Lecture Notes in Statistics

Edited by P. Bickel, P. Diggle, S. Fienberg, K. .
I. Olkin, N. Wermuth, S. Zeger

Springer

New York
Berlin
Heidelberg
Barcelona
Hong Kong
London
Milan
Paris
Singapore
Tokyo

David Ríos Insua
Fabrizio Ruggeri (Editors)

Robust Bayesian Analysis

David Ríos Insua
ESCET-URJC
Tulipan s/n
28933 Mostoles, Madrid
Spain

Fabrizio Ruggeri
CNR IAMI
Via Ampere 56
I-20131 Milano
Italy

Library of Congress Cataloging-in-Publication Data

Robust Bayesian analysis / editors, David Ríos Insua, Fabrizio Ruggeri.
p. cm. -- (Lecture notes in statistics ; 152)
Includes bibliographical references.
ISBN 0-387-98866-1 (softcover : alk. paper)
1. Bayesian statistical decision theory. I. Ríos Insua, David, 1964- II. Ruggeri, Fabrizio. III. Lecture notes in statistics (Springer-Verlag) ; v. 152.

QA279.5 .R64 2000
519.5'42--dc21

00-041912

Printed on acid-free paper.

Camera ready copy provided by the editors.
Printed and bound by Sheridan Books, Ann Arbor, MI.
Printed in the United States of America.

9 8 7 6 5 4 3 2 1

ISBN 0-387-98866-1 Springer-Verlag New York Berlin Heidelberg SPIN 10728773

Preface

Robust Bayesian analysis aims at overcoming the traditional objection to Bayesian analysis of its dependence on subjective inputs, mainly the prior and the loss. Its purpose is the determination of the impact of the inputs to a Bayesian analysis (the prior, the loss and the model) on its output when the inputs range in certain classes. If the impact is considerable, there is sensitivity and we should attempt to further refine the information available, perhaps through additional constraints on the incumbent classes and/or obtaining additional data; if the impact is not important, robustness holds and no further analysis and refinement would be required. Robust Bayesian analysis has been widely accepted by Bayesian statisticians; for a while it was even a main research topic in the field. However, to a great extent, their impact is yet to be seen in applied settings.

This volume, therefore, presents an overview of the current state of robust Bayesian methods and their applications and identifies topics of further interest in the area. The papers in the volume are divided into nine parts covering the main aspects of the field. The first one provides an overview of Bayesian robustness at a non-technical level. The paper in Part II concerns foundational aspects and describes decision-theoretical axiomatisations leading to the robust Bayesian paradigm, motivating reasons for which robust analysis is practically unavoidable within Bayesian analysis. Chapters in Part III discuss sensitivity to the prior, illustrating the key results in global and local robustness, along with their uses and limitations. Likelihood robustness is the topic of the paper in Part IV, whereas the papers in Part V address the issue of loss robustness, focussing on ranges of posterior expected losses, efficient sets and stability of Bayes decisions. The robust Bayesian approach is compared with other statistical methods in Part VI, specifically discussing sensitivity issues in Bayesian model selection and Bayesian nonparametrics and illustrating the Γ–minimax paradigm. Relevant algorithms in robust Bayesian analysis are presented in the papers in Part VII. Part VIII presents a discussion of case studies using a robust Bayesian approach, from medical decision making and statistics, economics and reliability. Finally, an extensive bibliography on Bayesian robustness concludes the volume.

We are grateful to all the contributors for their efforts in preparing chapters, which, we believe, provide a comprehensive illustration of the robust Bayesian approach. All the papers have been refereed by at least two people, mainly chosen among the contributors; we wish to acknowledge the work by Marek Męczarski, Antonio Pievatolo, Pablo Arias, Bruno Betrò, Alessandra

Guglielmi, Paul Gustafson, Juanmi Marin, Jacinto Martin, Elias Moreno, Peter Müller, Marco Perone Pacifico, Siva Sivaganesan, Cid Srinivasan, Luca Tardella and Mike Wiper in refereeing the papers.

We are also grateful to John Kimmel for all his fruitful comments and the Springer-Verlag referees for considering our proposal. Special thanks go to our families Susana and Anna, Giacomo and Lorenzo and our parents for their support and patience while preparing the volume.

David Ríos Insua and Fabrizio Ruggeri

MOSTOLES AND MILANO
(EUROPEAN UNION)

APRIL 2000

Contents

V Loss Robustness

VI Comparison With Other Statistical Methods

VII Algorithms

VIII Case Studies

Contributors

Pablo Arias, *jparias@unex.es*, Departamento de Matemáticas, Escuela Politécnica de Cáceres, Universidad de Extremadura, Cáceres, Spain.

Sanjib Basu, *basu@math.niu.edu*, Division of Statistics, Northern Illinois University, DeKalb, IL, USA.

James O. Berger, *berger@stat.duke.edu*, Institute of Statistics and Decision Sciences, Duke University, Durham, NC, USA.

Bruno Betrò, *bruno@iami.mi.cnr.it*, Consiglio Nazionale delle Ricerche, Istituto per le Applicazioni della Matematica e dell'Informatica, Milano, Italy.

Concha Bielza, *mcbielza@fi.upm.es*, Departamento de Inteligencia Artificial, Universidad Politecnica de Madrid, Boadilla del Monte, Madrid, Spain.

Enrico Cagno, *Enrico.Cagno@polimi.it*, Politecnico di Milano, Dipartimento di Meccanica, Milano, Italy.

Bradley P. Carlin, *brad@biostat.umn.edu*, Division of Biostatistics, School of Public Health, University of Minnesota, Minneapolis, MN, USA.

Franco Caron, *Franco.Caron@polimi.it*, Politecnico di Milano, Dipartimento di Meccanica, Milano, Italy.

Regino Criado, *rcriado@escet.urjc.es*, Department of Experimental Sciences and Engineering, Universidad Rey Juan Carlos, Mostoles, Madrid, Spain.

Dipak K. Dey, *dey@stat.uconn.edu*, Department of Statistics, University of Connecticut, Storrs, CT, USA.

Juan A. Fernández del Pozo, *jafernandez@fi.upm.es*, Departamento de Inteligencia Artificial, Universidad Politecnica de Madrid, Boadilla del Monte, Madrid, Spain.

Sandra Fortini, *sandra.fortini@uni-bocconi.it*, Istituto di Metodi Quantitativi, Università "L. Bocconi", Milano, Italy.

Manuel Gómez, *mgomez@fi.upm.es*, Departamento de Inteligencia Artificial, Universidad Politecnica de Madrid, Boadilla del Monte, Madrid, Spain.

Alessandra Guglielmi, *alessan@iami.mi.cnr.it*, Consiglio Nazionale delle Ricerche, Istituto per le Applicazioni della Matematica e dell'Informatica, Milano, Italy.

Paul Gustafson, *gustaf@stat.ubc.ca*, Department of Statistics, University of British Columbia, Vancouver, B.C., Canada.

Joseph B. Kadane, *kadane@stat.cmu.edu*, Department of Statistics, Carnegie Mellon University, Pittsburgh, PA, USA.

Michael Lavine, *michael@stat.duke.edu*, Institute of Statistics and Decision Sciences, Duke University, Durham, NC, USA.

Brunero Liseo, *brunero@pow2.sta.uniroma1.it*, Dipartimento di Studi Geoeconomici, Statistici e Storici per l'Analisi Regionale, Università di Roma, "La Sapienza", Roma, Italy.

Steven N. MacEachern, *snm@stat.ohio-state.edu*, Department of Statistics, Ohio State University, Columbus, OH, USA.

Mauro Mancini, *Mauro.Mancini@polimi.it*, Politecnico di Milano, Dipartimento di Meccanica, Milano, Italy.

Juan Miguel Marín, *j.m.marin@escet.urjc.es*, Department of Experimental Sciences and Engineering, Universidad Rey Juan Carlos, Mostoles, Madrid, Spain.

Jacinto Martín, *jrmartin@unex.es*, Departamento de Matemáticas, Escuela Politécnica de Cáceres, Universidad de Extremadura, Cáceres, Spain.

Athanasios Micheas, *atm97003@uconnvm.uconn.edu*, Department of Statistics, University of Connecticut, Storrs, CT, USA.

Elias Moreno, *emoreno@ugr.es*, Departamento de Estadística, Universidad de Granada, Granada, Spain.

Peter Müller, *pm@stat.duke.edu*, Institute of Statistics and Decision Sciences, Duke University, Durham, NC, USA.

María-Eglée Pérez, *eglee@cesma.usb.ve*, Departamento de Cómputo Científico y Estadística and Centro de Estadística y Software Matemático (CESMa), Universidad Simón Bolívar, Caracas, Venezuela.

Marco Perone Pacifico, *marcopp@pow2.sta.uniroma1.it*, Dipartimento di Statistica, Probabilità e Statistica Applicata, Università di Roma "La Sapienza", Roma, Italy.

David Ríos Insua, *drios@escet.urjc.es*, Department of Experimental Sciences and Engineering, Universidad Rey Juan Carlos, Mostoles, Madrid, Spain.

Sixto Ríos Insua, *srios@fi.upm.es*, Departamento de Inteligencia Artificial, Universidad Politecnica de Madrid, Boadilla del Monte, Madrid, Spain.

Fabrizio Ruggeri, *fabrizio@iami.mi.cnr.it*, Consiglio Nazionale delle Ricerche, Istituto per le Applicazioni della Matematica e dell'Informatica, Milano, Italy.

Gabriella Salinetti, *saline@pow2.sta.uniroma1.it*, Dipartimento di Statistica, Probabilità e Statistica Applicata, Università di Roma "La Sapienza", Roma, Italy.

N. D. Shyamalkumar, *shyamal@itam.mx*, Department of Actuarial Science and Insurance, Instituto Tecnologico Autonomo de Mexico, Mexico D.F., Mexico.

Siva Sivaganesan, *siva@math.uc.edu*, Department of Mathematical Sciences, University of Cincinnati, Cincinnati, OH, USA.

Cidambi Srinivasan, *srini@ms.uky.edu*, Department of Statistics, University of Kentucky, Lexington, KY, USA.

Luca Tardella, *tardella@pow2.sta.uniroma1.it*, Dipartimento di Statistica, Probabilità e Statistica Applicata, Università di Roma "La Sapienza", Roma, Italy.

Brani Vidakovic, *brani@stat.duke.edu*, Institute of Statistics and Decision Sciences, Duke University, Durham, NC, USA.

Simon P. Wilson, *simon.wilson@tcd.ie*, Department of Statistics, Trinity College, Dublin, Ireland.

Mike Wiper, *mwiper@est-econ.uc3m.es*, Departamento de Estadística y Econometría, Universidad Carlos III de Madrid, Getafe, Madrid, Spain.

1

Bayesian Robustness

James O. Berger, David Ríos Insua and Fabrizio Ruggeri

ABSTRACT An overview of the robust Bayesian approach is presented, primarily focusing on developments in the last decade. Examples are presented to motivate the need for a robust approach. Common types of robustness analyses are illustrated, including global and local sensitivity analysis and loss and likelihood robustness. Relationships with other approaches are also discussed. Finally, possible directions for future research are outlined.

Key words: prior/likelihood/loss robustness, algorithms, applications.

1.1 Introduction

Robust Bayesian analysis is concerned with the sensitivity of the results of a Bayesian analysis to the inputs for the analysis. The early 90s was the golden age of robust Bayesian analysis, in that many statisticians were highly active in research in the area, and rapid progress was being achieved. It was a major topic in many meetings, even some non-Bayesian meetings, and there were three meetings explicitly dedicated to the topic: the Workshop on Bayesian Robustness, held at Purdue University in 1989; and two International Workshops on Bayesian Robustness, held in Italy (Milano, 1992, and Rimini, 1995). The proceedings of the latter two conferences were published, respectively, in the *Journal of Statistical Planning and Inference* (vol. 40, no. 2 & 3), and in Berger et al. (1996), as an IMS Lecture Notes volume.

An issue of the journal *Test* (vol. 3, 1994) was mainly devoted to a review paper by Berger on robust Bayesian analysis, followed by discussions from many robust Bayesians. Several of these discussions were themselves extensive reviews of particular aspects of Bayesian robustness, so that the volume serves as a highly effective summary of the state of Bayesian robustness in 1994. Earlier reviews of Bayesian robustness were given in Berger (1984, 1985, 1990); these include considerable philosophical discussion of the approach, together with discussion of its history, including its origins in papers by Good in the 50s and, later, by Kadane and Chuang (1978). Another good source for information on Bayesian robustness is Walley (1991), although the book takes a somewhat different slant on the problem.

In the early 90s, there was an explosion of publications focusing on studying sensitivity to the prior distribution, in part because non-Bayesians often view this sensitivity to be the major drawback of Bayesian analysis. This work focused on replacing a single prior distribution by a class of priors and developing methods of computing the range of the ensuing answers as the prior varied over the class. This approach, called "global robustness", was soon supplemented by "local robustness" techniques, which focused on studying local sensitivity (in the sense of derivatives) to prior distributions. Interest naturally expanded into study of robustness with respect to the likelihood and loss function as well, with the aim of having a general approach to sensitivity towards all the ingredients of the Bayesian paradigm (model/prior/loss). Practical implementations of robust Bayesian ideas began to appear, including Godsill and Rayner (1996), Greenhouse and Wasserman (1996), Sargent and Carlin (1996) and Ríos Insua et al. (1999).

In the last half of the 90s, robust Bayesian analysis shifted from being a hot topic to being a mature field within Bayesian analysis, with continued gradual development, but with less of a sense of urgency. This change came about for several reasons. First, the initial flurry of fundamental theoretical advances naturally slowed, as will happen with any new field. Second, few Bayesians continued to question the need to view robustness or sensitivity as a serious issue, so that the philosophical excitement with the idea waned. Indeed, the consensus was that the time was now ripe to develop user-friendly implementations of the existing robust Bayesian methodology. The timing was not great, however, in that, coincidentally, this period also marked the explosion of interest in MCMC computational methods for Bayesian statistics. This had two serious effects on the field of Bayesian robustness. First, many of the researchers in robust Bayesian methods shifted their research into the MCMC arena. Second, the MCMC methodology was not directly compatible with many of the robust Bayesian techniques that had been developed, so that it was unclear how formal robust Bayesian analysis could be incorporated into the future "Bayesian via MCMC" world.

Paradoxically, MCMC has dramatically increased the need for consideration of Bayesian robustness, in that the modeling that is now routinely utilized in Bayesian analysis is of such complexity that inputs (such as priors) can be elicited only in a very casual fashion. It is now time to focus again on Bayesian robustness and to attempt to bring its ideas into the Bayesian mainstream.

This book aims at collecting contributions on the different aspects of Bayesian robustness, allowing statisticians and practitioners to access the wealth of research and the applied possibilities offered by the approach. New opportunities are offered by developments in algorithms (see Betrò and Guglielmi, 2000, Lavine et al., 2000, and MacEachern and Müller, 2000), the possibility of using MCMC methods and the need for sensitivity analysis in other fields, such as Bayesian model selection and Bayesian nonparametrics (see the papers by Liseo, 2000, and Basu, 2000). Bayesian robustness is playing a relevant role in SAMO (Sensitivity Analysis of Model Output), a group interested in investigating the relative importance of model input parameters on model predictions (see Saltelli et al., 2000) in many applied areas, from chemical engineering to econometrics.

In Section 2 of this paper, we present some motivating examples that indicate the importance of consideration of Bayesian robustness. Different approaches to Bayesian robustness are illustrated in Section 3. Description of "typical" robust analyses and ways to achieve robustness are presented, respectively, in Sections 4 and 5. Section 6 discusses connections between Bayesian robustness and other approaches, such as the frequentist approach of Huber (1973) and the approach based on lower probabilities of Walley (1991). Finally, directions for future research are discussed in Section 7.

1.2 Motivating examples

This section gives several examples and ideas to indicate the importance and uses of robustness in a Bayesian setting. The first example, concerned with p-values, demonstrates the use of robust Bayesian methods in identifying inadequacies in common statistical procedures. The next example considers the issue of imprecision in beliefs and preference assessment, which, in turn, points to foundational motivations for robust Bayesian analysis. We then give some examples that indicate potential dangers in approaching the problem from an overly simplified perspective.

1.2.1 Interpreting p-values

Robust Bayesian methods can be used to show that the common interpretations of certain classical procedures are not correct. Perhaps the most important example concerns p-values in testing a point-null hypothesis. Most statistical users believe that when the p-value is 0.05, the point null is very likely to be wrong, and when it is 0.01, it is almost certain to be wrong. However, robust Bayesian computations in Berger and Sellke (1987) show that these interpretations are not valid.

As an illustration, consider a long series of standard normal tests of, say, $H_0\colon \theta_i = 0$ versus $H_1\colon \theta_i \neq 0$, $i = 1, 2, \ldots$. (Similar results hold if the null hypotheses are small intervals, as discussed in Berger and Delampady, 1987.) Suppose 50% of the null hypotheses in this series are true. For each test, data is collected and the p-value computed, resulting in a corresponding series of p-values. Now consider the subset of these for which, say, $0.04 < p < 0.05$ (or $0.009 < p < 0.01$).

Fact: Among all the tests for which $0.04 < p < 0.05$ ($0.009 < p < 0.01$), at least 24% (7%) of the null hypotheses will be true.

The results are qualitatively similar for any initially specified proportion of true nulls; among those tests for which $0.04 < p < 0.05$ or $0.009 < p < 0.01$, a surprisingly large fraction (compared with the initial fraction) of the null hypotheses will be true.

This result is established by a robust Bayesian argument in which the parameters θ_i in the alternatives are allowed to assume any possible values (or, equivalently, any prior distribution for them is allowed). Then, conditional on the p-value being near 0.05, 24% is the *lower bound* on the proportion of true nulls (or posterior probability of the null), the lower bound being over all possible choices of the alternative hypotheses. Thus, in practice, one can expect the actual proportion of true nulls to be much higher than 24% when the p-value is 0.05 (and if 50% of the nulls were originally true). Further discussion of this phenomenon, from both parametric and nonparametric robust Bayesian perspectives, can be found in Sellke et al. (1999), which also gives a simple simulation program for explicitly exhibiting the phenomenon.

1.2.2 Elicitation in action

As suggested in many of the papers in this volume, the usual practical motivation underlying robust Bayesian analysis is the difficulty in assessing the prior distribution. Consider the simplest case in which it is desired to elicit a prior over a finite set of states $\theta_i, i = 1, \ldots, I$. A common technique to assess a precise $\pi(\theta_i) = p_i$, with the aid of a reference experiment, proceeds as follows: one progressively bounds $\pi(\theta_i)$ above and below until no further discrimination is possible and then takes the midpoint of the resulting interval as the value of p_i. Instead, however, one could directly operate with the obtained constraints $\alpha_i \le \pi(\theta_i) \le \beta_i$, acknowledging the cognitive limitations. Constraints might also come in different forms, such as knowledge that $\pi(\theta_i) \ge \pi(\theta_j)$.

The situation is similar in the continuous case; a common elicitation technique is to discretize the range Θ of the random variable and to assess the probabilities of the resulting sets. Again, however, one actually ends up only with some constraints on these probabilities. This is often phrased in terms of having bounds on the quantiles of the distribution. A related idea is having bounds on the distribution function corresponding to the random variable. Once again, qualitative constraints, such as unimodality, might also be added. Perhaps most common for a continuous random variable is to assess a parameterized prior, say a conjugate prior, and place constraints on the parameters of the prior; while this can be a valuable analysis, it does not serve to indicate robustness with respect to the specified form of the prior.

A similar situation holds in modeling preferences. One might, say, assess the loss of some consequences through the certainty equivalent method, and then fit a loss function. However, in reality, one only ends up with upper and lower constraints on such losses, possibly with qualitative features such as monotonicity and convexity, if preferences are increasing and risk averse. These constraints can often be approximated by an upper and a lower loss function, leading to the consideration of all loss functions that lie between these bounds (and that possibly satisfy the other qualitative constraints). If a parameterized loss function is assessed, say an exponential loss, the constraints are typically placed on the parameters of the loss, say the risk aversion coefficient.

In developing the model for the data itself there is typically great uncertainty, and a need for careful study of robustness or sensitivity. As this is appreciated by virtually all statisticians, we need not consider specific examples. It should be noted, however, that uncertainty in the model is typically considerably more important than uncertainty in the prior or loss.

A final comment concerning the limits of elicitation concerns the situation in which there are several decision makers (DMs) and/or experts involved in the elicitation. Then it is not even necessarily possible theoretically to obtain a single model, prior, or loss; one might be left with only classes of each, corresponding to differing expert opinions.

1.2.3 Foundations for robust analysis

One of the main arguments for the Bayesian approach is that virtually any axiomatic system leads to the subjective expected loss model, under which we are essentially led to the conclusions that: i) beliefs should be modeled with a probability distribution which, in the presence of additional information is updated through Bayes formula; ii) preferences and risk attitudes over consequences should be modeled with a loss function; and iii) preferences over alternatives should be modeled according to the minimum (subjective) expected loss principle. This framework provides the foundational underpinning for standard Bayesian inference and decision analysis. It is of considerable interest to understand the changes in this framework, and its conclusions, that result from the type of robust Bayesian considerations mentioned above.

We have argued that, in practical settings, it is not possible to precisely assess the prior, the model and the loss function. This contradicts the assumption, in the common axiom systems, of completeness in preferences and beliefs. Interestingly, however, relaxation of this assumption does not dramatically alter the conclusions of the foundations. Indeed, as outlined in Ríos Insua and Criado (2000) in this volume, one is essentially led to the same conclusion: imprecise beliefs and preferences can be modeled by a class of priors and a class of loss functions, so that preferences among alternatives can be represented by inequalities on the corresponding posterior expected losses. The web page `http://ippserv.rug.ac.be` contains considerable additional information on this subject.

1.2.4 Some views on simplified robust analyses

Before fully exploring key concepts in robust Bayesian analysis, it is useful to discuss what could be termed simplified robust analyses. The first, and most common, of these simplified analyses is informal sensitivity study, which consists of merely trying a few different inputs and seeing if the output changes significantly. A typical argument for this informal approach would be that Bayesian analysis in complex problems typically entails messy computations, and one cannot afford the additional computational burden that would be imposed by a formal robustness analysis. While informal analysis is certainly an important tool (and far better than ignoring robustness), a simple example shows its limitations and the need for undertaking more formal analyses.

Consider the no-data problem given by the decision table:

	θ_1	θ_2
a	c_1	c_2
b	c_2	c_3

Let $\pi(\theta_1) = p_1 = 1 - \pi(\theta_2)$. Assessment of this probability and the losses involved in the decisions results in the intervals $0.4 \leq p_1 \leq 0.6$, $-0.5 \leq L(c_1) \leq 0$, $-0.25 \leq L(c_2) \leq -.75$, and $-0.5 \leq L(c_3) \leq -1$. Consider the following four loss probability pairs associated with the bounds on losses and probabilities:

p_1	loss (l)	Exp. loss a	Exp. loss b
0.4	$L(c_1) = 0.$, $L(c_2) = -0.25$, $L(c_3) = -0.50$	-0.15	-0.40
0.4	$L(c_1) = -0.5$, $L(c_2) = -0.75$, $L(c_3) = -1$	-0.65	-0.90
0.6	$L(c_1) = 0.$, $L(c_2) = -0.25$, $L(c_3) = -0.50$	-0.10	-0.35
0.6	$L(c_1) = -0.5$, $L(c_2) = -0.75$, $L(c_3) = -1$	-0.60	-0.85

In all four cases, the expected loss of a is bigger than that of b, so this "simple" robustness analysis might lead one to conclude that alternative b is preferred to alternative a. Note, however, that this is not necessarily true: for instance, if $p_1 = 0.6$, $L(c_1) = -0.5$, $L(c_2) = -0.25$, and $L(c_3) = -0.5$, then the expected losses of a and b are, respectively, $-.4$ and $-.35$, so that a would be preferred.

Another type of simplified robustness analysis that is often encountered is the study of sensitivity of the outcome to one input of the problem, with the other inputs remaining fixed. It can happen, however, that a problem is insensitive to changes in only one input at a time, while being sensitive to simultaneous changes in inputs. This can also be illustrated using the above example.

Fix $p_1 = 0.5$ (the center of the interval) in the example above. Note that the difference in the expected loss of alternatives a and b is

$$.5L(c_1) + .5L(c_2) - .5L(c_2) - .5L(c_3) = .5(L(c_1) - L(c_3)).$$

Since $L(c_1) \geq L(c_3)$ for all values of L, it can be concluded that the problem is robust to changes only in L. Analogously, if we consider $L(c_1) = -.25$, $L(c_2) = -.5$ and $L(c_3) = -.75$ (the centers of the various intervals for the $L(c_i)$), the difference in the expected loss of a and b is

$$-p_1.25 - (1 - p_1).5 + p_1.5 + (1 - p_1).75 = .25.$$

Thus there is robustness against only changes in probabilities. However, as seen above, the problem is not robust with respect to joint changes in probabilities and losses.

1.3 Main issues in Bayesian robustness

In this section, we briefly outline the central issues in Bayesian robustness. Further discussion can be found in Berger (1984, 1985, 1990 and 1994), Wasserman (1992) and the papers in this volume.

1.3.1 The goal

As discussed earlier, robustness with respect to the prior stems from the practical impossibility of eliciting a unique distribution. Similar concerns apply to the other elements (likelihood and loss function) considered in a Bayesian analysis. The main goal of Bayesian robustness is to quantify and interpret the uncertainty induced by partial knowledge of one (or more) of the three elements in the analysis. Also, ways to reduce the uncertainty are studied and applied until, hopefully, robustness is achieved.

As an example, suppose the quantity of interest is the posterior mean; the likelihood is considered to be known, but the prior distribution is known only to lie in a given class. The uncertainty might be quantified by specifying the range spanned by the posterior mean, as the prior varies over the class. One must appropriately interpret this measure; for instance, one might say that the range is "small" if it spans less than 1/4 of a posterior standard deviation (suitably chosen from the range of possible posterior standard deviations). If the range is small, robustness is asserted, and the analysis would be deemed to be satisfactory. If, however, the range is not small, then some way must be found to reduce the uncertainty; narrowing the class of priors or obtaining additional data would be ideal.

1.3.2 *Different approaches*

We discuss the different possible approaches to Bayesian robustness for the simple situation in which the posterior mean is the quantity of interest and the uncertainty is in the prior distribution. Virtually any quantity of interest and any type of input uncertainty could have been used instead.

There are three main approaches to Bayesian robustness. The first is the *informal approach* already mentioned, in which a few priors are considered and the corresponding posterior means are compared. The approach is appealing because of its simplicity and can help, but it can easily "miss" priors that are compatible with the actually elicited prior knowledge and yet which would yield very different posterior means.

The second approach is called *global robustness* (see Moreno, 2000, for a thorough illustration). One (ideally) considers the class of all priors compatible with the elicited prior information and computes the range of the posterior mean as the prior varies over the class. This range is typically found by determining the "extremal" priors in the class that yield the maximum and minimum posterior means. Such computations can become cumbersome in multidimensional problems.

The third approach, called *local robustness*, is described by Gustafson (2000) and Sivaganesan (2000). It is interested in the rate of change in inferences, with respect to changes in the prior, and uses differential techniques to evaluate the rate. Local sensitivity measures are typically easier to compute in complicated situations than are global measures, but their interpretation (and calibration) is not always clear.

1.3.3 *Uncertainty modeling*

Classes of priors/likelihoods/loss functions are specified according to the available (partial) knowledge (e.g., quantiles or unimodality of the prior or convexity of the loss function). They can be classified into neighborhood or near-ignorance classes, using the classification proposed by Pericchi and Walley (1991). In the former case, one imagines that a "standard" Bayesian elicitation has yielded a specific prior (likelihood, loss), and considers neighborhoods of this prior (likelihood, loss) to study robustness. Various examples of such neighborhoods will be given later (and in other papers in this volume). Note, also, that the local approach to robustness implicitly operates on small perturbations within neighborhoods.

The near-ignorance class contains no baseline element and, instead, is based on some (few) features, such as quantiles of the prior distribution or values of the loss function at some points; see Martin et al. (1998). The names in this classification should not be taken at face value, since it is possible to have a near-ignorance class that is much smaller than a neighborhood class.

Desirable, but competing, properties of a class of priors include the following: computation of robustness measures should be as easy as possible; all reasonable priors should be in the class and unreasonable ones (e.g., discrete distributions in many problems) should not; and the class should be easily specified from elicited prior information. Similar properties can be defined for classes of losses and models.

1.3.4 Robustness measures

The most commonly used measure in global robustness is the range, that is, the difference between upper and lower bounds on the quantity of interest. Its value measures the variation caused by the uncertainty in the prior/likelihood/loss function. Its value may be deemed "satisfactorily small" or not, according to context-dependent criteria. It should always be recognized that the range depends strongly on the size of the classes used to represent the uncertain inputs. Thus a small range with an excessively small class is not really comforting, and a large range with an excessively large class is not necessarily damning.

When the range is too large, one ideally narrows the class further (through additional elicitation) or obtains additional data, hopefully resulting in a satisfactory range. If neither is possible, one must resort to an ad-hoc approach and should remain somewhat suspicious of the answer. A natural ad-hoc approach is to replace the class by a single input, obtained by some type of averaging over the class. For instance, if the class is a class of priors, one might choose a hyperprior on the class (perhaps using some default or noninformative-prior Bayesian method) and perform an ordinary Bayesian analysis. Numerous other ad-hoc criteria can be considered, however. An example is the Γ–minimax approach, described in Vidakovic (2000) for the problem of estimation. This typically conservative approach consists of choosing estimators minimizing the supremum, over the class, of either the posterior expected loss or the posterior regret.

While interpretation of the size of the range is usually done within the specific applied context, certain generic measures can be usefully introduced. For instance, Ruggeri and Sivaganesan (2000) consider a scaled version of the range, for a target quantity of interest $h(\theta)$, called *relative sensitivity* and defined as

$$R_\pi = \frac{(\rho_\Pi - \rho_0)^2}{V^\Pi},$$

where ρ_Π and ρ_0 equal $\mathbb{E}(h(\theta)|x)$ under Π and the baseline prior Π_0, respectively, and V^Π is the posterior variance of $h(\theta)$ with respect to Π. The idea is that, in estimation of $h(\theta)$, the posterior variance indicates the accuracy that is attainable from the experiment, so that one will often not worry if the (squared) range is small with respect to this posterior variance.

In the local sensitivity approach to Bayesian robustness, techniques from differential calculus are used and robustness is measured by the supremum of (functional) derivatives over a class.

1.3.5 *Algorithms*

Computation of robustness measures over a class is usually performed by considering the subset of extremal elements of the class and performing the computations over this subset. For example, in most robustness problems, the extremal elements of the class of all distributions are the Dirac (point-mass) measures, whereas the extremal elements of the class of all symmetric unimodal densities are the uniform densities. One of the first references to employ such variational mathematics in Bayesian robustness was Sivaganesan and Berger (1989).

Among the very large number of techniques that have been developed for robustness computations, the *linearization algorithm* deserves special mention because of its utility and generality of application. Introduced into the field by Lavine (1991), it allows determination of the range of a posterior expectation through a sequence of linear optimizations. A recent algorithmic development for generalized moment constraint classes can be found in Betrò and Guglielmi (2000).

1.4 A guided tour through Bayesian robustness

We illustrate typical steps in a robust Bayesian analysis in a "textbook" case, using one of the reference examples in the literature, due to Berger and Berliner (1986). We assume familiarity with Bayesian analysis; otherwise, we refer the reader to general textbooks, such as Berger (1985) or French and Ríos Insua (2000).

We analyze sensitivity to the prior in the example, following each of the three approaches presented above. (Later examples will illustrate likelihood and loss robustness.) The quantities of interest will be taken to be posterior means and posterior probabilities of credible sets.

Suppose a single observation, x, is observed from a normal distribution with unknown mean θ and variance 1, to be denoted $\mathcal{N}(\theta, 1)$. The classic (old) textbook Bayesian analysis would suggest selection of a conjugate, normal prior for θ, chosen to match some features determined by the elicitation process. Suppose, for instance, that $-.954$, 0, and $.954$ are the elicited quartiles of the distribution of θ. The unique, normal prior with these quartiles is $\Pi \sim \mathcal{N}(0, 2)$. The Bayes estimate of θ under squared error loss would be the posterior mean, given by $\delta_{\mathcal{N}}(x) = 2x/3$.

It is evident that the choice of a normal prior is mathematically convenient but otherwise arbitrary. It is thus natural to consider classes of priors that are compatible with the elicited quartiles, but are otherwise more general.

1.4.1 Informal approach

In the informal approach to robustness, one would try a few different priors, compatible with the elicited quartiles, and compare the answers. Three such priors are the normal $\mathcal{N}(0, 2)$, double exponential $\mathcal{DE}(0, \log 2/.954)$ and Cauchy distribution $\mathcal{C}(0, .954)$. Resulting posterior means, for different data x, are presented in Table 1.

x	.5	1	1.5	2	2.5	3	3.5	4	10
$\mathcal{N}$	0.333	0.667	1.000	1.333	1.667	2.000	2.333	2.667	6.667
$\mathcal{DE}$	0.292	0.606	0.960	1.362	1.808	2.285	2.776	3.274	9.274
$\mathcal{C}$	0.259	0.540	0.866	1.259	1.729	2.267	2.844	3.427	9.796

TABLE 1. Posterior means for different priors

For moderate or large values of x in Table 1, the posterior mean is reasonably consistent for the Cauchy and double exponential distributions but changes dramatically for the normal distribution. For small values of x, the posterior means are all reasonably consistent. The conclusion of this informal analysis would be that robustness likely obtains for smaller values of x, but not for larger values.

1.4.2 Global robustness

We will consider different neighborhood classes of priors, centered at the normal prior, $\mathcal{N}(0, 2)$.

1.4.2.1 ε-contamination neighborhoods

When a baseline prior Π_0 (often conjugate) is elicited by usual methods, a natural class of priors for studying sensitivity to the prior is the *ε-contamination* class

$$\Gamma_\varepsilon = \{\Pi : \Pi = (1-\varepsilon)\Pi_0 + \varepsilon Q,\ Q \in \mathcal{Q}\},$$

where $\mathcal{Q}$ is called the class of contaminations. (This class was first utilized in classical robustness studies by Huber, 1973.) The popularity of the class arises, in part, from the ease of its specification and, in part, from the fact that it is typically easily handled mathematically.

A variety of choices of $\mathcal{Q}$ have been considered in the literature. The most obvious choice is to let $\mathcal{Q}$ be the class of all distributions, as was considered in Berger and Berliner (1986) for the above example, with $\varepsilon = 0.2$. (The resulting class for the example is denoted Γ_A below.) Unfortunately, this class is often too large to yield useful robustness bounds.

One important class of refinements of the contaminating class involves the addition of shape constraints, such as unimodality and symmetry. Such constraints can often be readily elicited and can very significantly reduce the range of posterior quantities of interest.

Another class of refinements that is often considered is the addition of quantile constraints, since specification of probabilities of sets is typically easier than other elicitations. In the above example, for instance, Moreno and Cano (1991) considered the class Γ_A, but with, respectively, the median fixed to be that of Π_0 (class Γ_2) and the quartiles fixed at those of Π_0 (class Γ_4). An alternative quantile specification was considered in Betrò et al. (1994): because it is often easier to elicit features of the marginal distribution of X (i.e., features of the distribution of observables) than features of the prior, it is natural to consider Γ_{2m} and Γ_{4m}, the classes of priors in Γ_A which yield marginals for X that have, respectively, the same median and the same quartiles as the marginal distribution of X under Π_0.

Computation of the range of posterior functionals for these ε–contamination classes has been studied by many authors, and prototype computer codes are available (see Betrò and Guglielmi, 2000, for a review). The upper and lower bounds are actually achieved at discrete distributions, which greatly simplifies the analysis.

As in Betrò et al. (1994), we consider the robustness of the standard 95% credible interval, $C(x)$, computed assuming the conjugate normal prior Π_0 were true. For Γ_A and the various quantile classes given above, Table 2 gives the range of the posterior probability of $C(x)$ as the priors vary over the classes.

x	min	max	min	max	min	max	min	max	min	max
	$\Pi \in \Gamma_A$		$\Pi \in \Gamma_2$		$\Pi \in \Gamma_4$		$\Pi \in \Gamma_{2m}$		$\Pi \in \Gamma_{4m}$	
0.5	.818	.966	.842	.965	.906	.963	.844	.964	.906	.960
1.0	.773	.967	.823	.965	.876	.962	.834	.962	.907	.959
1.5	.707	.969	.787	.965	.861	.963	.810	.961	.892	.956
2.0	.615	.973	.728	.966	.827	.963	.746	.965	.856	.959
2.5	.496	.978	.640	.970	.755	.964	.652	.969	.795	.961
3.0	.363	.983	.522	.975	.666	.968	.525	.975	.735	.962
3.5	.235	.988	.377	.980	.534	.973	.377	.981	.624	.962
4.0	.135	.993	.237	.988	.377	.980	.237	.988	.469	.975

TABLE 2. Upper and lower bounds on the posterior probability of the 95% credible interval from the $\mathcal{N}(0,2)$ prior

As could be expected, the least robust situation is when $x = 4$. For the original Γ_A, the posterior probability of $C(4)$ varies over an enormous range. Utilizing the refined classes is of some help in reducing the range but, even for Γ_{4m}, the range would typically be viewed as unacceptably large. One could add additional constraints on the class, such as unimodality (as in Moreno and González, 1990), but even this would not help here; the basic problem is that the credible set from the conjugate prior is simply not a robust credible set when rather extreme data is observed. (For results concerning optimally robust credible sets, see Sivaganesan et al., 1993.)

A "nice" case in Table 2 is when $x = 1$ and Γ_{4m} is used; one might well be satisfied with the range $(.907, .959)$. We focused here on the range of the posterior probability of a credible interval, but we could have considered the posterior mean and variance, as in Sivaganesan and Berger (1989) and Sivaganesan (1988, 1989), where the prior Π_0 was contaminated by many classes of distributions: arbitrary, symmetric unimodal, unimodal,... Related results on hypothesis testing can be found in Moreno (2000). Finally, as indicated above in the comment about optimally robust credible sets, one can also search for the most robust posterior estimators among those commonly used, such as the mean, median and mode. See Sivaganesan (1991) for comparisons under the contaminating class of all bounded unimodal distributions having the same mode and median as Π_0.

1.4.2.2 Other neighborhood classes

One drawback of ε–contamination classes is that they are not true neighborhood classes in a topological sense. A variety of other classes have been considered that do have a formal interpretation in this sense, such as the class based on a concentration function and the class based on distribution bands. The former class consists of those distributions for which with the probability of each set is bounded by a given function of the set's probability under a baseline distribution. The example of Berger and Berliner (1986) has been investigated for concentration function neighborhoods; see Fortini and Ruggeri (2000) and references therein.

The distribution band class, described in Basu and DasGupta (1995) and Basu (1995), is defined as

$$\Gamma_{BDG} = \{F : F \text{ is a cdf and } F_L(\theta) \leq F(\theta) \leq F_U(\theta), \forall\theta\},$$

where F_L and F_U are given cdfs, with $F_L(\theta) \leq F_U(\theta)$. The class includes, as special cases, well-known metric neighborhoods of a fixed cdf, such as Kolmogorov and Lèvy neighborhoods. This class has the additional desirable property of being comparatively easy to elicit. For instance, in the normal example above one might specify the upper and lower cdfs to be those corresponding to the specified normal and Cauchy priors, if it were felt that the main uncertainty was in the tail of the prior. Additional "shape" requirements, such as symmetry and unimodality of the distributions, can also be added, as in Basu (1994).

1.4.2.3 Quantile classes

Often, there will be no baseline prior and the class of priors will simply be those that satisfy the features that are elicited. A particularly tractable class of features are generalized moment conditions (see Betrò and Guglielmi, 2000). The most common example of a generalized moment condition is specification of a quantile.

An example of a quantile class is that considered in Berger and O'Hagan (1988), O'Hagan and Berger (1988) and Moreno and Cano (1989). The elicitation proceeds by determining the prior probabilities of each of the intervals I_i, $i = 1, 6$, in Table 3. The probabilities indicated therein are the elicited values, and the quantile class, Γ_Q, consists of all distributions that are compatible with these assessments. Note that the $\mathcal{N}(0, 2)$ distribution is one such prior.

I_i	$(-\infty, -2]$	$(-2, -1]$	$(-1, 0]$	(0,1]	(1,2]	$(2, \infty)$
p_i	.08	.16	.26	.26	.16	.08

TABLE 3. Intervals and prior probabilities

Computation of ranges of a posterior quantity of interest is rather simple for a quantile class, since upper and lower bounds are achieved for discrete distributions giving mass to one point in each interval I_i (see, e.g., Ruggeri, 1990). But discrete distributions are typically quite unnatural, so that more realistic classes are often obtained by considering Γ_U, the unimodal priors in Γ_Q, or Γ_{BU}, the priors in Γ_U with known mode a_k and such that the density at a_k does not exceed a given threshold. Computations for Γ_U and Γ_{BU} are, however, quite involved. As a compromise, O'Hagan and Berger (1988) proposed the class Γ_{QU} of quasiunimodal priors, with $\Gamma_{BU} \subset \Gamma_U \subset \Gamma_{QU} \subset \Gamma_Q$, for which computations are relatively simple.

As an illustration of the effect of the various restrictions on the quantile class, Table 4, taken from Berger and O'Hagan (1988), considers the range of the posterior probability of the intervals I_i for the various classes, in a specific situation. Also reported is the posterior probability, p_i^*, for each I_i, under the single prior $\mathcal{N}(0, 2)$.

The main interest here is that imposing any of the shape constraints results in a considerable reduction in the ranges of the posterior probabilities, compared with the ranges for the raw quantile class, Γ_Q.

1.4.3 Local sensitivity

Local sensitivity analysis considers the behavior of posterior quantities of interest under infinitesimal perturbations from a specified baseline prior Π_0. Different approaches to local sensitivity are discussed in, e.g., Gustafson (2000) and Sivaganesan (2000), along with a discussion of their properties.

I_i	p_i^*	Γ_Q	Γ_{QU}	Γ_U	Π_{BU}
$(-\infty,-2]$	.0001	(0,.001)	(0,.0002)	(0,.0002)	(0,.0002)
$(-2,-1]$	.0070	(.001,.029)	(.006,.011)	(.006,.011)	(.006,.010)
$(-1,0]$	.1031	(.024,.272)	(.095,.166)	(.095,.166)	(.095,.155)
(0, 1]	.3900	(.208,.600)	(.320,.447)	(.322,.447)	(.332,.447)
(1, 2]	.3900	(.265,.625)	(.355,.475)	(.357,.473)	(.360,.467)
$(2,\infty)$	.1102	(0,.229)	(0,.156)	(0,.156)	(0,.154)

TABLE 4. Posterior ranges for I_i

As an illustration of the approach, consider the following example, studied in Basu (1996), which is similar to that analyzed in Section 1.4.2. Consider an $\mathcal{N}(\theta,1)$ model, for the data x, and a baseline prior Π_0, chosen to be $\mathcal{N}(0,\, 2.19)$; this prior has median zero and quartiles at -1 and 1. Consider "small" perturbations

$$\Pi_Q = (1-\varepsilon)\Pi_0 + \varepsilon Q = \Pi_0 + \varepsilon(Q-\Pi_0)$$

of Π_0, in the direction $Q-\Pi_0$, with Q in a class Γ of priors. As a sensitivity measure, consider

$$\overline{G}_{\rho_0} = \sup_{Q\in\Gamma} G_{\rho_0}(Q-\Pi_0),$$

where

$$G_{\rho_0}(Q-\Pi_0) = \frac{d}{d\varepsilon}\rho((1-\varepsilon)\Pi_0 + \varepsilon Q)_{\varepsilon=0} \tag{1}$$

is the Gâteaux derivative of the functional $\rho(\cdot)$ at Π_0 in the direction $Q-\Pi_0$.

We consider three classes (out of the five considered by Basu): Γ_A, Γ_U and Γ_{SU}, corresponding to arbitrary, unimodal (around 0) and symmetric, unimodal (around 0) distributions. We present Basu's results for the functionals ρ_1 (posterior mean), ρ_2 (posterior median) and ρ_4 (posterior variance), omitting the posterior mode, ρ_3, whose $\overline{G}_{\rho_0}$ is always ∞ for the three classes we consider when $x \neq 0$. This subset of results, from Basu (1996), is presented in Table 5.

As pointed out in Basu (1996), for the samples $x=0$ and $x=1$, the centers of Π_0 and the likelihood are nearly matching and the posterior mean is less sensitive than the posterior median to infinitesimal perturbations. The opposite happens for larger values of x. The table gives clear indications of the negative and positive aspects of local sensitivity. Observe that $\overline{G}_{\rho_0}$ is not scale invariant, so that interpreting its magnitude can be tricky; as an example, 0.5 could be large or small depending on the problem and the order of magnitude of the functionals. Scale-invariant measures could be

	$x=0$			$x=1$			$x=2$		
	ρ_1	ρ_2	ρ_4	ρ_1	ρ_2	ρ_4	ρ_1	ρ_2	ρ_4
Γ_A	1.08	1.86	1.23	1.69	2.17	1.79	3.47	3.47	4.33
Γ_U	0.81	1.86	1.23	0.78	1.73	1.06	1.25	1.05	1.08
Γ_{SU}	0.00	0.00	0.21	0.22	0.20	0.27	0.58	0.50	0.59
	$x=3$			$x=4$			$x=5$		
	ρ_1	ρ_2	ρ_4	ρ_1	ρ_2	ρ_4	ρ_1	ρ_2	ρ_4
Γ_A	9.38	7.59	13.83	33.72	22.65	57.84	161.88	92.41	318.92
Γ_U	3.17	2.48	3.66	10.64	7.45	14.93	46.92	29.05	78.55
Γ_{SU}	1.59	1.24	1.84	5.32	3.73	7.47	23.46	14.52	39.28

TABLE 5. $\overline{G}_{\rho_0}$ for the posterior mean (ρ_1), median (ρ_2) and variance (ρ_4)

obtained, for example by adapting the scaled range in Ruggeri and Sivaganesan (2000), but calibration remains a troublesome issue, as discussed in Sivaganesan (2000). Nonetheless, we can use $\overline{G}_{\rho_0}$ either to compare "similar" functionals, such as the mean and the median, in the same problem (same sample and class of priors), or to quantify the relative decrease in sensitivity when narrowing the class of priors and considering the same functional.

1.4.4 Interactive robustness

An effective robust Bayesian analysis should be interactive, asking for new information from different sources until robustness is achieved, for instance, the range of a posterior functional is below a fixed threshold. We have presented, in the previous sections, different ways to reduce the width of the class of priors, such as introduce new quantiles, unimodality and symmetry. Since the acquisition of new information has a cost, optimal ways of obtaining new information should be pursued.

Suppose, for instance, that one is developing a quantile class of priors. As suggested in Berger (1994), one useful way to proceed, when robustness has not been obtained for the current class, is to determine the "optimal" additional quantile to elicit, in the sense of optimally reducing the range of the posterior quantity of interest. Different implementations of this idea have been proposed in Liseo et al. (1996), Moreno et al. (1996) and Ríos Insua et al. (1999).

Moreno et al. (1996) consider, for each $j = 1, \ldots, n$, the class

$$\Gamma_j = \{\pi : \pi(\theta) = \pi_0(\theta) + \varepsilon_j \mathbf{1}(\theta)(A_j)(q(\theta) - \pi_0(\theta)), q \in \mathcal{Q}\}.$$

This class differs from the original class in that the contamination is allowed only on one set A_j at a time. Computing the ranges of the posterior functional over each class Γ_j, Moreno et al. select the interval leading to the largest range as that in which the new quantile is to be determined and they choose, as the splitting point of the interval, the minimizer of the range over an associated class.

Liseo et al. (1996) considered the quantity

$$\delta(A_j, \Gamma) = \sup_{\Pi \in \Gamma} \int_{A_j} h(\theta) l(\theta) \Pi(d\theta) - \inf_{\Pi \in \Gamma} \int_{A_j} h(\theta) l(\theta) \Pi(d\theta),$$

where h is the function of interest, l is the likelihood and $A_j, j = 1, \ldots, n$ and Γ are, respectively, the intervals and the class of priors determined by the quantiles. They computed $\delta(A_j, \Gamma)$ for all $j = 1, \ldots, n$, and split the interval A_j, with the largest of such values, into two intervals having (subjectively) the same probability content. The authors proved that their method leads to ranges as small as one wishes. The method can be easily implemented and its understanding for a practitioner is easy; new quantiles could be obtained by applying, for example, a betting scheme. The major drawback might be the slowness in the reduction of the range.

Ríos Insua et al. (1999) considered contaminations of a baseline prior, Π_0, with the signed measures $\delta = \varepsilon(\Pi - \Pi_0)$, Π being another probability measure and $\varepsilon \leq \varepsilon_0$, for a given ε_0. They computed the supremum of the Fréchet derivative, of the posterior functional of interest, over all δ's such that Π belongs to the quantile class under investigation. It can be proved (Ruggeri and Wasserman, 1993, and Martín and Ríos Insua, 1996) that the supremum is given by

$$\frac{\varepsilon_0}{\int l(\theta) \Pi_0(d\theta)} \max\{\sum_j p_j \overline{h_j}, -\sum_j p_j \underline{h_j}\},$$

where $\overline{h_j}$ ($\underline{h_j}$) is the supremum (infimum) of the function h over the interval A_j, $j = 1, \ldots, n$, associated with the quantiles, and p_j is the corresponding probability.

Ríos Insua et al. proposed splitting the interval which gives the largest contribution; in their paper the split was based on the experts' opinions. Pros and cons of this approach are similar to the ones for the method proposed by Liseo et al. (1996). In fact, the two approaches rely on very similar measures to choose the interval to split, although they are based, respectively, on local and global robustness measures. Other ideas may be seen in Ríos Insua (1990).

1.4.5 Likelihood and loss robustness

In this section, we briefly illustrate some issues regarding sensitivity with respect to the likelihood and loss function.

1.4.5.1 Likelihood robustness

An informal approach, similar to that discussed in Section 1.4.1, can be followed in analyzing likelihood robustness; one can simply try several models and see if the answer changes significantly. It is difficult to develop more formal approaches to likelihood robustness, for several reasons, one of which is that parameters often change meanings in going from one model to another; hence changing models also often requires changing the prior distributions. (This is one argument for performing prior elicitations for the marginal distribution of observables; such elicitations need only be done once and will induce appropriate priors for each of the models under consideration.)

Sometimes a parameter has a clear meaning, regardless of the model under consideration, and more formal methods can be employed. An example is that of the median of a symmetric distribution. For instance, Shyamalkumar (2000) discusses the situation in which there are two competing models (normal $\mathcal{N}(\theta, 1)$ and Cauchy $\mathcal{C}(\theta, 0.675)$), with classes of priors on the (well-defined) common parameter θ. The scale parameters of the two models are chosen to match the interquartile range. The priors on θ are specified to be in ε–contamination neighborhoods, where the contaminating classes are either Γ_A or Γ_{SU}, containing, respectively, arbitrary and symmetric, unimodal (around zero) distributions. Shyamalkumar considered $\varepsilon = 0.1$ and computed upper and lower bounds on the posterior mean $\mathbb{E}(\theta|x)$ of θ, as shown in Table 6.

Data	Likelihood	Γ_A		Γ_{SU}	
		$\inf \mathbb{E}(\theta\|x)$	$\sup \mathbb{E}(\theta\|x)$	$\inf \mathbb{E}(\theta\|x)$	$\sup \mathbb{E}(\theta\|x)$
$x = 2$	Normal	0.93	1.45	0.97	1.12
	Cauchy	0.86	1.38	0.86	1.02
$x = 4$	Normal	1.85	4.48	1.96	3.34
	Cauchy	0.52	3.30	0.57	1.62
$x = 6$	Normal	2.61	8.48	2.87	5.87
	Cauchy	0.20	5.54	0.33	2.88

TABLE 6. Bounds on posterior mean for different models and classes of priors

As noticed by Shyamalkumar, even though the widths of the ranges of the posterior means are similar for the two likelihoods, the centers of the ranges are quite different when x is larger.

As a further extension, one might consider a parametric class of models, such as that of Box and Tiao (1962), given by

$$\Lambda_{BT} = \{f(y|\theta,\sigma,\beta) = \frac{\exp\{-\frac{1}{2}\left|\frac{y-\theta}{\sigma}\right|^{\frac{2}{1+\beta}}\}}{\sigma 2^{(1.5+0.5\beta)}\Gamma(1.5+0.5\beta)};\ \text{any}\ \theta, \sigma > 0, \beta \in (-1,1]\}. \tag{2}$$

An application of this class is given in Shyamalkumar (2000), which reviews, as well, other current approaches to likelihood robustness.

1.4.5.2 Loss robustness

One of the motivations for comparing the robustness of the posterior mean and median, as done in Sivaganesan (1991), is that the two estimators are optimal Bayesian estimators for, respectively, the squared error and absolute error loss functions. More generally, a group of decision makers could have different ideas about the features of an optimal estimator, leading to different loss functions. We could be interested, for example, in quantifying the changes either in the posterior expected loss (e.g. in Dey et al., 1998) or in the optimal action (e.g., mean and median in Basu and DasGupta, 1995).

Various loss robustness measures have been proposed, most of which are discussed in Dey et al. (1998), which explores robustness properties of classes of LINEX losses under exponential and discrete power series families of distributions using conjugate priors. In particular, they consider the class of LINEX losses defined as

$$\Lambda = \{L_b : L_b(\theta, a) = \exp\{b(a-\theta)\} - b(a-\theta) - 1, b \neq 0, b_0 < b < b_1\},$$

where b_0 and b_1 are fixed. They consider a normal model for the observation and a conjugate prior for its mean, with specified variance, and compute the range of the posterior expected loss and the posterior regret (see Berger, 1985, for definition and properties). These results are reviewed in Dey and Micheas (2000).

Other approaches, based upon stability of decisions and nondominated actions, are thoroughly reviewed in Kadane et al. (2000) and Martín and Arias (2000), respectively.

1.4.5.3 Loss and prior robustness

Research on sensitivity jointly with respect to the prior and the loss is not very abundant. Note that, as indicated in Section 2, it is possible that the problem is robust with respect to the prior only and the loss only, but rather sensitive when both elements are jointly considered.

An outline of such general sensitivity analyses is given in Ríos Insua et al. (2000). Basu and DasGupta (1995) jointly considers priors in a distribution band and a finite number of loss functions. Martín and Ríos Insua (1996) used Fréchet derivatives to investigate local sensitivity for small perturbation in both the prior distribution and the loss function.

1.5 Inherently robust procedures

In the previous sections, we have considered ways to increase robustness by narrowing classes of priors and/or likelihood and/or loss functions. Now we focus on choices of priors/likelihood/loss which should be less sensitive to changes in the other components.

1.5.1 *Robust priors*

As thoroughly discussed in Berger (1985, 1994), flat-tailed distributions tend to be more robust than standard conjugate choices (e.g., Cauchy vs. normal distributions). The construction of a class of priors should take into account tail behavior in addition to symmetry, unimodality, etc. Tail modeling is straightforward for certain classes of priors, for instance the *density bounded* class

$$\Gamma_{DB} = \{\pi : L(\theta) \leq \pi(\theta) \leq U(\theta), \forall\theta\},$$

used by Lavine (1991). By appropriate choice of L and U, one can avoid priors with too-sharp tails. For instance, choosing L and U to be different multiples of a Cauchy density would preclude priors with normal-like tails. Additional constraints, such as unimodality, can be added to density bounded classes to further ensure that only reasonable prior densities are being considered.

Hierarchical priors are typically inherently robust in several senses, as discussed in Berger (1985). In part, this is because they are typically flat-tailed distributions. As an example, Moreno and Pericchi (1993) consider an ε-contamination neighborhood of an $\mathcal{N}(\mu, \tau)$ prior and show that, when the prior information concerning τ is vague, the range of the desired posterior quantity is large. However, if one places a prior on τ, which is equivalent to considering a hierarchical model, then the posterior range is drastically reduced.

1.5.2 *Robust models*

Many robust procedures in the literature are primarily concerned with obtaining resistance to outliers. For instance, the class (2), defined by Box and Tiao (1962), has been widely depicted as a robust class and its application for this purpose has been extensively discussed; see, e.g., Box and Tiao (1973). Another class of models, the extended power family, has been proposed by Albert et al. (1991). It can be viewed as a smooth alternative to the Box and Tiao power-series family, since its tails have similar functional forms but, unlike the power-series family, it is everywhere differentiable.

1.5.3 Robust estimators

As already discussed, different loss functions can lead to different Bayesian estimates. In particular, squared error and absolute error loss yield, respectively, the posterior mean and the posterior median as the optimal Bayesian estimator. It is interesting to study which such common estimators are more inherently robust. Sivaganesan (1991) investigated the behavior of the posterior mean, median and mode when the prior is in the ε–contamination class, restricted to all bounded unimodal distributions with the same mode and median as the baseline Π_0. He concludes that the posterior mean and posterior median are preferable to the posterior mode. Moreover, the posterior median seems to be worse, though only slightly, than the posterior mean.

Basu (1996) also studied this issue, for the example presented in Section 1.4.3. He found that the posterior mean is more (less) sensitive than the posterior median for large (small) samples.

1.5.4 Other robust procedures

Marin (2000) considers a robust version of the dynamic linear model which is less sensitive to outliers. This is achieved by modeling both sampling errors and parameters as multivariate exponential power distributions, instead of normal distributions.

Another possibility to enhance robustness is offered by the Bayesian non parametric approach; see Berger (1994). Parametric models can be embedded in nonparametric models, presumably achieving a gain in robustness. An example is discussed in MacEachern and Müller (2000), where mixtures of Dirichlet process are considered and an efficient MCMC scheme is presented.

1.6 Comparison with other approaches

Bayesian robustness has had interactions with other approaches, such as classical robustness, illustrated in Huber (1981), and upper and lower probabilities, presented in Walley (1991). We briefly review common points and differences.

1.6.1 Frequentist robustness

Certain notions developed in classical robustness have been utilized in Bayesian robustness. We have already mentioned one, ε–contamination classes of distributions, that were introduced in Huber (1973) for classical robustness problems. In the classical context, they are primarily used to model possible outliers.

The local sensitivity approach, discussed in Gustafson (2000), borrows the notion of an *influence function* from classical robustness theory (Hampel et al., 1986). Gustafson observes that the Gâteaux derivative can be written as

$$G_{\rho_0}(Q - \Pi_0) = \int I(z) d[Q - \Pi_0](z),$$

where $I(z)$ is the influence function. Thus plots of the influence function can be very useful in visually assessing sensitivity. For most posterior quantities, the influence function can be easily determined.

Peña and Zamar (1996) present an asymptotic approach to Bayesian robustness that allows use of classical robustness tools, such as the influence function and the maximum bias function. We refer to their paper for an illustration of the approach.

1.6.2 Imprecise probability

Imprecise probability is a generic term used to describe mathematical models that measure uncertainty without precise probabilities. This is certainly the case with robust Bayesian analysis, but there are many other imprecise probability theories, including upper and lower probabilities, belief functions, Choquet capacities, fuzzy logic, and upper and lower previsions; see Walley (1991) and the web page `http://ippserv.rug.ac.be`.

Some of these theories, such as fuzzy logic and belief functions, are only tangentially related to robust Bayesian analysis. Others are intimately related; for example, some classes of probability distributions that are considered in robust Bayesian analysis, such as distribution band classes, can also be interpreted in terms of upper and lower probabilities. Also, classes of probability distributions used in robust Bayesian analysis will typically generate upper and lower previsions as their upper and lower envelopes. Walley (1991) describes the connection between robust Bayesian analysis (in terms of sensitivity to the prior) and the theory of coherent lower previsions.

In a rough sense, the major difference between robust Bayesian analysis and these alternative theories is that robust Bayesian analysis stays with ordinary Bayesian intuition, considering, as ideal, classes (of priors, say) that consist only of those priors that are individually compatible with prior beliefs. In contrast, the alternative theories view the classes themselves (not the individual priors) as the basic elements of the theory. We do not take a philosophical position on this issue, but it is useful to note that the robust Bayesian approach is directly compatible with ordinary Bayesian analysis, and hence is more immediately accessible to Bayesian practitioners.

1.6.3 Hierarchical approaches

We have already mentioned that hierarchical modeling has certain inherent robustness properties. We further mentioned that one approach to dealing with a lack of robustness (with respect to, say, the prior) is to place a hyperprior on the class of priors, which is a type of hierarchical analysis. Indeed, if there were no possibility of obtaining additional information to deal with the lack of robustness, we would recommend this technique, with the hyperprior being chosen in some default fashion. Note, however, that this effectively means one would be working with a single prior (that arising from marginalizing over the hyperprior), so real robustness could not be claimed; in a sense, what has been accomplished is simply to use the inherent robustness of hierarchical modeling to develop a single prior that is hopefully reasonably robust. One could, of course, also embed the hyperprior in a class of priors and deal with robustness in that fashion; this is sensible if the hyperprior has clear intuitive meaning but otherwise can offer no real gain.

1.6.4 Reference and objective Bayes approaches

The primary applied form of Bayesian analysis, for over 200 years, has been the objective Bayes approach (often called the reference approach today), in which a single prior distribution is chosen in a default or noninformative fashion. A nice review of this approach is given in Kass and Wasserman (1996).

The objective Bayes approach is *not* related to the robust Bayesian approach through any type of formal attempt at being noninformative. (After all, if taken literally, *noninformative* would suggest that one consider the class of *all* possible prior distributions.) Rather, the objective Bayes approach is related in that it can be viewed as proposing a particular robust prior for default use of Bayesian methods.

There are two further relationships that should be mentioned. One has already been discussed, namely that, when robustness is not obtained, a good generic way to proceed is to determine a default hyperprior over the class of priors, and perform the (now single-prior) Bayesian analysis. Objective Bayesian techniques can be applied to determine a suitable default hyperprior. The second interesting relationship is the reverse: often there exist a number of possible default priors, and it is then natural to consider a Bayesian robustness analysis with respect to the class of default or noninformative priors.

1.7 Future work

We finish this paper with a brief discussion of potential topics for further research. Considerably more discussion of this issue can be found in the papers in the volume.

The most important challenge for the field of Bayesian robustness is to increase its impact on statistical practice; indeed, to make it a routine component of applied work. One avenue for doing so is simply to show, by example, the power of the methods. Thus we have included several papers in this volume that illustrate the uses of Bayesian robustness in applied settings. Marin (2000) applies a robust version of dynamic linear models to forecast unemployment and activity ratios. Carlin and Perez (2000) illustrates forward and backward robustness methods in epidemiological settings. Bielza et al. (2000) demonstrates the use of simulation methods and the expected value of perfect information to perform sensitivity analysis in a complex medical decision making problem. Cagno et al. (2000) applies robust methods to predict failures in gas pipelines, and Wilson and Wiper (2000) applies them in software reliability problems.

Perhaps the most important way to bring robust Bayesian methods to practitioners is to have these methods available in standard statistical software. There are several ongoing efforts to develop general Bayesian statistical software, and incorporation of robust Bayesian methodology into such software is clearly desirable. Indeed, as indicated in Section 1.4.4, Bayesian robustness can perhaps best be viewed as a component of general Bayesian software. Ideally, such software would proceed as follows:

- Elicit preferences and beliefs from the decision maker.
- Conduct a robustness study with classes compatible with the information elicited, and certain representative quantities of interest.
- If robustness appears to hold, conduct a standard Bayesian analysis with a representative prior/model/loss from the classes (or with respect to some hyperprior average); in part, performing the reported analysis with a single prior/model/loss would be for ease in communication.
- If robustness does not hold, elicit additional information to further constrain the classes (with the software perhaps suggesting which additional information would be best to elicit), or obtain additional data, and repeat.

The least understood aspect of this program is the possibility of interactive elicitation and is a topic that deserves much more research. Another aspect that requires clarification is the extent to which one can automate the process of determining whether or not robustness holds. Frequently, this will be a context-dependent decision, but generic calibration of robustness measures is often possible – some examples are given in McCulloch (1989) and Ríos Insua (1990) – and further efforts in this direction are clearly of interest.

In the introduction, we referred to the enormous impact that MCMC methods have had on Bayesian analysis, and we mentioned that Bayesian robustness methodology will need to be compatible with MCMC methods to become widely used. There are some useful results in this direction. For instance, once one has obtained an MCMC sample from the posterior associated with a baseline prior, obtaining an MCMC sample from the posterior of a "close" prior can be done via importance sampling, see for example Smith and Gelfand (1992). But much additional work needs to be done to more closely relate robust Bayesian techniques with MCMC.

Another area of enduring importance to the field is foundations. As will be seen in later papers, there is still interest in axiomatic models that lead to robust Bayesian analysis. For instance, a direct implication of most axiomatic systems is that nondominated actions or estimators should be considered. Indeed, a topic of considerable importance is that of further developing methods for approximating the set of nondominated estimators or actions, as explained in Martin and Arias (2000). On the other hand, axiomatics such as in Nau (1995), lead to somewhat different models that would require new computational developments.

To summarize, there are many challenging problems in robust Bayesian analysis from methodological, foundational and computational points of view. The field has a very exciting future.

Acknowledgments

Work supported by a HID-CICYT grant and by the National Science Foundation (USA), Grant DMS-9802261.

References

ALBERT, J., DELAMPADY, M. and POLASEK, W. (1991). A class of distributions for robustness studies. *Journal of Statistical Planning and Inference*, **28**, 291–304.

BASU, S. (1994). Variations of posterior expectations for symmetric unimodal priors in a distribution band. *Sankhya, A*, **56**, 320–334.

BASU, S. (1995). Ranges of posterior probability over a distribution band. *Journal of Statistical Planning and Inference*, **44**, 149–166.

BASU, S. (1996). Local sensitivity, functional derivatives and nonlinear posterior quantities. *Statistics and Decisions*, **14**, 405–418.

BASU, S. (2000). Bayesian robustness and Bayesian nonparametrics. In *Robust Bayesian Analysis* (D. Ríos Insua and F. Ruggeri, eds.). New York: Springer–Verlag.

BASU, S. and DASGUPTA, A. (1995). Robust Bayesian analysis with distribution bands. *Statistics and Decisions*, **13**, 333–349.

BERGER, J.O. (1984). The robust Bayesian viewpoint (with discussion). In *Robustness of Bayesian Analysis* (J. Kadane, ed.), Amsterdam: North-Holland.

BERGER, J.O. (1985). *Statistical Decision Theory and Bayesian Analysis* (2nd Edition). New York: Springer-Verlag.

BERGER, J.O. (1990). Robust Bayesian analysis: sensitivity to the prior. *Journal of Statistical Planning and Inference*, **25**, 303–328.

BERGER, J.O. (1994). An overview of robust Bayesian analysis. *TEST*, **3**, 5–58.

BERGER, J.O. and BERLINER, L.M. (1986). Robust Bayes and empirical Bayes analysis with ε-contaminated priors. *Annals of Statistics*, **14**, 461–486.

BERGER, J.O., BETRÒ, B., MORENO, E., PERICCHI, L.R., RUGGERI, F., SALINETTI, G. and WASSERMAN L. (1996). *Bayesian Robustness*, IMS Lecture Notes - Monograph Series, **29**. Hayward: IMS.

BERGER, J.O. and DELAMPADY, M. (1987). Testing precise hypotheses (with discussion). *Statistical Science*, **2**, 317–352.

BERGER, J.O. and O'HAGAN, A. (1988). Ranges of posterior probabilities for unimodal priors with specified quantiles. In *Bayesian Statistics 3*, (J.M. Bernardo, M.H. DeGroot, D.V. Lindley and A.F.M. Smith, eds.). Oxford: Oxford University Press.

BERGER, J.O. and SELLKE, T. (1987). Testing a point null hypothesis: the irreconcilability of p-values and evidence. *Journal of the American Statistical Association*, **82**, 112–122.

BETRÒ, B. and GUGLIELMI, A. (2000). Methods for global prior robustness under generalized moment conditions. In *Robust Bayesian Analysis* (D. Ríos Insua and F. Ruggeri, eds.). New York: Springer-Verlag.

BETRÒ, B., MĘCZARSKI, M. and RUGGERI, F. (1994). Robust Bayesian analysis under generalized moments conditions. *Journal of Statistical Planning and Inference*, **41**, 257–266.

BIELZA, C., RÍOS INSUA, S., GÓMEZ, M. and FERNÁNDEZ DEL POZO, J.A. (2000). Sensitivity analysis in IctNeo. In *Robust Bayesian Analysis* (D. Ríos Insua and F. Ruggeri, eds.). New York: Springer-Verlag.

BOX, G.E.P. and TIAO, G.C. (1962). A further look at robustness via Bayes theorem. *Biometrika*, **49**, 419–432 and 546.

BOX, G.E.P. and TIAO, G.C. (1973). *Bayesian Inference in Statistical Analysis.* New York: Addison-Wesley.

CAGNO, E., CARON, F., MANCINI, M. and RUGGERI, F. (2000). Sensitivity of replacement priorities for gas pipeline maintenance. In *Robust Bayesian Analysis* (D. Ríos Insua and F. Ruggeri, eds.). New York: Springer-Verlag.

CARLIN, B.P. and PÉREZ, M.E. (2000). Robust Bayesian analysis in medical and epidemiological settings. In *Robust Bayesian Analysis* (D. Ríos Insua and F. Ruggeri, eds.). New York: Springer-Verlag.

CUEVAS, A. and SANZ, P. (1988). On differentiability properties of Bayes operators. In *Bayesian Statistics 3* (J.M. Bernardo, M.H. DeGroot, D.V. Lindley and A.F.M. Smith, eds.), 569–577. Oxford: Oxford University Press.

DEY, D.K., GHOSH, S.K. and LOU, K. (1996). On local sensitivity measures in Bayesian analysis. In *Bayesian Robustness, IMS Lecture Notes - Monograph Series* (J.O. Berger, B. Betrò, E. Moreno, L.R. Pericchi, F. Ruggeri, G. Salinetti and L. Wasserman, eds.), 21–39. Hayward: IMS.

DEY, D., LOU, K. and BOSE, S. (1998). A Bayesian approach to loss robustness. *Statistics and Decisions*, **16**, 65–87.

DEY, D. and MICHEAS, A. (2000). Ranges of posterior expected losses and ϵ–robust actions. In *Robust Bayesian Analysis* (D. Ríos Insua and F. Ruggeri, eds.). New York: Springer-Verlag.

FORTINI, S. and RUGGERI, F. (2000). On the use of the concentration function in Bayesian robustness. In *Robust Bayesian Analysis* (D. Ríos Insua and F. Ruggeri, eds.). New York: Springer-Verlag.

FRENCH, S. and RÍOS INSUA, D. (2000). *Statistical Decision Theory.* London: Arnold.

GODSILL, S. J. and RAYNER, P.J.W. (1996). Robust treatment of impulsive noise in speech and audio signals. In *Bayesian Robustness, IMS Lecture Notes - Monograph Series*, (J.O. Berger, B. Betrò, E. Moreno, L.R. Pericchi, F. Ruggeri, G. Salinetti, L. Wasserman eds.), 41–62. Hayward: IMS.

GREENHOUSE, J. and WASSERMAN, L. (1996). A practical robust method for Bayesian model selection: a case study in the analysis of clinical trials (with discussion). In *Bayesian Robustness, IMS Lecture Notes - Monograph Series* (J.O. Berger, B. Betrò, E. Moreno, L.R. Pericchi, F. Ruggeri, G. Salinetti and L. Wasserman, eds.), 331–342. Hayward: IMS.

GUSTAFSON P. (2000). Local robustness in Bayesian analysis. In *Robust Bayesian Analysis* (D. Ríos Insua and F. Ruggeri, eds.). New York: Springer-Verlag.

HAMPEL, F.R., RONCHETTI, E.M., ROUSSEEUW, P.J. and STAHEL, W.A. (1986). *Robust Statistics: The Approach Based on Influence Functions.* New York: Wiley.

HUBER, P.J. (1973). The use of Choquet capacities in statistics. *Bullettin of the International Statistics Institute*, **45**, 181–191.

HUBER, P.J. (1981). *Robust Statistics.* New York: Wiley.

KADANE, J.B. and CHUANG, D.T. (1978). Stable decision problems. *Annals of Statistics*, **6**, 1095–1110.

KADANE, J.B., SALINETTI, G. and SRINIVASAN, C. (2000). Stability of Bayes decisions and applications. In *Robust Bayesian Analysis* (D. Ríos Insua and F. Ruggeri, eds.). New York: Springer-Verlag.

KASS, R. and WASSERMAN, L. (1996). The selection of prior distributions by formal rules. *Journal of the American Statistical Association*, **91**, 1343–1370.

LAVINE, M. (1991). Sensitivity in Bayesian statistics: the prior and the likelihood. *Journal of the American Statistical Association*, **86**, 396–399.

LAVINE, M., PERONE PACIFICO, M., SALINETTI, G. and TARDELLA, G. (2000). Linearization techniques in Bayesian robustness. In *Robust Bayesian Analysis* (D. Ríos Insua and F. Ruggeri, eds.). New York: Springer-Verlag.

LISEO, B. (2000). Robustness issues in Bayesian model selection. In *Robust Bayesian Analysis* (D. Ríos Insua and F. Ruggeri, eds.). New York: Springer-Verlag.

LISEO, B., PETRELLA, L. and SALINETTI, G. (1996). Bayesian robustness: An interactive approach. In *Bayesian Statistics 5* (J.O. Berger, J.M. Bernardo, A.P. Dawid and A.F.M. Smith, eds.), 661–666. Oxford: Oxford University Press.

MacEachern, S. and Müller, P. (2000). Efficient MCMC schemes for robust model extensions using encompassing Dirichlet process mixture models. In *Robust Bayesian Analysis* (D. Ríos Insua and F. Ruggeri, eds.). New York: Springer-Verlag.

Marin, J.M. (2000). A robust version of the dynamic linear model with an economic application. In *Robust Bayesian Analysis* (D. Ríos Insua and F. Ruggeri, eds.). New York: Springer-Verlag.

Martín, J. and Arias, J.P. (2000). Computing the efficient set in Bayesian decision problems. In *Robust Bayesian Analysis* (D. Ríos Insua and F. Ruggeri, eds.). New York: Springer-Verlag.

Martín, J. and Ríos Insua, D. (1996). Local sensitivity analysis in Bayesian decision theory (with discussion). In *Bayesian Robustness, IMS Lectures-Notes Monograph Series* (J.O. Berger, B. Betró, E. Moreno, L.R. Pericchi, F. Ruggeri, G. Salinetti and L. Wasserman eds.), Vol 29, 119–135. Hayward: IMS.

Martín, J., Ríos Insua, D. and Ruggeri, F. (1998). Issues in Bayesian loss robustness. *Sankhya, A*, **60**, 405–417.

McCulloch, R. (1989). Local model influence. *Journal of the American Statistical Association*, **84**, 473–478.

Moreno, E. (2000). Global Bayesian robustness for some classes of prior distributions. In *Robust Bayesian Analysis*, (D. Ríos Insua and F. Ruggeri, eds.). New York: Springer-Verlag.

Moreno, E. and Cano, J.A. (1989). Testing a point null hypothesis: Asymptotic robust Bayesian analysis with respect to the priors given on a subsigma field. *International Statistical Review*, **57**, 221–232.

Moreno, E. and Cano, J.A. (1991). Robust Bayesian analysis for ε-contaminations partially known. *Journal of the Royal Statistical Society, B*, **53**, 143–155.

Moreno, E. and González, A. (1990). Empirical Bayes analysis for ε-contaminated priors with shape and quantile constraints. *Brazilian Journal of Probability and Statistics*, **4**, 177–200.

Moreno, E., Martínez, C. and Cano, J.A. (1996). Local robustness and influence for contamination classes of prior distributions (with discussion). In *Bayesian Robustness, IMS Lecture-Notes Monograph Series* (J.O. Berger, B. Betró, E. Moreno, L.R. Pericchi, F. Ruggeri, G. Salinetti and L. Wasserman, eds.), Vol. 29, 137–154. Hayward: IMS.

MORENO, E. and PERICCHI, L.R. (1993). Bayesian robustness for hierarchical ε-contamination models. *Journal of Statistical Planning and Inference*, **37**, 159–168.

NAU, R.(1995). The shape of incomplete preferences. *Technical Report.* Duke University.

O'HAGAN, A. and BERGER, J.O. (1988). Ranges of posterior probabilities for quasi-unimodal priors with specified quantiles. *Journal of the American Statistical Association*, **83**, 503–508.

PEÑA, D. and ZAMAR, R.H. (1996). On Bayesian robustness: an asymptotic approach. In *Robust Statistics, Data Analysis and Computer Intensive Methods* (H. Rieder, ed.). New York: Springer-Verlag.

PERICCHI, L.R. and WALLEY, P. (1991). Robust Bayesian credible intervals and prior ignorance. *International Statistical Review*, **58**, 1–23.

RÍOS INSUA, D. (1990). *Sensitivity Analysis in Multiobjective Decision Making.* New York: Springer-Verlag.

RÍOS INSUA, D. and CRIADO, R. (2000). Topics on the foundations of robust Bayesian analysis. In *Robust Bayesian Analysis* (D. Ríos Insua and F. Ruggeri, eds.). New York: Springer-Verlag.

RÍOS INSUA, D., RUGGERI, F. and MARTIN, J. (2000). Bayesian sensitivity analysis: a review. To appear in *Handbook on Sensitivity Analysis* (A. Saltelli et al., eds.). New York: Wiley.

RÍOS INSUA, S., MARTIN, J., RÍOS INSUA, D. and RUGGERI, F. (1999). Bayesian forecasting for accident proneness evaluation. *Scandinavian Actuarial Journal*, **99**, 134–156.

RUGGERI, F. (1990). Posterior ranges of functions of parameters under priors with specified quantiles. *Communications in Statistics A: Theory and Methods*, **19**, 127–144.

RUGGERI, F. and SIVAGANESAN, S. (2000). On a global sensitivity measure for Bayesian inference. To appear in *Sankhya, A.*

RUGGERI, F. and WASSERMAN, L. (1993). Infinitesimal sensitivity of posterior distributions. *Canadian Journal of Statistics*, **21**, 195–203.

SALTELLI, A., CHAN, K. and SCOTT, M. (2000). *Mathematical and Statistical Methods for Sensitivity Analysis.* New York: Wiley.

SARGENT, D.J. and CARLIN, B.P. (1996). Robust Bayesian design and analysis of clinical trials via prior partitioning (with discussion). In *Bayesian Robustness, IMS Lecture Notes – Monograph Series* (J.O. Berger, B. Betrò, E. Moreno, L.R. Pericchi, F. Ruggeri, G. Salinetti, and L. Wasserman, eds.), 175–193. Hayward: IMS.

SELLKE, T., BAYARRI, M.J. and BERGER, J.O. (1999). Calibration of p-values for precise null hypotheses. *ISDS Discussion Paper 99–13,* Duke University.

SHYAMALKUMAR, N.D. (2000). Likelihood robustness. In *Robust Bayesian Analysis* (D. Ríos Insua and F. Ruggeri, eds.). New York: Springer-Verlag.

SIVAGANESAN, S. (1988). Range of posterior measures for priors with arbitrary contaminations. *Communications in Statistics A: Theory and Methods*, **17**, 1591–1612.

SIVAGANESAN, S. (1989). Sensitivity of posterior mean to unimodality preserving contaminations. *Statistics and Decisions*, **7**, 77–93.

SIVAGANESAN, S. (1991). Sensitivity of some standard Bayesian estimates to prior uncertainty – a comparison. *Journal of Statistical Planning and Inference*, **27**, 85–103.

SIVAGANESAN, S. (1993). Robust Bayesian diagnostics. *Journal of Statistical Planning and Inference*, **35**, 171–188.

SIVAGANESAN, S. (2000). Global and local robustness approaches: uses and limitations. In *Robust Bayesian Analysis*, (D. Ríos Insua and F. Ruggeri, eds.). New York: Springer-Verlag.

SIVAGANESAN, S. and BERGER, J. (1989). Ranges of posterior measures for priors with unimodal contaminations. *Annals of Statistics*, **17**, 868–889.

SIVAGANESAN, S., BERLINER, L.M. and BERGER, J. (1993). Optimal robust credible sets for contaminated priors. *Statistics and Probability Letters*, **18**, 383–388.

SMITH, A.F.M. and GELFAND, A.E. (1992). Bayesian statistics without tears: a sampling-resampling perspective. *American Statistician*, **46**, 84–88.

VIDAKOVIC, B. (2000). Γ-minimax: a paradigm for conservative robust Bayesians. In *Robust Bayesian Analysis* (D. Ríos Insua and F. Ruggeri, eds.). New York: Springer-Verlag.

WALLEY, P. (1991). *Statistical Reasoning With Imprecise Probabilities.* London: Chapman and Hall.

WASSERMAN, L. (1992). Recent methodological advances in robust Bayesian inference. In *Bayesian Statistics 4* (J.M. Bernardo, J.O. Berger, A. Dawid and A.F.M. Smith, eds.). Oxford: Oxford University Press.

WILSON, S.P. and WIPER, M.P. (2000). Prior robustness in some common types of software reliability models. In *Robust Bayesian Analysis* (D. Ríos Insua and F. Ruggeri, eds.). New York: Springer-Verlag.

2

Topics on the Foundations of Robust Bayesian Analysis

David Ríos Insua and Regino Criado

ABSTRACT One of the strengths of the Bayesian approach resides in the wealth and variety of axiomatic systems leading to the subjective expected loss model, which underpins Bayesian inference and decision analysis. This chapter surveys whether a similar situation holds for robust Bayesian analysis, overviewing foundational results leading to standard computations in robust Bayesian analysis.

Key words: Robust Bayesian analysis, foundations.

2.1 Introduction

One of the strengths of the Bayesian approach is the variety of axiomatic systems leading to the subjective expected loss model. Indeed, under several perspectives, we are essentially led to: i) model beliefs with a probability distribution, which in the presence of additional information is updated through Bayes' formula; ii) model preferences and risk attitudes over consequences with a loss function; and iii) model preferences over alternatives according to the minimum (subjective) expected loss principle. This framework underpins standard Bayesian inference and decision analysis modelling and computations. The aforementioned perspectives include Scott's axioms for probability; Von Neumann and Morgestern's axioms for expected utility; De Groot's two-stage axiomatisations for probabilities and utilities and Anscombe and Aumann's and Savage's joint axiomatisations (see, e.g., French and Ríos Insua (2000) for an introduction). The importance of these axiomatic systems stems not only because of their justification of the Bayesian approach, but also because, when applying the framework in a practical setting, if we need to make certain compromises with the foundations, we know which compromises are actually being made.

One example is robust Bayesian analysis. As argued in various chapters in Ríos Insua and Ruggeri (2000), one of its main motivations stems from the practical difficulty of assessing beliefs and preferences with precision, particularly in the presence of several decision makers (DMs) and/or experts. Some would even argue that priors, models and losses can never be quantified exactly. This suggests imprecision in judgments modelled by classes of losses, priors and models, the compromise with the foundations being on the assumption of completeness of beliefs and preferences.

In this chapter, we shall explore the computational implications of dropping completeness from standard Bayesian axioms, in an attempt to outline an appropriate axiomatic framework for robust Bayesian analysis. We shall see that developments are not as well rounded as in the precise (Bayesian) case, although various partial results exist which essentially lead to the same conclusion: imprecise beliefs and preferences may be modelled by a class of priors and a class of loss functions, so that preferences among alternatives may be represented by inequalities of the corresponding posterior expected losses. More technically, under certain conditions, we are able to model beliefs by a class Γ of priors π, which in the presence of additional information, are updated, one by one, via Bayes' theorem, and preferences by a class $\mathcal{L}$ of loss functions L, so that we find alternative a at most as preferred as b ($a \preceq b$) if and only if the posterior expected loss of a is greater than the posterior expected loss of b for any combination of losses and priors within the class, that is,

$$a \preceq b \iff \left(\left(\int L(a,\theta)\pi(\theta|x)d\theta \geq \int L(b,\theta)\pi(\theta|x)d\theta, \forall \pi \in \Gamma\right), \forall L \in \mathcal{L}\right).$$

The basic argument for such results assumes that the preference relation, instead of being a weak order (a complete, transitive binary relation), will be a quasi-order (a reflexive, transitive binary relation). Then, assuming conditions similar to the Bayesian axioms, we are able to obtain the representation sketched above.

We start by reviewing conditions similar to Scott's axioms for probability. Then, we provide several axiomatics which stem from modelling a class of linear functionals in a normed space and apply it to the axiomatisations of only probabilities. We then describe joint axiomatisations of beliefs and preferences, and the entailed computational problems, essentially leading to the computation of nondominated actions or estimators.

Since the nondominated set may be too big to reach a final conclusion, we explore foundations for Γ-minimax analyses, which lead to some departure from the robust Bayesian postulates. We could view the approach as follows: the statistician wants to behave as a Bayesian; he lacks information to obtain a prior, but he may obtain a class of priors; if this class does not lead to similar conclusions and he is not able to get additional information, he might be willing to break some Bayesian postulates to achieve a final answer.

2.2 Modelling beliefs

We start with the simplest case of modelling beliefs over a finite set $\Theta = \{1, \ldots, k\}$ of states ($k \geq 1$). The DM expresses his beliefs over $\mathcal{A}$, the algebra of subsets of Θ. They are modelled in terms of a binary relation $\preceq_\ell$ on $\mathcal{A}$ described as follows: "$B \preceq_\ell C$" for $B, C \in \mathcal{A}$ means he finds "B at most as likely to occur as C". We aim at modelling these beliefs by means of a probability distribution π on $(\Theta, \mathcal{A})$ giving a representation

$$B \preceq_\ell C \iff \pi(B) \leq \pi(C).$$

Scott's (1964) axioms provide such a representation. His axioms actually imply that $\preceq_\ell$ is a weak order.

We review here similar conditions leading to modelling beliefs by means of a class of probability distributions. We associate with each subset $B \in \mathcal{A}$ its indicator I_B. We have, see Ríos Insua (1990),

Proposition 1 . *Let $\Theta, \mathcal{A}, \preceq_\ell$ be as above. $(\mathcal{A}, \preceq_\ell)$ satisfies the conditions*

1. $\emptyset \prec_\ell \Theta$, $\emptyset \preceq_\ell B, \forall B \in \mathcal{A}$, *and*

2. *If for all integers $l, r \geq 1$ and for all $A_1, B_1, \ldots, A_{l+1}, B_{l+1} \in A$, such that $A_i \preceq_\ell B_i$, $i = 1, \ldots, l$ and*

$$I_{A_1} + \ldots + I_{A_l} + rI_{A_{l+1}} = I_{B_1} + \ldots + I_{B_l} + rI_{B_{l+1}},$$

then

$$B_{l+1} \preceq_\ell A_{l+1}$$

iff there is a (nonempty, finite) family Γ of probabilities π such that

$$B \preceq_\ell C \Leftrightarrow \pi(B) \leq \pi(C), \forall \pi \in \Gamma.$$

The proof uses standard results from solutions of linear inequalities; see Fishburn (1986) for other results. We may see that $(\mathcal{A}, \preceq_\ell)$ is a quasi-order.

We now describe conditional probabilities. As it happens with probability weak orders, non unicity may lead to problems while defining probabilities of events conditional on different events; see Fishburn (1986). However, in our context, it is enough to deal with probabilities conditional on the same event. $B|D$ represents the event B given D.

Proposition 2 . *Let $\Theta, \mathcal{A}, \preceq_\ell$ be as in Proposition 1. Assume, furthermore, that for any $B, C, D \in \mathcal{A}$, such that $\emptyset \prec_\ell D$,*

$$B|D \preceq_\ell C|D \iff B \cap D \preceq_\ell C \cap D.$$

Then, there is a family Γ *of probabilities such that for any* $B, C, D \in \mathcal{A}$, *with* $\emptyset \prec_\ell D$,

$$B|D \preceq_\ell C|D \iff \pi(B|D) \leq (\pi(C|D), \ \forall \pi \in \Gamma : \pi(D) > 0)\,,$$

where $\pi(B|D) = \pi(B \cap D)/\pi(D)$.

Albeit its limited context, the above results already describe the usual robust approach of modelling beliefs with a class of probability distributions, updated via Bayes' formula and performing various computational explorations within such a class, like computing ranges of posterior probabilities of certain sets or posterior expectations.

2.3 Towards joint models of beliefs and preferences

Most of the results we shall later provide stem from variations of the following lemma representing a quasi-order in a normed space (see, e.g., Ríos Insua (1990) for related proofs).

Lemma 1 . *Let* X *be a convex compact set in a normed real space* Y *and* $\preceq$ *a binary relation on it. Then, the following three conditions*

A1. $(X, \preceq)$ *is transitive and reflexive (quasi-order).*

A2. For $\alpha \in (0,1), x, y, z \in X$, $x \preceq y \iff \alpha x + (1-\alpha) z \preceq \alpha y + (1-\alpha) z$ *(independence condition).*

A3. For $x_n, y_n \in X$, *if* $x_n \to x$, $y_n \to y$, *and* $x_n \preceq y_n$, $\forall n$, *then* $x \preceq y$ *(continuity condition),*

are equivalent to the existence of a set W *of continuous linear functions* w *on* Y *such that,* $\forall x, y \in X$,

$$x \preceq y \iff (w(x) \leq w(y), \forall w \in W).$$

Other authors have provided related results. For example, Ríos Insua (1992) provides a result with a continuity condition alternative to A3, which Burtel (1997) shows is valid only in finite-dimensional spaces. This author also provides an alternative, more artificial, continuity condition.

Let us see the role of the previous type of results within robust Bayesian analysis. We start from a seminal theorem in Giron and Ríos (1980). They implicitly assume precision in preferences over consequences described by a (loss) function $L(a, \theta)$ to evaluate the consequence associated with alternative $a \in A$ and state $\theta \in \Theta$. Let

$$\mathcal{D} = \{h \mid \exists a \in A, h(\theta) = L(a, \theta), \forall \theta \in \Theta\}\,.$$

In principle, preferences $\preceq$ are established over these functions. However, they consider an extension of $\preceq$ over the set $C(\Theta)$ of continuous functions over Θ (they also consider the set of bounded functions). Their result (see also Walley, (1991)) states, under appropriate topological and measurability conditions, that:

Theorem 1 . *Suppose that* $(C(\Theta), \preceq)$ *satisfies the following conditions:*

GR1 $(C(\Theta), \preceq)$ *is a quasi-order.*

GR2 *For* $f, g, h \in C(\Theta)$, $\lambda \in (0,1)$, *then* $f \preceq g \iff \lambda f + (1-\lambda)h \preceq \lambda g + (1-\lambda)h$.

GR3 *For* $f_n, g, h \in C(\Theta)$, *if* $f_n \to f$ *and* $f_n \preceq g$, $h \preceq f_n$, $\forall n$, *then* $f \preceq g$, $h \preceq f$.

GR4 *If* $f(\theta) < g(\theta), \forall \theta \in \Theta$, *then* $g \prec f$.

Then, there exists a class Γ *of probability measures* π *on* Θ *such that, for* $a, b \in A$,

$$a \preceq b \iff \left(\int L(a,\theta)\pi(\theta)d\theta \geq \int L(b,\theta)\pi(\theta)d\theta, \forall \pi \in \Gamma \right).$$

Hence, we see that we model imprecision in beliefs by means of a class of probability distributions. We insist that, in this result, preferences over consequences are modelled with a single loss function. Interestingly enough, when $(C(\Theta), \preceq)$ is complete, that is, a weak order, Γ is a singleton. Under additional conditions, we may see that beliefs are updated applying Bayes' formula to each of the priors; see Giron and Ríos (1980) for a full discussion.

Observe that we have, therefore, the usual setting in sensitivity to the prior robust analyses. For example, if the loss is quadratic, since the optimal estimator is the posterior mean, we could explore how the posterior mean behaves as the prior varies in the class. We believe, though, that, conceptually, the appropriate computational aim would be the search for nondominated estimators, as later described.

A criticism to the previous result is the existence (by default) of a loss function. We can actually relax such assumption, adopting the richer framework in Anscombe and Aumann (1963). Consequences c of our decisions are in a space $\mathcal{C}$ of consequences. We consider $\mathcal{P}$, the set of distributions over $\mathcal{C}$ with finite support. The statistician has to choose among Anscombe–Aumann (AA) acts, which are measurable functions g from Θ to $\mathcal{P}$. The set of AA acts will be designated by $\mathcal{G}$. The set of AA constant acts will be $\mathcal{G}_c$. Preference among the elements of $\mathcal{G}$ will be denoted by $\preceq$. Note that $\mathcal{C}$ is embedded in $\mathcal{P}$, by identifying the consequence c with the probability distribution degenerate at c. Similarly, $\mathcal{P}$ is embedded in $\mathcal{G}$, by identifying the probability distribution p with the constant act with image p. Consequently, we induce preference relations in $\mathcal{P}$ and $\mathcal{C}$ in a natural way. We consider only simple Anscombe–Aumann acts, that is, acts which are measurable and take only finitely many values, their set being $\mathcal{G}_s$.

We next state the announced result. As a consequence of the suggested properties, we shall model the statistician's beliefs by means of a class of probability distributions and her preferences over consequences with a loss function, and acts will be ordered in a Pareto sense, based on comparisons of the corresponding expected losses. We describe first the required conditions and then state the result.

RB1a. Quasi-order $(\mathcal{G}_s, \preceq)$ is a quasi-order.

RB1b. Weak order for constants acts For constant acts $g_1, g_2 \in \mathcal{G}_c$,

$$g_1 \preceq g_2 \text{ or } g_2 \preceq g_1.$$

In words, preferences are transitive and reflexive. When we consider constant acts, preferences are complete. We do not require this for non-constant acts.

RB2. Independence As in GR2, applied to Anscombe–Aumann acts. In words, preferences depend only on parts of acts which are not common.

RB3. Continuity As in GR3.

RB4. Nontriviality There are two constant acts g_1, g_2 such that $g_1 \prec g_2$. We demand that the statistician find two constant acts, say $g_1(\theta) = p, g_2(\theta) = q, \forall\theta$, such that $p \prec q$.

RB5. Monotonicity For $B, D \in \mathcal{A},\ B \subset D,\ pI_{B^c} + qI_B \preceq pI_{D^c} + qI_D$. This is a dominance condition: the consequence obtained for the right-side act is preferred to the consequence obtained for the left act, hence we should prefer the right act to the left one.

We use p, q as in RB4 and RB5 to define, in a natural way, a likelihood relation $\preceq_\ell$ between events in $\mathcal{A}$:

$$B \preceq_\ell D \iff pI_{B^c} + qI_B \preceq pI_{D^c} + qI_D.$$

It is this relation $\preceq_\ell$ that we shall model with a class Γ of probability distributions.

The final axiom acknowledges the fact that preferences among acts should be based on the beliefs with which various consequences will be received. To do so, we shall need the concept of probability measure $y(g, r)$ induced by an act g and a probability measure r over Θ; see Fishburn (1982). We efine it by $y(g,r)(C) = \int g(\theta)(C)dr(\theta)$, where $g(\theta)(C)$ is the probability at act g gives to set C under state θ. If beliefs were precise, then it would natural to demand that $g_1 \preceq g_2 \iff y(g_1, r) \preceq y(g_2, r)$. Since there precision in beliefs, modelled by a class Γ of probability distributions, all demand

RB6. Compatibility of preferences under risk and under uncertainty

$$g_1 \preceq g_2 \iff (y(g_1, \pi) \preceq y(g_2, \pi), \forall \pi \in \Gamma).$$

Note that, for a constant act $g = qI_\Theta$, $y(g, \pi) = q, \forall \pi \in \Gamma$, so RB6 introduces no inconsistency in this case. Similarly, when $g = pI_{A^c} + qI_A$, $y(g, \pi) = (1 - \pi(A))p + \pi(A)q$, so RB6 introduces no inconsistency with $\preceq_\ell$.

We then have

Proposition 3 . *Let $\preceq$ be a binary relation in $\mathcal{G}_s$. Then:*

- *If $(\mathcal{G}_s, \preceq)$ satisfies assumptions RB1a, RB2–RB5, there is a closed convex class Γ of (finitely additive) probability distributions π such that*

$$A \preceq_\ell B \iff (\pi(A) \leq \pi(B), \forall \pi \in \Gamma).$$

- *If, in addition, RB1b and RB6 hold, there is a function L such that*

$$g_1 \preceq g_2 \iff \left(\int L(g_1(\theta))\pi(\theta)d\theta \geq \int L(g_2(\theta))\pi(\theta)d\theta, \forall \pi \in \Gamma \right).$$

Observe that if we define L in $\mathcal{C}$ by $L(c) = L(p_c)$, where p_c is the distribution degenerate at c, then any $p \in \mathcal{P}$ may be written

$$p = \sum_{i=1}^{n} p(c_i)p_{c_i},$$

and, by additivity,

$$L(p) = \sum p(c_i)L(p_{c_i}) = \sum p(c_i)L(c_i) = \int L(c)p(c)dc.$$

Hence, the result may be rewritten as

$$g_1 \preceq g_2 \iff$$

$$\left(\int \int L(c)g_1(\theta)(c)dc\, \pi(\theta)d\theta \geq \int \int L(c)g_2(\theta)(c)dc\, \pi(\theta)d\theta, \forall \pi \in \Gamma \right).$$

For the case often considered in which the lotteries $g(\theta)$ are degenerate for every θ (called pure horse lotteries in decision theory jargon), we then have, if (g, θ) designates the consequence obtained under act g and state θ,

$$g_1 \preceq g_2 \iff \left(\int L(g_1, \theta)\pi(\theta)d\theta \geq \int L(g_2, \theta)\pi(\theta)d\theta, \forall \pi \in \Gamma \right).$$

Again, by including an additional condition, we update beliefs via Bayes' formula applied to each distribution in the class.

As a consequence, in statistical problems, when there is information about θ provided by an experiment of interest with result x, we shall order acts according to

$$g_1 \preceq g_2 \iff \left(\int L(g_1, \theta)\pi(\theta|x)d\theta \geq \int L(g_2, \theta)\pi(\theta|x)d\theta, \forall \pi \in \Gamma \right).$$

Though we shall not discuss the issue of modelling preferences only, we shall just mention that dual results leading to a representation based on inequalities of expected losses, with a precise prior and imprecision in preferences modelled by a class of loss functions may be seen, for example, in Ríos Insua and Martin (1994).

The difficulty in assessing beliefs and preferences may lead us to wonder about joint axiomatisations of imprecise beliefs and preferences. This would provide models with a class of priors and a class of loss functions. We outline here some approaches to such problems.

Ríos Insua and Martin (1995) provide a joint axiomatisation based on the one given above. Essentially they show that removing RB1b from above leads to the model

$$g_1 \preceq g_2 \iff \left(\int L(g_1, \theta)\pi(\theta|x)d\theta \geq \int L(g_2, \theta)\pi(\theta|x)d\theta, \forall \pi \in \Gamma, \forall L \in \mathcal{L} \right).$$

Condition RB6 may seem too strong. Nau (1995) obtains a similar model by introducing the alternative condition of *partial substitution*. Such a condition requires that if

$$\alpha f + (1-\alpha)(I_E f' + I_{E^c} h) \preceq \alpha g + (1-\alpha)(I_E g' + I_{E^c} h)$$

for some $\alpha \in (0,1)$, where f', g' and h are constant lotteries and E is not potentially null, then

$$\alpha f + (1-\alpha)(pf' + (1-p)h) \preceq \alpha g + (1-\alpha)(pg' + (1-p)h)$$

for some $p \in (0,1]$. Nau's representation is actually in terms of a class of pairs of loss functions and probability distributions $(L_v, \pi_v)_{v \in V}$ such that

$$g_1 \preceq g_2 \iff$$

$$\int \int L_v(c) g_1(\theta)(c) dc\, \pi_v(\theta) d\theta \geq \int \int L_v(c) g_2(\theta)(c) dc\, \pi_v(\theta) d\theta, \forall (L_v, \pi_v)_{v \in V}.$$

Seidenfeld, Schervish and Kadane (1995) provide another interesting representation.

2.4 Γ-minimax approaches

The basic computational consequence of most previous results is that acts or alternatives are ordered in a Pareto sense as

$$a \preceq b \iff \left(\left(\int L(a,\theta)\pi(\theta|x)d\theta \geq \int L(b,\theta)\pi(\theta|x)d\theta, \forall \pi \in \Gamma \right), \forall L \in \mathcal{L} \right).$$

On the one hand, these models outline the typical context for robust Bayesian analyses in which we perturbate the prior and the loss within classes. On the other hand, they suggest that we should look for nondominated acts: $a \in A$ is nondominated if there is no other $b \in A$ such that $a \prec b$. Martin and Arias (2000) describe how the nondominated set may be approximated.

However, as shown in Martin and Arias (2000), the set of nondominated acts will not be a singleton. In the event that those acts differ a lot, in terms of their expected losses, we should try to get additional information. Strictly speaking, we should aim at reducing Γ and $\mathcal{L}$, so that we reduce the set of nondominated acts, and, even, differences in posterior expected losses between alternatives. However, it is conceivable that the statistician may not be able to obtain additional information. In that case, any nondominated act could be suggested as a solution for the problem. We could appeal to several heuristic procedures, to pick one nondominated solution. We concentrate here on the $\Gamma \times \mathcal{L}$-minimax procedure; see Vidakovic (2000) for details. Under such a criterion, we associate with each alternative its worst result, that is, its maximum posterior expected loss when the prior and the loss ranges in the classes, and pick the alternative with the best worst result, that is, with the minimum maximum posterior expected loss. We can easily see that, under certain conditions (see French and Ríos Insua (2000)) one of the $\Gamma \times \mathcal{L}$-minimax solutions is nondominated. Hence, this criterion does not lead to an unreasonable solution, given the current information.

On the other hand, note that, in general, the $\Gamma \times \mathcal{L}$-minimax ranking of acts does not correspond to a ranking according to posterior expected losses with respect to one of the distributions in Γ and one of the loss functions in $\mathcal{L}$. This means that this criteria is adding ad-hoc information. Interestingly enough, it is possible to provide foundations for these criteria, so that we are able to gain insights about it. We shall consider only imprecision in the prior, returning to the scenario in Proposition 3.

Since the adoption of the Γ-minimax criterion implies adding information, we are actually moving from $(\mathcal{G}_s, \preceq)$ to $(\mathcal{G}_s, \preceq')$. Of course, we shall demand that $\preceq \subseteq \preceq'$ in the sense that

$$g_1 \preceq g_2 \Rightarrow g_1 \preceq' g_2.$$

Note also that $(\mathcal{G}_c, \preceq') = (\mathcal{G}_c, \preceq)$ due to RB1b. The information that $\preceq'$

adds to $\preceq$ is not arbitrary though, in the sense that it follows some properties.

GM1. Weak order. $(\mathcal{G}_s, \preceq')$ is a weak order.
Essentially, $\preceq'$ adds comparisons to $\preceq$ so as to produce a complete relation, preserving transitivity.

GM2. Certainty independence. For all $g_1, g_2 \in \mathcal{G}_s, g_3 \in \mathcal{G}_c$ and $\alpha \in (0,1)$, then

$$g_1 \prec' g_2 \iff \alpha g_1 + (1-\alpha) g_3 \prec' \alpha g_2 + (1-\alpha) g_3.$$

The standard independence axiom RB2 is stronger than GM2 as it allows g_3 to be any act in $\mathcal{G}_s$. Note, though, that independence still holds for those pairs related by $\preceq$.

GM3. Continuity. For $g_1, g_2, g_3 \in \mathcal{G}_s$, if $g_1 \prec' g_2, g_2 \prec' g_3$, there are $\alpha, \beta \in (0,1)$ such that $\alpha g_1 + (1-\alpha) g_3 \prec' g_2$ and $g_2 \prec' \beta g_1 + (1-\beta) g_3$.
This is a continuity condition different to RB3.

GM4. Nontriviality. There are two acts g_1, g_2 such that $g_1 \prec' g_2$.

GM5. Monotonicity. If $g_1(\theta) \preceq' g_2(\theta), \forall \theta$, then $g_1 \preceq' g_2$.
This monotonicity condition is stronger than RB5.

GM6. Uncertainty aversion. $\forall g_1, g_2 \in \mathcal{G}_s$ and $\alpha \in (0,1)$,

$$g_1 \sim' g_2 \Rightarrow g_1 \preceq' \alpha g_1 + (1-\alpha) g_2.$$

We then have, following Proposition 3 and Gilboa and Schmeidler (1989):

Proposition 4 . *Let $\preceq'$ be a binary relation in $\mathcal{G}_s$. If $(\mathcal{G}_s, \preceq')$ satisfies assumptions GM1 GM6, we have that*

$$g_1 \preceq' g_2 \iff \left(\max_{\pi \in \Gamma} \int L(g_1(\theta)) \pi(\theta) d\theta \geq \max_{\pi \in \Gamma} \int L(g_2(\theta)) \pi(\theta) d\theta, \forall \pi \in \Gamma \right).$$

In the presence of additional information provided by a statistical experiment, we have that

$$g_1 \preceq g_2 \iff \max_{\pi \in \Gamma'} \int L(g_1(\theta)) \pi(\theta|x) d\theta \geq \max_{\pi \in \Gamma'} \int L(g_2(\theta)) \pi(\theta|x) d\theta, \forall \pi \in \Gamma'.$$

Therefore, we have a justification for the posterior Γ-minimax approach.

2.5 Discussion

The results have two basic implications. First, they provide a qualitative framework for robust Bayesian analysis, describing under what conditions we may undertake the standard and natural sensitivity analysis approach of perturbing the initial probability-loss assessments, within some reasonable constraints, thoroughly illustrated in Ríos Insua and Ruggeri (2000).

Second, they point to the basic solution concept of robust approaches, thus indicating the basic computational objective in sensitivity analysis, as long as we are interested in decision theoretic problems: that of non-dominated alternatives. Note the similarity with the admissibility concept. Both correspond to Pareto orderings of decision rules, see White (1982). Admissibility is based on inequalities on the risk function, therefore not making assumptions about the prior distribution and averaging over the data x (and assuming a precise loss function). Nondominance is based on inequalities on the posterior expected loss, therefore making assumptions over the prior distribution, using only the observed x (and, possibly, assuming imprecision in the loss function).

As we mentioned, the work on foundations is not as well rounded as that in standard Bayesian analysis, with various axiomatics leading to the subjective expected model. We have given an overview on some of the available results, but more work is still waiting. For example, it would be interesting to consider axiomatics for Savage's framework.

Acknowledgments

Work supported by a CICYT-HID project. We are grateful to discussions with Jacinto Martin.

References

ANSCOMBE, F. and AUMANN, R. (1963). A definition of subjective probability. *Annals of Mathematical Statistics*, **34**, 199–205.

BURTEL, D. (1997). A correction to a theorem by Ríos Insua. *Test*, **6**, 375–377.

FRENCH, S. and RÍOS INSUA, D. (2000). *Statistical Decision Theory*. London: Arnold.

FISHBURN, P.C. (1982). *The Foundations of Expected Utility*. Dordrecht: D. Reidel.

FISHBURN, P.C. (1986). The axioms of subjective probability. *Statistical Science*, **1**, 335–358.

GILBOA, I. and SCHMEIDLER, D. (1989). Maxmin expected utility with non-unique prior. *Journal of Mathematical Economics*, **18**, 141–153.

GIRON, F.J. and RÍOS, S. (1980). Quasi Bayesian behaviour: a more realistic approach to decision making? In *Bayesian Statistics* (J.M. Bernardo, M.H. De Groot, D. Lindley and A.F.M. Smith, eds.) Valencia: Valencia University Press.

MARTIN, J. and ARIAS, P. (2000). Computing the efficient set in Bayesian decision problems. In *Robust Bayesian Analysis* (D. Ríos Insua and F. Ruggeri, eds.). New York: Springer-Verlag.

NAU, R. (1992). Indeterminate probabilities on finite sets. *Annals of Statistics*, **20**, 4, 1737–1767.

NAU, R. (1995). The shape of incomplete preferences. *Technical Report*, Duke University.

RÍOS INSUA, D. (1992). The foundations of robust decision making: the simple case. *Test*, **1**, 69–78.

RÍOS INSUA, D. (1990). *Sensitivity Analysis in Multiobjective Decision Making.* Berlin: Springer-Verlag.

RÍOS INSUA, D. and MARTIN, J. (1994). On the foundations of robust decision making. In *Decision Theory and Decision Analysis: Trends and Challenges* (S. Ríos, ed.). Cambridge: Kluwer Academic Publishers.

RÍOS INSUA, D. and MARTIN, J. (1995). On the foundations of robust Bayesian statistics. *Technical Report*, Universidad Politecnica de Madrid.

RÍOS INSUA, D. and RUGGERI, F. (2000). *Robust Bayesian Analysis.* New York: Springer-Verlag.

SEIDENFELD, T., SCHERVISH, M. and KADANE, J. (1995). A representation of partially ordered preferences. *Annals of Statistics*, **23**, 2168–2217.

SCOTT, D. (1964). Measurement structures and linear inequalities. *Journal of Mathematical Psychology*, **1**, 233–247.

VIDAKOVIC B. (2000). Γ–minimax: a paradigm for conservative robust Bayesians. In *Robust Bayesian Analysis* (D. Ríos Insua and F. Ruggeri, eds.). New York: Springer-Verlag.

WALLEY, P. (1991). *Statistical Reasoning with Imprecise Probabilities.* London: Chapman and Hall.

WHITE, D.J. (1982). *Optimality and Efficiency.* New York: Wiley.

3

Global Bayesian Robustness for Some Classes of Prior Distributions

Elias Moreno

ABSTRACT Let θ represent the unobservable parameter of a sampling model $f(x|\theta)$ and $\varphi(\theta)$ the quantity of interest. In this article ranges of the posterior expectation of $\varphi(\theta)$, as the prior for θ varies over some commonly used classes of prior distributions, are studied. This kind of study is termed "global" Bayesian robustness of the class of priors for the quantity $\varphi(\theta)$. In this paper attention is paid to the methodology for finding these ranges and the prior distributions at which the extreme values are attained.

We shall consider contamination classes of prior distributions of a specified prior $\pi_0(\theta)$, corresponding to various contaminating classes. Contaminating classes include the class of priors having some "quantiles" as those of $\pi_0(\theta)$, the class with "shape" and quantile constraints compatible with $\pi_0(\theta)$, and the class with marginals as those of $\pi_0(\theta)$. Extensions of contamination classes –mainly proposed for modeling local uncertainty on $\pi_0(\theta)$, and for studying influence sets to guide an interactive robust analysis– are also reviewed.

The class of priors with given marginals plays an important role in multidimensional parameter spaces: this class contains any possible conditional distribution among the components, but maintains fixed the one-dimensional marginals. The main developments, which are based on the generalized moment constraints classes, are discussed.

The class of probability densities in a given "band" has interesting properties: it allows us to model with considerable freedom the tail behavior of θ and is the underlying class in the theory of precise measurement. The main results related to this class are also reviewed.

Key words: Bayesian robustness, ε-contamination classes, global robustness, influence set, interactive robustness, moment problem, priors with given marginals, probability bands, quantile classes.

3.1 Introduction

Let X be an observable random variable with density in a given parametric family $\{f(x|\theta), \theta \in \Theta\}$, where $\Theta \subset R^k$. To complete the Bayesian statistical model, a prior distribution on the unobservable parameter θ has to be subjectively elicited. The robust Bayesian approach is based on the assumption that subjective elicitation of a single prior for θ is in practice impossible, and instead a class of priors is the natural alternative. This implies an imprecision of the posterior inference on a given quantity of interest $\varphi(\theta)$, obtained as a consequence that the prior varies over the elicited class. This imprecision is measured by the range of posterior answers.

Although there is no general systematic procedure to elicit a class of prior distributions for θ, and the procedure will depend on the particular application considered, commonly used ways of eliciting such a class are the following.

(i) Some prior beliefs on the distribution of θ in the space Θ are established and a base prior $\pi_0(\theta)$ matching these beliefs is then chosen. The selection of this prior from the class of distributions that satisfies the prior beliefs, say Q, is typically dictated by its mathematical simplicity. Therefore, a probability ε reflecting our uncertainty on the chosen prior $\pi_0(\theta)$ must be specified and the class

$$\Gamma(Q, \varepsilon) = \{\pi(\theta) : \pi(\theta) = (1 - \varepsilon)\pi_0(\theta) + \varepsilon q(\theta), q(\theta) \in Q\} \tag{1}$$

is proposed. Notice that, by construction, $\Gamma(Q, \varepsilon) \subset Q$.

The class $\Gamma(Q, \varepsilon)$ is the well-known ε-contamination class of priors with base $\pi_0(\theta)$. Different types of ε-contamination classes have been studied by many authors: Berger (1984, 1985, 1990, 1994), Berger and Berliner (1986), Blum and Rosenblatt (1967), Bose (1994a), De la Horra and Fernández (1994), Huber (1973), Marazzi (1985), Moreno and Cano (1989, 1991, 1995), Moreno and Pericchi (1991, 1992a, 1993b), Sivaganesan (1988, 1989, 1993), Sivaganesan and Berger (1989), among others. In these papers the probability ε is associated with the chosen base prior π_0 and measures the whole uncertainty on this prior.

(ii) In some situations, however, the degree of our uncertainty on $\pi_0(\theta)$ would not be constant across the parameter space; for instance, we could assume more uncertainty in the "tails" of the distribution $\pi_0(\theta)$ than in the center. A formulation of the ε-contamination class to explicitly account for this can be given by replacing ε with an appropriate function $\varepsilon(\theta)$. A more flexible class, say $\Gamma(Q, \varepsilon(\theta))$, is then obtained. This class permits us to develop an analysis on influence sets which is useful to guide an interactive robust elicitation process (Berger, 1994; Liseo et al. 1996b; Moreno et al. 1996).

(iii) When prior beliefs are expressed by the probabilities of a collection of sets $\{C_i, i = 1, \ldots, n\}$ which forms a partition of the parameter space Θ (Berger 1985, 1990, 1994, Berliner and Goel, 1990, Kudō, 1967, Mansky, 1981, Cano et al., 1986, Wasserman and Kadane, 1990, 1992), the class Q

is then expressed as

$$Q_n = \{q(\theta) : \int_{C_i} q(\theta)d\theta = \alpha_i, i = 1, \ldots, n\}, \tag{2}$$

where $\sum_{i=1}^{n} \alpha_i = 1$. Although the elicitation made is not strictly on quantiles of the prior, the class Q_n is usually called a "quantile" class.

(iv) A more general "quantile" prior information might be of interest (Moreno and Cano, 1989, 1991). For example, suppose that for $\theta = (\theta_1, \theta_2)$ we are confident on the marginal $q(\theta_1)$ and the conditional probabilities satisfy

$$\int_{C_i} q(\theta_2|\theta_1)d\theta_2 = p_i.$$

Then the prior probability of any set in the σ-field $\mathcal{B}$ generated by $\{C \times C_i : C$ is a Borel set, $i = 1, \ldots, n\}$ is completely specified, and consequently Q becomes

$$Q_{\mathcal{B}} = \{q(\theta) : \int_{B} q(\theta)d\theta = \alpha(B), B \in \mathcal{B}\}. \tag{3}$$

This class processes a non-countable amount of prior information, say $\alpha(B), B \in \mathcal{B}$, while Q_n processes a countable amount of it, say α_i, $i = 1, \ldots, n$.

(v) Sometimes we are only able to elicit prior beliefs in terms of the probabilities of some sets on the component spaces of $\theta = (\theta_1, \ldots, \theta_k)$. For instance, suppose $\theta = (\theta_1, \theta_2) \in \Theta_1 \times \Theta_2$ and the prior beliefs $\{\int_{A_i} \pi_1(\theta_1)d\theta_1 = \alpha_i, i = 1, \ldots, n; \int_{B_j} \pi_2(\theta_2)d\theta_2 = \beta_j, j = 1, \ldots, m\}$, where $\{A_i, i = 1, \ldots, n\}$ and $\{B_j, j = 1, \ldots, m\}$ are partitions of Θ_1 and Θ_2, respectively, and $\{\alpha_i$, $i = 1, \ldots, n\}$, $\{\beta_j$, $j = 1, \ldots, m\}$ are specified probabilities. The class of prior distributions for θ compatible with these prior beliefs is given by $\Gamma = \cup_\gamma \Gamma(\gamma)$ where

$$\Gamma(\gamma) = \{\pi(\theta) : \int_{C_{ij}} \pi(\theta)d\theta = \gamma_{ij}, i = 1, \ldots, n, \ j = 1, \ldots, m\} \tag{4}$$

$C_{ij} = A_i \times B_j$, and $\gamma = \{\gamma_{ij} : \sum_{i=1}^{n} \gamma_{ij} = \beta_j, \sum_{j=1}^{m} \gamma_{ij} = \alpha_i, i = 1, \ldots, n, j = 1, \ldots, m\}$. Notice that $\{C_{ij}, i = 1, \ldots, n, \ j = 1, \ldots, m\}$ does form a partition of $\Theta_1 \times \Theta_2$, so that methods to deal with quantile classes are applicable to the class $\Gamma(\gamma)$, and consequently to Γ (Lavine et al., 1991, Moreno and Cano, 1995).

(vi) A natural extension of (4) is the class of priors with given marginals, that is

$$\Gamma = \{\pi(\theta_1, \ldots, \theta_k), \text{with given marginal} \pi_i(\theta_i), i = 1, \ldots, k\}. \tag{5}$$

Methods to deal with quantile classes are not applicable to this class and a separate analysis is needed (Liseo et al., 1996a, Moreno and Cano,

1995). We will see that a solution to the variational problems involved with this class can be obtained via the generalized moment class (Cambanis et al., 1976, Betró et al., 1994, Betró and Guglielmi, 2000, Dall'Aglio, 1995, Rachev, 1985, Tchen, 1980, Kemperman, 1987). This latter class is defined as

$$\Gamma = \{\pi(\theta) : \int g_i(\theta)\pi(\theta)d\theta = \mu_i, i = 1, \ldots, n\}, \tag{6}$$

where functions $g_i(\theta)$ and numbers μ_i, are a priori specified.

(vii) In situations where the elicitation is made by more than one elicitor, the prior beliefs might be given in terms of ranges of probabilities of a given partition $\{C_i, i = 1, \ldots, n\}$. The resulting class (DeRobertis, 1978, Moreno and Pericchi, 1992b) is written as

$$\Gamma(\alpha, \beta) = \{\pi(\theta) : \alpha_i \leq \int_{C_i} \pi(\theta)d\theta \leq \beta_i, \{i = 1, 2, \ldots, n\}, \tag{7}$$

where α_i, β_i are specified numbers such that $\sum_{i=1}^n \alpha_i \leq 1 \leq \sum_{i=1}^n \beta_i$. Notice that in this case no base prior can be proposed although this class is considered as "*the most natural elicitation mechanism*" (Berger, 1990, section 3.2).

(viii) An interesting extension of the class in (7) is that where the lower and upper probabilities α and β are specified for any set C in the Borel σ-field $\mathcal{B}$. This gives the class

$$\Gamma(L, U) = \{\pi(\theta) : L(C) \leq \int_C \pi(\theta)d\theta \leq U(C), C \in \mathcal{B}, \int_\Theta \pi(\theta)d\theta = 1\}, \tag{8}$$

where the measures L and U are such that $L(\Theta) \leq 1 \leq U(\Theta)$. This is an interesting class as long as it allows us to model the tail behavior with considerable freedom. As shown in Moreno et al. (1999b), this is the underlying one in the theory of precise measurement of Edwards et al. (1963). Related classes have been considered by Dickey (1976), Bose (1990, 1994b), De Robertis and Hartigan (1981), Lavine (1991), Perone-Pacifico et al. (1996) and Wasserman (1992).

Very little can be said, in general, on the question of which of the mentioned classes is the "right" one. In each specific application any one of them can be typically identified as the most appropriate one. In a general setting all of them are worthwhile to be considered.

The general problem to be solved is that of finding the range of

$$\rho(\pi) \equiv E^\pi(\varphi(\theta)|x) = \frac{\int \varphi(\theta)f(x|\theta)\pi(\theta)d\theta}{\int f(x|\theta)\pi(\theta)d\theta},$$

as the prior π varies over the given class Γ, where $\varphi(\theta)$ is the quantity of interest and $x = (x_1, \ldots, x_n)$ is a data set from $f(x|\theta)$. The posterior imprecision, say $R(\Gamma, \varphi) = \sup_{\pi \in \Gamma} E^\pi(\varphi(\theta)|x) - \inf_{\pi \in \Gamma} E^\pi(\varphi(\theta)|x)$, is the measure of robustness of Γ for the quantity $\varphi(\theta)$. When $R(\Gamma,\varphi)$ is not "small enough," Γ is said to be non-robust for the posterior inference on $\varphi(\theta)$, given x. In this case the next step would be either to elicit more features of the prior in order to reduce Γ or to increase the sample size to reduce the influence of the priors.

The quantity of interest $\varphi(\theta)$ may have different forms depending on the problem under consideration. For instance, in testing problems, $\varphi(\theta)$ might be the indicator function of the null hypothesis. In decision theory, $\varphi(\theta)$ might be a loss function associated with a given decision; in robustness analysis of the optimal decision under a quadratic loss function, the quantity of interest is $\varphi(\theta) = \theta$, etc.

The variational methods developed in Bayesian robustness for finding $R(\Gamma,\varphi)$ are essentially based on the assumption that $\rho(\pi)$ is a ratio of linear functionals of π and the class of priors Γ is a convex class; that is, for any two priors $\pi_1(\theta)$ and $\pi_2(\theta)$ in Γ, the convex combination $\lambda\pi_1(\theta) + (1-\lambda)\pi_2(\theta)$ is also in Γ, for any $0 \leq \lambda \leq 1$. Much more involved techniques, if available, are needed when these assumptions are not satisfied.

An important case where convexity of the class does not hold is when independence of the components of θ is assumed. For instance, when a base prior of the form $\pi_0(\theta_1,\ldots,\theta_k) = \prod_{i=1}^{k} \pi_i(\theta_i)$ is chosen, robustness to departure of $\pi_0(\theta)$ while maintaining the independence assumption can be modeled by considering some of the above classes on each of the components. The resulting class is not a convex class, and methods to deal with posterior robustness are now much more involved (Berger and Moreno, 1994). Very little is really known on the posterior behavior for those sorts of classes.

An example where $\rho(\pi)$ is not a ratio of linear functionals of $\pi$is the case where $\varphi(\theta)$ depends on π. For instance, for $\varphi(\theta,\pi) = [\theta - E^{\pi}(\theta|x)]^2$, $\rho(\pi)$ represents the posterior risk of the optimal decision under a quadratic loss (Sivaganesan and Berger, 1989). For this quantity and most of the mentioned classes, the extreme priors are not yet known.

This paper is organized as follows. Section 2 deals with ε-contamination classes for several types of contaminations, and the extension to the case where the uncertainty on the base prior $\pi_0(\theta)$ varies across the parameter space. In Section 3 classes with some specifications on the marginal spaces are considered. Section 4 deals with bands of probability densities, and some concluding remarks are given in Section 5.

3.2 ε-contamination classes

3.2.1 Contaminations with fixed quantiles

Let $\Gamma(Q_n,\varepsilon)$ be the class given by expressions (1) and (2). The supremum of the posterior expectation $E^{\pi}(\varphi(\theta)|x)$ as the prior ranges over $\Gamma(Q_n,\varepsilon)$, that is, $\sup_{\pi \in \Gamma(Q_n, \varepsilon)} E^{\pi}(\varphi(\theta)|x)$, is attained at a prior of the form

$$\pi^*(\theta) = (1-\varepsilon)\pi_0(\theta) + \varepsilon \sum_{i=1}^{n} \alpha_i \delta_{\theta_i^*}(\theta),$$

where θ_i^* is a point in C_i or a limit of a sequence of points in C_i. A similar statement is true for the infimum. These results follow from the following Lemma 1.

Lemma 1. For any functions $g_1(\theta)$ and $g_2(\theta)$, assumed to be integrable with respect to all priors in Q_n, we have that

(i)

$$\sup_{q \in Q_n} \int_{C_i} g_1(\theta) q(\theta) d\theta = \alpha_i \sup_{\theta \in C_i} g_1(\theta)$$

(ii)

$$\sup_{q \in Q_n} \frac{\int g_1(\theta) q(\theta) d\theta}{\int g_2(\theta) q(\theta) d\theta} = \sup_{\substack{\theta_i \in C_i \\ i=1,\ldots,n}} \frac{\sum_{i=1}^{n} \alpha_i g_1(\theta_i)}{\sum_{i=1}^{n} \alpha_i g_2(\theta_i)}.$$

Proof. See Theorem 1 in Moreno and Cano (1991). □

Letting $g_1(\theta) = \varphi(\theta) f(x|\theta)$ and $g_2(\theta) = f(x|\theta)$, the posterior range of $E^{\pi}(\varphi(\theta)|x)$ is obtained.

Notice that the variational problems associated with class Q_n are transformed in an optimization problem with n restrictions. The solution to this optimization problem may not be simple. The use of a linearization algorithm (Lavine et al. 1993, Lavine et al. 2000, Betró and Guglielmi, 1996), helps in this task. For instance, it can be seen that the sup in (ii) turns out to be the solution in λ to the equation

$$0 = \sup_{\substack{\theta_i \in C_i \\ i=1,\ldots,n}} \left\{ \sum_{i=1}^{n} \alpha_i g_1(\theta_i) - \lambda \sum_{i=1}^{n} \alpha_i g_2(\theta_i) \right\}.$$

The numerical optimization problem is avoided when the quantity of interest is an indicator function of a set, that is, when $\varphi(\theta) = 1_A(\theta)$. In this case, it can be seen that

$$\sup_{\pi \in \Gamma(Q_n, \varepsilon)} P^{\pi}(A|x) =$$

$$\frac{(1-\varepsilon) m(x|\pi_0)\beta_0 + \varepsilon \sum_i \alpha_i \sup_{\theta \in A \cap C_i} f(x|\theta)}{(1-\varepsilon) m(x|\pi_0) + \varepsilon \sum_{i \in K} \alpha_i \sup_{\theta \in C_i} f(x|\theta) + \varepsilon \sum_i \alpha_i \sup_{\theta \in A \cap C_i} f(x|\theta)},$$

where $m(x|\pi_0)$ is the marginal of the data under π_0, $\beta_0 = P^{\pi_0}(A|x)$ and $i \in K$ iff $A \cap C_i = \emptyset$. A similar expression is obtained for the infimum (see Moreno and Cano, 1991, Corollary 1).

Posterior ranges of $E^{\pi}(\varphi(\theta)|x)$ as π ranges over the class $\Gamma(Q_{\mathcal{B}}, \varepsilon)$, where $Q_{\mathcal{B}}$ is given in (3), are somewhat more involved and are expressed in terms of upper and lower integrals (Sacks, 1937). The reader is referred to Theorem 2 in Moreno and Cano (1991).

Example 1. Let X be a random variable $N(x|\theta, 1)$. Assume that $\pi_0(\theta) = N(\theta|0, 2)$ and $\varepsilon = .2$. The quantity of interest is the 95% π_0-Bayes credible interval denoted by $C_0(x)$. Consider the contaminating classes given by $Q_0 = \{\text{all distributions}\}$, $Q_1 = \{q(\theta) : \text{the median of } q(\theta) \text{ is as that of } \pi_0\}$ and $Q_2 = \{q(\theta) : \text{the quartiles of } q(\theta) \text{ are as those of } \pi_0\}$.

Let $\underline{P}_i = \inf_{\pi \in \Gamma(Q_i, \varepsilon)} P^\pi(C_0(x)|x)$ and $\overline{P}_i = \sup_{\pi \in \Gamma(Q_i, \varepsilon)} P^\pi(C_0(x)|x)$. For various values of x the resulting values of $(\underline{P}_i, \overline{P}_i)$ are displayed in the second and third columns in Table 1.

x	$(\underline{P}_0, \overline{P}_0)$	$(\underline{P}_1, \overline{P}_1)$	$(\underline{P}_2, \overline{P}_2)$
0.5	.82, .96	.84, .96	.91, .96
1.0	.77, .97	.82, .96	.87, .96
2.5	.71, .97	.79, .96	.86, .96
3.0	.36, .98	.52, .97	.66, .97
4.0	.13, .99	.20, .99	.50, .97

TABLE 1. Posterior ranges

The second column in Table 1 shows that robustness of the credible set $C_0(x)$ with respect to $\Gamma(Q_0, .2)$ strongly depends on the x observed. Robustness is present only when x is close to zero. Observations that are wholly incompatible with the underlying model, for instance $x = 4$, result in a lack of robustness. This is explained by observing that $\underline{P}_0 = .13$ is attained at the contamination $q(\theta = 4.27) = 1$, which is an unreasonable contamination.

By eliciting prior quartiles, as in the class $\Gamma(Q_2, .2)$, robustness increases considerably, as the fourth column in Table 1 shows. $\triangle$

3.2.2 Influence sets

An extension of the class $\Gamma(Q_n, \varepsilon)$ can be formulated as the class

$$\Gamma(Q_n, \varepsilon(\theta)) = \{\pi(\theta) : \pi(\theta) = [1 - \varepsilon(\theta)]\pi_0(\theta) + \varepsilon(\theta)q(\theta), q(\theta) \in Q_n\},$$

where $\varepsilon(\theta) \in [0, 1]$ expresses our uncertainty on $\pi_0(\theta)$ at the point θ. For an arbitrary $\varepsilon(\theta)$, priors in $\Gamma(Q_n, \varepsilon(\theta))$ do not necessarily satisfy the prior beliefs stated in (2). A necessary and sufficient condition for $\varepsilon(\theta)$ to satisfy the prior beliefs in (2) (Moreno et al., 1996) is that

$$\varepsilon(\theta) = \sum_{i=1}^{n} \varepsilon_i 1_{C_i}(\theta).$$

Each value $\varepsilon_i \in [0,1]$ can be interpreted as the uncertainty we have on the functional form of $\pi_0(\theta)$ on the set C_i, among all the possible priors in Q_n.

Formulae for finding the posterior imprecision of $E^\pi(\varphi(\theta)|x)$ as π ranges over $\Gamma(Q_n, \varepsilon(\theta))$ are given in Theorem 2 in Moreno et al. (1996).

Example 2. Let X be a random variable $N(x|\theta,1)$. Suppose we are interested in testing $H_0 : \theta \leq 0$. It has been elicited that the distribution of θ has first and third quartiles equal to -0.954 and 0.954, respectively. The elicited class is

$$Q_3 = \{q(\theta) : \int_{-\infty}^{-.954} q(\theta)d\theta = 0.25, \int_{-.954}^{.954} q(\theta)d\theta = 0.5\}.$$

The base prior $\pi_0(\theta) = N(\theta|0,2)$ that matches these quartiles is chosen. For $\varepsilon = .2$ and various values of x, the posterior imprecision of $P^\pi(H_0|x)$ as the prior π ranges over $\Gamma(Q_3,\varepsilon)$, say $R(\Gamma(Q_3,\varepsilon), H_0, x)$, is displayed in the second column of Table 2. The corresponding posterior imprecisions when $\varepsilon(\theta) = 0.5\{\mathbf{1}_{(-\infty,-.954)}(\theta) + \mathbf{1}_{(.954,\infty)}(\theta)\}$ are displayed in the third column in Table 2. The posterior imprecision is denoted by $R(\Gamma(Q_3, \varepsilon(\theta)), H_0, x)$.

x	$R(\Gamma(Q_3,\varepsilon), H_0, x)$	$R(\Gamma(Q_3,\varepsilon(\theta)), H_0, x)$
0.0	.21	.13
0.5	.18	.12
1.0	.14	.09
1.5	.09	.04

TABLE 2. Posterior imprecisions

The third column in Table 2 shows that a significant reduction of posterior imprecision is obtained only when uncertainty in the tails of $\pi_0(\theta)$ is considered, even when this uncertainty is as large as 0.5. We feel that the situation considered in the third column is a better reflection of our prior uncertainty than that in the second one. △

Notice that by taking $\varepsilon(\theta) = \sum_{i=1}^{n} \varepsilon_i 1_{C_i}(\theta)$, the uncertainty on $\pi_0(\theta)$ has been decomposed into local uncertainty on each of the elements of the partition $\{C_i, i \geq 1\}$. A natural question is to know which set C_i causes the largest effect on the posterior imprecision of our quantity of interest. A way to answer this question is as follows. Consider the class

$$\Gamma^i(Q_n) = \{\pi(\theta) : \pi(\theta) = \pi_0(\theta) + \varepsilon_i[q(\theta) - \pi_0(\theta)]1_{C_i}(\theta), q(\theta) \in Q_n\},$$

where only uncertainty on C_i is allowed. This class is derived from $\Gamma(Q_n, \varepsilon(\theta))$ by taking $\varepsilon_j = 0$ for any $j \neq i$.

Lemma 2. For $\varepsilon(\theta)$ as above and any observation x, the inequality

$$R(\Gamma(Q_n, \varepsilon(\theta)), \varphi, x) \geq \max_{i \geq 1} R(\Gamma^i(Q_n), \varphi, x)$$

holds.

Proof. Any $\pi(\theta)$ in $\Gamma^i(Q_n)$ can be written as

$$\pi(\theta) = \pi_0(\theta) + \sum_{i=1}^{n} \varepsilon_i[\tilde{q}(\theta) - \pi_0(\theta)]1_{C_i}(\theta),$$

where the contamination $\tilde{q}(\theta) = q(\theta)1_{C_i}(\theta) + \pi_0(\theta)1_{C_i^c}(\theta)$. Here C_i^c denotes the complement of C_i. Since $\pi_0(\theta) \in Q_n$, it follows that $\Gamma^i(Q_n) \subset \Gamma(Q_n, \varepsilon(\theta))$. This proves the assertion. □

Lemma 2 states that posterior robustness of the class $\Gamma(Q_n, \varepsilon(\theta))$ cannot be achieved when the classes $\Gamma^i(Q_n)$ are not robust. Thus, the elicitation efforts should be concentrated on the set C_k for which

$$R(\Gamma^k(Q_n), \varphi, x) = \max_{i \geq 1} R(\Gamma^i(Q_n), \varphi, x).$$

To find the set C_k is then of the utmost importance. The posterior imprecision in Q_n cannot be smaller than the threshold $R(\Gamma^k(Q_n), \varphi, x)$ even when we add an infinite number of quantiles in the set C_j $(\neq C_k)$.

Example 3. In the situation of Example 2, set $C_1 = (-\infty, -0.954)$, $C_2 = (-0.954, 0.954)$ and $C_3 = (0.954, \infty)$.

Values of $R(x) = R(\Gamma(Q_3, \varepsilon(\theta)), H_0, x)$ for $\varepsilon(\theta) = 0.2$ and $R_i(x) = R(\Gamma^i(Q_3), H_0, x)$ are computed and displayed in Table 3.

x	$R_1(x)$	$R_2(x)$	$R_3(x)$	$R(x)$
0.0	.03	.17	.03	.21
0.5	.02	.16	.03	.18
1.0	.01	.14	.02	.14
1.5	.00	.09	.01	.09

TABLE 3. Values of $R_i(x)$ and $R(x)$

Table 3 shows that the largest posterior imprecision for any of the sample points considered is that of $\Gamma^2(Q_3)$, which corresponds to contaminations on the set $C_2 = (-0.954, 0.954)$. Thus, our elicitation efforts should be concentrated in this set in order to reduce its posterior uncertainty. This explains the dramatic reduction on posterior uncertainty we obtained in Example 2 when we did not allow contaminations on C_2. △

3.2.3 Interactive robustness

The idea of using Bayesian robustness to guide the elicitation process was pointed out in the thoughtful paper by Berger (1994). He argued that "*since* *we are eliciting in terms of quantiles, this means that a new quantile θ^* must be chosen with the associated α^* (for the new interval created) being elicited."*

To this idea, the previous subsection adds the new issue of knowing on which set the new quantile θ^* must be located. The implementation of those two ideas was given in Moreno et al. (1996). Suppose the most influential set is $C_k = (\theta_{k-1}, \theta_k)$. If our confidence on $\pi_0(\theta)1_{C_k}(\theta)$ is not small, then it seems natural to impose the relationship between α^* and θ^*,

$$\alpha^* = \int_{\theta_{k-1}}^{\theta^*} \pi_0(\theta)d\theta.$$

If $R^*(x, \theta^*)$ denotes the posterior range of our quantity of interest as the prior varies over the class

$$\Gamma^* = \{\pi(\theta) : \pi(\theta) \in \Gamma(Q_n, \varepsilon(\theta)), \int_{\theta_{k-1}}^{\theta^*} \pi(\theta)d\theta = \int_{\theta_{k-1}}^{\theta^*} \pi_0(\theta)d\theta\},$$

the point θ^* should be determined in C_k as the most favorable point in the sense of robustness. That is, θ^* is such that

$$R^*(x, \theta^*) = \inf_{b \in C_k} R^*(x, b).$$

From Lemma 2 it is clear that the set where the new quantile is being elicited depends on the sample observation x. Therefore, it can be argued that the new class Γ^* is being designed as a function of the data, thereby implying that the prior elicitation is data-dependent. Interactive robustness entails jumps from samples to priors, and hence some dependence between samples and parameter is inherent in the idea.

Example 4. In Example 3 it is illustrated that the most influential set is C_2. Straightforward calculation gives that the new quantile is $\theta^* = .25$ with $\alpha^* = .25$ for all x considered in Table 3. The posterior imprecisions $R^*(x)$, $R_i^*(x)$, $i = 1, 2, 3, 4$, are displayed in Table 4.

x	$R_1^*(x)$	$R_2^*(x)$	$R_3^*(x)$	$R_4^*(x)$	$R(x)$
0.0	.03	.016	.016	.03	.087
0.5	.08	.032	.004	.03	.082
1.0	.01	.037	.008	.02	.081
1.5	.00	.031	.007	.01	.062

TABLE 4. Posterior imprecisions

Table 4 shows that after eliciting the new quantile the most influential set is now C_1^* or C_4^* for $x = 0$, and C_2^* for any other sample values. Therefore, if the desired degree of robustness is not yet achieved, a new quantile should be elicited either in C_2^* or C_1^* when the observation x is different from 0, or in C_4^* when $x = 0$. $\triangle$

For a slightly different interactive algorithm, see Liseo et al. (1996b).

3.2.4 Contaminations with shape constraints

When θ is a real parameter, symmetry and unimodality of the priors for θ are realistic assumptions in many applications (Berger and Berliner, 1986, Sivaganesan and Berger, 1989, Kadane et al., 1999). To simplify the notation and without loss of generality, we will assume that $\theta \in R^+$ and the mode $\theta_0 = 0$. Thus, the class Q_u is given by

$$Q_u = \{\text{all unimodal } q(\theta) \text{ with mode at } 0\}.$$

According to the Khintchine representation theorem (Dharmadhikari and Joag-Dev 1988), any $q \in Q_u$ can be written as the mixture of uniform distributions

$$q(\theta) = \int_0^\infty \frac{1}{z} 1_{(0,z)}(\theta) dF(z) = \int_\theta^\infty \frac{1}{z} dF(z),$$

where the mixing distribution F is some distribution on R^+. Hence, the expectation $E^q[\varphi(\theta)]$ can be written as

$$E^q[\varphi(\theta)] = \int_0^\infty \varphi(\theta) \left\{ \int_0^\infty \frac{1}{z} 1_{(0,z)}(\theta) dF(z) \right\} d\theta = \int_0^\infty G(z) dF(z),$$

where

$$G(z) = \frac{1}{z} \int_0^z \varphi(\theta) d\theta.$$

Therefore,

$$\inf_{q \in Q_u} E^q[\varphi(\theta)] = \inf_{F \in \mathcal{F}} E^F[G(z)],$$

where $\mathcal{F} = \{F(z) : \int_0^\infty dF(z) = 1\}$. In $\mathcal{F}$ the infimum is attained at a distribution that concentrates mass at a point, say z_0, and it implies that the extreme prior in Q_u is a uniform distribution, say $q(\theta) = U(\theta|0, z_0)$.

Similarly, it follows that

$$\inf_{q \in Q_u} E^q[\varphi(\theta)|x] = \inf_{q \in Q_u} \frac{\int_0^\infty \varphi(\theta) f(x|\theta) q(\theta) d\theta}{\int_0^\infty f(x|\theta) q(\theta) d\theta} = \inf_{z \geq 0} \frac{G_1(z;x)}{G_2(z;x)},$$

where

$$G_1(z;x) = \frac{1}{z} \int_0^z \varphi(\theta) f(x|\theta) d\theta, \text{and } G_2(z;x) = \frac{1}{z} \int_0^z f(x|\theta) d\theta.$$

Example 5 (Sivaganesan and Berger, 1989). Let $X|\theta \sim N(x|\theta,1)$, $\pi_0(\theta) = N(\theta|0,2)$ and $\varepsilon = 0.1$. When $x = 0.5$, the 95% HPD credible set is $C_0 = (-1.27, 1.93)$. We want to know how robust is the report $P^{\pi_0}(C_0|0.5) = .95$ in the class of priors

$$\Gamma = \{\pi(\theta) : \pi(\theta) = 0.9N(\theta|0,2) + 0.1q(\theta), q \in Q_{us}\},$$

where

$$Q_{us} = \{\text{all symmetric unimodal distributions with mode at } 0\}.$$

The range of $P^{\pi}(C_0|0.5)$ as π varies over Γ is given by

$$\inf_{\pi\in\Gamma} P^{\pi}(C_0|0.5) = .945, \quad \sup_{\pi\in\Gamma} P^{\pi}(C_0|0.5) = .958,$$

and these are attained when the contaminations are respectively $U(\theta| - 2.98, 2.98)$ and a point mass at 0. Thus, the posterior probability of C_0 is robust enough. $\triangle$

As the example shows, the supremum is attained at a degenerate prior concentrated on the mode. In most applications this is not an appealing fact and the way to avoid such a concentration is by adding quantile constraints to the unimodality condition (Berger and O'Hagan, 1988, Bertolino et al., 1995).

Lemma 3. Let Q_{uc} be the class

$$Q_{uc} = \{q(\theta) : q \text{ is unimodal at } 0, \int_{\theta_i}^{\theta_{i+1}} q(\theta)d\theta = \alpha_{i+1}, i = 0, \ldots, n-1\},$$

where $0 = \theta_0 < \ldots < \theta_n = \infty$, and $0 \le \alpha_i, \sum_{i=1}^{n} \alpha_i = 1$. Then any q in Q_{uc} can be written as

$$q(\theta) = \int_0^{\infty} \frac{1}{z} 1_{(0,z)}(\theta) dF(z),$$

for some distribution F in the class

$$\mathcal{F}_n = \{F(z) : \int_0^{\infty} g_i(z)dF(z) = \sum_{j=i+1}^{n} \alpha_j, i = 0, \ldots, n-1\},$$

where

$$g_i(z) = \frac{z - \theta_i}{z} 1_{(\theta_i,\infty)}(z), i = 0, \ldots, n-1.$$

Proof. The assertion follows from the equalities

$$\int_{\theta_i}^{\infty} q(\theta)d\theta = \sum_{j=i+1}^{n} \alpha_j = \int_{\theta_i}^{\infty} \frac{z - \theta_i}{z} dF(z), i = 0, \ldots, n-1.$$

Lemma 3 states that the variational problem related with the class Q_{uc} can be transformed in one related to the class $\mathcal{F}_n$. This class is of the form given in (6), that is, a class given by generalized moment constraints.

The variational problem related to this latter class can be transformed to an optimization problem in R^n (see the next section) whose solution may be complex (Kemperman, 1987, Lavine et al., 2000, Liseo *et al.*, 1996a, Salinetti, 1994, Betró and Guglielmi, 1996). This leads some authors to consider classes "close" to Q_{uc} but easier to analyze. For instance, quasi-unimodal priors were considered in O'Hagan and Berger (1988), and piecewise unimodal priors were introduced in Moreno and Gonzalez (1990).

3.3 Classes of priors with given marginals

When the parameter space Θ is multidimensional, prior elicitation of the joint distribution is difficult. The consideration of the class of priors with specified prior probabilities for some sets on the one-dimensional marginal spaces may alleviate this difficulty. When those sets form a partition of Θ, the variational problems arising in the robust Bayesian analysis have the solution given for the "quantile" class. However, a collection of sets that form a partition of the marginal spaces does not form a partition of the whole space Θ. Indeed, if $\Theta = \Theta_1 \times \ldots \times \Theta_k$ and $\{A_i,\ i = 1, \ldots, n\}$ is a partition of Θ_i, then prior specification of the probability $q(A_i)$ implies having the probability of the cylinder $A = \Theta_1 \times \ldots \times A_i \times \ldots \times \Theta_k$ with base A_i. The collection of cylinders overlaps and then Bayesian robustness methods for quantile classes do not apply.

On the other hand, it might be the case that we were able to completely specify the marginal densities $q_1(\theta_1), \ldots, q_k(\theta_k)$. This leads us to consider the class of priors $Q = \{q(\theta_1, \ldots, \theta_k) : \text{having marginals } q_1(\theta_1), \ldots, q_k(\theta_k)\}$. This situation corresponds to an extension of the above one when the partitions on the marginal spaces are as fine as possible. Unfortunately, even for $k = 2$, posterior robustness related to this class involves a very difficult variational problem (Monge–Kantorovich problem) that has not been solved yet (particular cases and approximations can be found in Cambanis et al., 1976, Tchen, 1980, Lavine et al., 1991, Moreno and Cano, 1995).

In this section we will consider a bidimensional parameter space, and the class of priors with given marginals will be taken as the contaminating class. The reasons are (i) to contaminate a given π_0 with priors having the same marginals as those of π_0 seems reasonable; notice that by taking $\varepsilon = 1$ in the ε-contamination class, π_0 is avoided; (ii) the class of priors with given marginals is so large that the resulting posterior ranges are useless.

As an illustration of the latter assertion we bring an example from Walley (1991, pp. 298–299). Suppose that a fair coin is tossed twice, in such a way that the second toss *may depend* on the outcome of the first. Let H_i

represent heads on toss i and T_i tails on toss i, $i = 1, 2$. Let Γ be the class of all distributions π with marginals $\pi_1(H_1) = \pi_2(H_2) = 1/2$. It can be seen that

$$\inf_{\pi \in \Gamma} P^{\pi}(H_2|H_1) = 0, \quad \sup_{\pi \in \Gamma} P^{\pi}(H_2|H_1) = 1,$$

and

$$\inf_{\pi \in \Gamma} P^{\pi}(H_2|T_1) = 0, \quad \sup_{\pi \in \Gamma} P^{\pi}(H_2|T_1) = 1.$$

This simple example shows that the class of priors with given marginals is useless to represent prior information.

3.3.1 Priors with marginals partially known

Let $\Theta = \Theta_1 \times \Theta_2$ be the parameter space and Γ the class of priors

$$\Gamma = \{\pi(\theta_1, \theta_2) : \pi(\theta_1, \theta_2) = (1 - \varepsilon)\pi_0(\theta_1, \theta_2) + \varepsilon q(\theta_1, \theta_2), q \in Q\},$$

where Q is given by

$$\begin{aligned} Q = \{q(\theta_1, \theta_2) &: \text{the marginals } q_1(\theta_1), q_2(\theta_2) \text{satisfy} \\ &\textstyle\int_{A_i} q_1(\theta_1) d\theta_1 = \alpha_i, \int_{B_j} q_2(\theta_2) d\theta_2 = \beta_j, i = 1, \ldots, n, j = 1, \ldots, m\}, \end{aligned}$$

$\{A_i, i = 1, \ldots, n\}$ and $\{B_j,\ j = 1, \ldots, m\}$, are partitions of Θ_1 and Θ_2, respectively, and $\{\alpha_i, i = 1, \ldots, n\}$ and $\{\beta_j,\ j = 1, \ldots, m\}$ are their corresponding probabilities. Conditions established on q must be derived from π_0 in order to preserve the prior information on which we are confident. In other words, Γ must be contained in Q.

In Lavine et al. (1991), Q was used as an approximation to the class of priors with marginals completely specified. They showed that the extreme values of $E^q(\varphi(\theta)|x)$ as q ranges over Q are obtained as the solution of a collection of linear programming problems for which there are efficient numerical algorithms.

In Moreno and Cano (1995) the extreme values of $E^q(\varphi(\theta)|x)$ as q ranges over Q were obtained as an extension of a quantile problem. This way the form of the extreme priors is exactly known. They considered the partition of Θ given as $C_{ij} = A_i \times B_j, i = 1, \ldots, n, j = 1, \ldots, m$. Then, any $q \in Q$ satisfies

$$\int_{C_{ij}} q(\theta) d\theta = \gamma_{ij},$$

for any matrix $G_{ij} = \{\gamma_{ij}, , i = 1, \ldots, n, j = 1, \ldots, m\}$ in the set of matrices

$$G = \{\gamma_{ij} : \gamma_{ij} \geq 0, \sum_{j=1}^{m} \gamma_{ij} = \alpha_i, \sum_{i=1}^{n} \gamma_{ij} = \beta_j, i = 1, \ldots, n, j = 1, \ldots, m\}.$$

Therefore, the problem of finding the extreme values of $E^q(\varphi(\theta)|x)$ as q ranges over Q is solved in two steps: first, as a quantile problem for each matrix G_{ij}; and second, G_{ij} is allowed to vary over G. This second step involves a numerical fractional linear programming problem. From this construction it is clear that the extreme priors $q^*(\theta_1, \theta_2)$ are discrete; that is,

$$q^*(\theta_1, \theta_2) = \sum \gamma_{ij}^* \delta_{(\theta_{1i}^*, \theta_{2j}^*)}(\theta_1, \theta_2),$$

for appropriate γ_{ij}^* in G and points $(\theta_{1i}^*, \theta_{2j}^*)$ in C_{ij}.

3.3.2 The case of a completely specified marginal

Extensions of the above result to the case where one of the marginals is completely specified and only one quantile is fixed on the other marginal are studied in Moreno and Cano (1995). They show that for Q defined as

$$Q = \{q : \text{the marginal } q_1(\theta_1) \text{ satisfies } \textstyle\int_A q_1(\theta_1) d\theta_1 = \alpha,$$
$$\text{and } q_2(\theta_2) \text{ is completely specified}\},$$

the extreme values of $E^q(\varphi|x)$ as q ranges over Q are attained at priors that concentrate their masses on curves in $\Theta_1 \times \Theta_2$.

Example 6. Nine patients were treated with ECMO (extracorporeal membrane oxygenation) and all of them survived. Ten patients were given standard therapy and six survived. Let p_1 be the probability of success under standard therapy and let p_2 be the probability of success under ECMO. Let $\eta_i = \log(p_i/(1 - p_i))$, $i = 1, 2$, and consider the parameters $\delta = \eta_2 - \eta_1$, $\gamma = (\eta_1 + \eta_2)/2$.

The prior suggested by Kass and Greenhouse (1989) for this problem was

$$\pi_0(\gamma, \delta) = \mathcal{C}(\gamma|0, 0.419^2)\mathcal{C}(\delta|0, 1.099^2),$$

where $\mathcal{C}$ represents a Cauchy density.

The π_0-posterior probability of $\delta > 0$, that is, the π_0-posterior probability that ECMO is a better treatment than standard therapy, turns out to be equal to 0.942. We want to know how robust this conclusion is.

A natural class to be considered is the ε-contamination class with base prior $\pi_0(\gamma, \delta)$, $\varepsilon = 0.2$ and contaminating class Q the class of priors $q(\gamma, \delta)$ with marginals $q_1(\gamma) = \mathcal{C}(\gamma|0, 0.419^2)$ and $q_2(\delta) = \mathcal{C}(\delta|0, 1.099^2)$. As an approximation of this class we consider

$$\Gamma_a = \{\pi(\gamma, \delta) : \pi(\gamma, \delta) = 0.8\pi_0(\gamma, \delta) + 0.2q(\gamma, \delta), q \in Q_a\},$$

where

$$Q_a = \{q(\gamma, \delta) : q_1(\gamma) \text{ is any prior and } q_2(\delta) = \mathcal{C}(\delta|0, 1.099^2)\}.$$

It can be seen that

$$\inf_{\pi \in \Gamma_a} P^\pi(\delta > 0|Data) = 0.855,$$

and we can conclude that the report $P^{\pi_0}(\delta > 0|Data) = 0.942$ is robust to departures from π_0 in Γ_a.

Notice that we can also consider the marginal of γ completely specified and the class of all priors for the marginal on δ. By so doing we obtain an infimum smaller than the one above. This implies that it is more important to specify the exact prior for δ than for γ. $\triangle$

A better approximation to the class of priors with given marginals is obtained by fixing one of the marginals and the probabilities of a partition on the other component. That is, consider the class of priors

$$Q = \{q(\theta_1, \theta_2) : q_2(\theta_2) \text{ fixed}, \int_{A_i} q_1(\theta_1)d\theta_1 = \alpha_i, i = 1, \ldots, n\},$$

where $\{A_i, i = 1, \ldots, n\}$ is an arbitrary partition of Θ_2 and α_i are their probabilities. Notice that $\int_{A_i} q_1(\theta_1)d\theta_1 = \alpha_i$ can be written as $\int 1_{A_i}(\theta_1)$ $q_1(\theta_1)d\theta_1 = \alpha_i$. These are constraints of the form given in (6).

The variational problem involved with this class (Liseo *et al.*, 1996a) can be solved by using the moment theory. Relevant references include Kemperman (1987) and Salinetti (1994).

For a given quantity of interest $\varphi(\theta_1, \theta_2)$, we use the following notation:
$\varphi_i(\theta_2) = \inf_{\theta_1 \in A_i} \varphi(\theta_1, \theta_2)$,
$d = (d_1, \ldots, d_{n-1}) \in R^{n-1}$,
$S_{ij} = S(d_i, d_j) = \{\theta_2 : \varphi_i(\theta_2) - \varphi_j(\theta_2) < d_i - d_j\}$,
$S_i = \left(\cap_{j=1}^{i-1} S_{ji}^c\right) \cap \left(\cap_{j=i+1}^{n} S_{ij}\right)$, $i = 1, \ldots, n$, where S_{ji}^c denotes the complement of S_{ji}.

The following result was proved in Liseo et al. (1996).

Lemma 4.

$$\inf_{q \in Q} \int \varphi(\theta_1, \theta_2) q(\theta_1, \theta_2) d\theta_1 d\theta_2 = \sup_{d \in R^{n-1}}$$
$$\left\{\int \sum_{i=1}^{n} \varphi_i(\theta_2) 1_{S_i}(\theta_2) q_2(\theta_2) d\theta_2 + \sum_{i=1}^{n-1} d_i(\alpha_i - \int_{S_i} q_2(\theta_2)d\theta_2)\right\}.$$

From Lemma 4 and the linearization algorithm (Lavine et al., 1993), the range of $E^q(\varphi(\theta_1, \theta_2)|x)$ as q varies over Q can be easily obtained.

Example 7. Let (X_1, X_2) be bivariate normal random variable $N(x_1, x_2| (\theta_1, \theta_2), I_2)$. Suppose the base prior $\pi_0(\theta_1, \theta_2) = N(\theta_1|0, 1)\ N(\theta_2|0, 1)$. We are interested in testing $H_0 : \theta_1 < \theta_2$ to departures from π_0 but preserving the marginals.

As an approximation we consider the ε-contamination class $\Gamma(Q, \varepsilon)$, where

$$Q = \{q(\theta_1, \theta_2) \in Q_1 : q_2(\theta_2) = N(\theta_2|0, 1), \int_{-\infty}^{-v} q_1(\theta_1)d\theta_1 =$$
$$\int_{v}^{\infty} q_1(\theta_1)d\theta_1 = \tfrac{3}{8}, \int_{-v}^{0} q_1(\theta_1)d\theta_1 = \int_{0}^{v} q_1(\theta_1)d\theta_1 = \tfrac{1}{8}\},$$

$\varepsilon = 0.2$, and $v = 0.3186$. The value for v is the 0.625-percentile of a standard normal distribution. For various values of $x = (x_1, x_2)$, the posterior ranges of $P^{\pi}(H_0|x)$ as π ranges over $\Gamma(Q, 0.2)$ are displayed in Table 5.

$x=(x_1,x_2)$	$\inf_{\pi\in\Gamma(Q,0.2)} P^\pi(H_0\|x)$	$\sup_{\pi\in\Gamma(Q,0.2)} P^\pi(H_0\|x)$
(0,0)	.378	.621
(0,3)	.821	.948
(3,0)	.025	.174

TABLE 5. Posterior ranges

Numbers in Table 5 show that $P^\pi(H_0|x)$ is reasonably robust to departures from π_0 when the marginals are maintained fixed. $\triangle$

3.4 Bands of probability measures

A simple and realistic class, especially when the elicitation is made by more than one elicitor, is that given in terms of ranges of probabilities of a partition $\{C_i,\ i=1,\ldots,n\}$ of the parameter space Θ. That is,

$$\Gamma(\alpha,\beta)=\{\pi(\theta):\alpha_i\leq\int_{C_i}\pi(\theta)d\theta\leq\beta_i, i=1,2,\ldots,n\},$$

where α_i, β_i are specified numbers such that $\sum_{i=1}^n \alpha_i \leq 1 \leq \sum_{i=1}^n \beta_i$. Expressions for finding ranges of posterior probabilities have been given in Moreno and Pericchi (1992b).

Lemma 5. For a given set C, we have

$$\inf_{\pi\in\Gamma(\alpha,\beta)} P^\pi(C|x)=\inf_{\gamma_i\in D}\frac{\sum_{i\in I}\gamma_i\inf_{\theta\in C_i}f(x|\theta)}{\sum_{i=1}^n\gamma_i\sup_{\theta\in C_i\cap C^c}f(x|\theta)+\sum_{i\in I}\gamma_i\inf_{\theta\in C_i}f(x|\theta)},$$

where I is a subset of indices of $\{1,\ldots,n\}$ defined by $i\in I$ if and only if $C_i\subset C$, $C^c=\Theta-C$, and $D=\{\gamma_i:\alpha_i\leq\gamma_i\leq\beta_i, i=1,\ldots,n,\sum_{i=1}^n\gamma_i=1\}$. For finding $\sup_{\pi\in\Gamma(\alpha,\beta)} P^\pi(C|x)$, use that $\sup_{\pi\in\Gamma(\alpha,\beta)} P^\pi(C|x)=1-\inf_{\pi\in\Gamma(\alpha,\beta)} P^\pi(C^c|x)$.

Example 8. Martz and Waller (1982) supposed that two engineers are concerned with the mean life θ of a proposed new industrial engine. The two engineers, A and B, quantify their prior beliefs about θ in terms of the probabilities given in Table 6.

In Berger and O'Hagan (1989), O'Hagan and Berger (1988) and Berliner and Goel (1990), this example was analyzed for the prior beliefs of engineers A and B, separately. Here we assume that any prior beliefs between those of the engineer A and B are deemed as reasonable. That is, we shall consider any prior $P(C_i)\in(\alpha_i,\beta_i)$, where $\alpha_i=\min(P(C_i|A),P(C_i|B))$, and $\beta_i=\max(P(C_i|A),P(C_i|B))$, $i=1,\ldots,6$. Those lower and upper prior probabilities are given in the third column of Table 7. Observe that these lower and upper probabilities are coherent in the sense that $\sum\alpha_i=.7<\sum\beta_i=1.3$.

i	Intervals C_i	$P(C_i\|A)$	$P(C_i\|B)$
1	[0,1000)	.01	.15
2	[1000,2000)	.04	.15
3	[2000,3000)	.20	.20
4	[3000,4000)	.50	.20
5	[4000,5000)	.15	.15
6	[5000,∞)	.10	.15

TABLE 6. Specified prior probabilities of sets

Suppose, as in O'Hagan and Berger (1988), that data become available in the form of two independent lifetimes that are exponentially distributed with mean θ. The observed lifetimes are 2000 and 2500 hours.

From Lemma 5 it follows that the extreme prior distributions are discrete. For instance, $\sup_{\pi\in\Gamma(\alpha,\beta)} P(C_3|x)$ is attained at the prior $\pi(\theta) = 0.15$ $1_{\{\theta=0\}}(\theta) + 0.15\ 1_{\{\theta=1000\}}(\theta) + 0.20\ 1_{\{\theta=2250\}}(\theta) + 0.20\ 1_{\{\theta=4000\}}(\theta) + 0.15$ $1_{\{\theta=5000\}}(\theta) + 0.15\ 1_{\{\theta=\infty\}}(\theta)$. The posterior ranges of each C_i, denoted by α_i^* and β_i^*, are given in the fourth column of Table 7.

i	Intervals C_i	(α_i, β_i)	(α_i^*, β_i^*)
1	[0,1000)	(.01, .15)	(.00, .11)
2	[1000,2000)	(.04, .15)	(.02, .25)
3	[2000,3000)	(.20, .20)	(.22, .40)
4	[3000,4000)	(.20, .50)	(.20, .60)
5	[4000,5000)	(.15, .15)	(.10, .22)
6	[5000,∞)	(.10, .15)	(.00, .15)

TABLE 7. Prior and posterior ranges for C_i

The fourth column in Table 7 shows that most of the sets C_i have large posterior ranges, motivating a search for more accurate prior information. Also, when the interest is on a set C, which is not generated by $\{C_i, i = 1, \ldots, n\}$, the posterior range of $P^\pi(C|x)$ as π ranges over $\Gamma(\alpha, \beta)$ is too wide. In fact, for any set B such that $B \cap C_i \neq \emptyset$, $B \cap C_i \neq C_i$ for all i, it turns out that the posterior range of B is $(0, 1)$ for any sample observation x. $\triangle$

More accurate prior information can be obtained by considering a shape constrain on the priors in $\Gamma(\alpha, \beta)$, as in O'Hagan and Berger (1989), Berger and O'Hagan (1988), Bose (1990), and Moreno and Pericchi (1991).

Another way of incorporating prior information to $\Gamma(\alpha, \beta)$ could be to elicit priors for any Borel set in R^+. Hence, two functions $L(\theta)$, $U(\theta)$ such that $L(\theta) \leq U(\theta)$ for any θ, and $\int L(\theta)d\theta \leq 1 \leq \int U(\theta)d\theta$, are elicited and the class $\Gamma(L, U)$ stated in (8) is then considered. If in $\Gamma(L, U)$ we suppress the condition $\int \pi(\theta)d\theta = 1$, the class of interval of measures considered by DeRobertis and Hartigan (1981) is obtained. This latter class will be denoted as Γ_{DH}.

The posterior range of a set A as π varies over $\Gamma(L, U)$ is determined by using the next lemma.

Lemma 6. Under mild conditions, $\sup_{\pi \in \Gamma(L,U)} P^{\pi}(A|x) = P^{\pi^*}(A|x)$, where $\pi^*(\theta)$ is

$$\pi^*(\theta) = U(\theta)1_B(\theta) + L(\theta)1_{B^C}(\theta),$$

and B is given by

$$B = \begin{cases} A \cap \{\theta : f(x|\theta) \geq z_A\}, \text{ if } \int_A U(\theta)d\theta + \int_{A^c} L(\theta)d\theta > 1, \\ A, \text{ if } \int_A U(\theta)d\theta + \int_{A^c} L(\theta)d\theta = 1, \\ A \cup \left(A^C \cap \{\theta : f(x|\theta) < z_A\}\right), \text{ if } \int_A U(\theta)d\theta + \int_{A^c} L(\theta)d\theta < 1, \end{cases}$$

z_A being such that $\int \pi^*(\theta)d\theta = 1$. $\inf_{\pi \in \Gamma(L,U)} P^{\pi}(A|x)$ is obtained by using the relationship $\inf_{\pi \in \Gamma(L,U)} P^{\pi}(A|x) = 1 - \sup_{\pi \in \Gamma(L,U)} P^{\pi}(A|x)$.

Proof. For a proof, see Moreno and Pericchi (1992b), Theorem 2.□

Example 9. Let $f(x_1, x_2|\theta_1, \theta_2) = N(x_1|\theta_1, 1)N(x_2|\theta_2, 1)$ and the classes of priors $\Gamma(L, U)$ and Γ_{DH} for $L(\theta_1, \theta_2) = 1/21 \times N(\theta_1|0, 1)N(\theta_2|0, 1)$ and $U(\theta_1, \theta_2) = 1/(2\pi)$.

Let $H_0 : \theta_1 \leq \theta_2$ be the quantity of interest. For several observations $x = (x_1, x_2)$ the second column in Table 8 gives ranges of $P^{\pi}(H_0|x)$ as π ranges over Γ_{DH} and, in the third column, the related ranges as π ranges over $\Gamma(L, U)$.

x	$P_{DH}(H_0\|x), P^{DH}(H_0\|x)$	$\underline{P}(H_0\|x), \overline{P}(H_0\|x)$
(0, 0)	.29, .71	.39, .61
(.5, 0)	.20, .61	.30, .51
(1, 0)	.11, .52	.23, .43
(1.5, 0)	.06, .44	.14, .36

TABLE 8. Posterior ranges of H_0

Table 8 shows that by simply adding to the class of bands of measures the restriction of being probability measures, posterior ranges are substantially reduced. Therefore, the use of the latter class is strongly recommended. △

The main question in eliciting a band of probability measures is how to choose the bounds $L(\theta)$ and $U(\theta)$. Some guidelines have been given in Moreno and Pericchi (1993a). A closely related problem is considered in the pathbreaking paper by Edwards et al. (1963). Under the title of "Principle of Stable Estimation" they studied conditions under which the prior distribution can be regarded as essentially uniform. What they really developed was a desiderata to define functions $L(\theta)$ and $U(\theta)$, and the approximation they refer to is for priors in the band specified by $L(\theta)$ and $U(\theta)$; the lower bound $L(\theta)$ was uniform (k_0) on a region B favored by the data, and zero otherwise, and the upper bound $U(\theta)$ was also uniform (k_1) on B and uniform (k_2) on B^c. They provided a system with three axioms containing indications for choosing k_0, k_1 and k_2. A similar formulation, tight to an "operational" prior $\pi_{OP}(\theta)$ instead of the uniform prior, was given in Dickey (1976). For a discussion on these classes, see Moreno et al. (1999b).

3.5 Conclusions

Global Bayesian robustness measures the posterior uncertainty of $\varphi(\theta)$ that arises from having partial prior information on the parameter. This partial prior information is modeled through classes of prior distributions Γ that satisfy the expert prior beliefs on the behavior of θ. Consequently, the degree of posterior uncertainty depends on the prior uncertainty described by the class of priors Γ, the quantity of interest $\varphi(\theta)$, the sampling model $f(x|\theta)$, and the available data set $x = (x_1, \ldots, x_n)$.

In a sense, a robust Bayesian analysis requires in its application a weaker set of inputs than those required in a conventional Bayesian analysis where the subjective prior is completely specified. The price to be paid in a robust Bayesian analysis is a certain degree of posterior uncertainty that should be quantified in the analysis.

For some classes of prior distributions, this quantification involves complex variational problems. In particular, multidimensional parameter spaces pose qualitatively much more involved variational problems to the robust Bayesian analysis than those encountered in one-dimensional parameter spaces. For instance, while a shape constraint, such as unimodality, might be a quite natural restriction in a one-dimensional parameter space, this concept does not have a clear meaning in a multidimensional parameter space. In this latter context, unimodality has to be specified for each of the possible existing directions, thus giving different types of unimodality (Dharmadhikari and Joag-Dev, 1988). A type of unimodality in a multidimensional setting (block unimodality) has been analyzed in Liseo *et al.*(1993).

Fortunately, in many applications realistic and accessible partial prior knowledge can be modeled with classes of priors for which the variational problems have a simple solution. In particular, ε-contamination classes with quantile restrictions pose simple variational problems that can be solved whatever the dimension of the parameter space is. In multidimensional settings an interesting alternative is the probability density band class that shares the simplicity of the variational problem with the ε-contamination classes.

A criticism to Bayesian robustness analysis is that here the sampling model is assumed to be fixed and the uncertainty considered is only on the prior distribution of the parameter in the sampling model. There is no reason for assuming such an asymmetric role for the sampling model and prior (Lavine, 1991). Instead, robustness should be studied with respect to both elements of the problem: the sampling model and the prior distribution. In this setting, however, the uncertainty a priori is typically so large that the posterior answer will be generally vacuous.

On the other hand, it is not clear that we are able to subjectively elicit on parameters that are unobservable in nature. In fact, some authors claim that the expert should be asked to assess only observable quantities (Kadane, 1980, Tversky and Kahneman, 1974, Winkler, 1980, Wolpert, 1989, among others). This means that elicitation should be on some features of the predictive distribution such as quantiles or shape constraint of the predictive distribution. For a formalization of this point of view that involves some robustness problems, see Moreno et al. (1999a).

References

BERGER, J.O. (1984). The robust Bayesian viewpoint (with discussion). In *Robustness of Bayesian Analysis* (J.B. Kadane, ed.), 63–144. Amsterdam: North-Holland.

BERGER, J.O. (1985). *Statistical Decision Theory and Bayesian Analysis.* New York: Springer-Verlag.

BERGER, J.O. (1990). Robust Bayesian analysis: sensitivity to the prior. *Journal of Statistical Planning and Inference* **25**, 303–328.

BERGER, J.O. (1994). An overview of robust Bayesian analysis (with discussion). *Test*, **3**, 5–124.

BERGER, J.O. and BERLINER, L.M. (1986). Robust Bayes and empirical Bayes analysis with ε-contaminated prior. *Annals of Statistics*, **14**, 461–486.

BERGER, J.O. and MORENO, E. (1994). Bayesian Robustness in bidimensional models: prior independence (with discussion). *Journal of Statistical Planning and Inference*, **40**, 161–176.

BERGER, J.O. and O'HAGAN, A. (1988). Ranges of posterior probabilities for unimodal prior with specified quantiles. In *Bayesian Statistics 3*, (J.M. Bernardo, M.H. DeGood, D.V. Lindley and A.F.M. Smith, eds.), 45–66. Oxford: Oxford University Press.

BERLINER, L.M. and GOEL, P.K. (1990). Incorporating partial prior information: ranges of posterior probabilities. In *Bayesian and Likelihood Methods in Statistics and Econometrics* (S. Geisser, J.S. Hodges and A. Zellner, eds.), 397–406. Amsterdam: North-Holland.

BERTOLINO, F., PICCINATO, L. and RACUGNO, W. (1995). Multiple Bayes factor for testing hypotheses. *Journal of the American Statistical Association*, **82**, 213–219.

BETRÓ, B., MECZARSKI, M. and RUGGERI, F. (1994). Robust Bayesian analysis under moment conditions. *Journal of Statistical Planning and Inference*, **41**, 257–266.

BETRÓ, B. and GUGLIELMI, A. (2000). Methods for global prior robustness under generalized moment conditions. In *Robust Bayesian Analysis* (D. Ríos Insua and F. Ruggeri, eds.). New York: Springer-Verlag.

BETRÓ, B. and GUGLIELMI, A. (1996). Numerical robust Bayesian analysis under generalized moment conditions (with discussion). In *Bayesian Robustness IMS Lectures-Notes Monograph Series*, (J.O. Berger, B. Betró, E. Moreno, L.R. Pericchi, F. Ruggeri, G. Salinetti and L. Wasserman, eds.), Vol 29, 3–20. Hayward: IMS.

BLUM, J.R. and ROSENBLATT, J. (1967). On partial prior information in statistical inference. *Annals of Mathematical Statistics*, **38**, 1671–1678.

BOSE, S. (1990). Bayesian robustness with shape-constrained priors and mixtures of priors. *Ph.D. Thesis*, Purdue University.

BOSE, S. (1994a). Bayesian robustness with more than one class of contaminations (with discussion). *Journal of Statistical Planning and Inference*, **40**, 177–188.

BOSE, S. (1994b). Bayesian robustness with mixture classes of priors. *Annals of Statistics*, **22**, 652–667.

CAMBANIS, S., SIMONS, G. and STOUT, W. (1976). Inequalities for $Ek[X, Y]$ when marginals are fixed. *Zeitschrift für Wahrscheinlichkeitstheorie und Verwandte Gebiete*, **36**, 285–294.

CANO, J.A., HERNÁNDEZ, A. and MORENO, E. (1986). Posterior measure under partial prior information. *Statistica*, XLVI, 119–230.

DALL'AGLIO, A. (19959 *Problema dei momenti e programmazione lineare semi-infinita nella robustezza bayesiana*, Ph.D. Dissertation, Universita "La Sapienza" di Roma.

DE LA HORRA, J. and FERNÁNDEZ, C. (1994). Bayesian analysis under ε-contaminated priors: a trade-off between robustness and precision. *Journal of Statistical Planning and Inference*, **38**, 13–30.

DEROBERTIS, L. (1978). The use of partial prior knowledge in Bayesian inference. *Ph. D. Thesis*, Yale University.

DEROBERTIS, L. and HARTIGAN, J.A. (1981). Bayesian inference using intervals of measures. *Annals of Statistics*, **9**, 235–244.

DHARMADHIKARI, S. and JOAG-DEV, K. (1988). *Unimodality, Convexity and Applications.* Boston: Academic Press.

DICKEY, J. (1976). Approximate posterior distributions. *Journal of the American Statistical Association*, **71**, 680–689.

EDWARDS, W., LINDMAN, H. and SAVAGE, L.J. (1963). Bayesian statistical inference for psychological research. *Psychological Review*, **79**, 193–242.

HUBER, P. (1973). The use of Choquet capacities in statistics. *Bulletin of the International Statistical Institute*, **45**, 181–191.

KADANE, J.B. (1980). Predictive and structural methods for eliciting prior distributions. In *Bayesian Analysis in Econometrics and Statistics* (A. Zellner, ed.). Amsterdam: North-Holland.

KADANE, J.B., MORENO, E., PÉREZ, M.E. and PERICCHI, L.R. (1999). Applying non-parametric robust Bayesian analysis to non-opinionated judicial neutrality. In *Proceedings of the First International Symposium on Imprecise Probabilities and Their Applications*(G. De Cooman, F.G. Cozman, S. Moral, P. Walley, eds.), 216–224.

KASS, R. and GREENHOUSE, J. (1989). Investigating therapies of potentially great benefit: a Bayesian perspective. Comments on Ware's paper. *Statistical Science*, **4**, 310–317.

KEMPERMAN, J. (1987). Geometry of the moment problem. *Proceedings of Symposia in Applied Mathematics*,**37**, 16–53.

KUDŌ, H. (1967). On partial prior information and the property of parametric sufficiency. *Proceedings of the Fifth Berkeley Symposium on Mathematical Statistics and Probability*, **1**, 251–265.

LAVINE, M. (1991). Sensitivity in Bayesian statistics: the prior and the likelihood. *Journal of the American Statistical Association*, **86**, 396–399.

LAVINE, M., WASSERMAN, L. and WOLPERT, R. (1991). Bayesian inference with specified prior marginals. *Journal of the American Statistical Association*, **66**, 964–971.

LAVINE, M., WASSERMAN, L. and WOLPERT, R. (1993). Linearization of Bayesian robustness problems. *Journal of Statistical Planning and Inference*, **37**, 307–316.

LAVINE, M., PERONE PACIFICO, M., SALINETTI, G. and TARDELLA, L. (2000). Linearization techniques in Bayesian Robustness. In *Robust Bayesian Analysis* (D. Ríos Insua and F. Ruggeri, eds.). New York: Springer-Verlag.

LISEO, B., PETRELLA, L. and SALINETTI, G. (1993). Block Unimodality for Multivariate Bayesian Robustness. *Journal of Italian Statistical Society*,**1**, 55–71.

LISEO, B., MORENO, E. and SALINETTI, G. (1996a). Bayesian robustness for classes of bidimensional priors with given marginals (with discussion). In *Bayesian Robustness IMS Lecture-Notes Monograph Series* (J.O. Berger, B. Betró, E. Moreno, L.R. Pericchi, F. Ruggeri, G. Salinetti and L. Wasserman, eds.), Vol. 29, 101–118. Hayward: IMS.

LISEO, B. PETRELLA and L. SALINETTI, G. (1996b). Bayesian Robustness: an Interactive Approach. In *Bayesian Statistics* 5, (J.O. Berger, J.M. Bernardo, A.P.,Dawid, A.F.M. Smith, eds.), 661–666. Oxford: Oxford University Press.

MANSKY, D.F. (1981). Learning and decision making when subjective probabilities have subjective domains. *Annals of Statistics*, **9**, 59–65.

MARAZZI, A. (1985). On constrained minimization of the Bayes risk for the linear model. *Statistics and Decisions*, **3**, 277–296.

MARTZ, H.F. and WALLER, R.A. (1982). *Bayesian Reliability Analysis.* New York: Wiley.

MORENO, E. and CANO, J.A. (1989). Testing a point null hypothesis: asymptotic robust Bayesian analysis with respect to the priors given in a subsigma field. *International Statistical Review*, **57**, 221–232.

MORENO, E. and CANO, J.A. (1991). Robust Bayesian analysis with ε-contaminations partially known. *Journal of the Royal Statistical Society Series B*, **53**,143–155.

MORENO, E. and CANO, J.A. (1995). Classes of bidimensional priors specified on a collection of sets: Bayesian robustness. *Journal of Statistical Planning and Inference*, **46**, 325–334.

MORENO, E. and GONZÁLEZ, A. (1990). Empirical Bayes analysis for ε-contaminated priors with shape and quantile constraints. *Brazilian Journal of Probability and Statistics*, **4**, 177–200.

MORENO, E., MARTÍNEZ, C. and CANO, J.A. (1996). Local robustness and influence for contamination classes of prior distributions (with discussion). In *Bayesian Robustness IMS Lectures-Notes Monograph Series*, (J.O. Berger, B. Betró, E. Moreno, L.R. Pericchi, F. Ruggeri, G. Salinetti and L. Wasserman, eds.),Vol 29, 139–156. Hayward: IMS.

MORENO, E. and PERICCHI, L.R. (1991). Robust Bayesian analysis for ε-contaminations with shape and quantile constraints. In *Proceedings of the Fifth International Symposium on Applied Stochastic Models and Data Analysis,* (R. Gutiérrez and M. Valderrama, eds.), 454–470. Singapore: World Scientific.

MORENO, E. and PERICCHI, L.R. (1992a). Subjetivismo sin dogmatismo: Análisis bayesiano robusto (with discussion). *Estadistica Española*, **34**, 5–60.

MORENO, E. and PERICCHI, L.R. (1992b). Bands of probability measures. In *Bayesian Statistics* 4, (J.O. Berger, J.M. Bernardo, A.P. Dawid, A.F.M. Smith, eds.), 607–713. Oxford: Oxford University Press.

MORENO, E. and PERICCHI, L.R. (1993a). Bayesian robustness for hierarchical ε-contamination models. *Journal of Statistical Planning and Inference*, **37**, 159–168.

MORENO, E. and PERICCHI, L.R. (1993b). Prior assessments for bands of probability measures: empirical Bayes analysis. *Test*, **1/2**, 101–110.

MORENO, E., BERTOLINO, F. and RACUGNO, W. (1999a). Inference under prior information on observable random variables. *Technical Report.* University of Granada.

MORENO, E., PERICCHI, L.R. and KADANE, J.B. (1999b). A robust Bayesian look at the theory of precise measurement. In *Decision Science and Technology: Reflections on the contributions of Ward Edwards* (J. Shanteau, B. Mellers and D. Schum, eds.), 171–178. Boston/Dordrecht/London: Kluwer Academic Publishers.

O'HAGAN, A. and BERGER, J.O. (1988). Ranges of posterior probabilities for quasi-unimodal priors with specified quantiles. *Journal of the American Statistical Association*, **83**, 503–508.

PERONE-PACIFICO, M., SALINETTI, G. and TARDELLA, L. (1996). Bayesian robustness on constrained density band classes. *Test*, **5**, 395–409.

SACKS, S. (1937). *Theory of the Integral.* Warsaw: Monografie Matematyezne.

SALINETTI, G. (1994) Comments on "An overview of Robust Bayesian Analysis" by J.O. Berger. *Test*, **3**, 5–124.

SIVAGANESAN, S. (1988). Range of posterior measures for priors with arbitrary contaminations. *Communications in Statistics*, **17**, 1591–1612.

SIVAGANESAN, S. (1989). Sensitivity of posterior mean to unimodality preserving contaminations. *Statistics and Decisions*, **7**, 77–93.

SIVAGANESAN, S. (1993). Range of the posterior probability of an interval for priors with unimodality preserving contaminations. *Annals of the Institute of Mathematical Statistics*, **45**, 171–188.

SIVAGANESAN, S. and BERGER, J.O. (1989). Ranges of posterior measures for priors with unimodal contaminations. *Annals of Statistics*, **17**, 868–889.

RACHEV, S.T. (1985). The Monge-Kantorovich mass transference problem and its stochastic applications. *Theory of Probability and Its Applications*, **29**, 647–671.

TCHEN, A.H. (1980). Inequalities distribution with given marginals. *Annals of Probability*, **8**, 814–817.

TVERSKY, A. and KAHNEMAN, D. (1974). Judgment under uncertainty: heuristic and biases. *Sciences*, **185**, 1124–1131.

WALLEY, P. (1991). *Statistical Reasoning with Imprecise Probabilities.* London: Chapman and Hall.

WASSERMAN, L. (1992). Recent methodological advances in robust Bayesian inference. In *Bayesian Statistics 4* (J.O. Berger, J.M. Bernardo, A.P. Dawid, A.F.M. Smith, eds.),483–502. Oxford: Oxford University Press.

WINKLER, R.L. (1980). Prior information, predictive distributions, and Bayesian model-building. In *Bayesian Analysis in Econometrics and Statistics* (A. Zellner, ed.). Amsterdam: North-Holland.

WOLPERT, R.L. (1989). Eliciting and combining subjective judgments about uncertainty. *International Journal of Technology Assessment Health Care*, **5**, 537–557.

4

Local Robustness in Bayesian Analysis

Paul Gustafson

ABSTRACT Whereas a global approach to prior robustness focusses on the range of inferences arising from a range of priors, the local approach is concerned with derivatives of posterior quantities with respect to the prior. The local approach has several advantages. First, in contrast to the linearization algorithm used for global robustness, an iterative scheme is not required. Second, it is typically straightforward to obtain a local analysis from Markov chain Monte Carlo posterior output. Third, local sensitivity analysis can be implemented in situations where the linearization algorithm for global analysis is not applicable because the posterior quantity of interest is not a ratio-linear functional of the distribution being perturbed. On the downside, however, it is hard to assess the accuracy of a local sensitivity analysis viewed as an approximation to a global analysis. In particular, summaries of local sensitivity often have nonsensical asymptotic behaviour, raising thorny questions about their calibration. In this article the local approach is broadly defined, ranging from differentiation of functions (e.g., assessing sensitivity to hyperparameters) to differentiation of functionals (e.g., assessing sensitivity to the prior distribution as a whole). In this article we review some of the basic formulations for local assessment of prior influence. We then discuss the use of local analysis to study sensitivity in contexts broader than merely prior uncertainty. Finally, we try to summarize the merits and demerits of local analysis as a robustness tool.

Key words: derivatives, model influence, prior influence.

4.1 Introduction

Questions of robustness are ubiquitous in statistical analysis. So often one cannot be sure that what goes into an analysis is exactly right; therefore, it seems imperative to check how strongly the answer coming out depends on the ingredients going in. For a Bayesian analysis these ingredients, or inputs, include the data, the model, the prior, and, perhaps implicitly, the loss function. The answer, or output, on the other hand, might be a scalar summary of the posterior distribution, a marginal posterior distribution, or even the entire posterior distribution. Thus the robustness question can be posed as: how *sensitive* is the output to the input? or equivalently as:

how *influential* is the input on the output?. While the term "Bayesian robustness" has become synonymous with prior influence, some attention has also been paid to the other inputs, the data, the model, and the loss function.

When prior influence is in question, we can distinguish at least three broad possibilities as to how sensitivity analysis might proceed. First, and most common, is *informal* sensitivity analysis in which the analysis is simply repeated for a few different prior distributions, and the resulting inferences are compared. Second, *global* sensitivity analysis aims for a more rigorous demonstration of robustness. Here the range of inferences arising from a large class of priors is determined, without explicitly carrying out the analysis for every prior in the class. The third possibility manifests itself if we paraphrase the sensitivity question slightly as: what is the rate of change in the inference with respect to change in the prior? Then *local* sensitivity analysis, which uses differential calculus to assess sensitivity, becomes an intuitively obvious and appealing option. Schematically, if μ represents or indexes the prior, and $T(\mu)$ is the posterior quantity of interest, then it is very natural to use $T'(\mu)$ to quantify prior influence.

This chapter will review and comment on Bayesian local sensitivity analysis. Section 2 attempts to give a reasonably thorough discussion of local prior influence, while Section 3 focusses on less-studied applications of local sensitivity analysis which go "beyond" the prior distribution. Comments on the utility of local robustness methods are saved for the discussion in Section 4. Note that a comparative review of the local and global approaches is given by Sivaganesan (2000).

4.2 Local prior influence

4.2.1 Various approaches

A variety of frameworks for assessing local prior influence have been proposed. Indeed, this variation gives the literature a fractured quality, though some of the differences are superficial. In broad generality, all the proposed methods involve differentiating a posterior quantity with respect to a changing prior. The derivative is evaluated at a *baseline* prior, giving a measure of posterior-to-prior sensitivity in the vicinity of the baseline prior.

One difference between methods lies in the way prior distributions are perturbed. Some authors take a parametric view and restrict the prior to a parametric family. For instance, say the parameter is θ, and the prior is restricted to be in a family of distributions $Q(\theta|\lambda)$ indexed by hyperparameter λ. Then the baseline prior is $P(\theta) = Q(\theta|\lambda)$, and the perturbed prior is taken to be $\tilde{P}_\mu(\theta) = Q(\theta|\lambda+\mu)$. We denote Bayesian updating with a superscript x, so that the posterior distributions arising from the baseline and perturbed priors are $P^x(\theta)$ and $\tilde{P}^x_\mu(\theta)$, respectively. The posterior quantity of interest is then a function of μ, say $T(\mu)$, and the derivative of $T(\cdot)$ evaluated at $\mu = 0$ reflects prior influence near the baseline prior.

Often it is argued that parametric families of distributions are not rich enough to fully assess prior influence. The most common nonparametric perturbation scheme is based on ϵ-contamination, which has its roots in the classical robustness literature. First we consider a very general *linear* perturbation under which the perturbed prior is

$$\tilde{P}_U = P + U,$$

with U a signed measure with total mass zero. We obtain ϵ-contamination by restricting U to be of the form $U = \epsilon(Q - P)$, where Q is a distribution. Typically Q might be assumed to belong to a rich nonparametric class of distributions. Now our posterior quantity of interest can be expressed as a functional of U, which we write as $T(U)$, and local sensitivity information resides in the derivative of T at $U = 0$, which we write as

$$\dot{T}(0)U = \left.\frac{\partial}{\partial \epsilon} T(\epsilon U)\right|_{\epsilon=0}.$$

While we will be primarily concerned with linear perturbations, we note that other schemes have been considered in the literature, such as *geometric* perturbations having densities of the form

$$\tilde{p}(\theta) = \frac{\{p(\theta)\}^{1-\epsilon}\{q(\theta)\}^{\epsilon}}{\int \{p(\theta)\}^{1-\epsilon}\{q(\theta)\}^{\epsilon}\, d\theta}.$$

Another way to grapple with the diversity of literature is to separate methods which assess sensitivity of posterior distributions from those which assess sensitivity of scalar posterior summaries. The former case requires some sort of divergence measure to quantify the discrepancy between two distributions. If $d(\cdot, \cdot)$ is such a measure, then in the case of linear perturbations, say, we may be interested in a functional of the form

$$T(U) = d(P^x, \tilde{P}^x_U), \qquad (1)$$

which is the divergence between baseline and perturbed posteriors. It is worth noting that for many divergence measures $\dot{T}(0)U = 0$ since $T()$ in (1) is minimized by taking $U = 0$. In such situations sensitivity analysis must be based on examination of second derivatives.

On the other hand, say we are interested in a scalar posterior summary such as the posterior mean of $h(\theta)$. Then, again assuming linear perturbations, the functional of interest is

$$T(U) = \int h(\theta) d\tilde{P}^x_U(\theta).$$

Of course, interest may centre on other posterior summaries, such as posterior quantiles, modes, or variances, with $T(U)$ defined accordingly.

With this preamble in mind, published articles on local prior influence are listed in Table 1, along with a crude classification scheme. In particular, the table indicates which posterior quantities are considered, and what kinds of prior perturbations are used. The final column in the table describes how worst-case sensitivity is measured (see subsection 2.3). Articles are included in Table 1 if they use local analysis based on derivatives specifically to assess sensitivity to the prior. Thus articles such as Diaconis and Freedman (1986) and Cuevas and Sanz (1988), which differentiate with respect to the prior for other reasons, are not included. It is interesting to note, however, that Diaconis and Freedman seemed to spur interest in the use of derivatives to measure prior influence. Articles which discuss concerns even more basic than differentiability, such as continuity and Lipschitz conditions with respect to the prior (Basu et al., 1998) are also not included in the table. Articles which extend sensitivity analysis beyond the prior distribution are also omitted from the table; this topic is discussed in Section 3. Other articles which are omitted because they do not readily fit into the classification scheme in Table 1 include Fortini and Ruggeri (1994), Fortini and Ruggeri (2000), Ruggeri and Wasserman (1995), and Meng and Sivaganesan (1996). The former two articles are based on *concentration functions.* The latter two involve first determining the form of the global range for a posterior quantity under prior uncertainty, and then applying local analysis by differentiating the endpoints of the range.

4.2.2 A specific approach

We now outline in more detail a specific approach to local prior influence. This approach, which takes elements from Ruggeri and Wasserman (1993), Sivaganesan (1993), and Gustafson (1996a), borrows the notion of an *influence function* from classical robustness theory (Hampel et al., 1986). We focus on a scalar posterior summary, denoted as $T(U)$ when the prior is $\tilde{P}_U = P + U$. Taking $U = \epsilon[Q - P]$ to obtain ϵ-contamination, the local sensitivity can be encapsulated with the influence function $I(\cdot)$, which satisfies

$$\frac{\partial}{\partial \epsilon} T(\epsilon[Q-P])\bigg|_{\epsilon=0} = \int I(z) d[Q-P](z). \tag{2}$$

Thus the influence function indicates sensitivity of the posterior quantity to "contamination" of the prior P by any other prior Q. Note that I is defined only up to an additive constant. It is convenient to standardize according to $\int I(z)dP(z) = 0$. Then, at least for one-dimensional priors, a plot of the influence function can be quite revealing. For instance, the effect of point-mass contamination can be eyeballed from the plot via $T(\epsilon[\delta_z - P]) \approx \epsilon I(z)$, while the effect of other contaminants Q can be assessed, quickly but roughly, by visual integration of $I(\cdot)$ with respect to Q.

Article	Posterior Quantity	Type of Perturb.	Worst-Case Analysis
Basu (1996)	var., med. mode	linear	absolute, various classes
Basu, Jammalamadaka and Liu (1996)	mean	para.	relative, Euclidean
Birmiwal and Dey (1993)	mean	linear	absolute, various classes
Delampady and Dey (1994)	dist.	linear	absolute, various classes
Dey and Birmiwal (1994)	dist.	linear, geom.	N/A
Dey, Ghosh and Lou (1996)	dist.	linear	absolute, various classes
Gelfand and Dey (1991)	dist.	geom.	N/A
Gustafson and Wasserman (1995)	dist.	linear, geom.	relative TV, ϕ
Gustafson (1996a)	mean	linear, nonlin.	relative L_1, L_2, L_∞
Gustafson (1996b)	mean	linear	relative L_2
McCulloch (1989)	dist.	para.	relative, KL
Moreno, Martínez and Cano (1996)	mean	linear	relative, quantile class
Peña and Zamar (1997)	mode	linear	N/A
Ruggeri and Wasserman (1993)	mean	linear	relative, TV
Sivaganesan (1993)	mean	linear	absolute, various classes

TABLE 1. Summary of the literature on local prior influence. The first column gives the citation. The second column indicates what posterior quantities are considered (abbreviations: dist. for distribution, var. for variance and med. for median). The third column indicates what kinds of prior perturbation are used (abbreviations: geom. for geometric and para. for parametric). The fourth column indicates whether worst-case sensitivity is measured in an absolute or relative sense, and what classes or discrepancy measures are considered (abbreviations: KL for Kullback–Leibler divergence, TV for total variation norm and ϕ for ϕ-divergence).

For most posterior quantities the influence function is easily determined. Take the case of $T(U) = \int h(\theta)d\tilde{P}^x_U(\theta)$, the posterior mean of $h(\theta)$. It is straightforward to determine that the corresponding influence function is

$$I(z) = \{h(z) - T(0)\} \left[\frac{dP^x}{dP}(z)\right]$$

For multidimensional priors, the influence function as defined in (2) is multidimensional, and therefore harder to visualize. But if we are interested in sensitivity to a one-dimensional prior marginal, then a one-dimensional influence function results. Say $\theta = (\theta_1, \theta_2)$, with the baseline prior $dP(\theta)$ expressed as the $\theta_2|\theta_1$ conditional $P_{2|1}$ and the θ_1 marginal P_1. We perturb P_1 according to $\tilde{P}_{1,U} = P_1 + U$ while keeping $P_{2|1}$ fixed, again letting $T(U)$ denote the posterior quantity of interest. The influence function is defined analogously to (2) as

$$\left.\frac{\partial}{\partial\epsilon}T(\epsilon[Q_1 - P_1])\right|_{\epsilon=0} = \int I_{P_1}(z)d[Q_1 - P_1](z), \qquad (3)$$

where E^x denotes a posteriori expectation. In the case that T is the posterior mean of $h(\theta)$, the standardized influence function is

$$I_{P_1}(z) = E^x\{h(\theta) - T(0)|\theta_1 = z\}\left[\frac{dP^x_1}{dP_1}(z)\right].$$

This opens up the possibility of assessing the sensitivity of various posterior expectations to various prior marginals.

As a very simple example of prior influence functions, consider a binomial model $X \sim \text{binomial}(n, \theta)$, with $n = 10$. The observed datum is $X = 7$, and the posterior quantity of interest is the posterior mean of θ. Five different baseline priors for θ are considered, with their densities plotted in the left-hand panels of Figure 1. The first four priors are Beta distributions, while the fifth is a mixture of two Beta distributions. The right-hand panels give the corresponding influence functions, with vertical lines to indicate the baseline posterior mean. Note that the influence function for the Beta(1,3) prior stands out from the others, with a large maxima around 0.8. This suggests that a prior giving a bit more weight to this region would substantially boost the posterior mean. Of course, this is in accord with intuition, since this particular prior is in conflict with the data.

4.2.3 Worst-case sensitivity

The influence function does a great service by summarizing a functional on signed measures by a function of a single variable. If still further summarization is required, it is customary to take the pessimistic view and consider worst-case sensitivity. In fact, some may criticize the robustness

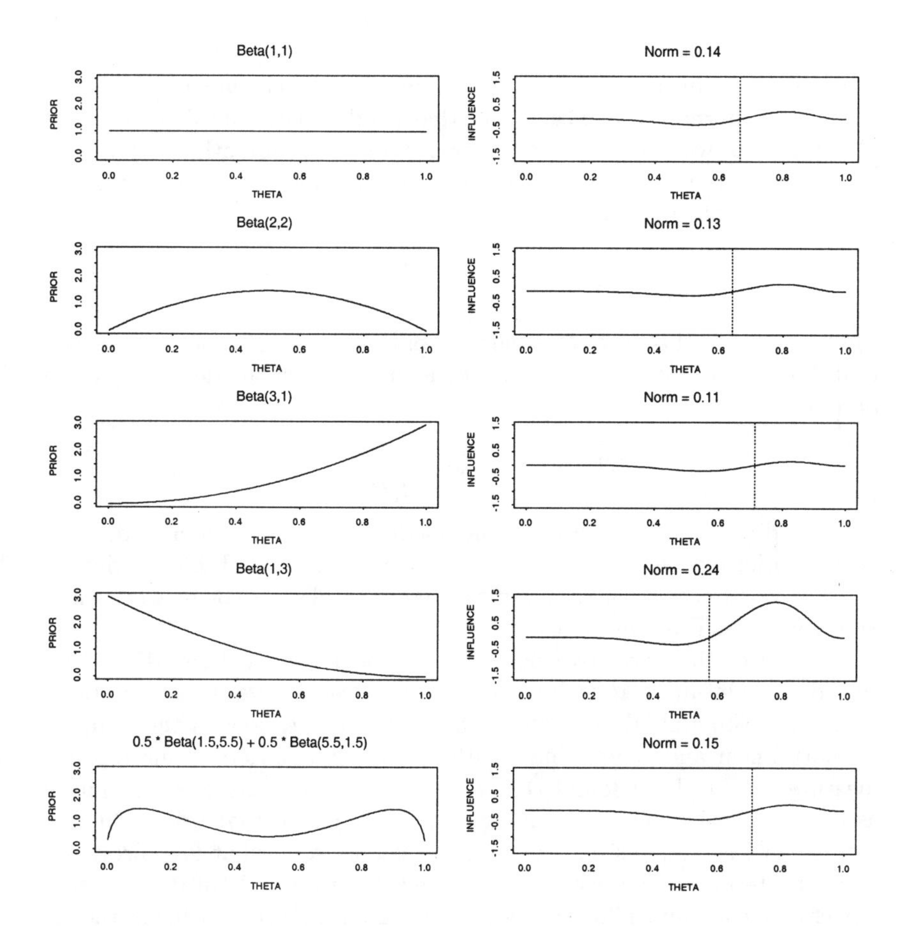

FIGURE 1. Prior influence functions for the posterior mean of θ, when $X \sim \text{Binomial}(n, \theta)$, with $n = 10$ and observed datum $X = 7$. The left-hand panels give five different baseline prior distributions for θ, while the right-hand panels give the corresponding influence functions. The vertical lines on the right-hand panels indicate the baseline posterior mean in each case. The value of the derivative norm is given above each influence function.

literature on the grounds of leaping to the worst case prematurely. For instance, most authors skip the influence function representation entirely.

Broadly speaking, one can quantify worst-case sensitivity in either an absolute or a relative sense. In the absolute case one might report

$$W = \sup_{U \in \Gamma} |\dot{T}(0)U| \tag{4}$$

as a worst-case measure of sensitivity, where $U \in \Gamma$ is a class of perturbing measures. Alternately, one can argue that (4) does not reflect the fact that $\tilde{P}_U = P + U$ will be closer to P for some U than for others. If we use a norm such that $d(P, \tilde{P}_U) = ||U||$ is a sensible measure of the size of a perturbation, then the derivative norm,

$$||\dot{T}(0)|| = \sup_{U \in \Gamma} \frac{|\dot{T}(0)U|}{||U||},$$

can be interpreted as a relative measure of local prior influence. In the case that T measures posterior discrepancy as in (1), then the derivative norm becomes

$$||\dot{T}(0)|| = \sup_{U \in \Gamma} \frac{d(P^x, \tilde{P}^x_U)}{d(P, \tilde{P}_U)}, \tag{5}$$

which is directly interpretable as the maximum ratio of posterior discrepancy to prior discrepancy. The literature classification in Table 1 indicates whether worst-case sensitivity is measured in an absolute or relative sense, and what class Γ is considered.

The use of the derivative norm as a measure of local sensitivity has considerable intuitive appeal; unfortunately, it can be problematic in practice. Gustafson and Wasserman (1995) studied expressions of the form (5) for both linear and geometric perturbations with a variety of discrepancy measures $d(\cdot, \cdot)$. They found that typically the norm diverges to infinity as the sample size increases, giving a false impression that prior influence increases with sample size. The situation is only somewhat less dire when scalar posterior summaries are considered. Gustafson (1996b) considered posterior means under linear prior perturbations, with $||U||$ being the L_p norm of $[dU/dP]$ with respect to measure P. When $p = 1$, the misleading dependence of the norm on the sample size persists. When $p = \infty$, the norm is a poor measure of sensitivity for a different reason: its dependence on the observed data is very weak. However, when $p = 2$, the norm appears to behave adequately as a measure of prior influence. In the case of absolute worst-case sensitivity, Sivaganesan (1996) studies the asymptotics of W and gives conditions on the contaminating classes under which W vanishes as the sample size increases.

Say we do use the L_2 norm,

$$||U|| = \left\{ \int [dU/dP]^2 dP \right\}^{1/2},$$

to measure the magnitude of a linear prior perturbation. Then in the framework of subsection 2.2 it is a simple matter to verify that

$$\|\dot{T}(0)\| = \left[\int \{I(z)\}^2 dP(z)\right]^{1/2} \tag{6}$$

is the derivative norm. Consequently, (6) can be regarded as a worst-case summary of the influence function. For the examples in Figure 1, the norm is displayed above each influence function. We see that the norm for the Beta(3,1) prior is about twice as big as the norm for the other four priors, reinforcing the notion that this prior is considerably more influential than the others.

Finally, we mention that much has been made in the literature about whether derivatives with respect to the prior are weak or strong. In fact, arguably the local sensitivity literature is too hung-up on this point. Without being very careful about the formalism, the derivative is strong (a Fréchet derivative) if the Taylor approximation

$$T(U) = T(0) + \dot{T}(0)U + o(\|U\|)$$

holds uniformly in U. If the approximation holds only pointwise, then we have only a weak (Gâteaux) derivative. Typically an innocuous condition such as boundedness of the likelihood is enough to ensure the derivative is strong (Ruggeri and Wasserman, 1993). The practical implication of a strong derivative is that the derivative norm, which we have already considered as a sensitivity measure, can be used to approximate a global robustness analysis. In particular, a strong derivative implies

$$\sup_{\{U \in \Gamma : \|u\| < \epsilon\}} T(U) = T(0) + \epsilon \|\dot{T}(0)\| + o(\epsilon), \tag{7}$$

so that the local analysis provides a valid "small-class" approximation to a global analysis. Of course, the approximation is not uniform in n, as the norm can diverge to infinity as n increases, while the width of the global range shrinks to zero.

4.3 Beyond the prior

4.3.1 Decision theory

Several authors have suggested broadening the notion of local sensitivity in a decision-theoretic context. Ramsay and Novick (1980) introduce PLU (prior, likelihood, utility) robustness, suggesting that the influence of each should be measured via an influence function. More recently, Martín and Ríos Insua (1996) consider posterior expected loss as the target quantity and study sensitivity by differentiating this target with respect to the prior and the loss function jointly. Kadane and Srinivasan (1996) and Kadane et al. (2000) also treat posterior expected loss as the target quantity, though they only consider variation in the prior, through the notion of *stability*, which dates back to Kadane and Chuang (1978) and is also considered by Salinetti (1994). Kadane and Srinivasan (1996) draw specific connections between stability and weak differentiability. In a different vein, Rukhin (1993) uses differentiation to study the influence of the prior on the frequentist risk of the Bayes rule. One intriguing question in the decision theory context is whether sensitivity of the expected loss or sensitivity of the optimal action is more relevant. One might argue that both should be assessed simultaneously.

4.3.2 Hierarchical models

Consider the following simple prototype of a hierarchical model. Observables $X_1, \ldots, X_n$ are modelled as independent, with X_i having density $f(x_i|\theta_i)$. For concreteness we might think of θ_i as a random effect associated with the ith sampling unit. In turn, $\theta_1, \ldots, \theta_n$ are modelled as independent and identically distributed with density $f(\cdot|\lambda)$, where λ is also an unknown parameter. Thus the specification is completed by assigning a prior distribution to λ. Note that there are different ways to view such a structure. One might view $f(x|\theta, \lambda) = f(x|\theta)$ as the model, and $f(\theta, \lambda) = f(\theta|\lambda)f(\lambda)$ as the prior. Or one might integrate out the random effects to obtain $f(x|\lambda)$ as the model, and $f(\lambda)$ as the prior.

An obvious question of interest is how sensitive inferences in this sort of model are to the form of the distribution for $\theta|\lambda$. To be specific, say that the random effect distribution is centered at zero, via $\theta_i = \lambda Z_i$, with $Z_1, \ldots, Z_n$ independent and identically distributed with density g. The baseline prior might correspond to $g \equiv N(0, 1)$. Further, say interest lies in the posterior mean of $h(\theta, \lambda)$, which we express as $T(g)$. Conceptually, at least, we could subject $T(\cdot)$ to either a local or global sensitivity analysis in order to study the influence of g.

If we attempt a global analysis in this context, we quickly run into trouble. Virtually all global robustness analyses rely upon the quantity of interest being a *ratio-linear* functional of the distribution being perturbed (Lavine, 1991, Lavine et al., 1993). In the present setting, however, the density under perturbation appears n times multiplicatively in the posterior density. That is,

$$T(g) = \frac{\int h(\theta, \lambda) \left\{\prod_{i=1}^n f(x_i|\theta_i)\lambda^{-1} g(\theta_i/\lambda)\right\} f(\lambda)\, d\theta\, d\lambda}{\int \left\{\prod_{i=1}^n f(x_i|\theta_i)\lambda^{-1} g(\theta_i/\lambda)\right\} f(\lambda)\, d\theta\, d\lambda},$$

which is clearly not ratio-linear in g. This effectively makes global analysis intractable for any rich nonparametric class of distributions. On the other hand, local analysis for this sort of problem has been investigated by Gustafson (1996a) and turns out to be not much more difficult than for single prior problems as in Section 2. In particular, it can be shown that the influence function for perturbation of g as a whole is the sum of individual influence functions corresponding to perturbation of g for single sampling units. Arguably it is this kind of situation where the merits of local analysis are manifested, as global analysis is not feasible.

We have painted an oversimplified picture of sensitivity analysis in hierarchical models. If $\theta|\lambda$ is viewed as contributing to the integrated out sampling model $f(x|\lambda)$, then some of the difficulties described in the next subsection come into play.

4.3.3 The sampling model

Say we model data $X_1, \ldots, X_n$ as being independent and identically distributed with density $f(\cdot|\theta)$, and assign a prior density $f(\theta)$. Of course, both the model and prior specifications will be approximate in some sense, and in most scenarios inferences will depend much more heavily on the model than on the prior. So model influence may be of more import than prior influence, despite the fact that the literature on Bayesian robustness has almost exclusively focussed on the prior rather than the model.

Right away we see that prior influence and model influence cannot be treated in exactly the same way. The prior density $f(\theta)$ is a marginal density, whereas the model density $f(x_i|\theta)$ is a conditional density. Inherently this makes it more difficult to assess model influence than prior influence. In some situations we might be willing to infuse enough structure so that model perturbations can be applied to a marginal. For instance, say the parameter is $\theta = (\mu, \sigma)$, and the baseline model has location-scale structure

$$X_i = \mu + \sigma Z_i,$$

where $Z_1, \ldots, Z_n$ are independent and identically distributed with distribution P. For sensitivity analysis we can then replace P with $\tilde{P}_U = P + U$ in the usual way, and we can express the posterior quantity of interest as $T(U)$. Again, $T(\cdot)$ will not be a ratio-linear functional of U, so global analysis is hopeless. But at least computationally it should be feasible to study T locally. Unfortunately, though, such an analysis may not convey useful information.

There are two fundamental problems with using local analysis to assess sampling model influence. The first is that data usually have something to say about how well various models fit them. For instance, the data may make it clear that the model $P + \epsilon(Q_1 - P)$ fits the data better than $P + \epsilon(Q_2 - P)$, but $|T(\epsilon[Q_2 - P]) - T(0)|$ might be much larger than $|T(\epsilon[Q_1 - P]) - T(0)|$. Then a straightforward worst-case sensitivity analysis would focus on contamination of P by Q_2, even though contamination by Q_1 is better supported by the data.

The second problem is that model perturbations may render parameters uninterpretable. For instance, in the location-scale case μ and σ may be interpreted as the mean and standard deviation of the data-generating distribution, for the purposes of both prior specification and inference. However, unless we are careful to restrict the forms of perturbation to the distribution of Z, these interpretations will be lost. That is, without restrictions on the perturbations U, μ and σ will not be the mean and standard deviation of $\tilde{P}_U$.

In the case of location-scale families, Gustafson (1996d) tries to obviate these problems with the following scheme. First, parameters of interest are regarded as functions of θ *and* U, so that consistent interpretations can be maintained for all perturbed distributions. Second, two influence functions are determined: one for the posterior quantity of interest, and one for the Bayes factor comparing the perturbed model to the baseline model. Arguably, non-robustness obtains only if the same contaminating distribution can produce a large change in the posterior quantity and a large increase in the Bayes factor. Viewing the two influence functions simultaneously can give a rough idea of whether these two conditions can be met by the same contaminating distribution. A third feature of this methodology is that the sampling distribution is not perturbed directly. Rather, the baseline distribution is expressed as a mixture, and the mixing distribution is perturbed. This engenders smoother and more realistic perturbations to the baseline sampling distribution. Ostensibly the combination of these features does give a legitimate scheme for assessing local model influence, at least in simple models. However, the methodology is probably too complicated for routine use in practice. The notion of assessing the sensitivity of Bayes factors to sampling distributions is also investigated by Carota (1996), albeit in quite a different sense.

Clarke and Gustafson (1998) also tackle local model influence, in a much broader setting. In particular, these authors attempt to quantify the overall worst-case sensitivity of inferences to the prior, model and data jointly. Furthermore, this worst-case sensitivity is apportioned to see what fraction is attributable to the prior, what fraction is attributable to the model, and what fraction is attributable to the data. The use of local sensitivity is clearly vital; there is no hope of addressing this question using global analysis. While the qualitative findings are interesting, the methodology is again probably too complicated to have practical value as a diagnostic tool. In particular, the authors were only able to study simple models.

4.3.4 Local sensitivity in asymptotia

Local sensitivity techniques can also play a role in the study of asymptotics. Since Bayesian and maximum likelihood estimates share the same asymptotics, this is not a purely Bayesian endeavor. Gustafson (1996e) considered the asymptotics of hierarchical models, as per subsection 3.2, in situations where the integrated model $p(x|\lambda)$ has a closed form under the baseline specification. A difficulty in studying sensitivity is that the closed form is usually lost as soon as $p(\theta_i|\lambda)$ is perturbed. Asymptotically, inferences about λ concentrate at the value of λ for which the specified model $p(x|\lambda)$ is closest to the true density of X in terms of Kullback–Leibler divergence. Roughly speaking, if the true distribution of $\theta|\lambda$ is an ϵ-contamination of the postulated distribution, then the asymptotic value of λ can be expressed as $\lambda(\epsilon)$. The lack of a closed form for the distribution of $X|\lambda$ under contamination makes evaluation of $\lambda(\epsilon)$ difficult for $\epsilon > 0$. But evaluation of $\lambda'(0)$ is relatively straightforward, because $\epsilon = 0$ gives rise to closed-form calculations. Using this approach, it is possible to assess large-sample sensitivity for some stock hierarchies, such as the beta-binomial, the gamma-Poisson, and the normal-normal, at least when inference about the moments of $\theta_i|\lambda$ is of interest. Again, it is not clear that there is any other tractable approach to this kind of analysis.

4.4 Discussion

We close by weighing the pros and cons of local sensitivity analysis and pondering the future of these methods.

4.4.1 Pros

For prior influence, local analysis is easier to compute than global analysis. Typically an influence function and a worst-case summary are readily computed. When MCMC algorithms are used to obtain baseline posterior summaries, obtaining these local sensitivity quantities is not much extra effort. We have not discussed computational details, but Gustafson (1996a) addresses efficient use of MCMC methods to compute influence functions. Whereas the local quantities can be computed in one go, a global prior influence analysis requires iterative evaluation of posterior quantities to drive the linearization algorithm.

For prior influence, local analysis seems plausible for comparative purposes. For instance, say an analyst wants to know which parts of the posterior are sensitive to a particular prior marginal, in a complicated model. Comparison of derivative norms for various posterior expectations under perturbation of this marginal provide a reasonably satisfying answer to the question. Similarly, the analyst might compare the sensitivity of a particular posterior quantity under perturbations of different parts of the prior. In fact, this leads to a *sensitivity matrix* summarizing the sensitivity of various posterior quantities to various prior perturbations. Examples of such matrices can be found in Gustafson (1996a, f). Comparisons of sensitivity for various baseline priors can also be carried out, as in Section 2, for example.

Local sensitivity analysis has had some success in aiding qualitative understanding which transcends particular data sets and models. For instance, Gustafson (1996b) delineates a sense in which Jeffreys' prior is least influential amongst all priors, and Gustafson (1996a, c) provides somewhat general results about how sensitivity "propagates" in hierarchical models.

As discussed in Section 3, local sensitivity analysis can be a useful tool in broadening the study of sensitivity and influence beyond the prior. In contrast, global analysis seems to come to a crashing halt once the quantity of interest is no longer ratio-linear in the distribution being perturbed. Since the Bayesian robustness literature as a whole has been criticized for its excessive and perhaps inward focus on the prior, this speaks to the potential value of local analysis.

4.4.2 Cons

The chief drawback of local sensitivity analysis seems to be that rightly or wrongly, a local analysis lacks credibility. While influence functions might be reasonably interpretable, reported derivative norms may seem somewhat arbitrary. Some authors view this as an issue of *calibration* and try to provide meaningful scales on which to interpret the magnitude of a worst-case sensitivity measure (e.g., McCulloch, 1989). It is not clear that calibration schemes can be developed which do not seem overly contrived.

To some extent the concern about calibration is very valid in light of the results on asymptotic behaviour cited in Section 2. If a measure of prior influence increases to infinity as the sample size grows, then it is probably of little value, except perhaps for comparative purposes, as mentioned in the previous subsection. If we are assessing sensitivity of scalar posterior summaries rather than posterior distributions, then we can choose a framework for sensitivity assessment which has more sensible asymptotic behaviour. However, we are still not likely to get exact asymptotic agreement between corresponding local and global analyses, and so (7) must be taken with a grain of salt. In particular, if one really wants a global analysis, it is probably worth actually doing it (providing it is feasible), rather than approximating it with a local analysis using (7).

4.4.3 Wither Bayesian robustness?

As far as local analysis goes, the cons seem to outweigh the pros, in that the local approach has received virtually no attention from the wider statistical community. This is not really a criticism of the local approach compared to the global approach, as the latter has also not caught on in any practical sense. In terms of Bayesian statistical practice, both local and global robustness schemes are clear losers compared to informal sensitivity analysis, which is used in applications all the time. It seems there are two reasons for this. First, the proponents of both local and global sensitivity analyses have not clearly demonstrated what added value they provide over an informal analysis. Whereas the informal analysis might be viewed as skimpy, formal analysis, with very rich classes of distributions and an emphasis on the a posteriori worst case, might be viewed as unduly heavy and pessimistic. Second, modern Bayesian analysis involves very rich models, often with priors that function as automatic Occam's razors, cutting back to simpler models whenever warranted. With such models there are simply fewer assumptions to be violated, and sensitivity analysis is necessarily less important.

Having said all that, both the local and global approaches to robustness are motivated by the very honest and important question of input-to-output sensitivity. There may well be a middle ground between the ad-hoc informal sensitivity analysis, and the heavily mathematized and abstracted approaches both local and global, found in the Bayesian robustness literature. Ideally researchers will find new ways to meld the theory and practice of sensitivity analysis. Given the fundamental role of the derivative as a measure of rate of change, perhaps local analysis will have a role to play in such hybrid enterprises.

References

BASU, S. (1996). Local sensitivity, functional derivatives and nonlinear posterior quantities. *Statistics and Decisions*, **14**, 405–418.

BASU, S., JAMMALAMADAKA, S.R. and LIU, W. (1996). Local posterior robustness with parametric priors: maximum and average sensitivity. In *Maximum Entropy and Bayesian Methods* (G. Heidlbreder, ed.), 97–106. Netherlands: Kluwer.

BASU, S., JAMMALAMADAKA, S.R. and LIU, W. (1998). Qualitative robustness and stability of posterior distributions and posterior quantities. *Journal of Statistical Planning and Inference* **71** 151–162.

BIRMIWAL, L.R. and DEY, D.K. (1993). Measuring local influence of posterior features under contaminated classes of priors. *Statistics and Decisions*, **11**, 377–390.

CAROTA, C. (1996). Local robustness of Bayes factors for nonparametric alternatives. In *Bayesian Robustness* (J.O. Berger, B. Betrò, E. Moreno, L.R. Pericchi, F. Ruggeri, G. Salinetti and L. Wasserman, eds.), 283–291. IMS Lecture Notes - Monograph Series Volume 29. Hayward, CA: Institute of Mathematical Statistics.

CLARKE, B. and GUSTAFSON, P. (1998). On the overall sensitivity of the posterior distribution to its inputs. *Journal of Statistical Planning and Inference*, **71**, 137–150.

CUEVAS, A. and SANZ, P. (1988). On differentiability properties of Bayes operators. In *Bayesian Statistics 3* (J.M. Bernardo, M.H. DeGroot, D.V. Lindley and A.F.M. Smith, eds.), 569–577. Oxford: Oxford University Press.

DELAMPADY, M. and DEY, D.K. (1994). Bayesian robustness for multiparameter problems. *Journal of Statistical Planning and Inference*, **40**, 375–382.

DEY, D.K. and BIRMIWAL, L.R. (1994). Robust Bayesian analysis using entropy and divergence measures. *Statistics and Probability Letters*, **20**, 287–294.

DEY, D.K., GHOSH, S.K. and LOU, K.R. (1996). On local sensitivity measures in Bayesian analysis (with discussion). In *Bayesian Robustness* (J.O. Berger, B. Betrò, E. Moreno, L.R. Pericchi, F. Ruggeri, G. Salinetti and L. Wasserman, eds.), 21–39. IMS Lecture Notes - Monograph Series Volume 29. Hayward, CA: Institute of Mathematical Statistics.

DIACONIS, P. and FREEDMAN, D. (1986). On the consistency of Bayes estimates. *Annals of Statistics*, **14**, 1–26.

FORTINI, S. and RUGGERI, F. (1994). Concentration functions and Bayesian robustness (with discussion). *Journal of Statistical Planning and Inference*, **40**, 205–220.

FORTINI, S. and RUGGERI, F. (2000). On the use of the concentration function in Bayesian robustness. In *Robust Bayesian Analysis* (D. Ríos Insua and F. Ruggeri, eds.). New York: Springer-Verlag.

GELFAND, A.E. and DEY, D.K. (1991). On Bayesian robustness of contaminated classes of priors. *Statistics and Decisions*, **9**, 63–80.

GUSTAFSON, P. and WASSERMAN, L. (1995). Local sensitivity diagnostics for Bayesian inference. *Annals of Statistics*, **23**, 2153–2167.

GUSTAFSON, P. (1996a). Local sensitivity of inferences to prior marginals. *Journal of the American Statistical Association*, **91**, 774–781.

GUSTAFSON, P. (1996b). Local sensitivity of posterior expectations. *Annals of Statistics*, **24**, 174–195.

GUSTAFSON, P. (1996c). Aspects of Bayesian robustness in hierarchical models (with discussion). In *Bayesian Robustness* (J.O. Berger, B. Betrò, E. Moreno, L.R. Pericchi, F. Ruggeri, G. Salinetti and L. Wasserman, eds.), 63–80. IMS Lecture Notes - Monograph Series Volume 29. Hayward, CA: Institute of Mathematical Statistics.

GUSTAFSON, P. (1996d). Model influence functions based on mixtures. *Canadian Journal of Statistics*, **24**, 535–548.

GUSTAFSON, P. (1996e). The effect of mixing-distribution misspecification in conjugate mixture models. *Canadian Journal of Statistics*, **24**, 307–318.

GUSTAFSON, P. (1996f). Robustness considerations in Bayesian analysis. *Statistical Methods in Medical Research*, **5**, 357–373.

HAMPEL, F.R., RONCHETTI, E.M., ROUSSEEUW, P.J. and STAHEL, W.A. (1986). *Robust Statistics: The Approach Based on Influence Functions.* New York: Wiley.

KADANE, J.B. and CHUANG, D.T. (1978). Stable decision problems. *Annals of Statistics*, **6**, 1095–1110.

KADANE, J.B., SALINETTI, G. and SRINIVASAN, C. (2000). Stability of Bayes decisions and applications. In *Robust Bayesian Analysis* (D. Ríos Insua and F. Ruggeri, eds.). New York: Springer-Verlag.

KADANE, J.B. and SRINIVASAN, C. (1996). Bayesian robustness and stability (with discussion). In *Bayesian Robustness* (J.O. Berger, B. Betrò, E. Moreno, L.R. Pericchi, F. Ruggeri, G. Salinetti and L. Wasserman, eds.), 81–99. IMS Lecture Notes–Monograph Series Volume 29. Hayward, CA: Institute of Mathematical Statistics.

LAVINE, M. (1991). Sensitivity in Bayesian statistics: the prior and the likelihood. *Journal of the American Statistical Association*, **86**, 396–399.

LAVINE, M., WASSERMAN, L. and WOLPERT, R. (1993). Linearization of Bayesian robustness problems. *Journal of Statistical Planning and Inference*, **37**, 307–316.

MARTÍN, J. and RÍOS INSUA, D. (1996). Local sensitivity analysis in Bayesian decision theory (with discussion). In *Bayesian Robustness* (J.O. Berger, B. Betrò, E. Moreno, L.R. Pericchi, F. Ruggeri, G. Salinetti and L. Wasserman, eds.), 119–135. IMS Lecture Notes - Monograph Series Volume 29. Hayward, CA: Institute of Mathematical Statistics.

McCulloch, R.E. (1989). Local model influence. *Journal of the American Statistical Association*, **84**, 473–478.

Meng, Q. and Sivaganesan, S. (1996). Local sensitivity of density bounded priors. *Statistics and Probability Letters*, **27**, 163–169.

Moreno, E., Martínez, C. and Cano, J.A. (1996). Local robustness and influence for contamination classes of prior distributions (with discussion). In *Bayesian Robustness* (J.O. Berger, B. Betrò, E. Moreno, L.R. Pericchi, F. Ruggeri, G. Salinetti and L. Wasserman, eds.), 137–152. IMS Lecture Notes - Monograph Series Volume 29. Hayward, CA: Institute of Mathematical Statistics.

Peña, D. and Zamar, R. (1997). A simple diagnostic tool for local prior sensitivity. *Statistics and Probability Letters*, **36**, 205–212.

Ramsay, J.O. and Novick, M.R. (1980). PLU robust Bayesian decision theory: point estimation. *Journal of the American Statistical Association*, **75**, 901–907.

Ruggeri, F. and Wasserman, L. (1993). Infinitesimal sensitivity of posterior distributions. *Canadian Journal of Statistics*, **21**, 195–203.

Ruggeri, F. and Wasserman, L. (1995). Density based classes of priors: infinitesimal properties and approximations. *Journal of Statistical Planning and Inference*, **46**, 311–324.

Rukhin, A. L. (1993). Influence of the prior distribution on the risk of a Bayes rule. *Journal of Theoretical Probability*, **6**, 71–87.

Salinetti, G. (1994). Stability of Bayesian decisions. *Journal of Statistical Planning and Inference*, **40**, 313–320.

Sivaganesan, S. (1993). Robust Bayesian diagnostics. *Journal of Statistical Planning and Inference*, **35**, 171–188.

Sivaganesan, S. (1996). Asymptotics of some local and global robustness measures (with discussion). In *Bayesian Robustness* (J.O. Berger, B. Betrò, E. Moreno, L.R. Pericchi, F. Ruggeri, G. Salinetti and L. Wasserman, eds.), 195–209. IMS Lecture Notes - Monograph Series, Volume 29. Hayward, CA: Institute of Mathematical Statistics.

Sivaganesan, S. (2000). Global and local robustness approaches: uses and limitations. In *Robust Bayesian Analysis* (D. Ríos Insua and F. Ruggeri eds.). New York: Springer-Verlag.

5

Global and Local Robustness Approaches: Uses and Limitations

Siva Sivaganesan

ABSTRACT We provide a comparative review of the global and local approaches to Bayesian robustness. Here, we focus on the issue of assessing robustness on a (more) practical level. Specifically, we address the issues of interpretation, calibration, and the limitations of these measures. The important issue of how one may proceed with a follow-up investigation, when robustness is found to be lacking, is also reviewed, including how both the global and local robustness measures, together, can be useful in such a follow-up study. We then briefly summarize the asymptotics, and end with a discussion.

Key words: Calibration, asymptotics, measures of robustness.

5.1 Introduction

Several approaches to investigation of robustness in Bayesian analysis can be found in the literature, many of which are rather ad hoc. Berger (1984) introduced a formal approach to robustness investigation, now known as the (global) robust Bayesian approach. The approach is general in scope in the sense that it provides a framework to investigate robustness with respect to any aspect of the assumptions in the statistical analysis, whether it is the prior, the sampling model, or the loss function. Any such input is typically hard to specify accurately, given the usual practical limitations, and is subject to uncertainty. The robust Bayesian approach is to address this uncertainty by specifying a class, for each input, consisting of all plausible choices for that input, and then finding the ranges of possible answers resulting from the use of all possible combinations of these inputs. If the range is small, one concludes that robustness occurs. Suppose that we are interested in the robustness with respect to a specified prior π_0, which is our main focus here, and that a class Γ of priors is specified. Then, the sensitivity or robustness of a posterior functional $\psi(\pi)$ is assessed by calculating the bounds $\underline{\psi} = \inf_{\pi \in \Gamma} \psi(\pi)$ and $\bar{\psi} = \sup_{\pi \in \Gamma} \psi(\pi)$, yielding a range $(\underline{\psi}, \bar{\psi})$, for $\psi(\pi)$. In particular, a small range is an indication of robustness w.r.t. the prior. More details on the development and computation can be found in other chapters in this book.

Some difficulties with global robustness have led to the emergence of the local robustness approach. While the overall goal is similar, this approach is mainly suited to assess sensitivity to especially small (local) deviations from a specified input, for example, a prior π_0. Here, as in the global robustness approach, plausible deviations from a specified prior are elicited using a class Γ. Then, a suitable derivative of $\psi(\pi)$ w.r.t. π, evaluated at the base prior π_0, is used to assess the sensitivity of $\psi(\pi)$ to small deviations from a specified prior π_0. The choices of derivatives, their properties and related issues are covered in Gustafson (2000).

This chapter focuses on the issues of assessing robustness using these two robustness measures, global and local, with emphasis on interpretation and calibration. Our main focus is on robustness w.r.t. the prior, although we will have a few remarks later.

5.2 Classes of priors

The use of the robustness measures to evaluate robustness, global or local, is closely tied with the class of priors being used. In this regard, it is important to keep in mind how a class is elicited and what it means as a reflection of prior uncertainty. Here we briefly address this topic. More details on the actual elicitation, the type of classes commonly used, computational issues, and other related discussions are available in other chapters in this book.

A class Γ of priors is typically chosen by first specifying a single prior π_0, and then choosing Γ as a suitable neighborhood to reflect one's uncertainty about π_0. Many forms of classes are considered in the literature. One is the ϵ-contamination class

$$\Gamma = \{\pi = (1-\epsilon)\pi_0 + \epsilon q : q \in Q\}, \tag{1}$$

where Q is a set of probability distributions representing plausible deviations from π_0. Some others are density bounded classes, density ratio classes, and mixture classes.

Ideally, the class Γ would be the right "size" by including all priors that are a priori plausible, that is, compatible with one's prior information, and excluding all priors that are not. But practical limitations and mathematical convenience often require the use of classes that do not conform to the above ideal. As such, the classes often used are either too large, consisting of priors that are a priori not plausible, or too small, excluding priors that are plausible, or both. While this may be unavoidable in practice, it is important to know where a specified class stands, in this respect, and to be mindful about it in interpreting the answers, such as ranges, derived using this class.

5.3 Global approach

Suppose that, as described before, a class Γ has been chosen to reflect the uncertainty about a specified prior π_0. Then, the global robustness measure of a posterior functional $\psi(\pi)$ of interest is the range $I(\psi) = (\underline{\psi}, \bar{\psi})$. Alternatively, the sensitivity of $\psi(\pi)$ to deviations from π_0 is $S(\psi) = \max\{\bar{\psi} - \psi_0, \psi_0 - \underline{\psi}\}$, where $\psi_0 = \psi(\pi_0)$. The range $I(\psi)$ (or $S(\psi)$) has a simple interpretation. When Γ reasonably reflects the uncertainty in π_0, that is, of the right size, a "small" range $I(\psi)$ indicates robustness w.r.t. the prior. In this case, one may use, for example, ψ_0, as the Bayesian answer with the satisfaction that it is robust w.r.t. possible misspecification of the prior. On the other hand, a "large" range is an indication that there is lack of robustness w.r.t. the prior. Here, what is meant by "small"(or "large") is context dependent and is best determined by the investigator or the end-user. (Note also that these robustness measures are not scale-free.)

5.3.1 Robust Bayesian analysis: an iterative process

Thus, the range, as a global robustness measure, is easy to interpret and use in assessing robustness, provided the class is of the right size. Otherwise, the use of the range to assess robustness is more complicated, and depends on how well the class reflects one's uncertainty about the prior information.

If the class is considered large (in the sense of containing all plausible priors, and some that are not), then a small range is evidence of robustness w.r.t. the prior. But if the class is suspected to be too small, then further evaluation using a larger class is needed to arrive at a conclusion. In this regard, it may be worthy to use a large and computationally convenient class as a first step in a robustness analysis. If the range for this large class itself is small, then the range for the "correct" class would also have to be small, and hence one can conclude robustness without the need for a more careful and lengthy process of attempting to elicit the right class. For example, ϵ-contamination with arbitrary contaminations may be such a candidate.

On the other hand, when the range is large, it may be for two reasons: (i) the class is of the right size, and the result is sensitive to uncertainty about the prior; or (ii) the class is unduly large (containing priors that are a priori not plausible), which results in a misleadingly large range giving a false indication of the extent of non-robustness. (The true range from the correct size class may even be small enough to conclude robustness.)

Fortunately, which of the above two is the cause for the large range may not be difficult to ascertain in a typical robust Bayesian analysis. Often, the extremal priors, say π_L and π_U, at which the lower and upper bounds are attained, are known from the computation of the bounds. One can then subjectively judge whether these priors are a priori plausible. If plausible, it means the large range is indeed a true indication of non-robustness. If, on the other hand, these priors are deemed not plausible, one cannot conclusively decide on the nature of robustness and thus needs to pursue the investigation by refining the class and recomputing the range. The refinement may be done by excluding some priors which are deemed not plausible either by specifying more quantiles or by putting additional shape constraints provided such finer prior elicitations are possible and justified. The fact that the extremal priors π_L and π_U above are some of the inappropriate priors which need to be excluded would be valuable information in the process of refining a class. Indeed, the robust Bayesian investigation is an iterative process involving repeated subjective assessment and computation of the range. For details on refining a class by specifying more quantiles, see Liseo et al. (1996) and Ríos Insua et al. (1999).

5.4 Limitations of the global robustness approach

As with most statistical methods, the global approach also has its limitations. Before we discuss these limitations, we feel that this is a good place to stress the merits and advantages of the global robustness approach. The global robustness approach was first put forward formally in Berger (1984). In this book, Berger provided the formal approach, which consists of using a class to describe one's uncertainty about the inputs and of using the ranges of the posterior quantity of interest as a final measure of sensitivity, along with the required foundational underpinning for this approach. There is full agreement among researchers, as exemplified by the large number of articles that appeared over a fairly short period of time, that the approach is foundationally and conceptually sound, that it is intuitively very sensible, and that the *range* is very easy to understand and interpret as a robustness measure.

Everyone agrees that when the range is small, there is robustness with respect to uncertainty about the input(s) specified by the class(es). The limitations discussed here only relate to scenarios where the range is large that one could not conclude robustness, and further analysis may be needed to resolve the robustness issue. Moreover, we feel that some of these limitations only rarely present themselves in practice, and, when they occur, they can be resolved with further careful evaluation of the prior elicitation, as illustrated in examples given below, while others relate to the more general issue of prior elicitation in Bayesian inference.

5.4.1 Commonly known limitations

While conceptually simple, the global robustness approach is difficult to implement in many practical situations. Two well-known difficulties are

(i) specification of reasonable (non-parametric) class of priors in multidimensional parameter problems, and (ii) calculation of ranges with classes of multidimensional priors and with non-parametric classes of models f that preserve the common assumption that sample values are independent and identically distributed (i.i.d.). The difficulty with the elicitation of classes of priors for multiimensional parameters probably lies more with the Bayesian paradigm than with the robust Bayesian paradigm in particular. It is not very common to find articles where priors are carefully elicited for multidimensional parameters, either due to lack of prior information or due to the difficult and time-consuming task of prior elicitation. Two articles where the issue of prior elicitation for multidimensional parameters is addressed are Kadane et al. (1980) and Garthwite and Dickey (1992). When a multidimensional prior is specified, envisioning the uncertainty in such a prior is likely to be conceptually very difficult. Using a convenient neighborhood of the specified prior as a class, on the other hand, usually results in excessively large ranges. One may have to put many restrictions on the class to obtain ranges that are not excessively large, but that raises the question of whether such restrictions are indeed a priori justifiable. For such issues, see Lavine et al. (1991), Berger and Moreno (1994), and Sivaganesan (1995). Mixture classes appear to be a good choice in this context, as they tend to exclude priors with such features as sharp discontinuities and lack of smoothness. But computation with this class can (also) be intractable; for example, when certain components of multidimensional parameters are a priori thought to be i.i.d., as in hierarchical models.

Although prior robustness may be considered to be a more immediate concern to Bayesians, a full robustness investigation would involve study of sensitivity to other inputs such as the sampling model. The numerical computation involved in calculating the ranges is mostly tractable when parametric classes are used to express uncertainty in a specified sampling model. But when non-parametric classes of sampling models with i.i.d. assumption for the observations are used, finding the ranges of a posterior measure is a very difficult task, except in some special situations, such as Bayarri and Berger (1998).

5.4.2 *Limitations of the global robustness measure*

As discussed in the previous section, use of the range to assess robustness is very closely tied to the question of how compatible the class under consideration is with the prior information, that is, whether it is of the right size, too large, or too small. But this should perhaps not be regarded as a criticism of the approach. The idea of using a class which is too large is to simplify the whole investigation including the elicitation, by making it a stepwise process involving repeated elicitation and computation, which goes further only when a definite conclusion concerning robustness is not reached.

However, there is a difficulty that goes beyond the question of whether the class of priors under consideration is compatible with the prior information. It is possible that the range is large, giving a false indication of non-robustness, not necessarily because some priors in the class are incompatible with the prior information, but due to the possibility that certain priors are in conflict with the data. This phenomenon is typically reflected by a (relatively) very small or negligible value for the marginal of the data x:

$$m(x|\pi) = \int f(x|\theta)d\pi(\theta).$$

This is an important issue in trying to resolve the robustness question when the range is large but no further refinement is feasible. In order to motivate the need to consider $m(x|\pi)$ in assessing robustness, imagine putting a "flat" prior on the class of priors and proceeding with a fully Bayesian approach. Then if $\psi(\pi)$ is a posterior expectation of a function of θ, the fully Bayesian answer is the overall expectation of $\psi(\pi)$ taken w.r.t. the "posterior" of π, which can be regarded as a weighted average of $\psi(\pi)$, with the weights being proportional to the marginals $m(x|\pi)$. Thus, such a fully Bayesian answer would tend to down-weight the priors π's with relatively very small values for $m(x|\pi)$. It is therefore sensible, in assessing robustness, to down-weight or exclude from consideration those priors with little or no data support, as measured by the value of the marginals. To illustrate this, consider the following simple example.

Example 1. Let $X \sim N(\theta, 1)$ and the class of priors for θ be

$$\Gamma = \{N(\mu, 1) : -4 < \mu < 8\}.$$

Suppose $x = 2$ is observed. Then the range of the posterior mean of θ is $(-1, 5)$, and $m(x|\mu) = 0.282e^{-(2-\mu)^2/4}$. Note that while the largest value of $m(x|\mu)$ is 0.282, its value corresponding to the prior giving the upper bound $\overline{\psi} = 5$ is $0.282e^{-9}$ (which is also the value corresponding to the lower bound $\underline{\psi} = -1$). Thus it is clear that there is very little data support to these bounds, and hence the range $(-1, 5)$ is not a reasonable indication of the extent of robustness. Indeed, a more realistic indicator of robustness would be a range that is narrower than the above, with reasonable data support for both bounds. $\triangle$

Two approaches have been suggested in the literature to overcome the above phenomenon.

Approach 1: Using a Hierarchical Prior

Once a class Γ is elicited, one possibility is to proceed in the usual Bayesian way and express the uncertainty about the prior by putting a hierarchical prior (or hyper-prior) on Γ, say $G(d\pi)$. Or, instead of choosing a single hyper-prior, one may choose a class Γ^* of hyper-priors, expressing the uncertainty at this level of the hierarchy. To illustrate this, consider Example 1 again.

Example 1 (Continued). Suppose that the uncertainty in μ is approximately expressed by the class of priors for μ, given by

$$\Gamma^* = \{N(4, \tau^2) : 0 < \tau^2 < 4\}.$$

Then the posterior mean of θ for a given μ is given by

$$E(\theta|x, \mu) = \psi(\mu, x) = \frac{1}{2}\mu + \frac{1}{2}x,$$

and the marginal of X for a given μ, $m(x|\mu)$, is $N(\mu, 2)$. Using a prior from Γ^*, the posterior mean of μ is

$$E(\mu|x) = \frac{\tau^2}{\tau^2 + 2}x + \frac{8}{\tau^2 + 2},$$

and hence the posterior mean of θ, for fixed τ^2, is

$$E(\theta|x) = \frac{x}{2} + \frac{1}{2}\{\frac{\tau^2}{\tau^2 + 2}x + \frac{8}{\tau^2 + 2}\},$$

and the corresponding marginal is

$$m(x|\tau^2) = \frac{1}{\sqrt{2\pi(2 + \tau^2)}} \exp\{-\frac{(x - 4)^2}{2(2 + \tau^2)}\}.$$

Thus, when $x = 2$, the new range for the posterior mean, over the priors defined by Γ and Γ^*, is (2.33, 3). Note that, unlike the earlier range over the class Γ, the bounds here have much larger data support, the marginals be ng $0.282e^{-1} = 0.1037$ and $0.163e^{-1/3} = 0.1167$, corresponding respectively to the lower and upper bounds. $\triangle$

Some remarks are in order with regards to the above example. The purpose of the example is merely to illustrate the idea of using a hyper-prior on the original class of priors, Γ, in cases where the bounds resulting from the use of Γ have very little data support and so do not reflect the true nature of robustness. Such an elicitation of a second class Γ^* would require the availability of additional prior information. But such prior information may not be available if, as is the case, all available prior information has been exhausted in the elicitation of Γ. Moreover, when Γ is a non-parametric class, for example, a mixture class or the ϵ-contamination class with unimodal contaminations, specifying a hyper-prior (or a class of hyper-priors Γ^*) on Γ may be difficult to accomplish. Thus, although conceptually attractive, the above approach can be difficult to implement in practice.

Here is another related example, where a second-stage prior is used to obtain more sensible ranges.

Example 2 (Moreno and Pericchi, 1993). Here, the idea of using a hierarchical prior is used to obtain more sensible ranges for the posterior probability of a set. In their paper, the authors show that when the ϵ-contamination class

$$\Gamma(Q) = \{\pi(\theta) = (1-\epsilon)\pi_0(\theta|\mu,\tau) + \epsilon q(\theta) : q \in Q\},$$

where Q is the class of all unimodal distributions with mode and median at μ, is considered, the bounds for the posterior probability of certain sets are highly sensitive to the value of τ. In particular, when prior information is vague and τ is allowed to take all positive values, or when τ is very large, the range becomes almost vacuous, that is, nearly (0, 1). To overcome the sensitivity to the specification of τ, the authors propose using a hyper-prior $h(d\tau)$ for τ, the uncertainty of which, in turn, is expressed by using a class of distributions H. Thus, the authors propose the use of the class $\Gamma(H,Q)$ for prior distributions of θ, by combining the above two classes $\Gamma(Q)$ and H, to get

$$\Gamma(H,Q) = \{\pi(\theta) = (1-\epsilon)\int \pi_0(\theta|\tau)h(d\tau) + \epsilon q(\theta) : h \in H, q \in Q\}.$$

For example, when $\pi_0(\theta|\mu,\tau)$ is $N(\theta|0,\tau)$, with $\mu = 0$, $\epsilon = 0.2$, and $x = 2$, the range of the posterior probability of the set $(0,\infty)$ over $\Gamma(Q)$ is (0, 1) when τ is allowed to take any positive value. But when the uncertainty about τ is expressed by the class

$$H = \{h(\tau) : \int_0^1 h(\tau)d\tau = 0.26, \int_1^\infty h(\tau)d\tau = 0.74\},$$

the range over $\Gamma(H,Q)$ is (0.36, 0.57).

Note that using $\Gamma(H,Q)$ is similar in spirit to putting a hierarchical prior on a class of priors for θ, indexed by τ. This process, as in Example 1, again gives a range that is more sensible in terms of the data support. △

Approach 2: Using Credible or Likelihood Regions

While the above hierarchical approach of putting a prior (or priors) on the class of priors Γ may be appealing in principle, it is at best very difficult to carry out in practice. One reason is that, in more complex problems (needless to say, we are only dealing with very simple problems in Examples 1 and 2), the idea of putting a prior on priors is hard to envision. Another is that, as remarked at the end of Example 1 above, typically, one would have exhausted almost all available prior information in the elicitation of the original class Γ, and consequently, putting an informative/subjective proper prior on Γ may not be feasible. The following two approaches attempt to overcome these difficulties associated with the hierarchical approach.

Using Credible Regions

One way of appropriately accounting for $m(x|\pi)$ is to use a non-informative hyperprior on Γ, and use only those priors which belong to an HPD-credible region for π of size, say, 0.95. Thus, if $G(d\pi)$ is a non-informative prior on Γ, we may let $\Gamma^* = \{\pi \in \Gamma : m(x|\pi) > K\}$, where K is chosen so that the posterior probability of Γ^* calculated using $G(d\pi)$ is 0.95. Here, instead of directly using $m(x|\pi)$, a convenient non-informative prior is used to obtain a smaller class Γ^*. Note that only a non-informative prior is used and so there is no need for additional prior information. Moreover, since it is used only as a means to obtain a suitable scaling of $m(x|\pi)$ using HPD regions, its effect on the answers may be minimal. Note that using an HPD region to select the priors π for inclusion in Γ^* may be intuitively more appealing than directly working with $m(x|\pi)$. As an example, consider again Example 1.

Example 1 (Continued). Suppose we use the non-informative prior $\pi(\mu) = 1$. Then, the posterior for μ is $N(\mu|x, 2)$, and hence the 95% HPD region for μ, when $x = 2$, is $(-0.83, 4.83)$. The resulting range for the posterior mean is $(0.565, 3.415)$. $\triangle$

Example 3 (Sivaganesan and Berger, 1993). The above approach was used to obtain the ranges of the posterior mean of the binomial parameter in a binomial empirical Bayes problem. In this case, only a finite number of moments of the prior π (for the binomial parameters $\theta_i, i = 1, \ldots, N$) are relevant in determining the posterior mean. Using this fact, the authors use a non-informative hyper-prior on the moments of the prior π and obtain $100\gamma\%$ HPD regions for the moments, for values of γ ranging from 0.05 to 0.99. For each γ, the corresponding HPD region defines a (reduced class) Γ^*. The ranges of the posterior mean, denoted ψ, of the (current) θ, namely θ_N, for various γ are obtained and plotted against γ, as in Fig. 1, which is obtained using data consisting of 40 past observations, x_i's $(i = 1, \ldots, N - 1 = 40)$, where x_i is from $Bin(n_i = 5, \theta_i)$ distribution, and a current observation $x_N = 2$ from $Bin(n_N = 5, \theta_N)$. Note that the range is fairly stable for moderate to large γ (up to 0.95), followed by a rapid widening, as γ increases from 0.95 to 1. It may therefore be reasonable to conclude that the size of the range in the stable region is a more appropriate indicator of the extent of robustness than the usual range (at $\gamma = 1$). $\triangle$

Using Likelihood Regions

Alternatively, one can directly use the marginal $m(x|\pi)$ to define a smaller class

$$\Gamma^* = \{\pi \in \Gamma : m(x|\pi) > K\} \tag{2}$$

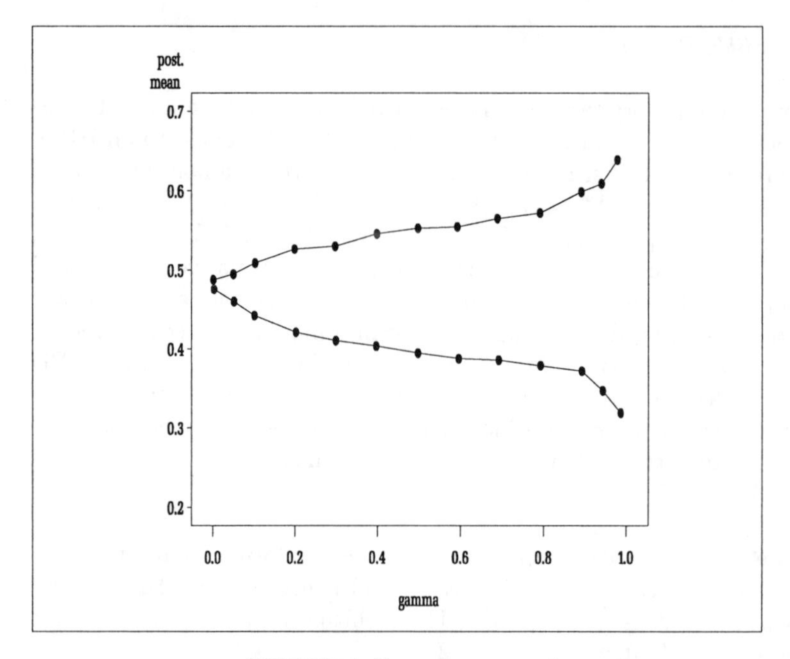

FIGURE 1. Posterior mean ψ versus γ

for some $K > 0$. Then the ranges over Γ^* can be reported for different choices of K chosen directly without the use of a hyper-prior and credible region, as in the previous approach. This approach may be attractive when it is difficult to envision a hyper-prior on Γ. The idea is that by looking at these ranges, one may be able to judge the extent of robustness, taking into account the data support associated with such ranges. It may turn out that for all reasonable choices of K, the ranges are small enough to conclude robustness, while the full range over Γ is not. (Note that we are only discussing scenarios where the full range is large.)

Example 1 (Continued). Suppose we decide to consider

$$\Gamma^* = \{\pi \in \Gamma : \frac{m(x|\pi)}{\sup_{\pi \in \Gamma} m(x|\pi)} > 0.05\}.$$

Here, we have scaled the marginal by its largest value, so that the (relative) values of the marginal are scaled between 0 and 1. Thus, the class Γ^* consists only of those priors whose relative marginal is at least 0.05. This, in turn, means that the prior mean μ must satisfy the inequality $0 < \mu < 5.46$ for those priors in Γ^*, yielding a (smaller) range $(0.27, 3.73)$ for the posterior mean of θ, which may well be considered small enough to conclude robustness. $\triangle$

In the same vein, Sivaganesan (1999) proposed a measure of data support for each possible value, called rΓ-likelihood, of the posterior mean (say, ρ)

given by

$$Rl_\Gamma(\rho) = \frac{l_\Gamma(\rho)}{\sup_{\pi\in\Gamma} m(x|\pi)},$$

where

$$l_\Gamma(\rho) = \sup_{\pi\in\Gamma:\psi(\pi)=\rho} m(x|\pi).$$

Note that $Rl_\Gamma(\rho)$ is the maximum value of (the scaled version of) $m(x|\pi)$ among all priors that give the value ρ for the functional $\psi(\pi)$. Thus, $Rl_\Gamma(\rho)$ is a measure of data support for the value ρ. A plot of $Rl_\Gamma(\rho)$ versus ρ can be a useful additional tool for assessing robustness. Such a plot readily gives the range for any subclass of the form (2). Moreover, it was shown in Sivaganesan (1999) that, in most cases, the same computations used to obtain the ranges are sufficient to produce this plot.

Example 4 (Sivaganesan, 1999). The above approach was used in the context of the ECMO example studied in Ware (1989) and Kass and Greenhouse (1989). Here the quantity of interest is the difference δ in the log-odds ratio (LOR) of ECMO and a standard therapy based on data from a clinical trial, while τ, the average LOR, is a nuisance parameter. See Moreno (2000) for more background details on this example. Kass and Greenhouse (1989) considered, among many other priors, the prior $\pi_1(\delta) = \text{Cauchy}(0, 1.099^2)$ for δ, and the prior $\pi_2(\tau)=$ Cauchy$(0, .419^2)$ for τ, assuming independence of δ and τ. In Lavine et al. (1991), the authors investigated the sensitivity of the posterior probability, $\rho = P(\delta > 0|data)$, to departures from the independence assumption by considering priors with fixed marginals as above. In their analysis without any independence assumption, they found that the range of ρ, over the class

$$\Gamma = \{\pi(\delta, \tau) : \text{ marginals are } \pi_1(\delta) \text{ and } \pi_2(\tau)\},$$

to be almost the entire range (0, 1), indicating a severe lack of robustness with respect to the class Γ. They observed that the maximizing (minimizing) prior puts most of its mass in regions where the likelihood is relatively small.

Using the approach outlined above, a plot of the relative Γ-likelihood, $Rl_\Gamma(\rho)$, versus ρ is obtained; see Fig.2. The figure reveals that for ρ smaller than 0.5, there is relatively little data support, suggesting a much reduced range for ρ. From this figure, one may conclude that the values of ρ that are reasonably supported by the data are certainly larger than 0.5, indicating the superiority of ECMO. $\triangle$

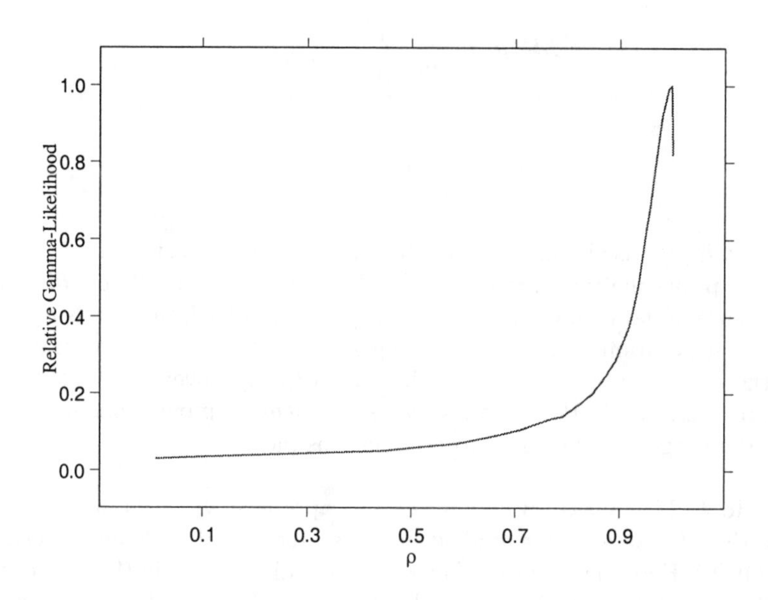

FIGURE 2. Relative Γ-likelihood versus ρ

Remarks.

(1) The above approaches of using the credible or likelihood regions, while attractive as they do not require careful and difficult elicitation of hyper-priors, are only diagnostic in nature. Moreover, whether these approaches would actually resolve the robustness issue depends on the specific model(s) and data at hand. But we believe that the method based on likelihood regions can easily be used without additional effort and can be valuable in gaining more insight, which can help resolve the robustness issue.

(2) Although the motivation given at the beginning of this section for considering the marginals $m(x|\pi)$ and hence the need for the above approach (or a similar one) based on likelihood regions may be compelling, there are situations where the above approach is not appropriate. This is illustrated by an example in Bayarri and Berger (1994), which is outlined below.

Example 5 (Bayarri and Berger, 1994). Here, the authors considered the question of testing $H_0 : \theta = 0$ versus $H_1 : \theta \neq 0$, where based on $X|\theta \sim N_p(\theta, I)$ and a prior, $\theta \sim N_p(0, \tau^2 I)$, where τ^2 is not specified. Using a class

$$\mathcal{G} = \{N_p(0, \tau^2 I) : \tau^2 > 0\},$$

as prior distributions for θ. Suppose that the quantity of interest is the Bayes factor. Then, the lower bound for the Bayes factor (over $\mathcal{G}$) for H_0

versus H_1 is

$$\underline{B} = (1 + \hat{\tau}^2)^{p/2} \exp\{-\frac{p}{2}\hat{\tau}^2\},$$

where $\hat{\tau}$ is the value maximizing the marginal of X. Now, consider a full Bayesian approach, using a prior distribution $g(\tau^2)$ for τ^2 of the form

$$g(\tau^2) \propto c(\tau^2)^a,$$

for some $c > 0$ and $a \geq 0$. This leads to a Bayes factor, say $B(g)$, which, for large p, satisfies

$$\frac{B(g)}{\underline{B}} = Kp^{(a+1)/2}(1 + o(1)),$$

where $K > 0$ is a constant. Note that $\underline{B}$, which is usually regarded as a "robust Bayesian answer" in the testing problem, corresponds to the ML-II prior, namely, the prior with the largest value for the marginal. Note that $\underline{B}$ is clearly not appropriate, since for large classes of reasonable priors (on τ^2) the discrepancy between the full Bayesian answer and the "robust Bayesian answer" $\underline{B}$ is huge for large p, while the latter has the maximum data support among all members of the class $\mathcal{G}$. $\triangle$

The phenomenon observed in the above example illustrates that an automatic use of the approaches described in this section can lead to misleading or incorrect answers, and one should be cautious in applying the above approaches as an aid in a robustness study. On the other hand, note that, as p grows, the range of the Bayes factor also grows very rapidly, and hence a reduced range using only those τ^2 with marginal value above a reasonable bound would still result in a huge range for the Bayes factor, forcing one to conclude a severe lack of robustness. This should effectively prevent one from (incorrectly) using $\underline{B}$ as a satisfactory robust Bayesian answer. In such a scenario, a natural robust Bayesian remedy is to use a hierarchical approach (as indicated earlier) and use a class of priors for τ^2 in the second stage.

Accounting for Posterior Variance

When the range is large, one may get some insight into the extent or the source of sensitivity by considering the size of the range relative to the posterior standard deviation corresponding to the "true" prior (or the base prior). For instance, when the posterior standard deviation w.r.t. the true prior is small, a large range would be regarded as an indication of sensitivity, whereas the same range may not be regarded as indicating sensitivity when the posterior standard deviation w.r.t. the true prior is small. Thus, it would be desirable to appropriately scale the range using the posterior standard deviation. This notion is taken up in Ruggeri and Sivaganesan (2000), where a scaled version of the range is proposed as an additional robustness tool.

5.5 Local approach

Local robustness approach to Bayesian robustness may be thought of as an offshoot of the global approach, partly due to the difficulty in the calculation of ranges in complex problems.

Suppose $\psi(\pi_f, f)$ is a posterior quantity of interest, and we are interested in the sensitivity of this quantity to *small* deviations from some specified (base) choices π_0 and f_0, respectively, of the prior π_f and the sampling model f. In the local robustness approach, one uses suitable derivatives of $\psi(\pi_f, f)$ w.r.t. π and f evaluated at the base choices π_0 and f_0, respectively, as measures of sensitivity to *small* deviations from π_0 and f_0. Other forms of local robustness measures involving the whole posterior distribution have also been studied; more details can be seen in Gustafson (2000).

Uses and Advantages

It appears that the basic idea of using some form of derivatives in robust Bayesian investigation has been recommended by different authors with at least three somewhat different goals, which we list below.

(i) To obtain approximations to the global bounds $\underline{\psi}$ and $\bar{\psi}$.

(ii) To use as a diagnostic tool in the global approach, when the range is large, in order to determine which aspect of the specified model (for example, sampling model or prior) is most influential, so that further (elicitational) effort may be directed on that part of the model.

(iii) To use as a robustness measure when the global approach is difficult to implement.

In Srinivasan and Truszczynska (1990), quite good approximations to global bounds are derived using Fréchet derivatives. In Sivaganesan (1993), derivatives of posterior quantities w.r.t. ϵ (using $\epsilon-$contamination classes to model uncertainties in priors and sampling models) are used as diagnostics to determine which aspect of the overall model is most influential. For example, in a hierarchical model, derivatives were used to determine which stage prior is most influential; this could help decide where to focus further elicitational efforts. In Ríos Insua et al. (1999), Fréchet derivative is used, in the same vein, to decide which (additional) quantile of the prior to specify in order to improve robustness. In Gustafson (1996), the local approach is used to assess sensitivity to prior specifications in hierarchical models. There are many other related articles that use the local approach to assess sensitivity; for references, see Gustafson (2000).

The main underlying motivation for the use of the local approach is the ease of computation, especially when assessing sensitivity due to uncertainty in a common distribution, where the model involves a random sample (i.i.d.) of quantities from this common distribution. Important examples of this type include: (i) evaluation of sensitivity due to uncertainty in a specified sampling distribution $f_0(x|\theta)$, where a random sample of observations X_i, $i = 1, \ldots, n$, is assumed to come from a common sampling distribution in a non-parametric neighborhood of $f_0(x|\theta)$, and (ii) in hierarchical models, it is common to assume, at a given stage, that parameters $\theta_1, \ldots, \theta_k$ are exchangeable with a common prior distribution. It would be of interest to assess the sensitivity to changes in the elicited common prior distribution, while preserving exchangeability. Using the local approach in these types of problems is usually very easy, unlike the global approach; for instance see Gustafson (1996) and Sivaganesan (1993).

Calibration: A Major Concern

While the local approach is relatively easy to use regardless of whether we are interested in robustness w.r.t. a prior or a sampling model, there is a major concern. The local robustness measure, say $W(\psi)$, is the value of an appropriate derivative of a functional of interest evaluated at the "base" choice of the prior, π_0, or the sampling model, f_0. Intuitively, therefore, a larger value of $W(\psi)$ would mean more sensitivity to "small" deviations from the base choice. But this apparently simple interpretation quickly runs into difficulties when we extend the notion a little further and ask, for example, (i) What does the exact value of $W(\psi)$ really tell us? (ii) How small is small when we say "small" deviations from the base choice? Does it matter? (iii) If $W(\psi)$ for one class of priors is twice as large as $W(\psi)$ for another, can we (approximately) quantify the relative sensitivity w.r.t. these classes, besides saying there is more sensitivity w.r.t. the former than the latter?

Specific answers to these questions are difficult to obtain, even in specific examples. Yet these are certainly reasonable questions in the context of Bayesian robustness, especially in view of the global robustness measure, *range*, for which one can indeed attempt to give reasonable answers to similarly posed questions. This is viewed as a major disadvantage of the local approach. We do not quite know how to interpret or calibrate the local robustness measure.

In a more general context of robustness, however, including frequentist robustness where the idea of using a derivative was readily accepted, one might be quite happy to use the local robustness measure as a sensitivity diagnostic. (Note that the use of influence functions in frequentist robustness, while well received, probably has the same limitations in terms of their interpretation or their calibration.) This is probably the motivation for the local robustness approach, as a diagnostic tool to study sensitivity in complex models (see Gustafson, 1996, and Sivaganesan, 1993), and *not* (at least not yet) as an approximation to global robustness approach.

In Sivaganesan (1995), a slightly modified version of the local sensitivity measure was proposed in the context of ϵ-contamination classes , see (1). One form of local sensitivity measure of the posterior quantity ψ, in this context, is given by

$$W(\psi) = \sup_{q \in Q} \frac{m(x|q)}{m(x|\pi_0)} (\psi(q) - \psi(\pi_0)), \tag{3}$$

where Q is the class of contaminations.

Now, let q^* be that q at which the sup above is attained, and define

$$\psi^* = \psi(\pi^*) \text{ and } W^*(\psi) = \psi^* - \psi(\pi_0),$$

where $\pi^* = (1-\epsilon)\pi_0 + \epsilon q^*$. $W^*(\psi)$ was proposed as a measure of calibration of the local sensitivity $W(\pi)$. The value of $W(\pi)$ is interpreted as causing a change (in the quantity of interest $\psi(\pi)$ of the size $W^*(\psi)$ from its base value $\psi(\pi_0)$. In fact, it was suggested that $W^*(\psi)$ itself could be regarded as a local robustness measure. It was shown that $W^*(\psi)$ has many desirable asymptotic behavior. For instance, when Q is the set of all contaminations, the measure $W^*(\psi)$ is of the order $O(n^{-1/2})$ as the sample size n grows. It was further argued that $W^*(\psi)$ is more resistant (than the global bound, say $\bar{\psi}$) to the "adverse" effects of those contaminations that are unreasonable, or, more specifically, that have little data support. The following result, given in that paper, lends some credence to this argument.

Result 1. Let $\bar{m}$ (resp., m^*) be the value of the marginal $m(x|\pi)$ corresponding to the π for which $\bar{\psi}$ (resp., ψ^*) is attained. Then,

(i) $\bar{m} < m^*$, and

(ii) $\sup_{\{\pi \in \Gamma : \psi(\pi) = \psi^*\}} m(x|\pi) = m^*$.

5.6 Asymptotics

It is well known that in most cases of interest the influence of the prior on the posterior distribution diminishes as the sample size increases. One would therefore expect a measure of sensitivity with respect to prior uncertainty to converge to zero, as sample size n goes to ∞. In view of this, many authors have investigated the asymptotic behavior of various robustness measures. Gustafson (2000), Sivaganesan (1996) and Ruggeri and Sivaganesan (2000) are some of the more recent references which address this issue and contain many other references on this topic. The limit and the rate of convergence, in general, depend on the robustness measure used, the class used, and the posterior summary for which robustness is considered.

The global robustness measure of a posterior expectation, namely the range, has been shown to converge to 0, although at different rates for different classes. For most classes consisting of priors which have uniformly bounded densities w.r.t. Lebesgue measure, such as ϵ-contamination classes with unimodal contaminations with bounded height at the mode, the range converges to 0 at the rate of $n^{-1/2}$. For classes which allow point masses, for example ϵ-contamination class with arbitrary contaminations or quantile class, the rate of convergence of the range is $\sqrt{\log n/n}$; see, e.g., Sivaganesan (1988). One instance where the range does not have the desired asymptotic limit is when considering the range of the posterior probability of the frequentist confidence interval for the normal mean when the variance is known. Here the inf of the posterior coverage probability converges to 0 rather than to, say, 0.95 or 1. Ruggeri and Sivaganesan (2000) proposed a global relative sensitivity measure that is a scaled version of the range. They showed that this global robustness measure does not converge to 0 as n goes to ∞ for ϵ-contamination class with arbitrary contaminations, but does so for many other classes.

The asymptotic behavior of local robustness measures is summarized in the article by Gustafson (2000), so we avoid the details here. In summary, the local robustness measures do not all have satisfactory limits as the sample size grows. As explained in Gustafson (2000) and the references therein, certain local robustness measures for certain classes of priors diverge to ∞, or have limits that do not depend on the data in a sensible manner. However, we may find sensible local robustness measures and classes of priors for which there is satisfactory asymptotic behavior, see Sivaganesan (1996).

In summary, the global robustness measure range converges to 0 for all classes of priors as the sample size grows to ∞, while the rate of convergence to 0 is rather slow when the priors are allowed to have point masses. The asymptotic behavior of local robustness measures is not as satisfactory. In particular, when point masses are allowed for priors, the local robustness measure has the counter-intuitive behavior of diverging to ∞ as the sample size grows. At this juncture, it may be worth noting that the modified local sensitivity measure W^* proposed in the previous section (see Sivaganesan, 1995) does not suffer from such anomalous behavior.

5.7 Discussion

There are several approaches to Bayesian robustness investigation. Three of them are the global robustness approach, the local robustness approach, and the "informal" approach. In the "informal" approach, one tries a few different priors (and/or sampling models) to see if the answers vary substantially. If not, one is fairly happy that there is robustness. This third approach is probably most widely used by practitioners. It is reasonable to say that the global robustness approach has extended this informal approach to a formal one with a sound foundation, via the robust Bayesian viewpoint advanced in Berger (1984). Most, if not all, other Bayesian robustness methods found in the literature are, to varying degrees, informal approaches. Many of them are meant to be used as diagnostics, and some other approaches suggest the use of inherently robust priors, an idea that may not be in line with one's subjective elicitation of the priors. In any case, these methods do not provide clear and meaningful guidelines to determine whether or not there is robustness. The global robustness approach does. Despite its appeal, however, the global robustness approach has not been widely embraced. This is due, partly, to three reasons: (i) the enormous research on this topic has so far not provided many new and useful insight into the issue of robustness; (ii) the elicitation of priors in realistic multidimensional problems is, in itself, a very difficult task, making the specification of the uncertainty in a prior using a class of priors even more difficult; (iii) the computation of bounds in many important problems is very difficult to carry out.

In spite of the above limitations, the global robustness approach remains very appealing, more so than other approaches, for the reasons given above, and therefore ought to remain an important goal of future robust Bayesian research. Until the elicitational and computational difficulties with the global approach are overcome, the informal method of trying a few priors is likely to be dominant in Bayesian applications. Among the several other diagnostic and similar approaches, the local robustness approach seems very appealing for the following reasons: it also requires the elicitation of classes of priors/sampling models; it allows the use of many forms of classes including non-parametric ones; it is numerically tractable even in many complex problems; it is flexible enough to allow one to study the robustness of the overall posterior or a posterior expectation to a prior or any other input; it has some intuitive appeal and, in terms of its interpretation, the local robustness measure is probably not inferior to many other ad-hoc measures suggested in the literature including the frequentist literature; and there is also some evidence from examples, although few, suggesting that it approximates the global bounds well and also adjusts for the data–prior conflict reasonably well (see Sivaganesan, 1995). For these reasons, we feel that the local robustness approach will continue to be a useful and attractive method in robust Bayesian analysis.

References

BAYARRI, M.J. and BERGER, J.O. (1994). Application and limitations of robust Bayesian bounds and type-II MLE. In *Statistical and Decision Theory and Related Topics V* (S.S. Gupta and J.O. Berger, eds.), 121–134. Berlin: Springer-Verlag.

BAYARRI, M. J. and BERGER, J.O. (1998). Robust Bayesian analysis of selection models. *Annals of Statistics*, **26**, 645–659.

BERGER, J.O. (1984). The robust Bayesian viewpoint (with discussion). In *Robustness of Bayesian Analysis* (J.B. Kadane, ed.), 63–134. Amsterdam: North-Holland.

BERGER, J.O. and MORENO, E. (1994). Bayesian robustness in bidimensional models: prior independence. *Journal of Statistical Planning and Inference*, **40**, 161–176.

GARTHWITE, P.H. and DICKEY, J.M. (1992). Elicitation of prior distributions for variable selection in regression. *Annals of Statistics*, **20**, 1697–1710.

GUSTAFSON, P. (1996). Aspects of Bayesian robustness in hierarchical models (with discussion). In *Bayesian Robustness* (J. O. Berger, B. Betrò, E. Moreno, L. R. Pericchi, F. Ruggeri, G. Salinetti, and L. Wasserman eds.), IMS-Lecture Notes-Monograph Series, Volume 29, 63–80. Hayward: IMS.

GUSTAFSON, P. (2000). Local robustness in Bayesian analysis. In *Robust Bayesian Analysis*, (D. Ríos Insua and F. Ruggeri, eds.). New York: Springer-Verlag.

KADANE, J.B., DICKEY, J.M., WINKLER, R.L., SMITH, W.S. and PETERS, S.C. (1980). Interactive elicitation of opinion for a normal linear model. *Journal of the American Statistical Association*, **75**, 845–854.

KASS, R. and GREENHOUSE, J. (1989). Investigating therapies of potentially great benefit: a Bayesian perspective. Comment on: "Investigating therapies of potentially great benefit: ECMO," by J. H. Ware. *Statistical Sciences*, **4**, 310–317.

LAVINE, M., WASSERMAN, L. and WOLPERT, R. (1991). Bayesian inference with specified marginals. *Journal of the American Statistical Association*, **86**, 964–971.

LISEO, B., PETRELLA, L. and SALINETTI, G. (1996). Bayesian robustness: An interactive approach. In *Bayesian Statistics 5*, (J.O. Berger, J.M. Bernardo, A.P. Dawid and A.F.M. Smith, eds.), 661–666. Oxford: Oxford University Press.

MORENO, E. (2000). Global Bayesian robustness for some classes of prior distributions. In *Robust Bayesian Analysis* (D. Ríos Insua and F. Ruggeri, eds.). New York: Springer-Verlag.

MORENO, E. and PERICCHI, L.R. (1993). Bayesian robustness for hierarchical ϵ-contamination models. *Journal of Statistical Planning and Inference*, **37**, 159–168.

RÍOS INSUA, S., MARTIN, J., RÍOS INSUA, D. and RUGGERI, F. (1999). Bayesian forecasting for accident proneness evaluation. *Scandinavian Actuarial Journal*, **99**, 134–156.

RUGGERI, F. and SIVAGANESAN, S. (2000). On a global sensitivity measure for Bayesian inference. To appear in *Sankhya*.

SIVAGANESAN, S. (1988). Range of posterior measures for priors with arbitrary contaminations. *Communications in Statistics: Theory and Methods*, **17**, 1591–1612.

SIVAGANESAN, S. (1995). multidimensional Priors: Global and Local Robustness. *Technical Report*, University of Cincinnati.

SIVAGANESAN, S. (1996). Asymptotics of some local and global robustness measures (with discussion). In *Bayesian Robustness* (J.O. Berger, B. Betrò, E. Moreno, L.R. Pericchi, F.Ruggeri, G.Salinetti, and L.Wasserman, eds.), IMS Lecture Notes - Monograph Series, Volume 29, 195–209. Hayward: IMS.

SIVAGANESAN, S. (1999). A likelihood based robust Bayesian summary. *Statistics & Probability Letters*, **43**, 5–12.

SIVAGANESAN, S. and BERGER, J.O. (1993). Robust Bayesian analysis of the binomial empirical Bayes problem. *Canadian Journal of Statistics*, **21**, 107–119.

SRINIVASAN, C. and TRUSZCZYNSKA, H. (1990). Approximations to the Range of a Ratio Linear Posterior Quantity. *Technical Report*, **289**, University of Kentucky.

WARE, J.H. (1989). Investigating therapies of potentially great benefit: ECMO. *Statistical Sciences*, **4**, 298–349.

6

On the Use of the Concentration Function in Bayesian Robustness

Sandra Fortini and Fabrizio Ruggeri

ABSTRACT We present applications of the concentration function in both global and local sensitivity analyses, along with its connection with Choquet capacities.

Key words: global sensitivity, local sensitivity, classes of priors.

6.1 Introduction

In this paper, we expose the main properties of the concentration function, defined by Cifarelli and Regazzini (1987), and its application to Bayesian robustness, suggested by Regazzini (1992) and developed, mainly, by Fortini and Ruggeri (1993, 1994, 1995a, 1995b, 1997).

The concentration function allows for the comparison between two probability measures Π and Π_0, either directly by looking at the range spanned by the probability, under Π, of *all* the subsets with a given probability under Π_0 or by considering summarising indices. Such a feature of the concentration function makes its use in Bayesian robustness very suitable.

Properties of the concentration function are presented in Section 2. Some applications of the concentration function are illustrated in the paper; in Section 3 it is used to define classes of prior measures, whereas Sections 4 and 5 deal with global and local sensitivity, respectively. An example in Section 6 describes how to use the results presented in previous sections. Finally, Section 7 illustrates connections between the concentration function and 2-alternating Choquet capacities, described in Wasserman and Kadane (1990, 1992).

6.2 Concentration function

Cifarelli and Regazzini (1987) defined the concentration function (c.f.) as a generalisation of the well-known Lorenz curve, whose description can be found, for example in Marshall and Olkin (1979, p. 5): "Consider a population of n individuals, and let x_i be the wealth of individual $i, i = 1, \dots, n$. Order the individuals from poorest to richest to obtain $x_{(1)}, \dots, x_{(n)}$. Now plot the points $(k/n, S_k/S_n), k = 0, \dots, n$, where $S_0 = 0$ and $S_k = \Sigma_{i=1}^k x_{(i)}$ is the total wealth of the poorest k individuals in the population. Join these points by line segments to obtain a curve connecting the origin with the point $(1, 1)$.... Notice that if total wealth is uniformly distributed in the population, then the Lorenz curve is a straight line. Otherwise, the curve is convex and lies under the straight line."

The classical definition of concentration refers to the discrepancy between a probability measure Π (the "wealth") and a uniform one (the "individuals"), say Π_0, and allows for their comparison, looking for subsets where the former is much more concentrated than the latter. The definition can be extended to non-uniform discrete distributions; we use data analysed by DiBona et al. (1993), who addressed the issue of racial segregation in the public schools of North Carolina, USA. The authors proposed a method to check if students tend to be uniformly distributed across the schools in a district or, otherwise, if they tend to be segregated according to their race. The proposed segregation index allowed the authors to state that segregation was an actual problem for all grades (K–12) and it had increased from 1982 to 1992. Lorenz curves are helpful in analysing segregation for each grade in a school.

	Native Americans	Asians	Hispanics	Blacks	Whites
Durham (82)	0.002	0.012	0.002	0.332	0.652
School 332	0.011	0.043	0.022	0.403	0.521
Ratio (S/D)	5.500	3.580	11.000	1.210	0.799
Durham (92)	0.002	0.026	0.009	0.345	0.618
School 332	0.007	0.106	0.013	0.344	0.530
Ratio (S/D)	3.500	4.070	1.440	0.997	0.858

TABLE 1. Public Kindergartens in Durham, NC

In Table 1 we present the distribution of students, according to their race, in a school (labelled as 332) in the city of Durham and compare it with the race distribution in the city public school system. We consider the ratios between percentages in the school (S_i) and the city (D_i), ordering the races according to their (ascending) ratios. Similarly to Marshall and Olkin, we plot (Fig. 1) the straight line connecting $(0, 0)$ and the points $\left(\Sigma_{j=1}^i D^{(j)}, \Sigma_{j=1}^i S^{(j)}\right), i = 1, \dots, 5$, where $D^{(j)}$ and $S^{(j)}$ correspond to the race with the jth ratio (in ascending order). The distance between the straight line and the other two denotes an unequal distribution of students in the school with respect to (w.r.t.) the city, and its largest value is one of the proposed segregation indexes. Moreover, the 1992 line lies above the 1982 one up to 0.9 (approx.), denoting an increase in adherence to the city population among the kids from the largest groups (White and Black) and, conversely, a decrease among other groups. Therefore, segregation at school 332 is decreasing over the 10 year period, in contrast with the general tendency in North Carolina, at each grade (see Di Bona et al., 1993, for more details).

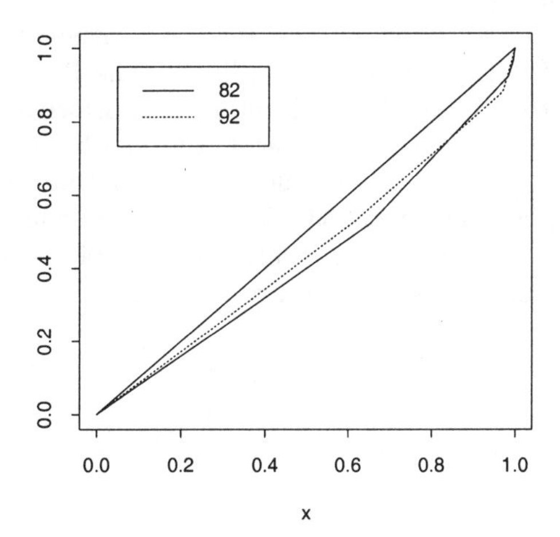

FIGURE 1. Lorenz curve for races at school 332 vs. Durham schools

As an extension of the Lorenz curve, Cifarelli and Regazzini (1987) defined and studied the c.f. of Π w.r.t. Π_0, where Π and Π_0 are two probability measures on the same measurable space $(\Theta, \mathcal{F})$. According to the Radon–Nikodym theorem, there is a unique partition $\{N, N^C\} \subset \mathcal{F}$ of Θ and a nonnegative function h on N^C such that

$$\Pi(E) = \Pi_a(E \cap N^C) + \Pi_s(E \cap N), \ \forall E \in \mathcal{F},$$

where Π_a and Π_s are, respectively, the absolutely continuous and the singular part of Π w.r.t. Π_0, that is, such that

$$\Pi_a(E \cap N^C) = \int_{E \cap N^C} h(\theta)\Pi_0(d\theta), \Pi_0(N) = 0, \Pi_s(N) = \Pi_s(\Theta).$$

Set $h(\theta) = \infty$ all over N and define $H(y) = \Pi_0(\{\theta \in \Theta : h(\theta) \leq y\})$, $c_x = \inf\{y \in \Re : H(y) \geq x\}$ and $c_x^- = \lim_{t \to x^-} c_t$. Finally, let $L_x = \{\theta \in \Theta : h(\theta) \leq c_x\}$ and $L_x^- = \{\theta \in \Theta : h(\theta) < c_x\}$.

Definition 1 *The function* $\varphi_\Pi : [0,1] \to [0,1]$ *is the concentration function of* Π *with respect to* Π_0 *if* $\varphi_\Pi(x) = \Pi(L_x^-) + c_x[x - H(c_x^-)]$ *for* $x \in (0,1)$, $\varphi_\Pi(0) = 0$ *and* $\varphi_\Pi(1) = \Pi_a(\Theta)$.

Observe that $\varphi_\Pi(x)$ is a nondecreasing, continuous and convex function, such that $\varphi_\Pi(x) \equiv 0 \Longrightarrow \Pi \perp \Pi_0$, $\varphi_\Pi(x) = x, \forall x \in [0,1] \Longleftrightarrow \Pi = \Pi_0$, and

$$\varphi_\Pi(x) = \int_0^{c_x} [x - H(t)]dt = \int_0^x c_t dt. \tag{1}$$

It is worth mentioning that $\varphi_\Pi(1) = 1$ implies that Π is absolutely continuous w.r.t. Π_0 while $\varphi_\Pi(x) = 0, 0 \le x \le \alpha$, means that Π gives no mass to a subset $A \in \mathcal{F}$ such that $\Pi_0(A) = \alpha$.

We present two examples to illustrate how to compute c.f.s. Consider the c.f. of a normal distribution $\mathcal{N}(0,1)$ w.r.t. a Cauchy $\mathcal{C}(0,1)$. The Radon–Nikodym derivative $h(\theta)$ is plotted in Fig. 2, a horizontal line is drawn and the subset of Θ with Radon-Nikodym derivative below the line becomes L_x (for an adequate x). This procedure is equivalent to computing and ordering ratios as in the example about school 332. The c.f. is obtained by plotting the points $(x, \varphi_\Pi(x))$, where x and $\varphi_\Pi(x)$ are the probabilities of L_x under the Cauchy and the normal distributions, respectively.

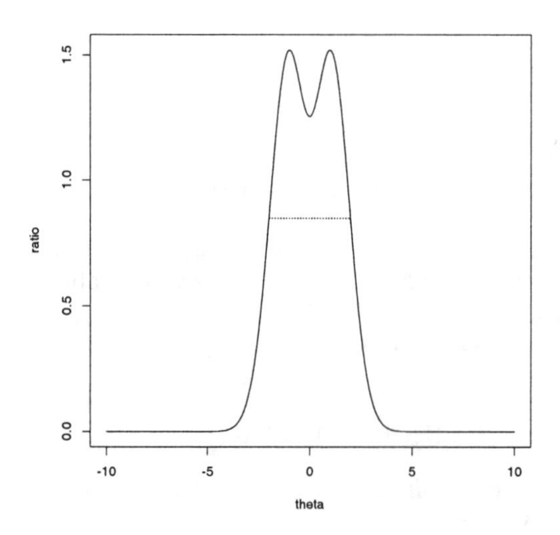

FIGURE 2. Radon–Nikodym derivative of $\mathcal{N}(0,1)$ vs. $\mathcal{C}(0,1)$

As another example, consider a gamma distribution $\Pi \sim \mathcal{G}(2,1)$ and an exponential one $\Pi_0 \sim \mathcal{E}(1)$. Their densities on $\Re^+$ are, respectively, $\pi(\theta) = \theta e^{-\theta}$ and $\pi_0(\theta) = e^{-\theta}$, so that $h(\theta) = \theta$, $\theta \ge 0$. For any $x \in [0,1]$, we compute the c.f. by finding the value y such that $x = \Pi_0(\{\theta \in \Theta : h(\theta) \le y\})$.

It follows that $y = -\log(1-x)$ since $x = \int_0^y e^{-\theta} d\theta = 1 - e^{-y}$. Finally, we get

$$\begin{aligned}\varphi_\Pi(x) &= \Pi(\{\theta \in \Theta : h(\theta) \leq -\log(1-x)\}) \\ &= \int_0^{-\log(1-x)} \theta e^{-\theta} d\theta \\ &= 1 - (1-x)(1 - \log(1-x)).\end{aligned}$$

The comparison of probability measures in a class is made possible by the partial order induced by the c.f. over the space $\mathcal{P}$ of all probability measures, when considering c.f.s lying above others. Total orderings, consistent with the partial one, are discussed in Regazzini (1992); they are achieved when considering synthetic measures of concentration as Gini's (1914) concentration index $C_{\Pi_0}(\Pi) = 2\int_0^1 \{x - \varphi_\Pi(x)\}dx$ and Pietra's (1915) index $G_{\Pi_0}(\Pi) = \sup_{x \in [0,1]}(x - \varphi_\Pi(x))$. The latter coincides with the total variation distance between Π and Π_0, as proved by Cifarelli and Regazzini (1987).

The following theorem, proved in Cifarelli and Regazzini (1987), states that $\varphi_\Pi(x)$ substantially coincides with the minimum value of Π on the measurable subsets of Θ with Π_0-measure not smaller than x.

Theorem 1 *If $A \in \mathcal{F}$ and $\Pi_0(A) = x$, then $\varphi_\Pi(x) \leq \Pi_a(A)$. Moreover, if $x \in [0,1]$ is adherent to the range of H, then there exists a B_x such that $\Pi_0(B_x) = x$ and*

$$\varphi_\Pi(x) = \Pi_a(B_x) = \min\{\Pi(A) : A \in \mathcal{F} \text{ and } \Pi_0(A) \geq x\}. \tag{2}$$

If Π_0 is nonatomic, then (2) holds for any $x \in [0,1]$.

This theorem is relevant when applying the c.f. to robust Bayesian analysis: for *any* $x \in [0,1]$, the probability, under Π, of all the subsets A with measure x under Π_0, satisfies

$$\varphi_\Pi(x) \leq \Pi(A) \leq 1 - \varphi_\Pi(1-x). \tag{3}$$

As an example, we can consider the c.f. of $\Pi \sim \mathcal{G}(2,2)$ w.r.t. $\Pi_0 \sim \mathcal{E}(1)$, showing that $[0.216, 0.559]$ is the range spanned by the probability, under Π, of the sets A with $\Pi_0(A) = 0.4$ (see Fig3).

Finally, we mention that the c.f., far from substituting other usual distribution summaries, e.g. the mean, furnishes different information about probability measures. As an example, consider two measures concentrated on disjoint, very close sets in $\Re$: their means are very close, their variances might be the same, but their c.f. is 0 in $[0,1)$.

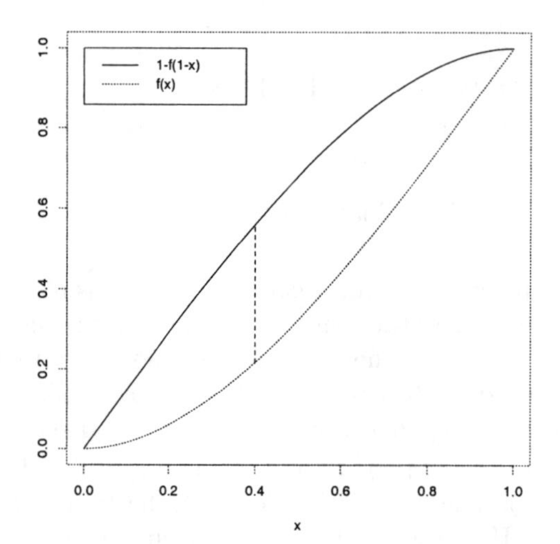

FIGURE 3. Range of $\Pi(A)$ spanned by A s.t. $\Pi_0(A) = .4$ ($\mathcal{G}(2,2)$ vs. $\mathcal{E}(1)$)

6.3 Classes of priors

Fortini and Ruggeri (1995a) presented a method, based upon the c.f., to define neighbourhoods of probability measures and applied it in robust Bayesian analyses in Fortini and Ruggeri (1994). Their approach allows the construction of probability measures Π with functional forms close to a nonatomic baseline measure Π_0. In particular, they defined neighbourhoods of Π_0 by imposing constraints on the probability of all measurable subsets, such as requiring $|\Pi_0(A) - \Pi(A)| \leq \Pi_0(A)(1 - \Pi_0(A))$, for any $A \in \mathcal{F}$. By observing that the above relation can be written $\Pi(A) \geq g(\Pi_0(A))$, with $g(x) = x^2$, Fortini and Ruggeri gave the following definitions.

Definition 2 *A function $g : [0,1] \to [0,1]$ is said to be compatible if g is a monotone nondecreasing, continuous, convex function, with $g(0) = 0$.*

Definition 3 *If g is compatible, then the set*

$$K_g \doteq \{\Pi : \Pi(A) \geq g(\Pi_0(A)), \ \ \forall A \in \mathcal{F}\}$$

will be a g-neighbourhood of Π_0.

Observe that, if $\Pi \in K_g$, then $g(\Pi_0(A)) \leq \Pi(A) \leq 1 - g(1 - \Pi_0(A))$, for any $A \in \mathcal{F}$. The requirement $g(0) = 0$ is needed to avoid $\Pi(\Theta) \leq 1 - g(0) < 1$, while monotonicity, continuity and convexity are thoroughly discussed in Fortini and Ruggeri (1995).

As proved in Fortini and Ruggeri (1995), $\{K_g\}$ generates a topology since it becomes a fundamental system of neighbourhoods of Π_0, when g belongs to an adequate class G of compatible functions.

The definition of a g–neighbourhood of Π_0 can be reformulated by means of the c.f. w.r.t. Π_0, as stated in the following,

Theorem 2 *The set* $K_g = \{\Pi : \varphi_\Pi(x) \geq g(x), \forall x \in [0,1]\}$ *is a* g*-neighbourhood of* Π_0.

Fortini and Ruggeri (1995) proved that any compatible g is a c.f.

Theorem 3 *Given a function* $g : [0,1] \to [0,1]$*, there exists at least one measure* Π *such that* g *is the c.f. of* Π *w.r.t.* Π_0 *if and only if* g *is compatible.*

6.3.1 Main results

Consider the space $\mathcal{P}$ of all probability measures on Θ endowed with the weak topology. $\mathcal{P}$ can be metrized as a complete separable metric space. Consider the set F_g of extremal points of K_g, that is, the probability measures $\Pi \in K_g$ such that

$$\Pi = \alpha\Pi_1 + (1-\alpha)\Pi_2, \Pi_1 \in K_g, \Pi_2 \in K_g, 0 < \alpha < 1 \Longrightarrow \Pi = \Pi_1 = \Pi_2.$$

The following results were proved by Fortini and Ruggeri (1995).

Theorem 4 $F_g \subseteq E_g$*, where* $E_g = \{\Pi : \varphi_\Pi(x) = g(x), \forall x \in [0,1]\}$*. If* $g(1) = 1$*, then* F_g *coincides with* E_g.

Furthermore, every probability measure whose c.f. is greater than g can be represented as a mixture of probability measures having g as c.f., applying Choquet's Theorem (Phelps, 1966).

Theorem 5 *Let the function* $g : [0,1] \to [0,1]$ *be compatible. Then, for any probability measure* $\tilde{\Pi} \in K_g$*, there exists a probability measure* $\mu_{\tilde{\Pi}}$ *on* $\mathcal{P}$ *such that* $\mu_{\tilde{\Pi}}(F_g) = 1$ *and* $\tilde{\Pi} = \int_{\mathcal{P}} \Pi \mu_{\tilde{\Pi}}(d\Pi)$.

The supremum (and infimum) of ratio-linear functionals of Π is found in E_g, as shown in

Theorem 6 *Let* f *and* m *be real-valued functions on* Θ *such that* $\int_\Theta |f(\theta)|\Pi(d\theta) < \infty$ *and* $0 \leq \int_\Theta m(\theta)\Pi(d\theta) < \infty$ *for any* $\Pi \in K_g$*. Then*

$$\sup_{\Pi \in K_g} \frac{\int_\Theta f(\theta)\Pi(d\theta)}{\int_\Theta m(\theta)\Pi(d\theta)} = \sup_{\Pi \in E_g} \frac{\int_\Theta f(\theta)\Pi(d\theta)}{\int_\Theta m(\theta)\Pi(d\theta)}.$$

Computations of bounds on prior expectations are simplified by taking in account

Theorem 7 *Let* $H_f(y) = \Pi_0(\{\theta \in \Theta : f(\theta) \leq y\})$*,* $c_f(x) = \inf\{y : H_f(y) \geq x\}$*. Then* $\sup_{\Pi \in K_g} \int_\Theta f(\theta)\Pi(d\theta) = \int_0^1 c_f(x)g'(x)dx$.

The result can be applied to find bounds on posterior expectations, too, using the linearization technique presented by Lavine (1988) and Lavine et al. (2000). Finally, the result was used in Ruggeri (1994) to compute bounds on the posterior probability of sets.

Corollary 1

$$\sup_{\Pi \in K_g} \Pi(A|x) = \left\{ 1 + \frac{-\int_0^{\Pi_0(A^C)} c_{-l_1}(x)c(x)dx}{\int_{\Pi_0(A^C)}^1 c_{l_2}(x)c(x)dx} \right\}^{-1},$$

where I_A is the indicator function of the subset A, $l_x(\theta)$ is the likelihood function and, for any $\theta \in \Theta$, $l_1(\theta) = l_x(\theta)I_{A^C}(\theta)$ and $l_2(\theta) = l_x(\theta)I_A(\theta)$.

6.3.2 Classes of priors as concentration function neighbourhoods

Fortini and Ruggeri (1994) considered classes of prior measures K_g such that their c.f.s w.r.t. a nonatomic base one, say Π_0, are pointwise not smaller than a specified compatible function g. The function g gives the maximum concentration of a measure w.r.t. a base one which is deemed compatible with our knowledge. Note that, assuming Π_0 nonatomic, the discrete measures can be ruled out or not by choosing $g(1) = 1$ or < 1, respectively. The posterior expectation of any function $f(\theta)$, say $E^*(f)$, can be maximised all over K_g applying Theorems 6 and 7. The results are consistent with those found in the literature.

Here we review the collection of classes, including some that are well known, defined by Fortini and Ruggeri (1994). Note that F_g can be a proper subset of E_g, as in the cases of $\varepsilon-$contamination and total variation neighbourhoods.

$\varepsilon-$CONTAMINATIONS. The ε-contamination class $\Gamma_\varepsilon = \{\Pi_Q = (1-\varepsilon)\Pi_0 + \varepsilon\Pi, \Pi \in \mathcal{P}\}$ is defined by $g(x) = (1-\varepsilon)x$, $\forall x \in [0,1]$. The sets E_g and F_g are obtained, respectively, considering singular (w.r.t. Π_0) and Dirac contaminating measures. As shown in Berger (1990), $E^*(f)$ is maximised by contaminating Dirac measures, i.e. over F_g.

DENSITY BOUNDED CLASS. Given a probability measure Π_0 and $k > 0$, consider the class

$$\Gamma_k^B = \{\Pi : (1/k)\Pi_0(A) \le \Pi(A) \le k\Pi_0(A), \forall A \in \mathcal{F}\},$$

studied by Ruggeri and Wasserman (1991). This class, a special case of the density bounded classes defined by Lavine (1991), is a c.f. neighbourhood K_g, with $g(x) = \max\{x/k, k(x-1)+1\}$.

DENSITY RATIO CLASS. Density ratio classes were defined by DeRobertis and Hartigan (1981); Ruggeri and Wasserman (1995) considered Γ_k^{DR}, the density ratio neighbourhood around Π_0 (with density $\pi_0(\theta)$), given by all the probability measures whose densities $\pi(\theta)$ are such that there exists $c > 0$ so that $\pi_0(\theta) \le c\pi(\theta) \le k\pi_0(\theta)$ for almost all θ. It can be shown that Γ_k^{DR} is the class of the probability measures such that their c.f.s w.r.t. Π_0 are inside any triangle with vertices $(0,0)$, $(1,1)$ and a point on the curve $g(x) = x/(k-(k-1)x)$, $0 < x < 1$.

TOTAL VARIATION NEIGHBOURHOOD. A class Γ^T is said to be a total variation neighbourhood of a probability measure Π_0 if it contains all the probability measures Π that satisfy $\sup_{A\in\mathcal{F}} |\Pi(A) - \Pi_0(A)| \le \varepsilon$, given a fixed $\varepsilon \in [0,1]$. Assuming Π_0 nonatomic, then the class is a c.f. neighbourhood K_g with $g(x) = \max\{0, x - \varepsilon\}$.

OTHER NEIGHBOURHOODS. As discussed earlier, we can consider many neighbourhoods, like the class of all the probability measures Π satisfying

$$|\Pi_0(A) - \Pi(A)| \le \Pi_0(A)(1 - \Pi_0(A)), \forall A \in \mathcal{F}.$$

In this case, the neighbourhood K_g is given by $g(x) = x^2$.

6.4 Global sensitivity

As discussed by Moreno (2000), global sensitivity addresses the issue of computing ranges for quantities of interest as the prior measure varies in a class. Usually, quantities like posterior means and set probabilities have been considered, whereas less attention has been paid to changes in the functional form of the posterior measures (see Boratynska, 1996, for the study of the "radius" in the class of posterior measures, endowed with the total variation metric). C.f.s have been used in such a context, and the main reference is the paper by Fortini and Ruggeri (1995b), who considered $\varepsilon-$contaminations and compared the c.f.s of the posterior probability measures w.r.t. a base posterior measure Π_0^*. In computing, pointwise, the infimum $\varphi_{\hat{\Pi}}(x)$ of the c.f., their interest was twofold: providing a measure of the distance between the distributions in the class and Π_0^* and checking if the probability of all measurable sets would satisfy bounds like those used in the previous section to define classes of measures.

Consider a class Γ of probability measures Π and a base prior Π_0, as in the $\varepsilon-$contamination class given by

$$\Gamma_\varepsilon = \{\Pi_Q = (1-\varepsilon)\Pi_0 + \varepsilon Q, Q \in \mathcal{Q}\},$$

where $\mathcal{Q} \subseteq \mathcal{P}$ and $0 \le \varepsilon \le 1$.

Let Π^* denote the posterior measure corresponding to the prior Π. Consider the class

$$\Psi = \{\varphi_\Pi : \varphi_\Pi \text{ is the c.f. of } \Pi^* \text{ w.r.t. } \Pi_0^*, \Pi \in \Gamma\}.$$

From Theorem 1 and (3), it follows, for any $\Pi \in \Gamma$ and $A \in \mathcal{F}$ with $\Pi_0^*(A) = x$, that

$$\widehat{\varphi}(x) \leq \Pi^*(A) \leq 1 - \widehat{\varphi}(1-x),$$

where $\widehat{\varphi}(x) = \inf_{\Pi \in \Gamma} \varphi_\Pi(x)$, for any $x \in [0,1]$.

The interpretation of $\widehat{\varphi}$, in terms of Bayesian robustness, is straightforward: the closest $\widehat{\varphi}(x)$ and $1 - \widehat{\varphi}(1-x)$ are for all $x \in [0,1]$, the closest the posterior measures are. It is then possible to make judgments on robustness by measuring the distance between $\widehat{\varphi}(x)$ and the line $y = x$, for example, by Gini and Pietra's indices as in Carota and Ruggeri (1994) and Fortini and Ruggeri (1995b).

Fortini and Ruggeri (1995b) proved, for ε-contaminations, the following:

Theorem 8 *If φ and φ_0 denote the c.f.'s of Π_Q^* and Q^* w.r.t. Π_0^*, respectively, then it follows that*

$$\varphi(x) = \lambda_Q x + (1 - \lambda_Q)\varphi_0(x),$$

where

$$\lambda_Q = (1-\varepsilon)D_0/[(1-\varepsilon)D_0 + \varepsilon D_Q],$$

with $D_0 = \int_\Theta l_x(\theta)\Pi_0(d\theta)$ and $D_Q = \int_\Theta l_x(\theta)Q(d\theta)$.

They were able to find $\varepsilon-$contaminations of a nonatomic prior Π_0 leading to the lowest c.f., when considering arbitrary contaminations and those given by generalised moment conditions; they found the lowest c.f. in the unimodal case when $\sup_Q D_Q \leq l(\theta_0)$ holds.

A similar approach was followed by Carota and Ruggeri (1993), who considered the class of mixtures of probability measures defined on disjoint sets with weights known to vary in an interval. The class is suitable to describe, with some approximation, the case of two (or more) populations, depending on the same parameter θ, which are strongly concentrated in disjoint subsets.

Finally, it is worth mentioning that Fortini and Ruggeri (1995b) used c.f.s and compatible functions g in checking posterior robustness as well. They considered g as a threshold function, denoting how much the posterior set probabilities were allowed to vary (for example, $\Pi^*(A) \geq g(\Pi_0^*(A))$, for any $A \in \mathcal{F}$). Therefore, robustness is achieved when $\widehat{\varphi}(x) \geq g(x)$ for all $x \in [0,1]$.

6.5 Local sensitivity

Fortini and Ruggeri (1997) studied functional derivatives of the c.f. and mentioned they could be used in Bayesian robustness to perform local sensitivity analysis (see Gustafson, 2000, on the latter). An example is presented in the next section. Here we present some results based on Gâteaux differentials.

Definition 4 *Let X and Y be linear topological spaces. The Gâteaux differential in the direction of $h \in X$ and at a point x_0 of a mapping $f : X \to Y$ is given by*

$$\lim_{\lambda \to 0^+} \frac{f(x_0 + \lambda h) - f(x_0)}{\lambda}$$

if the limit exists.

Fortini and Ruggeri (1993) extended the definition of c.f. given by Cifarelli and Regazzini (1987), considering the c.f. between a signed measure and a probability. The extended version of the c.f. allows for the computation of the limit

$$\mathcal{L}(x, \Delta) = \lim_{\lambda \to 0} \frac{\varphi((\Pi_0 + \lambda\Delta)^*, x) - \varphi(\Pi_0^*, x)}{\lambda},$$

where $\varphi(\Pi, \cdot)$ denotes the c.f. of Π w.r.t. a baseline measure and Δ is a signed measure such that $\Delta(\Theta) = 0$ and $||\Delta|| \leq 1$ for a suitable norm $|| \cdot ||$.

This limit coincides with the differential $\psi'_\Delta(\Pi_0, x)$ of the functional $\psi(\Pi) = \varphi(\Pi^*(\Pi), x)$ in Π_0 in the direction of Δ. The following theorem gives an explicit expression for $\psi'_\Delta(\Pi_0, x)$.

Theorem 9

$$\psi'_\Delta(\Pi_0, x) = \begin{cases} \frac{D_\Delta}{D_{\Pi_0}}(\varphi(\Delta^*, x) - x) & \text{if} D_\Delta > 0 \\ 0 & \text{if} D_\Delta = 0 \\ -\frac{D_\Delta}{D_{\Pi_0}}(\varphi(\Delta^*, 1 - x) - \varphi(\Delta^*, 1) + x) & \text{if} D_\Delta < 0, \end{cases}$$

where Δ^* is defined by $\Delta^*(B) = \int_B l_x(\theta)\Delta(d\theta)/D_\Delta$, for any $B \in \mathcal{F}$ and $D_\Delta = \int_\Theta l_x(\theta)\Delta(d\theta) \neq 0$.

Proof. Along the lines of the proof of Theorem 2 in Fortini and Ruggeri (1993), it can be shown that, for any real λ such that $\lambda D_\Delta \geq 0$, $\varphi((\Pi_0 + \lambda\Delta)^*, x) = D_{\Pi_0}/(D_{\Pi_0} + \lambda D_\Delta)x + (\lambda D_\Delta/(D_{\Pi_0} + \lambda D_\Delta))\varphi(\Delta^*, x)$. Otherwise, λ is taken so that $-D_{\Pi_0} < \lambda D_\Delta < 0$, and it follows, from Lemma 1 in Fortini and Ruggeri (1993), that

$$\varphi((\Pi_0 + \lambda\Delta)^*, x) = \frac{D_{\Pi_0}}{D_{\Pi_0} + \lambda D_\Delta} x - \frac{\lambda D_\Delta}{D_{\Pi_0} + \lambda D_\Delta} [\varphi(\Delta^*, 1 - x) - \varphi(\Delta^*, 1)].$$

Applying the definition of Gâteaux differential, then $\psi'_{\Delta^*}(\Pi_0, x)$ is easily computed. □

Given $\varepsilon \in [0,1]$ and a probability measure Q, the choice $\Delta = \varepsilon(Q - \Pi_0)$ implies that $\Pi_0 + \lambda\Delta$ is a contaminated measure for any $\lambda \in [0,1]$. In this case, the Gâteaux differential is given by

$$\psi'_{\varepsilon(Q-\Pi_0)}(\Pi_0, x) = \varepsilon \frac{D_Q}{D_{\Pi_0}} \{\varphi(Q^*, x) - x\}.$$

The previous Gâteaux differential mainly depends on three terms: ε, D_Q/D_{Π_0} and $\varphi(Q^*, x)$. Because of their interpretation, they justify the use of the Gâteaux differential to measure the sensitivity of the concentration function to infinitesimal changes in the baseline prior. In fact, the first term measures how contaminated the prior is with respect to the baseline one, while the third says how far the contaminating posterior is from the baseline one. Besides, the second term can be interpreted as the Bayes factor of the contaminating prior with respect to the baseline one. Hence, the Gâteaux differential stresses any possible aspect which might lead to nonrobust situations.

We consider $||\psi'_{\varepsilon(Q-\Pi_0)}(\Pi_0)|| = \sup_{0\le x\le 1} \mid \psi'_{\varepsilon(Q-\Pi_0)}(\Pi_0, x) \mid$ as a concise index of the sensitivity of Π_0 to contaminations with Q.

Theorem 10 *Given Q and Π_0 as above, it follows that*

$$||\psi'_{\varepsilon(Q-\Pi_0)}(\Pi_0)|| = \varepsilon \frac{D_Q}{D_{\Pi_0}} G_{\Pi_0^*}(Q^*),$$

where $G_{\Pi_0^*}(Q^*) = \sup_{0\le x\le 1} \{x - \varphi(Q^*, x)\}$ is the Pietra concentration index.

When contaminating Π_0 with the probability measures in a class $\mathcal{Q}$, we can assume

$$||\psi'(\Pi_0, \mathcal{Q})|| = \sup_{Q\in\mathcal{Q}} ||\psi'_{\varepsilon(Q-\Pi_0)}(\Pi_0)|| \tag{4}$$

as a measure of local robustness.

The index (4) can be found analytically in some cases. Let Π_0 be absolutely continuous w.r.t. the Lebesgue measure on $\Re$. If the contaminating class is the class $\mathcal{Q}_{\mathrm{a}}$ of all the probability measures over Θ, then $||\psi'(\Pi_0, \mathcal{Q}_{\mathrm{a}})|| = \varepsilon(l_x(\hat{\theta})/D_{\Pi_0}$, where $\hat{\theta} \in \Theta$ is the maximum likelihood estimator of θ. Considering the class $\mathcal{Q}_{\mathrm{q}}$ of all probability measures sharing $m-1$ given quantiles, then $||\psi'(\Pi_0, \mathcal{Q}_{\mathrm{q}})|| = \varepsilon \sum_{i=1}^m q_i l_x(\hat{\theta}_i)/D_{\Pi_0}$, where $\hat{\theta}_i \in I_i$ is the maximum, for $l_x(\theta)$, over any interval I_i of the partition $\{I_i\}$ of Θ determined by the quantiles, while q_i is the probability of I_i, $i = 1, \ldots, m$.

6.6 Example

Consider the model $P_\theta \sim \mathcal{N}(\theta, 1)$, the prior $\Pi_0 \sim \mathcal{N}(0, 2)$ and an ε–contamination class of probability measures around Π_0. Let $\varepsilon = 0.1$. This example has been widely used in Bayesian robustness by Berger and Berliner (1986) and many other authors since then.

6.6.1 *Global sensitivity*

Consider Π_0 contaminated either by the class $\mathcal{Q}_{1/2}$ of probability measures which have the same median as Π_0 or by $\mathcal{Q}$, the class of the probability measures which are either $Q_k \sim \mathcal{U}(\theta_0 - k, \theta_0 + k)$, $k > 0$, or $Q_\infty \equiv \delta_{\theta_0}$, the Dirac measure at θ_0. Observe the sample $s = 1$. In the former case, the lowest c.f. is $\varphi \equiv 0$, given by the contamination $(\delta_0 + \delta_1)/2$, whereas, in the latter case, the lowest c.f. is $\hat{\varphi}(x) = 0.879x$. Should we decide to compare $\hat{\varphi}(x)$ with, say, the function $g(x) = x^2$, it is evident that $\hat{\varphi}(x) < g(x)$ for some x and, therefore, robustness is not achieved. A different choice of $g(x)$, which allows for discrete contaminations (i.e., such that $g(1) < 1$), might have led to a different situation.

6.6.2 *Local sensitivity*

Consider Π_0 to be contaminated either by the class $\mathcal{Q}_{\text{a}}$ or by $\mathcal{Q}_{1/2}$. It can be easily shown that $\|\psi^{'}\|$ is achieved for a Dirac prior concentrated at the sample s in the former case and for a two-point mass prior, which gives equal probability to 0 and s, in the latter. The values of $\|\psi^{'}(\Pi_0, \mathcal{Q}_{\text{a}})\|$ and $\|\psi^{'}(\Pi_0, \mathcal{Q}_{1/2})\|$ are shown in Table 2, for different samples s's. As expected, the class $\mathcal{Q}_{1/2}$ induces smaller changes than $\mathcal{Q}_{\text{a}}$. It is worth noting that the difference is negligible for small values of s while it increases when observing larger values of s (in absolute value). The finding is coherent with Table 1 in Betrò et al. (1994). While their table was obtained by numerical solution of a constrained nonlinear optimisation problem, here the use of the Gâteaux differential requires just a simple analytical computation.

This example shows that sometimes local sensitivity analysis can give information on the global problem as well, but favoured by simpler computations.

Notice that

$$|\psi^{'}(\Pi_0, \mathcal{Q}_{\text{a}})| = \varepsilon \frac{D_{\delta_s}}{D_{\Pi_0}} = \sup_{Q \in \mathcal{Q}_{\text{a}}} \varepsilon \frac{D_Q}{D_{\Pi_0}}, \tag{5}$$

so that the Pietra index does not seem to have an important part in (4). The same happens when $\mathcal{Q}_{1/2}$ is considered. As shown in the following example, there are contaminating classes for which (5) does not hold. Consider, for example, the class $\mathcal{Q}_{\text{N}} = \{\text{N}(0, \tau^2) : 1 \leq \tau \leq 2\}$. The values of

$\sup_{Q \in \mathcal{Q}_N} \varepsilon D_Q / D_{\Pi_0}$, are shown in Table 2 for different samples s's. They are quite large, especially if compared with those of $\|\psi'(\Pi_0, \mathcal{Q}_N)\|$.

s	$\|\psi'(\Pi_0, \mathcal{Q}_a)\|$	$\|\psi'(\Pi_0, \mathcal{Q}_{1/2})\|$	$\|\psi'(\Pi_0, \mathcal{Q}_N)\|$	$\sup_{Q \in \mathcal{Q}_N} \varepsilon \frac{D_Q}{D_{\Pi_0}}$
[1ex] 0.5	0.1805	0.1606	0.0127	0.1199
1.0	0.2046	0.1399	0.0170	0.1126
1.5	0.2520	0.1392	0.0202	0.1019
2.0	0.3373	0.1717	0.0218	0.1023
2.5	0.4908	0.2458	0.0225	0.1174
3.0	0.7762	0.3881	0.0317	0.1411
3.5	1.3342	0.6671	0.0453	0.1752
4.0	2.4927	1.2463	0.0656	0.2250

TABLE 2. Gâteaux differentials

6.7 Connections with Choquet capacities

We conclude the paper by observing that $g-$neighbourhoods can be considered as an example of 2-alternating Choquet capacities. We refer to Wasserman and Kadane (1990, 1992) for a thorough description of the properties of the latter, their application in Bayesian robustness and their links with other notions, like special capacities (see, for example, Bednarski, 1981, and Buja, 1986) and upper and lower probabilities in Walley's (1991) approach. Details on capacities can be found in Choquet (1955) and Huber and Strassen (1973).

Let $\mathcal{Q}$ be a nonempty set of prior probability measures on $(\Theta, \mathcal{F})$. We define upper and lower prior probability functions by

$$\overline{\Pi}(A) = \sup_{\Pi \in \mathcal{Q}} \Pi(A) \quad \text{and} \quad \underline{\Pi}(A) = \inf_{\Pi \in \mathcal{Q}} \Pi(A),$$

for any $A \in \mathcal{F}$.

The set $\mathcal{Q}$ is said to be *2-alternating* if

$$\overline{\Pi}(A \cup B) \leq \overline{\Pi}(A) + \overline{\Pi}(B) - \overline{\Pi}(A \cap B),$$

for any A, B in $\mathcal{F}$. The set $\mathcal{Q}$ is said to generate a *Choquet capacity* if $\overline{\Pi}(C_n) \downarrow \overline{\Pi}(C)$ for any sequence of closed sets $C_n \downarrow C$. It can be shown that $\mathcal{Q}$ generates a Choquet capacity if and only if the set $\mathcal{C} = \{\Pi : \Pi(A) \leq \overline{\Pi}(A), \forall A \in \mathcal{F}\}$ is weakly compact. We say that $\mathcal{Q}$ is *m-closed* (or *closed w.r.t. majorisation*) if $\mathcal{C} \subseteq \mathcal{Q}$.

Consider a $g-$neighbourhood K_g around a nonatomic probability measure Π_0. Let g be such that $g(1) = 1$. For any set $A \in \mathcal{F}$, we show that

$$\overline{\Pi}(A) = 1 - g(1 - \Pi_0(A)) \quad \text{and} \quad \underline{\Pi}(A) = g(\Pi_0(A)).$$

From the properties of $g-$neighbourhoods and the definition of upper and lower probability functions, it follows that

$$g(\Pi_0(A)) \leq \underline{\Pi}(A) \leq \Pi(A) \leq \overline{\Pi}(A) \leq 1 - g(1 - \Pi_0(A)),$$

for any $\Pi \in \mathcal{Q}$. We show that both upper and lower bounds are actually achieved for probability measures in $\mathcal{Q}$, so that they coincide with upper and lower probability functions, respectively.

Consider

$$\pi_A(\cdot) = \frac{g(\Pi_0(A))}{\Pi_0(A)}\pi_0(\cdot)I_A(\cdot) + \frac{1 - g(\Pi_0(A))}{1 - \Pi_0(A)}\pi_0(\cdot)I_{A^C}(\cdot),$$

where π_A and π_0 are the densities, w.r.t. some dominating measure, of the probability measures Π_A and Π_0, respectively.

Note that Π_A differs from Π_0 because of two different multiplicative factors (the one on A is smaller than 1, whereas the one on A^C is bigger than 1). The c.f. of Π_A w.r.t. Π_0 is made of two segments joining on the curve g at $(\Pi_0(A), g(\Pi_0(A)))$. Because of the convexity of g, the c.f. is above g, so that $\Pi_A \in K_g$. Besides, $\Pi_A(A) = g(\Pi_0(A)) = \underline{\Pi}(A)$ follows from the construction of Π_A. Since the upper probability function is obtained in a similar way, we have proved that upper and lower probability functions in K_g can be expressed by means of g (i.e., respectively, as $1 - g(1 - x)$ and $g(x)$, $x \in [0, 1]$).

Fortini and Ruggeri (1995a) proved that the set K_g is compact in the weak topology; therefore, its definition (see (3)) implies that it generates a Choquet capacity, besides being m-closed.

Using the equation $\overline{\Pi}(A) = 1 - \underline{\Pi}(A^c)$, the 2-alternating property can be rewritten as

$$\underline{\Pi}(A \cup B) \geq \underline{\Pi}(A) + \underline{\Pi}(B) - \underline{\Pi}(A \cap B),$$

for any A, B in $\mathcal{F}$.

In our case, the property becomes

$$g(\Pi_0(A) + \Pi_0(B) - \Pi_0(A \cap B)) \geq g(\Pi_0(A)) + g(\Pi_0(B)) - g(\Pi_0(A \cap B)),$$

which is satisfied because the convexity of g implies

$$\frac{g(x_1 + x_2 - x_3) + g(x_3)}{2} \geq \frac{g(x_1) + g(x_2)}{2},$$

for $x_3 \leq x_1 \leq x_2$; note that $x_2 \leq x_1 + x_2 - x_3$.

Therefore, K_g is an $m-$closed, 2-alternating Choquet capacity, and results in Wasserman and Kadane (1990) apply to it.

References

BEDNARSKI, T. (1981). On solutions of minimax test problems for special capacities. *Zeitschrift für Wahrscheinlichkeitstheorie und Verwandte Gebiete*, **58**, 397–405.

BERGER, J.O. (1985). *Statistical Decision Theory and Bayesian Analysis* (2nd edition). New York: Springer-Verlag.

BERGER, J.O. (1990). Robust Bayesian analysis: sensitivity to the prior. *Journal of Statistical Planning and Inference*, **25**, 303–328.

BERGER, J.O. (1994). An overview of robust Bayesian analysis. *TEST*, **3**, 5–58.

BERGER, J.O. and BERLINER, L.M. (1986). Robust Bayes and empirical Bayes analysis with ε–contaminated priors. *Annals of Statistics*, **14**, 461–486.

BETRÒ, B., MĘCZARSKI, M. and RUGGERI, F. (1994). Robust Bayesian analysis under generalized moments conditions. *Journal of Statistical Planning and Inference*, **41**, 257–266.

BORATYNSKA, A. (1996). On Bayesian robustness with the ε–contamination class of priors. *Statistics & Probability Letters*, **26**, 323–328.

BUJA, A. (1986). On the Huber-Strassen theorem. *Probability Theory and Related Fields*, **73**, 149–152.

CAROTA, C. and RUGGERI, F. (1994). Robust Bayesian analysis given priors on partition sets. *TEST*, **3**, 73–86.

CHOQUET, G. (1955). Theory of capacities. *Annales de l'Institute Fourier*, **5**, 131–295.

CIFARELLI, D.M. and REGAZZINI, E. (1987). On a general definition of concentration function. *Sankhyā, B*, **49**, 307–319.

DEROBERTIS, L. and HARTIGAN, J. (1981). Bayesian inference using intervals of measures. *Annals of Statistics*, **9**, 235–244.

DI BONA, J., PARMIGIANI, G. and RUGGERI, F. (1993). Are we coming together or coming apart? Racial segregation in North Carolina schools 1982–1992. *Technical Report*, **93.6**, CNR-IAMI.

FORTINI, S. and RUGGERI, F. (1993). Concentration function and coefficients of divergence for signed measures. *Journal of the Italian Statistical Society*, **2**, 17–34.

FORTINI, S. and RUGGERI, F. (1994). Concentration function and Bayesian robustness. *Journal of Statistical Planning and Inference*, **40**, 205–220.

FORTINI, S. and RUGGERI, F. (1995a). On defining neighbourhoods of measures through the concentration function. *Sankhyā, A*, **56**, 444–457.

FORTINI, S. and RUGGERI, F. (1995b). Concentration function and sensitivity to the prior. *Journal of the Italian Statistical Society*, **4**, 283–297.

FORTINI, S. and RUGGERI, F. (1997). Differential properties of the concentration function, *Sankhyā, A*, **59**, 345–364.

GINI, C. (1914). Sulla misura della concentrazione della variabilità dei caratteri. *Atti del Reale Istituto Veneto di S.L.A., A.A. 1913–1914, parte II*, **73**, 1203–1248.

GUSTAFSON, P. (2000). Local robustness in Bayesian analysis. In *Robust Bayesian Analysis* (D. Ríos Insua and F. Ruggeri, eds.). New York: Springer-Verlag.

HUBER, P.J. and STRASSEN, V. (1973). Minimax tests and the Neyman-Pearson lemma for capacities. *Annals of Statistics*, **1**, 251–263.

LAVINE, M. (1988). Prior influence in Bayesian statistics. *Technical Report*, **88–06**, ISDS, Duke University.

LAVINE, M. (1991). Sensitivity in Bayesian statistics: the prior and the likelihood. *Journal of the American Statistical Association*, **86**, 396–399.

LAVINE, M., PERONE PACIFICO, M., SALINETTI, G. and TARDELLA, G. (2000). Linearization techniques in Bayesian robustness. In *Robust Bayesian Analysis* (D. Ríos Insua and F. Ruggeri, eds.). New York: Springer-Verlag.

MARSHALL, A.W. and OLKIN, I. (1979). *Inequalities: Theory of Majorization and Its Applications.* New York: Academic Press.

MORENO, E. (2000). Global Bayesian robustness for some classes of prior distributions. In *Robust Bayesian Analysis* (D. Ríos Insua and F. Ruggeri, eds.). New York: Springer-Verlag.

PHELPS, R.R. (1966). *Lectures on Choquet's Theorem.* Princeton: Van Nostrand Company.

PIETRA, G. (1915). Delle relazioni tra gli indici di variabilità. *Atti del Reale Istituto Veneto di S.L.A. A.A. 1914-1915, parte II*, **74**, 775-792.

REGAZZINI, E. (1992). Concentration comparisons between probability measures. *Sankhyā, B*, **54**, 129–149.

RUGGERI, F. (1994). Local and global sensitivity under some classes of priors. In *Recent Advances in Statistics and Probability* (J.P. Vilaplana and M.L. Puri, eds). Ah Zeist: VSP.

RUGGERI, F. and WASSERMAN, L.A. (1991). Density based classes of priors: infinitesimal properties and approximations. *Technical Report*, **91.4**, CNR-IAMI.

RUGGERI, F. and WASSERMAN, L.A. (1995). Density based classes of priors: infinitesimal properties and approximations. *Journal of Statistical Planning and Inference*, **46**, 311–324.

WALLEY, P. (1991). *Statistical Reasoning with Imprecise Probabilities.* London: Chapman Hall.

WASSERMAN, L.A. (1992). Recent methodological advances in robust Bayesian inference. *Bayesian Statistics 4* (J.M. Bernardo, J.O. Berger, A.P. Dawid and A.F.M. Smith, eds.). Oxford: Oxford University Press.

WASSERMAN, L.A. and KADANE, J. (1990). Bayes' theorem for Choquet capacities. *Annals of Statistics*, **18**, 1328–1339.

WASSERMAN, L.A. and KADANE, J. (1992). Symmetric upper probabilities. *Annals of Statistics*, **20**, 1720–1736.

7

Likelihood Robustness

N.D. Shyamalkumar

ABSTRACT Most of the research in the theory of Bayesian robustness has concerned the sensitivity of the posterior measures of interest to imprecision solely in the prior, the primary reason being that the prior is perceived to be the weakest link in the Bayesian approach. Another reason is that the operator which maps the likelihood to its posterior is not ratio-linear, making the problem of global robustness with respect to the likelihood, for interesting nonparametric neighborhoods, not very mathematically tractable. Despite these reasons the problem retains its importance, and there have been some interesting studies, which we review here.

Initial research has concerned itself with embedding the likelihood in a parametric class or a discrete set of likelihoods; see for example Box and Tiao (1962). Such research suggested methodologies which, though computationally easy, were not considering sufficiently rich neighborhoods of the likelihood. A deviation from this approach is that of Lavine (1991), where the class of neighborhoods chosen was nonparametric. Apart from these, there have been some interesting studies on restricted problems; for instance, imprecision of weights in weighted distributions (Bayarri and Berger, 1998) and the case of regression functions (Lavine, 1994). Until now we were implicitly considering the problem of global robustness, where it is clear that the problem is quite difficult, except for some restricted problems. We conclude by discussing the local sensitivity approach to likelihood robustness where the mathematical problem becomes more tractable (see, e.g., Sivaganesan, 1993), but of course at the cost of ease of interpretation that the global robustness approach entails.

Key words: Bayesian robustness, global robustness, likelihood robustness.

7.1 Introduction

Robust Bayesian analysis is the study of sensitivity of posterior measures of interest with respect to some or all of the *uncertain* inputs. Most of the literature in the area has focused on sensitivity of the posterior answers solely with respect to the prior, the primary reason being that the prior is perceived to be the weakest link in the Bayesian approach (Berger, 1985). Moreover, as far as the likelihood is concerned, the technical difficulty of the variational problems arising from the study of posterior robustness with respect to the likelihood was also a main hindrance. In any case, carrying out such a sensitivity study is very important and in many problems could prove to be more relevant than sensitivity with respect to the prior, as would be shown through some examples.

In the following $X_1, X_2, \ldots, X_n$ will denote independent random variables taking values in $(\mathcal{S}, \mathcal{B}_\mathcal{S})$ which, unless specified otherwise, could safely be assumed as a subset of the Euclidean space with its associated Borel σ-field. We shall assume that there is always a base model, which we will denote by $\{P_\theta, \theta \in \Theta\}$. In robustness studies, this base model may be considered as the initial *guess* of the experimenter. We would further assume that the probability measures in the model are dominated by a σ-finite measure λ on $(\mathcal{S}, \mathcal{B}_\mathcal{S})$ and its Radon–Nikodym derivative $dP_\theta/d\lambda$ will be denoted by $f_P(\cdot|\theta)$. The parameter space would be a subset of a Euclidean space and $(\Theta, \mathcal{B}_\Theta)$ will be the associated Borel space. We will denote by π_0 the initial prior *guess*, a probability measure on $(\Theta, \mathcal{B}_\Theta)$. It is sometimes convenient to think of it in terms of the product space $(\mathcal{S} \times \Theta, \mathcal{B}_\mathcal{S} \otimes \mathcal{B}_\Theta)$ with P_θ as the regular conditional probability conditioned on θ and π_0 as the marginal of θ.

The function of interest will be denoted by $\phi(\cdot)$, a real-valued function defined on $(\Theta, \mathcal{B}_\Theta)$. For example, ϕ could be defined as $\phi(\theta) = \theta$ and would be estimated by the posterior mean. Then a study of global likelihood robustness of this function of the unknown parameter would seek the variational bounds of

$$\frac{\int \phi(\theta) \prod_{i=1}^n f_Q(x_i|\theta) \mathrm{d}\pi(\theta)}{\int \prod_{i=1}^n f_Q(x_i|\theta) \mathrm{d}\pi(\theta)},$$

where $\{Q_\theta,\ \theta \in \Theta\}$ and π would be some model and prior *close* to the base model and prior, respectively. Some studies focus on varying only the likelihood or model by fixing the prior (maybe non-informative); we classify these too under likelihood robustness studies. Note the asymmetry in the above ratio when considered as a function of either the model or the prior. When considered as a function of the prior, the above is a ratio of two linear functionals of the prior, or as is commonly termed in the literature, *ratio-linear*. This is not true when considered as a function of the model, and in fact the complexity increases with the sample size unless either one introduces a nice structure through an appropriate definition of *closeness* or when the nature of the problem imposes such a structure on the form of the likelihood. In the following, we will review the literature in the area which has attempted such likelihood robustness studies.

7.2 Likelihood classes

One way of classifying the robustness studies concerning likelihood is whether they study variational problems as described above, that is global robustness, or whether they study local sensitivity as we shall describe in a later section. Another way of classification is through the *dimensionality* or cardinality of the likelihood classes considered in the studies. This usually does translate into the complexity of the variational problem. It is necessary to observe that increasing complexity of the class does not always imply increasing pertinence of the study or results in a practical problem. What is necessary is that the class of likelihoods considered corresponds to the knowledge or uncertainty in the inputs concerning the likelihood. In this section, we have used this latter classification scheme and consider classes which have a finite cardinality or which are infinite and isomorphic to an Euclidean subset or which are nonparametric in the sense that they do not have a natural isomorphism to a finite-dimensional space. In later sections we will look at the literature on robustness studies in more specialized problems.

7.2.1 Finite classes

One way to proceed with likelihood robustness studies is to consider the likelihood to be one among a finite class. One attraction of the finite class approach is that it leads to simple variational problems. The issue is that it might not be a good representation of the uncertainty in the likelihood. But even in the case that it only approximately models the uncertainty, it is a good way to start robustness studies as it might help in defining more representative classes by pointing out a direction where sensitivity is more pronounced. The example below is taken from Pericchi and Perez (1994) and Mortera (1994) and was first considered in Berger (1990).

Example 1. Let us consider $X \backsim N(\theta, 1)$ with θ unknown and the prior for θ as $\pi_0 = N(0,1)$. As is usually the case, neither the likelihood nor the prior is free of uncertainty. We consider two classes for the uncertainty in the prior π_0, $\Gamma^A_{0.1}$ and $\Gamma^{SU}_{0.1}$; $\Gamma^A_{0.1}$ denotes the ϵ-contamination class with $\epsilon = 0.1$ and arbitrary contaminants, that is,

$$\Gamma^A_{0.1} = \{\pi : \pi = 0.9\pi_0 + 0.1q, q \text{ arbitrary,}\}$$

and $\Gamma^{SU}_{0.1}$ the ϵ-contamination class with $\epsilon = 0.1$ and contaminants restricted to unimodal probabilities symmetric around zero, namely,

$$\Gamma^{SU}_{0.1} = \{\pi : \pi = 0.9\pi_0 + 0.1q, q \text{ symmetric unimodal around zero}\}.$$

Clearly, $\Gamma^{SU}_{0.1} \subset \Gamma^A_{0.1}$. The extreme points of both sets are well known: those with degenerate and symmetric uniform contaminants, respectively. Likelihood uncertainty is modeled by considering the class

$$\mathcal{M} = \{N(\theta, 1), C(\theta, 0.675)\},$$

the choice of the scale for Cauchy being such that the interquartile ranges matches those of the normal likelihood. The quantity of interest is θ itself, and hence we study the sensitivity of its Bayes estimate, $\mathbb{E}(\theta|x)$. The table below gives the results.

Data	Likelihood	$\Gamma^A_{0.1}$		$\Gamma^{SU}_{0.1}$	
		$\inf \mathbb{E}(\theta\|x)$	$\sup \mathbb{E}(\theta\|x)$	$\inf \mathbb{E}(\theta\|x)$	$\sup \mathbb{E}(\theta\|x)$
$x = 2$	normal	0.93	1.45	0.97	1.12
	Cauchy	0.86	1.38	0.86	1.02
$x = 4$	normal	1.85	4.48	1.96	3.34
	Cauchy	0.52	3.30	0.57	1.62
$x = 6$	normal	2.61	8.48	2.87	5.87
	Cauchy	0.20	5.54	0.33	2.88

Notice that even though the widths of the ranges of the posterior expectations are similar, the centers are quite different between those arising from the normal and Cauchy likelihoods. △

Even though simple, this example is interesting in its results. The computational complexity does not increase significantly with increasing sample size. This latter property is the main attraction of the finite class approach to likelihood robustness.

7.2.2 Parametric classes

A parametric class, while not increasing the computational complexity too much with respect to finite classes, may capture much more of the uncertainty in the likelihood if suitably chosen. One of the first such analyses was that of Box and Tiao (1962). In their analysis, the normal location-scale family is embedded in the parametric class of location-scale family of likelihoods,

$$\frac{e^{\frac{-1}{2}\left|\frac{y-\theta}{\sigma}\right|^{\frac{2}{1+\beta}}}}{\sigma 2^{(1.5+0.5\beta)}\Gamma(1.5+0.5\beta)}; \quad y,\theta \in (-\infty,\infty), \sigma \in (0,\infty), \beta \in (-1,1].$$

Clearly, when $\beta = 0$, we have the normal likelihood and when $\beta = 1$ we have the double exponential likelihood. Also, observe that, as $\beta \to -1$, the likelihood approaches that of a symmetric uniform. For intermediate values of β, the likelihood is an intermediate symmetric distribution; platykurtic for $\beta < 0$ and leptokurtic for $\beta > 0$. In some sense, β can be thought of as a non-normality parameter. Note that the assumption of normality is equivalent to assuming that $\beta = 0$. O'Hagan (1979) shows that this family is similar to the normal, in the sense that they are both *outlier-resistant* and that for $\beta > 1$ we get likelihoods that are *outlier-prone*.

This is an appropriate place to observe that the distinction between likelihood and prior is, in fact, somewhat artificial. Above, as an example, some part of what was previously the likelihood could now be considered as part of the prior; that is, working with the normal likelihood is equivalent to having the above *expanded* likelihood with parameters (θ, σ, β) and a prior on β which is a unit mass at zero. Extending this argument to its extreme, we could look at the distribution of the observations as the parameter and the prior to be on the space of all distributions on the sample space. In this latter case, we could look at the usual likelihood and a prior on a (finite dimensional) parameter as a way of defining a probability on the space of all distributions on the sample space with the support as a subset of the closure of the set of probability measures defined by $\{P_\theta, \theta \in \Theta\}$.

We give below the gist of the Bayesian analysis of Darwin's data as in Box and Tiao (1962).

Example 2. The data here is the Darwin's paired data on the heights of self and cross-fertilized plants. The parameter of interest is the location parameter θ of the distribution of the difference of the heights. In usual analysis the difference of the heights is modeled as normal. The robustness study here focuses on this normality assumption. The 15 ordered differences of the heights which form the sample are given below.

#	Value of Obsn.	#	Value of Obsn.	#	Value of Obsn.
1	−67	6	16	11	41
2	−48	7	23	12	49
3	6	8	24	13	56
4	8	9	28	14	60
5	14	10	29	15	75

The importance of the normality assumption in a classical analysis is shown by the conclusions in a testing problem of $\theta = 0$ against the alternative of $\theta > 0$. It is seen that even though the t-test criterion, which is what one would use under normal assumption and classical theory, is robust to variations in the likelihood from normal to symmetric uniform, changing the significance level of the above test from 2.485% to 2.388%, the inference problem is not: if, in fact, the likelihood was symmetric uniform, one would not use the t-test but instead use one based on the first and the last order statistic, and in the latter case the significance level changes to 23.215% (from the previous 2.388%), leading to very different conclusions. $\triangle$

The robustness of the former kind is called *criterion* robustness and that of the latter as *inference* robustness by Box and Tiao (1964). In the robust Bayesian analysis we *automatically* deal with *inference* robustness. In this respect, we should mention the, by now well-known, discrepancy between the p-values and conditional and Bayes' measures of evidence; see Berger and Sellke (1987) and other references in Berger (1994).

Example 2 (continued). We return to the Bayesian analysis of the problem. For this, let us fix for the priors on θ and σ the non-informative priors

$$\pi_\theta(\theta) \propto k \text{ and } \pi_\sigma(\sigma) \propto \frac{1}{\sigma}.$$

With these independent priors for the location and scale parameters and the above likelihood for fixed β, we see that the posterior for θ as a function of β, $\pi(\theta|\beta,\text{data})$ varies significantly with β. This happens because as we vary β the likelihood attains its maximum at different measures of location and these measures for the given data are quite disparate, for example, the sample mid-range = 4.0 and the sample median = 24.0. Hence, we see that the posterior answers would vary as the likelihood varies over the above class of exponential power distributions or, in other words, posterior robustness is not achieved.

Instead of the above global robustness study with respect to β, we could do a more complete Bayesian analysis with a prior for β, using, for example, the symmetric beta family given below,

$$\pi_\beta(\beta) \propto (1-\beta^2)^{a-1}, \quad \beta \in (-1, 1]$$

where the parameter a, representing concentration around normality ($\beta = 0$), takes values in $[1, \infty)$. The posterior of β is found to concentrate away from -1, and since the sensitivity of the posterior answers was pronounced near $\beta = -1$, this more complete Bayesian analysis results in tighter bounds with respect to a. This latter part would be looked at as a prior robustness analysis and reconfirms our observation above that the distinction between likelihood and prior is nothing but artificial. $\triangle$

In the above example, note that we could extend the analysis to include the extended family with $\beta \in (-1, \infty)$. Also, we could allow for a richer class of priors for the location and scale parameters instead of the non-informative priors chosen above. This latter type of analysis should still be very computationally accessible by using suitable prior classes which have proven algorithms, as can be found in this volume. In case one is interested in studying robustness with respect to symmetry, a suitable parametric class can be found in Azzalini (1985). For other examples and discussion of the above kind of Bayesian analysis, we refer to Box (1980) and references therein.

7.2.3 Nonparametric classes

One of the first approaches to the likelihood robustness problem with non-parametric classes was that of Lavine (1991). DeRobertis (1978) mentioned also a nonparametric neighborhood for likelihoods.

Let $\mathcal{P}$ denote the set of all probability measures on $(\mathcal{S}, \mathcal{B}_\mathcal{S})$. The approach of Lavine (1991) begins with a probability on a finite-dimensional space $(\Theta, \mathcal{B}_\Theta)$, with the conditional distribution constrained to assign probability one to a neighborhood $N(\theta)$ of P_θ. The complete class of probability measures on $\mathcal{P}$ is given by

$$\Gamma = \{\pi = \pi^m \times \pi^c : \pi^m \in \Gamma^m, \pi^c \in \Gamma^c\},$$

where Γ^m could be any of the well-studied classes found in the literature on prior robustness and $\Gamma^c = \{\pi^c : \pi^c(N(\theta)|\theta) = 1\}$. The family of neighborhoods $\{N(\theta) : \theta \in \Theta\}$ could be any of the topological or other standard neighborhoods, like ϵ-contaminations, such that $P_\theta \in N(\theta), \forall\theta \in \Theta$. With this structure, it follows that the extrema of the posterior quantities over the class Γ would be attained at the extreme points of the set of conditionals, which is the set of degenerate conditionals. Hence, we can think of this structure in the sense of the usual likelihood robustness setup with the class of likelihood being $\prod_{\theta\in\Theta} N(\theta)$,that is,

$$\{\{f(\cdot|\theta)|\theta \in \Theta\} : f(\cdot|\theta) \in N(\theta), \forall\theta \in \Theta\}.$$

The computational problem uses the linearization algorithm; see Lavine et al. (1993). There are two important steps in the computational problem: the optimization in $\prod_{\theta\in\Theta} N(\theta)$ and the optimization in Γ^m. For the first problem, the specific posterior functionals of interest, one could work out the extrema using simple arguments, especially when the $N(\theta)$ are density bounded classes. The second stage is handled in the same way as in the prior robustness literature. It can be shown that when the $N(\theta)$ do not bound the densities away from both 0 and ∞, the bounds could be trivial. This rules out usual topological neighborhoods (weak convergence), which do not bound densities, and the ϵ-contamination neighborhoods, which do not bound the density from above. The strength of the above approach is its computational ease, but this comes at the cost of the class of likelihoods being large. We will return to this approach again in the next section.

One important variational problem encounters in likelihood robustness involving nonparametric classes of likelihoods is that of obtaining the extrema of the expectation of a real-valued function defined on $\mathbb{R}^n$ over the class of probability measures on $\mathbb{R}^n$ which make the coordinate mappings i.i.d. Since our functionals are ratio-linear, we could equivalently consider the class of all (infinitely) exchangeable probabilities. In general, this problem cannot be simplified, as each probability measure with i.i.d. coordinate mappings is an extreme point of this subset of probability measures. A possible approach is to consider a larger class whose set of extreme points is simple enough to facilitate evaluation of the extrema in this larger class and small enough for these extrema to be a good approximation of the extrema we sought in the first place. One such class is that of finitely exchangeable probability measures with sampling without replacement as the extreme probabilities; see Diaconis and Freedman (1980). This is the approach taken in Shyamalkumar (1996a), where one could find the convergence results of the computational scheme which uses the result that finitely exchangeable probabilities approach infinitely exchangeable probabilities in total variation norm. The upside of this approach is that we are able to consider very realistic classes of likelihoods, especially in cases where there is an implied structure for, e.g., location families or location-scale families. This restriction is not imposed by classes of the form $\prod_{\theta\in\Theta} N(\theta)$. The downside of this approach is that the dimension of the optimization problem grows with the sample size, which is expected when one imposes a reasonable restriction on the class of likelihoods. Nevertheless, this method would be useful when one has small sample sizes or to understand the kind of uncertainties that could affect the results significantly. Below is an example of the type of problems that could be solved using this approach.

Example 3. Suppose that (X_1, X_2, X_3, X_4) is a vector of i.i.d. observations from a symmetric unimodal location model. The base model is assumed to be normal, and uncertainty in the likelihood is modeled by considering the following class:

$$\{0.9N(\theta, 1) + 0.1Q_\theta,\ Q_\theta \text{ symmetric unimodal}\}.$$

We work with the usual noninformative prior for the location parameter, the Lebesgue measure on the line. The posterior quantity of interest is the probability of coverage of the 95% confidence interval for θ, denoted by $C = \{\bar{x} - 0.98, \bar{x} + 0.98\}$, determined using the base model. Clearly, what is of prime interest is the infimum of the posterior probability as the likelihood varies over the above class. This is given by

$$\inf_P \frac{\int I_C(\theta) \prod_{i=1}^4 U_{(-\beta_i, \beta_i)}(x_i - \theta) \mathrm{d}\theta \prod_{i=1}^4 \mathrm{d}P(\beta_i)}{\int \prod_{i=1}^4 U_{(-\beta_i, \beta_i)}(x_i - \theta) \mathrm{d}\theta \prod_{i=1}^4 \mathrm{d}P(\beta_i)}.$$

Note that above, the vector $(\beta_1, \beta_2, \beta_3, \beta_4)$ is supposed to have some i.i.d. distribution, which we relax to finitely exchangeable laws to compute lower bounds for the above infimum. Below we give this infimum relaxing this assumption of i.i.d. to just independence and compare this to the infimum found by using the above method.

$\mathbf{x} = (x_1, x_2, x_3, x_4)$	Independent	I.I.D. (lower bounds)
$(-1.0, -0.5, 0.5, 1.0)$	0.886	0.916
$(-2.0, -1.0, 1.0, 2.0)$	0.742	0.757
$(-1.0, 0.0, 0.1, 3.0)$	0.670	0.718

$\triangle$

An alternative computational method for the robustness problems with mixture classes, like the one in Example 3, is based on the steepest descent method; see Basu (1999). In this method, one works with the Gâteaux derivative and finds the direction of steepest descent at every iteration. Given this direction, one then finds the convex combination of the estimate on the last iteration and the direction of steepest descent which attains the infimum, which is the next estimate. As above, if the extreme points of the class of likelihoods considered is parametrized in a low-dimensional space, then the computations are facilitated. This algorithm makes even problems with large sample sizes possible, but the downside is that the speed of convergence of the algorithm cannot be universally guaranteed to be efficient. This approach could also potentially entertain likelihood and prior robustness simultaneously.

7.3 Regression analysis

In a parametric Bayesian regression model we could think of likelihood robustness studies concerning sensitivity to either the regression function or the error distribution. In the former case, we could let the unknown parameter θ be specified by a function $s(\cdot,\cdot)$ from $\mathcal{B} \times \mathcal{Z}$ to Θ, where $\mathcal{B}$ is a finite-dimensional Euclidean parameter space and $\mathcal{Z}$ is the covariate space, fix the functional form of the likelihood and study sensitivity with respect to the function s. An example of such a study is Lavine (1994) which we shall discuss below. We shall then discuss some studies regarding sensitivity with respect to the error distribution with a fixed linear or nonlinear regression function and prior, in a multivariate regression setting.

7.3.1 Regression function

Suppose that $X_1, X_2, \ldots, X_n$ are independent random variables with laws $P_{\theta_1}, P_{\theta_2}, \ldots, P_{\theta_n}$, $\theta_i \in \Theta$ is known to be of the form $\theta_i = s(\beta, x_i)$, $i = 1, 2, \ldots, n$ where the x_i's are known covariates. In the robust Bayesian analysis of this subsection, we would entertain uncertainty in both the prior and the functional form of the function $s(\cdot,\cdot)$. The approach here is similar to the one discussed in Lavine (1991) and followed in Lavine (1994). Hence we do not discuss the computational procedure, which is similar to the one above, but mention an example in Lavine (1994).

Example 4. The following data are from Racine et al. (1986). A probit regression model was considered for such data.

Dose (mg/ml)	ln(Dose)	# of Animals	# of Deaths
422	6.045	5	0
744	6.612	5	1
948	6.854	5	3
2069	7.635	5	5

The probability measure is binomial with $n = 5$ and θ is supposed to have the base functional form $\Phi(\beta_0 + \beta_1 \ln(x))$, where x is the dosage of the toxic and Φ is the standard normal c.d.f. The prior π_0 for (β_0, β_1) was elicited and modeled as a bivariate normal with mean $(-17.31, 2.57)$, standard deviations $(32.461, 4.821)$ and correlation coefficient -0.99976. Uncertainty in the prior is modeled as a density bounded class, the bounds being $\pi_0 2^{-0.5}$ and $\pi_0 2^{0.5}$. The neighborhood $N(\beta)$ for the regression function at β, that is, for the function $s(\beta, \cdot)$, is taken to be the set of all functions bounded between $\Phi(\beta_0 + \beta_1 \ln(x2^{-0.5}))$ and $\Phi(\beta_0 + \beta_1 \ln(x2^{0.5}))$. We give below the bounds for different covariate values.

ln(Dose)	5.5	6.0	6.5
inf, sup	0.00, 0.28	0.00, 0.54	0.01, 0.97
ln(Dose)	7.0	7.5	8.0
inf, sup	0.26, 1.00	0.54, 1.00	0.72, 1.00

$\triangle$

Note that some of the bounds are quite wide apart. This is because of entertaining a large class of regression functions ($\prod_{\beta \in \mathcal{B}} N(\beta)$). Nevertheless, this approach could be part of understanding the sensitivity to uncertainties by looking at the regression functions at which the bounds are attained.

7.3.2 *Error distribution*

In this subsection, we study likelihood robustness in a multivariate regression setting with respect to the distribution of the error by working with a fixed prior. We state a result from Fernández et al. (1997) which summarizes the kind of results that are available in these kinds of studies. See Osiewalski and Steel (1993a) for further references in this area.

The key idea is to construct a probability measure on $\mathbb{R}^n$ by using the mapping from $\mathcal{Y} \times \mathbb{R}_+$ to $\mathbb{R}^n - \{0\}$ given by $ry \rightarrow z$, where $\mathcal{Y}$ is an $(n-1)$-dimensional manifold. For example, $\mathcal{Y}$ could be taken as the unit sphere S^{n-1} and $r = \|z\|_2$. Then, having fixed such a $\mathcal{Y}$ and any probability measure P on $\mathcal{Y}$, we could construct a class of probability measures on $\mathbb{R}^n$, say $\mathcal{C}$, given by

$$\mathcal{C} = \{\mathcal{L}(Z) : Z = RY \text{ with } \mathcal{L}(Y) = P\},$$

that is, by varying the joint distribution on (R, Y) having fixed the marginal of Y. This family is quite large and contains the l_q spherical models of Osiewalski and Steel (1993a), elliptical errors of Osiewalski and Steel (1993b) and the family generated with i.i.d. errors from the exponential power distribution discussed above.

Consider the regression setup

$$X_i = s_i(\beta) + \sigma_i \epsilon_i, \; i = 1, 2, \dots, p,$$

where $\epsilon_1, \epsilon_2, \dots, \epsilon_p$ are i.i.d. n-dimensional random vectors, $\sigma_i > 0, i = 1, 2, \dots, p$ and $s_i(\cdot)$ are known functions from $\mathcal{B} \subseteq \mathbb{R}^m$ to $\mathbb{R}^n$ with $m \leq n$. In this setup, it can be shown that if we adopt a prior on $(\beta, \sigma_1, \sigma_2, \dots, \sigma_p)$ as a product measure of the noninformative prior $\prod_{i=1}^p \sigma_i^{-1}$ and any prior (σ finite) for β, then the joint distribution of $(X_1, X_2, \dots, X_p, \beta)$ is the same for all choices of error distributions from $\mathcal{C}$. This, in particular, implies the invariance of the posterior of β, with respect to the particular choice of distribution from $\mathcal{C}$.

7.4 Weighted distribution models

Let $\{P_\theta^T, \theta \in \Theta\}$ be a parametrized family of probability measures on $(\mathcal{S}, \mathcal{B}_\mathcal{S})$. In the case when the observations $X_1, X_2, \ldots, X_n$ arise from $\{P_\theta, \theta \in \Theta\}$, which is related to $\{P_\theta^T, \theta \in \Theta\}$ by the relation

$$f_P(\cdot|\theta) \propto w(\cdot) f_{P^T}(\cdot|\theta),$$

the base model is called a weighted distribution model. Observe that for the above to make sense, w must be a nonnegative, real-valued function which is integrable with respect to each probability in $\{P_\theta^T, \theta \in \Theta\}$. The function w is referred to as the weight function. When w is bounded above, without loss of generality by one, we could think of w as the probability of selecting the observation arising from $\{P_\theta^T, \theta \in \Theta\}$. In the special case when this probability is either zero or one, the model is called a selection model. The weighted distribution model arises in many practical situations (see, e.g., Rao, 1985, Bayarri and DeGroot, 1992, and references therein). Quite frequently there is considerable uncertainty in the specification of the weight function. We review in this section the robust Bayesian analysis carried out with respect to the weight function in Bayarri and Berger (1998).

As before, let ϕ be the function of the parameter of interest. Then, the posterior quantity of interest would be

$$\frac{\int \phi(\theta)(\nu_w(\theta))^{-n} \prod_{i=1}^n f_{P^T}(x_i|\theta) \mathrm{d}\pi_0(\theta)}{\int (\nu_w(\theta))^{-n} \prod_{i=1}^n f_{P^T}(x_i|\theta) \mathrm{d}\pi_0(\theta)},$$

where

$$\nu_w(\theta) = \int w(x) f_{P^T}(x|\theta) \mathrm{d}\lambda(x)$$

is the normalizer. Observe that, in this case, the class of likelihoods is equivalently specified by a class of weight functions. We shall consider the following two classes of weight functions:

$$\begin{aligned} \mathcal{W}_1 &= \{w : w_1(x) \le w(x) \le w_2(x)\}, \\ \mathcal{W}_2 &= \{\text{nondecreasing } w : w_1(x) \le w(x) \le w_2(x)\}. \end{aligned}$$

Note that the posterior is dependent on w only through ν_w.

One could find bounds for the above classes using prior robustness techniques in DeRobertis and Hartigan (1981) and Bose (1990). In order to apply the latter method, we would use the fact that the class of ν_w induced by $\mathcal{W}_1$ is contained in

$$\Gamma_1 = \{\nu_w : \nu_{w_1}(\theta) \le \nu_w(\theta) \le \nu_{w_2}(\theta)\}$$

and that induced by $\mathcal{W}_2$ is contained in

$$\Gamma_2 = \{\text{nondecreasing } \nu_w : \nu_{w_1}(\theta) \leq \nu_w(\theta) \leq \nu_{w_2}(\theta)\}$$

when the likelihood of $\{P_\theta^T, \theta \in \Theta\}$ has the monotone likelihood ratio property (MLR). The resulting computational problem is unidimensional irrespective of the sample size. But, as we see in the example below, these bounds do not suffice.

Example 5. Suppose that $f_{P^T}(x|\theta) = \theta^{-1}\exp\{-x/\theta\}$, $x \geq 0$ and $\theta \geq 0$. We shall work with $\pi_0(\theta) = \theta^{-1}$, the usual noninformative prior. We assume a selection model setup with $w_i(x) = I_{[\tau_i,\infty)}(x)$, with $\tau_2 < \tau_1$. Our function of interest is $\phi(\theta) = \theta$. The table below gives the bounds when $\tau_1 = 0.8$ and $\tau_2 = 1.2$ for different sample sizes.

	$n = 5$	$n = 10$	$n = 50$
Bounds for Γ_1	$(0.813, 1.711)$	$(0.610, 1.681)$	$(0.421, 1.976)$
Bounds for Γ_2	$(1.117, 1.711)$	$(0.864, 1.681)$	$(0.582, 1.976)$

Note that the bounds widen with increasing sample sizes. This suggests that the bounds can be significantly improved. One possible reasoning goes as follows. Using the memoryless property of the exponential, it can be shown that if X has a distribution for a w between w_1 and w_2, then $X - \tau_1$ is stochastically larger than the exponential and $X - \tau_2$ is stochastically smaller than the exponential. Using the MLR property we can conclude that the likelihood would concentrate around a neighborhood of $\overline{x}_n - \tau_1$ and $\overline{x}_n - \tau_2$.So a widening of the range does not fit with our intuition. $\triangle$

The above implies that finding exact bounds would improve results significantly. For this, the concept of variation-diminishing transformations is used; see Brown et al. (1981). One set of assumptions on ϕ and $f_{P^T}(x|\theta)$ are monotonicity and strictly decreasing monotone likelihood ratio, respectively. If we relax the condition on ϕ to be monotonicity on two halves of the real line, then we require that $f_{P^T}(x|\theta)$ satisfies a stronger property of strictly variation reducing with order 3. Many distributions, including those in the exponential family satisfy these properties. As usual, we do require some integrability conditions. With these conditions it can be shown that the extrema are attained by members of four classes of functions or their limiting cases, and these classes are either unidimensional or bidimensional. Hence, even for exact bounds the computational complexity is independent of the sample size. For details, we refer to Bayarri and Berger (1998).

Example 5 (continued). Using the computational procedure above, it can be shown that for $\mathcal{W}_2$ the extremes are attained for w, an indicator function of an interval of the form (τ, ∞). For such a weight function, it can be seen that the posterior is an inverse gamma, and this gives a closed-form analytic expression for the posterior mean as $n(\overline{x}_n - \tau)/(n-1)$. Hence the bounds are as given below.

	$n = 5$	$n = 10$	$n = 50$
Bounds for $\mathcal{W}_2$	(1.0, 1.5)	(0.889, 1.333)	(0.816, 1.224)

Note that this agrees with our intuition on the asymptotics of the bounds. Also, note that the width of the bounds would not converge to zero, and this makes a robust Bayesian analysis all the more important. $\triangle$

7.5 Discussion

We have concentrated only on studies that find bounds of posterior quantities when the likelihood is allowed to vary over a class which represents uncertainty in the likelihood. Another approach is to find the extrema of the functional derivative of the posterior expectation, with respect to the likelihood at the *guess* model, in all directions represented by a class of likelihoods. This is the approach taken, for example, in Cuevas and Sanz (1988), Sivaganesan (1993) and Dey et al. (1996) and references therein. The main advantage of this approach is the computational ease that comes as a result of the derivative being a linear functional of the likelihood. For illustrative purposes, let us consider the simple situation when the base model is $\{f(x|\theta), \theta \in \Theta\}$, the prior on θ is π_0, the parametric function of interest is ρ and the uncertainty in the model is modeled by the ϵ-contamination class,

$$\{g(x|\cdot) : g(x|\theta) = (1-\epsilon)f(x|\theta) + \epsilon p(x|\theta), \forall\theta \in \Theta;\ p \in \mathcal{P}\}.$$

In the above case, local sensitivity studies would consider, for example, the supremum of the squared derivative with respect to ϵ at $\epsilon = 0$. Then for a random sample, $x_1, x_2, \ldots, x_n$, it can be seen easily that the above supremum is equal to

$$\sup_{p\in\mathcal{P}} \left(\frac{\int (\rho(\theta) - \rho_0(\mathbf{x})) \left(\prod_{i=1}^n f(x_i|\theta)\right) \sum_{j=1}^n \left(\frac{p(x_j|\theta)}{f(x_j|\theta)}\right) d\pi_0(\theta)}{m(\mathbf{x})} \right)^2,$$

where m and ρ_0 are the marginal and posterior expectation of ρ, calculated under base model, respectively. The linearity of the above integral with respect to p is what makes local sensitivity much more computationally tractable; this comes at the cost of ease in interpretation of results. The chapters on local robustness provide further discussion.

Earlier in Example 2, we saw that robustness was not achieved with respect to β, but when β was given a prior, robustness with respect to this prior was achieved. This, in fact, is a limitation of global robustness studies, see Berger (1994). The above phenomenon arises because global robustness does not look at the marginal of the data, which might support or rule out some of the likelihoods in the considered class. One approach to overcome this limitation is that taken in Example 2, that is, to use a prior. Another approach is that of Sivaganesan and Berger (1993), where the marginal is used to define the class of likelihoods, that is the set of likelihoods considered is taken to be $\{f \in \Gamma : m(x|f, \pi_0) > c\}$.

In prior robustness studies, when the class of priors satisfies certain conditions (see Shyamalkumar, 1996b), it can be shown that the effect of the prior vanishes asymptotically. As we saw earlier, this is not the case in the selection model problem when uncertainty was entertained in the weight function. In fact, let us suppose that our sample is arising from a likelihood h_{θ_0}. For a likelihood f, if we denote by Θ_f the set of θ which minimizes the Kullback–Leibler distance between h_{θ_0} and f_θ, then under some regularity conditions it can be shown that the posterior bounds would asymptotically be equal to bounds of φ on the set $\cup_{f \in \Gamma} \Theta_f$. The argument for the latter would be similar to Berk (1966). This shows that the study of likelihood robustness would still be important in large samples.

Many studies above entertained uncertainty in the likelihood but assumed a fixed prior. Sometimes the algorithm does not easily allow uncertainty in both the likelihood and prior and, in others, the complexity increases substantially. An interesting example is in the selection model where the first algorithm, though producing crude bounds, does entertain easily both the prior and likelihood but the more refined algorithm does not. An argument in support for considering uncertainty in only the likelihood could be that for large sample sizes the likelihood would have a more pronounced effect than the prior. It is not uncommon that in these cases the choice is a noninformative prior appropriate for the problem. An interesting question could be that, for results where for a fixed or a restricted class of priors one obtains invariance of the posterior with respect to the likelihood in a suitably chosen class, what could be said when the prior is allowed to vary in one of the standard classes of priors?

It should be mentioned that in many practical studies, even if a formal Bayesian analysis is not carried out, checking a couple of models and priors does often lead to valuable insight. Simple theoretical results, such as in DasGupta (1994), could also be derived for the problem in hand to further develop the understanding of the sensitivity to assumptions.

References

AZZALINI, A. (1985). A class of distributions which includes the normal ones. *Scandinavian Journal of Statistics*, **12**, 171-178.

BASU, S. (1999). Posterior sensitivity to the sampling distribution and the prior. *Annals of the Institute of Statistical Mathematics*(to appear).

BAYARRI, M.J. and BERGER, J. (1998). Robust Bayesian analysis of selection models. *Annals of Statistics*, **26**, 645–659.

BAYARRI, M.J. and DEGROOT, M. (1992). A "BAD" view of weighted distributions and selection models. In *Bayesian Statistics 4* (J.M. Bernardo, J.O. Berger, A.P. Dawid and A.F.M. Smith, eds.), 17–33. Oxford: Oxford University Press.

BERGER, J. (1985). *Statistical Decision Theory and Bayesian Analysis*, second edition. New York: Springer–Verlag.

BERGER, J. (1990). Robust Bayesian analysis: sensitivity to the prior, *Journal of Statistical Planning and Inference*, **25**, 303-328.

BERGER, J. (1994). An overview of robust Bayesian analysis. *Test*, **3**, 5–124.

BERGER, J.O. and SELLKE, T. (1987). Testing a point null hypothesis: the irreconcilability of significance levels and evidence. *Journal of the American Statistical Association*, **82**, 112–122.

BERK, R.H. (1966). Robust limiting behaviour of posterior distributions when the model is incorrect. *Annals of Mathematical Statistics*, **37**, 51–58.

BOSE, S. (1990). Bayesian robustness with shape-constrained priors and mixture of priors. *Ph.D. Dissertation*, Dept. of Statistics, Purdue University.

BOX, G.E.P. (1980). Sampling and Bayes inference in scientific modelling and robustness (with discussion). *Journal of the Royal Statistical Society Series A*, **143**, 383–430.

BOX, G.E.P. and TIAO, G.C. (1962). A further look at robustness via Bayes theorem. *Biometrika*, **49**, 419–432 and 546.

BOX, G.E.P. and TIAO, G.C. (1964). A note on criterion robustness and inference robustness. *Biometrika*, **51**, 169–173.

BROWN, L.D., JOHNSTONE, I.M. and MACGIBBON, K.B. (1981). Variation diminishing transformations: a direct approach to total positivity and its statistical applications. *Journal of the American Statistical Association*, **76**, 824–832.

CUEVAS, A. and SANZ, P. (1988). On differentiability properties of Bayes operators. In *Bayesian Statistics 3* (J.M. Bernardo, M.H. DeGroot, D.V. Lindley and A.F.M. Smith, eds.), 569–577. Oxford: Oxford University Press.

DasGupta, A. (1994). Discussion of "An overview of robust Bayesian analysis" by J.O. Berger. *Test*, **3**, 83–89.

DeRobertis, L. (1978). The use of partial prior knowledge in Bayesian inference. *Ph.D. Dissertation*, Dept. of Statistics, Yale University.

DeRobertis, L. and Hartigan, J. (1981). Bayesian inference using intervals of measures. *Annals of Statistics*, **9**, 235–244.

Dey, D.K., Ghosh, S.K. and Lou, K.R. (1996). On local sensitivity measures in Bayesian analysis. *Bayesian Robustness*, IMS Lecture Notes - Monograph Series, **29**, 21–39.

Diaconis, P. and Freedman, D. (1980). Finite exchangeable sequences. *Annals of Probability*, **8**, 745–764.

Fernández, C., Osiewalski, J. and Steel, M.F.J. (1997). Classical and Bayesian inference robustness in multivariate regression models. *Journal of the American Statistical Association*, **92**, 1434–1444.

Lavine, M. (1991). Sensitivity in Bayesian statistics: the prior and the likelihood. *Journal of the American Statistical Association*, **86**, 396–399.

Lavine, M. (1994). An approach to evaluating sensitivity in Bayesian regression analysis. *Journal of Statistical Planning and Inference*, **40**, 233–244.

Lavine, M., Wasserman, L. and Wolpert, R. (1993). Linearization of Bayesian robustness problems. *Journal of Statistical Planning and Inference*, **37**, 307–316.

Mortera, J. (1994). Discussion on "Posterior robustness with more than one sampling model" by L.R. Pericchi and M.E. Perez. *Journal of Statistical Planning and Inference*, **40**, 291–293.

O'Hagan, A. (1979). On outlier rejection phenomena in Bayes inference. *Journal of the Royal Statistical Society Series B*, **41**, 358–367.

Osiewalski, J. and Steel, M.J. (1993a). Robust Bayesian inference in l_q models. *Biometrika*, **80**, 456–460.

Osiewalski, J. and Steel, M.J. (1993b). Robust Bayesian inference in elliptical regression models. *Journal of Econometrics*, **57**, 345–363.

Pericchi, L.R. and Perez, M.E. (1994). Posterior robustness with more than one sampling model. *Journal of Statistical Planning and Inference*, **40**, 279–291.

RACINE, A., GRIEVE, A.P., FLUHLER, H. and SMITH, A.F.M. (1986). Bayesian methods in practice: experiences in the pharmaceutical industry. *Applied Statistics*, **35**, 93–150.

RAO, C.R. (1985). Weighted distributions arising out of methods of ascertainment: what population does a sample represent? In *A Celebration of Statistics: The ISI Centenary Volume* (A.G. Atkinson and S. Fienberg, eds.). New York: Springer-Verlag.

SHYAMALKUMAR, N.D. (1996a). Bayesian robustness with respect to the likelihood. *Technical Report*, **96–23**, Department of Statistics, Purdue University.

SHYAMALKUMAR, N.D. (1996b). Bayesian robustness with asymptotically nice classes of priors. *Technical Report*, **96–22**, Department of Statistics, Purdue University.

SIVAGANESAN, S. (1993). Robust Bayesian diagnostics. *Journal of Statistical Planning and Inference*, **35**, 171–188.

SIVAGANESAN, S. and BERGER, J.O. (1993). Robust Bayesian analysis of the binomial empirical Bayes problem. *Canadian Journal of Statistics*, **21**, 107–119.

8

Ranges of Posterior Expected Losses and ϵ-Robust Actions

Dipak K. Dey and Athanasios C. Micheas

ABSTRACT Robustness of the loss functions is considered in a Bayesian framework. Variations of the posterior expected loss are studied when the loss functions belong to a certain class. Ranges of the posterior expected loss, influence function and minimax regret principles are adopted to capture the robustness and optimum choice of the loss functions. Applications include the continuous exponential family under the class of Linex loss functions, weighted squared error and Hellinger distance loss. Methods are applied to standard examples, including normal distributions. We also illustrate the use of power divergence in this framework with an application to the normal case.

Key words: Bayesian robustness, continuous exponential family, Hellinger distance loss, influence approach, Linex loss function, minimax regret action, posterior expected loss.

8.1 Introduction

The problem of robustness has always been an important element of the foundation of statistics. However, it has only been in recent decades that attempts have been made to formalize the problem beyond ad-hoc measures towards a theory of robustness. Huber (1964) and Hampel (1971) concentrated on some approaches which try to capture the essential features of real life robustness problems within a mathematical framework. Robustness studies also have recently received considerable interest among Bayesians. A detailed review of the subject and its literature from a Bayesian viewpoint can be found in Berger (1984, 1985 and 1995). Traditionally, Bayesian robustness is concerned with sensitivity analysis regarding choice of priors. This can be achieved through marginal distributions, through posterior expected loss and through Bayes risk. In recent years, there has been an explosion of literature relating to the variation of posterior features when the prior belongs to a certain class. A review of this subject can be found in Berger (1984, 1994).

Most of the research efforts in the study of Bayesian robustness avoid the choice of loss function. In a decision-theoretic framework, specification of a loss function can be even more severe than that of specifying a prior. Of course, reporting the entire posterior distribution avoids this problem, since a loss function can be developed later and can be combined with any posterior distribution. It is important to note that a decision problem will generally be sensitive to the specification of the loss function for small or moderate errors. For example, if a weighted squared error loss such as $w(\theta)(\theta - a)^2$ is used, an optimum decision rule, say Bayes action, is highly sensitive to $w(\theta)$ in the same way that it is sensitive to the prior.

There are only a few papers relating to loss robustness. The delay is probably due to the fact that robustness problems in the prior distribution are likely more important than loss robustness problems. Furthermore, when performing a statistical analysis, there might time pressure or cost constraints that do not allow us to consider repeating the analysis under a different loss. A well-performed analysis, though, requires the examination of different loss functions in order to determine the loss that will allow the experimenter to make the optimum decision.

The most important papers in the literature on loss robustness include Brown (1975), Kadane and Chuang (1978), Ramsay and Novick (1980), Makov (1994), Dey et al. (1998) and Martin et al. (1998). In the paper of Dey et al. (1998) the problem is formulated towards a theory of loss robustness with formal definitions of measures. They explore robustness properties of Linex (linear-exponential) loss under exponential and discrete power-series families of distributions using the conjugate priors. On the contrary, Martin et al. (1998) described the scenario in which imprecision in preferences is modeled with a class of loss functions. Rather than analyzing the local or global behavior of a posterior or predictive expectation, they have considered foundations of robust Bayesian analysis and suggested that a main computational objective would be the obtainment of the set of nonrandomized alternatives.

In this chapter, we review the literature by considering various measures of loss robustness via posterior expected loss (PEL). The format of the chapter is as follows: Section 2 is devoted to the development of the measures that help us assess the loss robustness. In Section 3, we consider the famous Linex loss function as introduced in Klebanov (1972) and studied extensively by Klebanov (1974, 1976) and Varian (1975). We then study the loss robustness for continuous exponential family and discrete power series distributions. The use of power divergence is investigated in the this context, for the simple case of the normal distribution. We also discuss loss robustness for some well-known classes in the literature, including weighted squared error loss. Section 4 is devoted to a few concluding remarks.

8.2 Measures of posterior loss robustness

Let $\mathcal{A}$ denote the action space, Θ denote the parameter space and $\mathcal{L}$ denote a class of loss functions. The loss in taking action $a \in \mathcal{A}$ when $\theta \in \Theta$ will be denoted by $L(\theta, a) \in \mathcal{L}$. In the following, we let π be the prior measure. Also, we will assume the likelihood function $f(x|\theta)$ is held fixed. Furthermore, we will, unless otherwise stated, assume that $x \in \mathcal{X}$, the sample space, is held fixed and hence suppress the dependence on x of various quantities.

For $L \in \mathcal{L}$, define the posterior expected loss (PEL) of an action a by

$$P_L(a) = E^{\pi(\theta|x)}[L(\theta, a)] = \int_{\Theta} L(\theta, a)\pi(\theta|x)d\theta,$$

where $\pi(\theta|x)$ is the posterior distribution of θ given x and is given by $(\pi(\theta)f(x|\theta))/m(x)$, where $m(x) = \int_{\Theta} f(x|\theta)\pi(\theta)d\theta$ is the marginal distribution of x.

Several approaches for measuring loss robustness based on PEL, have appeared in the literature. For assessment of the sensitivity of the class $\mathcal{L}$, we define the range of the posterior expected loss as

$$R(x) = \sup_{L \in \mathcal{L}} P_L(a) - \inf_{L \in \mathcal{L}} P_L(a). \tag{1}$$

The class $\mathcal{L}$ will be robust if for given data x, $R(x)$ is sufficiently small. A plot of $R(x)$ against x can indicate the degree of robustness for different values of x.

A natural question after obtaining robustness is an appropriate choice of the loss function within the class $\mathcal{L}$. Following Dey et al. (1998), the first approach in selecting a loss will be called the minimax regret approach, which is defined as follows. Define the posterior regret of an action $a \in \mathcal{A}$ by

$$\mathcal{G}_L(a) = P_L(a) - \inf_{a' \in \mathcal{A}} P_L(a') = P_L(a) - P_L(a_b), \tag{2}$$

where a_b is the Bayes action under the loss $L \in \mathcal{L}$. Our objective is to determine the minimax regret action a_0 defined by the condition

$$\sup_{L \in \mathcal{L}} \mathcal{G}_L(a_0) = \inf_{a \in \mathcal{A}} \sup_{L \in \mathcal{L}} \mathcal{G}_L(a). \tag{3}$$

The loss that we select is defined by

$$L_o = \arg \sup_{L \in \mathcal{L}} \mathcal{G}_L(a_0).$$

The second approach is based on the influence of the posterior expected loss. Following Ramsay and Novick (1980), we define the influence of a loss function as

$$\frac{\partial}{\partial x}\inf_a P_L(a) = \frac{\partial}{\partial x} P_L(a_b), \tag{4}$$

where a_b is the Bayes action. In view of this we can consider the following ordering in $\mathcal{L}$.

Makov (1994) suggests two ways for ordering loss functions in a class, using influence of the PEL. First, we say L_1 is preferred to L_2 if

$$\sup_x |\inf_a P_{L_1}(a)| < \sup_x |\inf_a P_{L_2}(a)|.$$

In addition, L_1 is preferred to L_2 if

$$E^x[P_{L_1}(a)] < E^x[P_{L_2}(a)],$$

where the expectation is taken with respect to $m(x)$, the marginal distribution of x.

Using the influence approach, Dey et al. (1998) suggest measuring the posterior loss robustness by the quantity

$$D = \sup_x |\frac{\partial}{\partial x} P_L(a_b)| - \inf_x |\frac{\partial}{\partial x} P_L(a_b)|. \tag{5}$$

Again, large values of D indicate lack of robustness. Finally, in terms of an optimum choice of a loss function, we say $L_o \in \mathcal{L}$ is posterior loss robust if

$$\sup_x |\frac{\partial}{\partial x} P_{L_o}(a_b)| = \inf_{L \in \mathcal{L}} \sup_x |\frac{\partial}{\partial x} P_L(a_b)|.$$

Martin et al. (1998) argue that the best choice of a loss function in a class $\mathcal{L}$ is the loss L_o that minimizes the PEL, that is,

$$P_{L_o}(a) = \inf_{L \in \mathcal{L}} P_L(a).$$

The above measures give a guideline for the optimum choice of a loss within a class of loss functions.

An alternative approach towards the construction of measures that assess loss robustness would be to incorporate maximum a posteriori estimates (MAP) instead of computing measures based on PEL. This approach will simplify calculations significantly and provide a basis for the construction of a large variety of loss robustness measures. This approach is pursued in Micheas and Dey (1999).

8.3 Loss robustness under different classes of loss functions

There is a vast literature about different loss functions and their applications. In the context of loss robustness, one is faced with the task of computing measures as those introduced in Section 2, over a class of loss functions. Among the most important such classes are Linex, Hellinger and weighted squared error loss functions. In what follows, we consider measuring loss robustness for these classes of loss functions.

8.3.1 *Linex loss*

The Linex loss function is an asymmetric loss first introduced by Klebanov (1972) and used by Varian (1975) in the context of real estate assessment. It is defined as

$$L_b(\theta, a) = \exp\{b(a-\theta)\} - b(a-\theta) - 1, \tag{6}$$

where $b \neq 0$. The constant b determines the shape of the loss function. Klebanov (1976) characterizes the Linex loss functions from the Rao–Blackwell condition. Zellner (1986) discusses this loss in connection with normal estimation problems. Both Zellner (1986) and Varian (1975) discuss the behavior of the loss functions and their various applications. When $b > 0$, the loss (6) increases almost linearly for negative error and almost exponentially for positive error. Therefore, overestimation is a more serious mistake than underestimation. When $b < 0$, the linear exponential increases are interchanged, where underestimation is more serious than overestimation. For small values of b, the loss function is close to the squared error loss. For application of Linex loss, see Zellner (1986) and Kuo and Dey (1990).

8.3.1.1 Loss robustness for continuous exponential family

Consider a random variable X having density from exponential family of the form

$$f(x|\theta) = \exp\{x\theta - \beta(\theta) - M(x)\}, -\infty < x < \infty$$

where $\beta(\theta) = ln \int_{-\infty}^{\infty} \exp\{x\theta - M(x)\}dx < \infty$. The natural conjugate prior, following Diaconis and Ylvisaker (1979), is given as

$$\pi(\theta|n_0, x_0) = \exp\{n_0 x_0 \theta - n_0 \beta(\theta) - k(n_0, x_0),\}$$

where x_0 and n_0 are hyperparameters chosen such that

$$k(n_0, x_0) = \ln \int \exp\{n_0 x_0 \theta - n_0 \beta(\theta)\} d\theta < \infty.$$

It follows immediately that the posterior pdf is given by

$$\begin{aligned} \pi(\theta|x, n_0, x_0) \quad = \quad & \exp\{(x + n_0 x_0)\theta - (n_0+1)\beta(\theta) \\ & - k(n_0+1, (x+n_0 x_0)/(n_0+1))\}. \end{aligned}$$

Now in order to calculate the posterior expected loss we need to calculate the posterior Laplace transform. Direct calculation shows that the posterior Laplace transform is

$$E^{\theta|x}[\exp(-b\theta)] = \exp\{k(n_0+1, \frac{x-b+n_0x_0}{n_0+1}) - k(n_0+1, \frac{x+n_0x_0}{n_0+1})\}, \tag{7}$$

and therefore the posterior mean is

$$E^{\theta|x}[\theta] = k'(n_0+1, \frac{x+n_0x_0}{n_0+1}), \tag{8}$$

where k' is the derivative of k with respect to x.

From (6), it follows that the posterior expected loss for the action a is

$$P_{L_b}(a) = \exp(ba)E^{\theta|x}\{\exp(-b\theta)\} - b\{a - E^{\theta|x}(\theta)\} - 1. \tag{9}$$

Thus, substituting (7) and (8) it follows from (9) that

$$\begin{aligned} P_{L_b}(a) &= \exp\{ba + k(n_0+1, \tfrac{x-b+n_0x_0}{n_0+1}) - k(n_0+1, \tfrac{x+n_0x_0}{n_0+1})\} \\ &- \; b[a - k'(n_0+1, \tfrac{x+n_0x_0}{n_0+1})] - 1. \end{aligned} \tag{10}$$

Now in order to calculate the range of the posterior expected loss $P_{L_b}(a)$, we need to know the behavior of P_{L_b} when $b_0 < b < b_1$. It follows after lengthy algebraic manipulation that $\partial P_{L_b}/\partial b|_{b=0} = 0$ and $\partial^2 P_{L_b}/\partial b^2 \geq 0$, $\forall b$. Thus, $P_{L_b}(a)$ is a convex function in b having minimum at $b \to 0$ (since $b \neq 0$). Finally, it follows from (1) that the range

$$R(x) = \begin{cases} P_{L_{b_1}} - P_{L_{b_0}} & \text{if } 0 < b_0 < b_1, \\ P_{L_{b_0}} - P_{L_{b_1}} & \text{if } b_0 < b_1 < 0. \end{cases} \tag{11}$$

Let us now obtain the minimax regret action as defined in (3). First we need to find the Bayes rule and the posterior regret of an action. Now differentiating $P_{L_b}(a)$ in (10) with respect to a and equating to 0, we get the Bayes rule as

$$a_b = \frac{1}{b}\{k(n_0+1, \frac{x+n_0x_0}{n_0+1}) - k(n_0+1, \frac{x-b+n_0x_0}{n_0+1})\}. \tag{12}$$

Now it can be shown (Dey et al., 1998, Lemmas 3.3 and 3.4) that a_b is a decreasing function of b and for $b > 0$, $E_b(\theta) < a_b$. Thus, the posterior expected loss evaluated at the Bayes rule is

$$\inf_{a' \in \mathcal{A}} P_L(a') = P_{L_b}(a_b) = b[k'(n_0+1, \frac{x+n_0x_0}{n_0+1}) - a_b],$$

and the posterior regret $\mathcal{G}_b(a)$ (we put subscript b because $L \in \mathcal{L}_b$) as defined in (2) is

$$\mathcal{G}_b(a) = \exp\{b(a-a_b)\} - b(a-a_b) - 1,$$

where a_b is the Bayes action under loss L_b. One can write

$$\mathcal{G}_b(a) = \exp(f_a(b)) - f_a(b) - 1,$$

where $f_a(b) = b(a - a_b)$. Let us define the function $g(y) = e^y - y - 1$. Note that $\mathcal{G}_b(a) = g(f_a(b))$. Let $S = S_a = \{f_a(b) : b_0 \leq b \leq b_1\}$, $M_1 = \inf\{y \in S : y < 0\}$ and $M_2 = \sup\{y \in S : y > 0\}$. Following Dey et al. (1998), the minimax regret action is a_{b_m}, where b_m satisfies the equation $g(M_1(b_m)) = g(M_2(b_m))$ and a_b is given in (12).

Now in order to find the minimax regret action a_{b_m}, one may find b_m by using any numerical method for finding the root of the function $h(b') = g(M_1(b')) - g(M_2(b'))$. Then the root of $h(b') = 0$ is $b' = b_m$. For more details on the various cases involved, we refer the reader to Dey et al. (1998).

To obtain the influence, we follow the definition given in (4). After some algebraic manipulation it can be shown that

$$\begin{aligned} \frac{\partial}{\partial x} P_{L_b}(a_b) &= k'(n_0+1, \frac{x - b + n_0x_0}{n_0+1}) + bk''(n_0+1, \frac{x+n_0x_0}{n_0+1}) \\ &\quad - k'(n_0+1, \frac{x+n_0x_0}{n_0+1}). \end{aligned} \tag{13}$$

Exact evaluation of the supremum and infimum of (13) involves likelihood specification.

Next, we consider an example from Dey et al. (1998), to demonstrate the loss robustness studies.

Example 1. Let x be an observation from a normal distribution with mean θ and variance 1. Then it follows from the basic representation of the exponential family density that $\beta(\theta) = \theta^2/2$ and $M(x) = x^2/2 + 1/2\log(2\pi)$. The corresponding conjugate prior is

$$\pi_{n_0,x_0}(\theta) = \exp\{n_0x_0\theta - n_0\theta^2/2 - k(n_0, x_0)\},\ n_0 > 0$$

where

$$k(n_0, x_0) = n_0x_0^2/2 + \frac{1}{2}\log(2\pi/n_0).$$

Therefore, the posterior density is given as

$$\pi_{n_0,x_0}(\theta|x) = \exp\{(x+n_0x_0)\theta - (n_0+1)\beta(\theta) - k(n_0+1, \frac{x+n_0x_0}{n_0+1})\},$$

where $n_0 > 0$ and

$$k(n_0+1, \frac{x+n_0x_0}{n_0+1}) = \frac{1}{2}\frac{(x+n_0x_0)^2}{(n_0+1)} + \frac{1}{2}\log(2\pi/(n_0+1)).$$

It follows that

$$k'(n_0+1, \frac{x+n_0x_0}{n_0+1}) = (x+n_0x_0)/(n_0+1),$$

and thus the posterior expected loss is

$$\begin{aligned} P_{L_b}(a) &= \exp[\frac{b}{n_0+1}\{b/2-(x+n_0x_0)+(n_0+1)a\}] \\ &-b\{a-\frac{x+n_0x_0}{n_0+1}\}-1. \end{aligned}$$

From (11), it follows that the range is given as

$$\begin{aligned} R(x) \quad &= \exp[\tfrac{b_1}{n_0+1}\{\tfrac{b_1}{2}-(x+n_0x_0)+(n_0+1)a\}] \\ &- \exp[\tfrac{b_0}{n_0+1}\{\tfrac{b_0}{2}-(x+n_0x_0)+(n_0+1)a\}] \\ &- (b_1-b_0)(a-\tfrac{x+n_0x_0}{n_0+1}), \text{ if } 0<b_0<b_1. \end{aligned}$$

From (12), the Bayes rule is given as

$$a_b = \frac{x+n_0x_0-\frac{b}{2}}{n_0+1}.$$

To obtain the posterior regret, first it can be shown that

$$\inf_{a'\in\mathcal{A}} P_L(a') = P_{L_b}(a_b) = \frac{b^2}{2(n_0+1)}$$

and thus the posterior regret is given as

$$\begin{aligned} \mathcal{G}_b(a) &= \exp[\frac{b}{n_0+1}\{\frac{b}{2}-(x+n_0x_0)+(n_0+1)a\}] \\ &-b\{a-\frac{x+n_0x_0}{n_0+1}\}-1-\frac{b^2}{2(n_0+1)}. \end{aligned}$$

Next we find the minimax regret action for the following sets of b_0, b_1 values. In each case, let $n_0=1$.

i) Suppose $b_0=1$ and $b_1=2$. Then $b_m=1.679$, and the minimax regret is 0.0136.

ii) Suppose $b_0=1$ and $b_1=2.5$. Then $b_m=2.097$, and the minimax regret is 0.0345.

iii) Suppose $b_0=1$ and $b_1=3$. This time $b_m=2.529$ and the minimax regret is 0.0703.

Finally, the influence for this problem is

$$\frac{\partial}{\partial x}P_{L_b}(a_b) = \frac{x-b+n_0x_0}{n_0+1} + \frac{b}{n_0+1} - \frac{x+n_0x_0}{n_0+1} = 0.$$

Hence, all $L\in\mathcal{L}$ are equivalent, D as defined in (5) is zero and from the influence point of view, any $L\in\mathcal{L}$ is posterior loss robust. $\triangle$

8.3.2 *Hellinger-power divergence loss*

Cressie and Read (1984) defined using the convex function $v(u) = 2(u^r - u)/(r(r-1))$, $0 < r < 1$, the power-divergence family of loss functions as

$$L_\varphi(\theta, a) = \int_X \varphi\left(\frac{f(x|a)}{f(x|\theta)}\right) f(x|\theta)dx,$$

the loss in choosing $a \in A$. They used this quantity for the unified presentation, exploration and comparison of the maximum likelihood and Pearson-goodness-of-fit tests.

We will consider a simple modification of this convex function

$$\varphi(u) = \frac{u^r}{r(r-1)}, \quad 0 < r < 1.$$

8.3.2.1 Loss robustness for continuous exponential family

For the continuous exponential family as defined in Section 8.3.1.1, the loss in choosing $a \in A$ is reduced to

$$L_\varphi(\theta, a) = \int_X \varphi(\exp\{x(a-\theta) - [\beta(a) - \beta(\theta)]\}) \exp\{x\theta - \beta(\theta) - M(x)\}dx.$$

The loss function induced by the modified Cressie and Read power-divergence will be of the form

$$\begin{aligned} L_\varphi(\theta, a) &= \frac{1}{r(1-r)} \int_X \exp\{x[r(a-\theta) + \theta] - (r[\beta(a) - \beta(\theta)] + \beta(\theta)) \\ &\quad - M(x)\}dx. \end{aligned}$$

Let

$$\gamma(\theta, a) = \log\left(\int_X \exp\{x[r(a-\theta) + \theta] - M(x)\}dx\right).$$

Then it follows that

$$L_\varphi(\theta, a) = \frac{1}{r(1-r)} \exp\{-r\beta(a) - (1-r)\beta(\theta) + \gamma(\theta, a)\}.$$

Clearly the loss involves evaluation of an integral through $\gamma(\theta, a)$, which cannot be expressed in general, in a closed form. Thus we consider the normal example, where the loss function has a closed analytical expression.

Example 2. We turn now to the case of the normal distribution. We have $\beta(\theta) = \theta^2/2$, $M(x) = x^2/2 + 1/2\log(2\pi)$ and $k = k(n_o+1, (x+n_o x_o)/(n_o+1)) = (x+n_o x_o)^2)/(2(n_o+1)) + 1/2\ \log((2\pi)/(n_o+1))$. Hence the posterior distribution of $\theta|x$ is $N((x+n_o x_o)/(n_o+1), 1/(n_o+1))$.

Now we can easily see that $\exp\{\gamma(\theta, a)\} = \exp\{1/2[r(a-\theta)+\theta]^2\}$. The loss in taking action $a \in A$ after some algebra will be reduced to

$$L_\varphi(\theta, a) = \frac{1}{r(1-r)} e^{-\frac{1}{2}r(1-r)\{\theta-a)^2}.$$

The posterior expected loss under φ−divergence is defined as

$$\begin{aligned} P^{\varphi}_{L_r}(a) &= E^{\theta|x}(L_\varphi(\theta, a)) = E^{\theta|x}(\frac{1}{r(1-r)}\exp\{-\frac{1}{2}r(1-r)(\theta-a)^2\}) \\ &= \frac{1}{r(1-r)} E^{\theta|x}(\exp\{-\frac{1}{2}r(1-r)(\theta-a)^2\}), \end{aligned}$$

and since $\theta|x \backsim N((x+n_o x_o)/(n_o+1), 1/(n_o+1))$ we have after some algebraic manipulations

$$P^{\varphi}_{L_r}(a) = \frac{\sqrt{n_o+1}}{r(1-r)\sqrt{r(1-r)+n_o+1}} \exp\{-\frac{1}{2}\{r(1-r)a^2 + \frac{(x+n_o x_o)^2}{n_o+1} - \frac{[ar(1-r)+x+n_o x_o]^2}{r(1-r)+n_o+1}\}\}. \tag{14}$$

We first retrieve the Bayes rule. By differentiating $P^{\varphi}_{L_r}(a)$ w.r.t. a,

$$\begin{aligned} \frac{\partial P^{\varphi}_{L_r}(a)}{\partial a} &= \frac{-\sqrt{n_o+1}[2r(1-r)a - \frac{2(r(1-r)a+x+n_o x_o)r(1-r)}{r(1-r)+n_o+1}]}{2r(1-r)\sqrt{r(1-r)+n_o+1}} \\ &\quad \exp\{-\frac{1}{2}\{r(1-r)a^2 + \frac{(x+n_o x_o)^2}{n_o+1} \\ &\quad - \frac{[ar(1-r)+x+n_o x_o]^2}{r(1-r)+n_o+1}\}, \end{aligned}$$

and hence the Bayes rule is given by

$$a_b = \frac{x+n_o x_o}{n_o+1}.$$

Recall that the Bayes rule under Linex loss was given by

$$a_b = \frac{x+n_o x_o - \frac{b}{2}}{n_o+1}.$$

In order to measure robustness we calculate the range of the posterior expected loss. Let

$$g(r,a,x,n_o,x_o) = \exp\{-\frac{1}{2}[r(1-r)a^2 + \frac{(x+n_ox_o)^2}{n_o+1} - \frac{[ar(1-r)+x+n_ox_o]^2}{r(1-r)+n_o+1}]\}$$

and

$$h(r,a,x,n_o,x_o) = \frac{[ar(1-r)+x+n_ox_o]^2}{r(1-r)+n_o+1}.$$

We have from (14) after some algebra that

$$\frac{\partial P^{\varphi}_{L_r}(a)}{\partial r} = \frac{\sqrt{n_o+1}g(r,a,x,n_o,x_o)}{2r(1-r)\sqrt{r(1-r)+n_o+1}}[-a^2(1-2r) + \frac{\partial h(r,a,x,n_o,x_o)}{\partial r} - 2\frac{1-2r}{r(1-r)} - \frac{1-2r}{r(1-r)+n_o+1}].$$

Thus $\partial P^{\varphi}_{L_r}(a)/\partial r = 0$ if and only if

$$\frac{\partial h(r,a,x,n_o,x_o)}{\partial r} - 2\frac{1-2r}{r(1-r)} - \frac{1-2r}{r(1-r)+n_o+1} - a^2(1-2r) = 0.$$

After lengthy algebraic manipulations, we have that $\partial P^{\varphi}_{L_r}(a)/\partial r = 0$ if and only if $r(1-r) = t$, where t satisfies

$$3t^2 - t[(x+n_ox_o)(2a(n_o+1) - x - n_ox_o) - (n_o+1)(5+a^2(n_o+1))] + 2(n_o+1)^2 = 0. \tag{15}$$

Now let

$$\Delta = [(x+n_ox_o)(2a(n_o+1) - x - n_ox_o) - (n_o+1)(5+a^2(n_o+1))]^2 - 24(n_o+1)^2.$$

Then for appropriate selection of the hyperparameters, $\Delta > 0$ and the roots of (15) are given by

$$t_1 = \frac{(x+n_ox_o)(2a(n_o+1) - x - n_ox_o) - (n_o+1)(5+a^2(n_o+1)) - \sqrt{\Delta}}{6}$$

and

$$t_2 = \frac{(x+n_ox_o)(2a(n_o+1) - x - n_ox_o) - (n_o+1)(5+a^2(n_o+1)) + \sqrt{\Delta}}{6}.$$

Hence $\partial P^{\varphi}_{L_r}(a)/\partial r = 0$ if and only if

$$r = r_t^{(1)} = \frac{1}{2} + \frac{\sqrt{1-4t}}{2} \quad \text{or}$$
$$r = r_t^{(2)} = \frac{1}{2} - \frac{\sqrt{1-4t}}{2}$$

for all t such that $t \leq 1/4$.

Let $r_{t_1}^{(1)}, r_{t_2}^{(1)}, r_{t_1}^{(2)}, r_{t_2}^{(2)}$ denote all the possible roots. Assuming all four cases are valid for some n_o, x_o, we can easily find the range for the posterior expected loss in taking action $a \in \mathcal{A}$. We have

$$\begin{aligned} &\inf_{L_r \in \mathcal{L}} P_{L_r}^{\varphi}(a) = \inf_{r \in (0,1)} P_{L_r}^{\varphi}(a) \\ &= \min\{P_{L_{r_{t_1}^{(1)}}}^{\varphi}(a), P_{L_{r_{t_2}^{(1)}}}^{\varphi}(a), P_{L_{r_{t_1}^{(2)}}}^{\varphi}(a), P_{L_{r_{t_2}^{(2)}}}^{\varphi}(a)\} \end{aligned} \tag{16}$$

and

$$\begin{aligned} &\sup_{L_r \in \mathcal{L}} P_{L_r}^{\varphi}(a) = \sup_{r \in (0,1)} P_{L_r}^{\varphi}(a) \\ &= \max\{P_{L_{r_{t_1}^{(1)}}}^{\varphi}(a), P_{L_{r_{t_2}^{(1)}}}^{\varphi}(a), P_{L_{r_{t_1}^{(2)}}}^{\varphi}(a), P_{L_{r_{t_2}^{(2)}}}^{\varphi}(a)\}. \end{aligned} \tag{17}$$

Hence from (1), (16) and (17), the PEL range is given by

$$R(x) = \sup_{L_r \in \mathcal{L}} P_{L_r}^{\varphi}(a) - \inf_{L_r \in \mathcal{L}} P_{L_r}^{\varphi}(a). \ \triangle$$

8.3.3 Other classes of loss functions

We now discuss different classes of loss functions that can be used when trying to assess loss robustness with different measures. For what follows, we assume the setup of Section 8.3.1.1. The first class we consider is the weighted squared error loss defined by Whittle and Lane (1967), where the weights are given by $I^F(\theta)$, the Fisher information about θ. Their study concentrated on the case where a sequential estimation procedure is nonsequential. The loss is defined as

$$L_F(\theta, a) = I^F(\theta)(\theta - a)^2.$$

Shapiro and Wardrop (1980) extended this loss function to a larger family of loss functions, when studying Bayesian sequential estimation. Their loss is given by

$$L_F^{r,s}(\theta, a) = \exp\{\theta r - \beta(\theta)s\} I^F(\theta)(\theta - a)^2,$$

where r and s are parameters that allow a greater variety of shapes for the loss function.

Makov (1994) studied these two classes of weighted squared error loss functions, namely,

$$L_w(\theta, a) = w(\theta)(\theta - a)^2$$

and

$$L_w^{r,s}(\theta, a) = \exp\{\theta r - \beta(\theta)s\}w(\theta)(\theta - a)^2$$

in the context of loss robustness. He showed that, from an influence standpoint, $L_F(\theta, a)$ as defined above, is posterior loss robust. He also considered the extended class of loss functions $L_w^{r,s}(\theta, a)$ and found under which conditions for r and s, $L_F^{r,s}(\theta, a)$ will be approximately posterior loss robust, again from an influence viewpoint.

8.4 Conclusions

In this chapter we have considered loss robustness problems. We investigated the measures that help us in assessing loss robustness for a class of loss functions. The most important such classes were considered, including Linex, Hellinger and squared error loss functions. Similar studies can be performed on other classes of loss functions. In most cases, it is assumed that the loss function depends upon the difference $(a-\theta)$. The loss functions we have seen so far are extensions of this difference. Martin et al. (1998) argue that these extensions should be carefully determined when one uses the range of the PEL to assess robustness. They consider the following: suppose, for example, that the class of loss functions includes functions of the form $L(\theta, a) + k$, where $k \in [k_o, k_1]$; for example, the universal class studied by Hwang (1985) includes such functions. Notice that all these loss functions are strategically equivalent, therefore leading to the same optimal decision, hence having a robust problem. Should we insist in computing the range of the PEL, we would find it to be greater than $k_1 - k_0$, which if large, may suggest that we lack robustness. The same problem arises in the case where the class has loss functions of the form $L(\theta, a) + k$, with $k \in [k_o, k_1]$.

In order to solve this problem, Martin et al. (1998) turn to the foundations of robust Bayesian analysis. Foundational results suggest that preferences among loss functions in a class $\mathcal{L}$ will follow a Pareto order with respect to the class of expected losses. Therefore, they consider the following rule when selecting the appropriate loss function: $L_1 \in \mathcal{L}$ is at most as preferred as $L_2 \in \mathcal{L}$, if and only if $P_{L_1}(a) \geq P_{L_2}(a)$. If strict inequality holds, they say that L_2 dominates L_1. Obviously, we are interested in the set of nondominated loss functions.

This approach is closely related to the classical concept of admissibility. In this case we use posterior expected losses, for various loss functions, whereas in the case of admissibility, we use the risk function for various priors.

Martin et al. (1998) continue with an extensive study of the nondominated sets of actions. They find conditions for the existence of the nondominated loss function and investigate its relationship with the Bayes and $\mathcal{L}$-minimax loss functions, also providing ways of computing the nondominated set of loss functions.

The methods introduced in this chapter were applied to the exponential family of distributions with conjugate prior, which includes a large variety of models for the likelihood function. The methods can easily be extended to the power series family of distributions. Finally, several other measures can be introduced to measure the loss robustness, which will be pursued elsewhere.

References

BERGER, J.O. (1984). The robust Bayesian viewpoint (with discussion). In *Robustness of Bayesian Analysis* (J. Kadane, ed.). Amsterdam: North Holland.

BERGER, J.O. (1985). *Statistical Decision Theory and Bayesian Analysis.* NewYork: Springer-Verlag.

BERGER, J.O. (1994). An overview of robust Bayesian analysis. *Test*, **3**, 5–59.

BROWN, L.D. (1975). Estimation with incompletely specified loss functions. *Journal of the American Statistical Association*, **70**, 417–427.

CRESSIE, N. and READ, T.R.C. (1984). Multinomial goodness-of-fit tests. *Journal of the Royal Statistical Society Series B*, **46**, 440–464.

DEY, D., LOU, K. and BOSE, S. (1998). A Bayesian approach to loss robustness. *Statistics and Decisions*, **16**, 65–87.

DIACONIS, P. and YLVISAKER, D. (1979). Conjugate priors for exponential families. *Annals of Statistics*, **7**, 269–281.

HAMPEL, F.R. (1971). A general qualitative definition of robustness. *Annals of Mathematical Statistics,* **42**, 1887–1896.

HUBER, P.J. (1964). Robust estimation of a location parameter. *Annals of Mathematical Statistics,* **35**, 73–101.

HWANG, J.T. (1985). Universal domination and stochastic domination: estimation simultaneously under a broad class of loss functions. *Annals of Statistics,* **13**, 295–314.

KADANE, J.B. and CHUANG, D.T. (1978). Stable decision problems. *Annals of Statistics,* **6**, 1095–1110.

KLEBANOV, L.B. (1972). Universal loss function and unbiased estimation. *Dokl. Akad. Nauk SSSR Soviet Math. Dokl. t. 203*, **N6**, 1249–1251.

KLEBANOV, L.B. (1974). Unbiased estimators and sufficient statistics. *Probability Theory and Applications*, **19**, 2, 392–397.

KLEBANOV, L.B. (1976). A general definition of unbiasedness. *Probability Theory and Applications*, **21**, 571–585.

KUO, L. and DEY, D. (1990). On the admissibility of the linear estimators of the Poisson mean using LINEX loss functions. *Statistics and Decisions,* **8**, 201–210.

MAKOV, U.E. (1994). Some aspects of Bayesian loss robustness. *Journal of Statistical Planning and Inference*, **38**, 359–370.

MARTÍN, J., RÍOS INSUA D. and RUGGERI, F. (1998). Issues in Bayesian loss robustness. *Sankhya: The Indian Journal of Statistics, Series A*, **60**, 405–417.

MICHEAS, A. and DEY, D. (1999). On measuring loss robustness using maximum a posteriori estimate. *Technical Report*, **99–12**, Department of Statistics, University of Connecticut.

RAMSAY, J.O. and NOVICK, M.R. (1980). PLU robust Bayesian decision. Theory: point estimation. *Journal of the American Statistical Association*, **75**, 901–907.

SHAPIRO, C.P. and WARDROP, R.L. (1980). Bayesian sequential estimation for one-parameter exponential families. *Journal of the American Statistical Association,* **75**, 984–988.

VARIAN, H.R. (1975). A Bayesian approach to real estate assessment. In *Studies in Bayesian Econometrics and Statistics in Honor of Leonard J. Savage* (S.E. Feinberg and A. Zellner eds.), 195–208. Amsterdam: North Holland.

WHITTLE, P. and LANE, R. O. D. (1967). A class of situations in which a sequential estimation procedure is non-sequential. *Biometrika,* **54**, 229–234.

ZELLNER, A. (1986). Bayesian estimation and prediction using asymmetric loss functions. *Journal of the American Statistical Association,* **81**, 446–451.

9

Computing Efficient Sets in Bayesian Decision Problems

Jacinto Martín and J. Pablo Arias

ABSTRACT Previous work in robust Bayesian analysis has concentrated mainly on inference problems, either through local or global analysis. Foundational approaches to robust Bayesian analysis suggest that, when addressing decision theoretic issues, computations should concentrate on comparing pairs of alternatives, checking for dominance between them. We study this problem for several important classes of priors and losses. Then, we propose an approximation scheme to compute the efficient set. Finally, we mention some screening methods.

Keywords. Bayesian analysis, robustness, classes of priors, dominance, classes of loss functions.

9.1 Introduction

Most previous work on robust Bayesian analysis and sensitivity analysis has concentrated on inference problems and prior sensitivity, especially, on computing the range of the posterior expected loss when the prior varies in a certain class. Berger (1994) provides an excellent review.

From a decision theory perspective, various foundations lead to a class of prior distributions modelling beliefs and a class of loss functions modelling preferences, suggesting ordering alternatives in a Pareto sense(see e.g. Nau et al., 1996). For example, using a class Γ of priors over the set of states Θ, and a class $\mathcal{L}$ of loss functions over the set of consequences $\mathcal{C}$, we have

$$b \preceq a \iff \left(\frac{\int_\Theta L(a,\theta) f(x|\theta) d\pi(\theta)}{\int_\Theta f(x|\theta) d\pi(\theta)} \leq \frac{\int_\Theta L(b,\theta) f(x|\theta) d\pi(\theta)}{\int_\Theta f(x|\theta) d\pi(\theta)}, \forall \pi \in \Gamma, \forall L \in \mathcal{L} \right),$$

where L is the loss function, x is the result of an experiment with likelihood $f(x|\theta)$, $\preceq$ is the preference relation between alternatives and $\pi(\theta|x)$ denotes the posterior when π is the prior.

As a consequence, the solution concept is the efficient set, that is, the set of nondominated alternatives. If $\mathcal{A}$ designates the set of alternatives:

Definition 1 *Let $a, b \in \mathcal{A}$ such that $a \neq b$. b dominates a if $a \prec b$ (that is, $a \preceq b$ and $\neg(b \preceq a)$).*

Definition 2 *$a \in \mathcal{A}$ is nondominated if there is no $b \in \mathcal{A}$ such that b dominates a.*

This chapter deals with the problem of approximating the efficient set. This task has been undertaken in simple decision Analytic problems (see, e.g., Ríos Insua and French, 1991) and references quoted therein. Clearly, in order to do that we need procedures to check dominance between different alternatives. We study this problem in three contexts: first, for several classes of priors; second, for several classes of losses; and, finally, for classes of priors and losses simultaneously. This problem leads to the study of the operator (Ríos Insua and Martín, 1994b):

$$T(\pi, L; a, b) = \frac{\int_\Theta (L(a,\theta) - L(b,\theta)) f(x|\theta) d\pi(\theta)}{\int_\Theta f(x|\theta) d\pi(\theta)},$$

since

$$\text{if} \sup_{\pi\in\Gamma, L\in\mathcal{L}} T(\pi, L; a, b) \le 0, \text{ then } b \preceq a. \tag{1}$$

This clearly helps us to determine whether a dominates b. When the alternatives are fixed, we use $T(\pi, L)$ (suppressing a and b), and similarly if π or L is fixed.

Note that we are mainly interested in knowing whether the supremum (1) is not positive rather than knowing its specific value. This will be useful when we are only able to bound it. Moreover, we shall consider only distributions $\pi \in \Gamma$ (and data x) such that $0 < \int_\Theta f(x|\theta)d\pi(\theta) < \infty$. Hence, we may ignore denominators since

$$b \preceq a \iff \sup_{\pi\in\Gamma, L\in\mathcal{L}} N(\pi, L; a, b) \le 0,$$

with

$$N(\pi, L; a, b) = \int_\Theta (L(a,\theta) - L(b,\theta)) f(x|\theta) d\pi(\theta).$$

We shall compute $\sup_{\pi\in\Gamma, L\in\mathcal{L}} N(\pi, L; a, b)$ for several important classes Γ and $\mathcal{L}$ appearing in the robust Bayesian literature. Note that our problem is intimately related with the computation of bounds for linear functionals in robust Bayesian analysis, see Berger (1990). There are, however, two main differences: first, as we have mentioned, we are mainly interested in the sign of the supremum, rather than its value. Moreover, in some cases, the computations are more complicated since they refer to comparisons of two alternatives.

After checking dominance between alternatives, we study the problem of computing the efficient set. We provide some examples where the computations are straightforward. We describe also an approximation scheme based on the discretisation of the set of alternatives. Finally, for problems with a "big" efficient set, we mention several screening methods.

9.2 Prior robustness

In this section, we assume that there is precision in preferences and we have a unique loss function L. Since a and b will remain fixed, we shall use the notation

$$\begin{aligned} h(\theta) &= (L(a,\theta) - L(b,\theta))f(x|\theta), \text{ and} \\ N(\pi) &= \int h(\theta)d\pi(\theta). \end{aligned} \tag{2}$$

For computational reasons, we shall assume that the loss function L is normalized between 0 and 1.

9.2.1 *ε-contaminated class*

ε-contaminated classes are very popular in robust Bayesian analysis: see Berger and Berliner (1986), among others. This class defines a neighborhood around a distribution π_0 as follows:

$$\Gamma^{\varepsilon}(\pi_0,\ \mathcal{Q}) = \{\pi : \ \pi = (1-\varepsilon)\pi_0 + \varepsilon q \ : \ q \in \mathcal{Q}\},$$

where ε represents imprecision over π_0 and $\mathcal{Q}$ is a class of distributions, typically including π_0.

The computation of the supremum of $N(\pi)$ when $\pi \in \Gamma^{\varepsilon}(\pi_0,\ \mathcal{Q})$ leads to the computation of the supremum of $N(q)$ when $q \in \mathcal{Q}$:

$$\begin{aligned} \sup_{\pi \in \Gamma^{\varepsilon}(\pi_0,\ \mathcal{Q})} N(\pi) &= \sup_{q \in \mathcal{Q}} \int h(\theta) d((1-\varepsilon)\pi_0 + \varepsilon q)(\theta) \\ &= (1-\varepsilon)N(\pi_0) + \varepsilon \sup_{q \in \mathcal{Q}} N(q). \end{aligned}$$

Therefore,

$$\sup_{\pi \in \Gamma^{\varepsilon}(\pi_0,\ \mathcal{Q})} N(\pi) \leq 0 \iff \sup_{q \in \mathcal{Q}} N(q) \leq -\frac{1-\varepsilon}{\varepsilon} N(\pi_0).$$

Hence, dominance studies for ε-contaminated classes lead to studies over the contaminating class $\mathcal{Q}$, which could be, for example, any of the ones we study below.

9.2.2 *Parametric class*

Parametric classes are also popular in robust Bayesian analysis (see, e.g., Berger, 1990, 1994). If $\pi_\lambda(\theta)$ designates the density function or the probability function, we define the parametric class as:

$$\Gamma_\lambda = \{\pi : \ \pi \text{ has d. f. (or p. f.) } \pi_\lambda(\theta),\ \lambda \in \Lambda\}.$$

Typical examples are the class of normal distributions with constraints on the mean and standard deviation, exponential distributions with constraints on the parameter, and so on. These classes are not convex in general, which conflicts with results in foundations (see, e.g., Ríos Insua and Martín, 1994a). However, if we use the convex hull, we obtain the same supremum, so we need to consider only Γ_λ.

If we have a primitive of (2) for each $\pi \in \Gamma_\lambda$, we shall have a nonlinear programming problem. If we lack that primitive, we may use the following Monte Carlo importance sampling strategy to approximate

$$\sup_{\pi \in \Gamma_\lambda} N(\pi) = \sup_{\lambda \in \Lambda} \int h(\theta)\pi_\lambda(\theta)d\theta (= N(\lambda)).$$

Let $g(\theta)$ be a density with support Θ:

Step 1. Generate n i.i.d. vectors $\theta_1, \ldots, \theta_n \sim g$.

Step 2. Approximate $N(\lambda)$ by

$$N_n(\lambda) = \frac{1}{n} \sum_{i=1}^{n} h(\theta_i) \frac{\pi_\lambda(\theta_i)}{g(\theta_i)}.$$

Step 3. Approximate $\sup_{\lambda \in \Lambda} N(\lambda)$ by

$$\hat{N}_n = \sup_{\lambda \in \Lambda} N_n(\lambda). \tag{3}$$

Shao (1989) used a similar scheme to approximate maximum expected utility alternatives in standard decision theoretic problem.

Since we are actually interested in the sign of the supremum, we may use an interval estimate of $\sup_{\lambda \in \Lambda} N(\lambda)$ to determine the sample size n. For example, if λ_n designates the optimum of (3) and $\widehat{V(\lambda_n)}$ an estimate of the variance of $\hat{N}_n$, we could stop sampling whenever

$$0 \notin \left[\hat{N}_n - 2\sqrt{\widehat{V(\lambda_n)}}, \hat{N}_n + 2\sqrt{\widehat{V(\lambda_n)}}\right].$$

If $\hat{N}_n + 2\sqrt{\widehat{V(\lambda_n)}} < 0$, we would suggest $b \prec a$. We could also stop sampling when the interval is small enough, suggesting no dominance. If none of these conditions holds, we should sample more from g. We use this method in Section 9.4.1, where we study imprecision in beliefs and preferences.

9.2.3 Quantile class

Quantile classes are fairly popular in robustness studies. Consider a measurable partition $A_1, \ldots, A_n$ of Θ and define

$$\Gamma_Q = \left\{\pi : \underline{p}_i \leq \pi(A_i) \leq \bar{p}_i,\ i = 1, \ldots, n\right\},$$

where $\sum_{i=1}^n \underline{p}_i \leq 1 \leq \sum_{i=1}^n \bar{p}_i$ and $\underline{p}_i \geq 0$, $i = 1, \ldots, n$. This class, or variants thereof, has been studied, among others, by Moreno and Pericchi (1992) and Ruggeri (1992). A particular case is in Berliner and Goel (1986), with $\bar{p}_i = \underline{p}_i$, $i = 1, \ldots, n$. Denote this class by $\Gamma_{\vec{p}}$ where $\vec{p} = (p_1, \ldots, p_n)$, that is,

$$\Gamma_{\vec{p}} = \left\{ \pi : \pi(A_i) = p_i,\ i = 1, \ldots, n,\ \sum_{i=1}^n p_i = 1 \right\}.$$

Now we need to solve now the optimisation problem

$$\begin{aligned} &\sup \int h(\theta) d\pi(\theta) \\ &s.t. \\ &\underline{p}_i \leq \pi(A_i) \leq \bar{p}_i,\ i = 1, \ldots, n \\ &\sum_{i=1}^n \pi(A_i) = 1 \end{aligned}$$

The next result suggests its solution with one linear programming problem and n nonlinear programming problems:

Proposition 1

$$\begin{aligned} \sup_{\pi \in \Gamma_Q} N(\pi) = \ & \max \sum_{i=1}^n p_i \sup_{\theta \in A_i} h(\theta) \\ & s.t. \\ & \underline{p}_i \leq p_i \leq \bar{p}_i,\ i = 1, \ldots, n \\ & \sum_{i=1}^n p_i = 1 \end{aligned}$$

Proof. For $\Gamma_{\vec{p}}$, we have

$$\sup_{\pi \in \Gamma_{\vec{p}}} N(\pi) = \sup_{\pi \in \Gamma_{\vec{p}}} \sum_{i=1}^n \int_{A_i} h(\theta) d\pi(\theta) = \sum_{i=1}^n p_i \sup_{\theta \in A_i} h(\theta).$$

Decompose Γ_Q in subclasses $\Gamma_{\vec{p}}$, where $\underline{\mathbf{p}} \leq \vec{p} \leq \bar{\mathbf{p}}$, $\underline{\mathbf{p}} = (\underline{p}_1, \ldots, \underline{p}_n)$, $\bar{\mathbf{p}} = (\bar{p}_1, \ldots, \bar{p}_n)$ and $\sum_{i=1}^n p_i = 1$, with $\leq$ the element by element inequality. Γ_Q is the union of these classes. Then,

$$\sup_{\pi \in \Gamma_Q} N(\pi) = \sup_{\underline{\mathbf{p}} \leq \vec{p} \leq \bar{\mathbf{p}}} \ \sup_{\pi \in \Gamma_{\vec{p}}} N(\pi),$$

leading to the suggested result. □

9.2.4 Probability bands class

When Θ is not finite, a usual criticism to the quantile class is that the supremum is achieved at a discrete distribution, as shown above. One way of mitigating against this is by adding a unimodality condition to the probability distributions in Γ_Q; see Section 9.2.5. Alternatively, we may use a class of probability bands.

Let L and U be two measures over (Θ, β), with β a σ-field, such that $L(A) \leq U(A), \forall A \in \beta$, and $L(\Theta) \leq 1 \leq U(\Theta)$. Let

$$\Gamma(L,U) = \{\pi : \; L(A) \leq \pi(A) \leq U(A), \; \forall A \in \beta, \; \pi(\Theta) = 1\}.$$

This class has been considered and studied, among others, by Berger (1994), Wasserman et al. (1993) and Moreno and Pericchi (1992). Proposition 2 is based on results from the last authors.

Proposition 2 *Let* $A_z = \{\theta \in \Theta : \; h(\theta) \leq z\}$ *and* $B_z = \{\theta \in \Theta : \; h(\theta) < z\}$. *If there is* z *such that*

$$L(A_z) + U(A_z^c) = 1, \text{ or} \tag{4}$$

$$L(B_z) + U(B_z^c) = 1, \tag{5}$$

then

$$\sup_{\pi \in \Gamma(L,U)} N(\pi) = N(\pi^*) = \int_{B_z} h(\theta) dL(\theta) + \int_{A_z^c} h(\theta) dU(\theta) + z(1 - L(B_z) - U(A_z^c)),$$

with

$$\pi^*(d\theta) = L(d\theta) I_{A_z} + U(d\theta) I_{A_z^c}$$

if (4) *holds, and*

$$\pi^*(d\theta) = L(d\theta) I_{B_z} + U(d\theta) I_{B_z^c}$$

if (5) *holds.*

Proof.

Suppose (4) holds. If there is $\pi_1 \in \Gamma(L,U)$ such that $N(\pi^*) < N(\pi_1)$, then

$$\int_{A_z} h(\theta) d\pi_1(\theta) + \int_{A_z^c} h(\theta) d\pi_1(\theta) > \int_{A_z} h(\theta) dL(\theta) + \int_{A_z^c} h(\theta) dU(\theta)$$

$$\Longrightarrow \int_{A_z} h(\theta) d(\pi_1 - L)(\theta) > \int_{A_z^c} h(\theta) d(U - \pi_1)(\theta)).$$

Since $\pi_1 - L$ and $U - \pi_1$ are measures,

$$z(\pi_1(A_z) - L(A_z)) > z(U(A_z^c) - \pi_1(A_z^c)). \tag{6}$$

Inequality (6) cannot hold for $z = 0$, hence we consider only two cases:

- $z > 0$. Then, since (4) holds,

$$\pi_1(\Theta) = \pi_1(A_z) + \pi_1(A_z^c) > L(A_z) + U(A_z^c) = 1,$$

which is a contradiction.

- $z < 0$. Then

$$\pi_1(\Theta) = \pi_1(A_z) + \pi_1(A_z^c) < L(A_z) + U(A_z^c) = 1,$$

obtaining, as well, a contradiction.

The proof is similar if (5) holds. □

Let $E_z = \{\theta \in \Theta : \ h(\theta) = z\}$. Equations (4) or (5) will hold when $L(E_z) = U(E_z) = 0$, $\forall z$. Cases in which the hypothesis in Proposition 2 does not hold are the following:

- There is z such that

$$U(E_z) > 0, \ L(A_z) + U(A_z^c) < 1 \text{ and } L(B_z) + U(B_z^c) > 1. \tag{7}$$

- There is z such that

$$L(E_z) > 0, \ L(A_z) + U(A_z^c) > 1 \text{ and } \forall z' > z, \ L(B_{z'}) + U(B_{z'}^c) < 1 \tag{8}$$

We solve both cases with the following result:

Proposition 3 *If* (7) *or* (8) *holds, then*

$$\sup_{\pi \in \Gamma(L,U)} N(\pi) = N(\pi^*) = \int_{B_z} h(\theta) dL(\theta) + \int_{A_z^c} h(\theta) dU(\theta) + z(1 - L(B_z) - U(A_z^c)),$$

where

$$\pi^*(d\theta) = L(d\theta) I_{B_z} + U(d\theta) I_{A_z^c} + \pi_0^*(d\theta) I_{E_z}$$

with π_0^* a probability distribution in $\Gamma(L,U)$ such that $\pi_0^*(E_z) = 1 - L(B_z) - U(A_z^c)$.

Proof. We prove the result when (7) holds. Suppose there is $\pi_1 \in \Gamma(L,U)$ such that $N(\pi^*) < N(\pi_1)$. Suppose that $\pi^*(E_z) \leq \pi_1(E_z)$. The proof would be analogous in the other case. Then,

$$\begin{aligned}
& \int_{B_z} h(\theta) d\pi_1(\theta) + \int_{A_z^c} h(\theta) d\pi_1(\theta) + \int_{E_z} h(\theta) d\pi_1(\theta) \\
> \ & \int_{B_z} h(\theta) dL(\theta) + \int_{A_z^c} h(\theta) dU(\theta) + \int_{E_z} h(\theta) d\pi_0^*(\theta) \\
\Longrightarrow \ & \int_{B_z} h(\theta)(d\pi_1(\theta) - dL(\theta)) + z[(\pi_1 - \pi_0^*)(E_z)] \\
> \ & \int_{A_z^c} h(\theta)(dU(\theta) - d\pi_1(\theta)).
\end{aligned}$$

Since $\pi_1 - L$ and $U - \pi_1$ are measures, we have

$$z(\pi_1(B_z) - L(B_z) + \pi_1(E_z) - \pi_0^*(E_z)) > z(U(A_z^c) - \pi_1(A_z^c)),$$

and, therefore, we are in cases similar to those in the proof of Proposition 2.

The proof is similar if (8) holds. □

Obviously, we need a procedure to compute z in Propositions 2 and 3. We provide an approximation method based on the bisection method of nonlinear programming (see e.g. Bazaraa and Shetty, 1979). Compute, with the aid of two nonlinear programs:

$$z_0 = \frac{\sup_{\theta\in\Theta} h(\theta) + \inf_{\theta\in\Theta} h(\theta)}{2},$$

$$H = \frac{|\sup_{\theta\in\Theta} h(\theta)| + |\inf_{\theta\in\Theta} h(\theta)|}{2}.$$

Let $\varphi(z) = L(A_z) + U(A_z^c)$ and $\varphi'(z) = L(B_z) + U(B_z^c)$, then

Step 1 $i = 0,\ z = z_0$.

Step 2 Compute $\varphi(z)$ and $\varphi'(z)$. $i = i + 1$.

Step 3 If $0 < \varphi(z) - 1 < \varepsilon$, stop (apply Proposition 2, (4)).
If $0 < 1 - \varphi'(z) < \varepsilon$, stop (apply Proposition 2, (5)).
If $\varphi(z) < 1 < \varphi'(z)$, stop (apply Proposition 3).
If $\varphi'(z) < 1 - \varepsilon$, do $z = z - 2^{-i}H$. Go to 2.
If $\varphi(z) > 1 + \varepsilon$, do $z = z + 2^{-i}H$. Go to 2.

Note that, in order to ensure convergence of the algorithm, we apply Proposition 2 approximately, the accuracy of the approximation depending on the value of ε in **Step 3**. It is easy to prove the validity of the algorithm.

9.2.5 Unimodal distributions class

We consider now the following classes over $\Theta \subseteq \mathbb{R}$:

$$\Gamma_U(\theta_0) = \{\pi :\ \pi \text{ is unimodal in } \theta_0\}$$

and

$$\Gamma_{US}(\theta_0) = \{\pi : \pi \text{ is unimodal in } \theta_0 \text{ and symmetric}\}.$$

These classes, either as contaminators in ε-contaminations, or per se, have been used, among others, by Sivaganesan and Berger (1989) and Dall'-Aglio and Salinetti (1994). We compute the supremum of $N(\pi)$ over classes $\Gamma_U(\theta_0)$ and $\Gamma_{US}(\theta_0)$. The result for the corresponding ε-contaminated class follows immediately, as we saw in Section 9.2.1. Let $\Gamma_1 \subseteq \Gamma_{US}(\theta_0)$ and $\Gamma_2 \subseteq \Gamma_U(\theta_0)$ be defined as follows:

$$\begin{aligned}\Gamma_1 &= \{\pi : \pi \sim U(\theta_0 - z, \theta_0 + z) \text{ for some } z > 0\},\\ \Gamma_2 &= \{\pi : \pi \sim U(\theta_0 - z, \theta_0) \text{ or } \pi \sim U(\theta_0, \theta_0 + z) \text{ for some } z > 0\},\end{aligned}$$

with $U(m,n)$ the uniform distribution over (m,n). Results in Sivaganesan and Berger (1989) show that the supremum of $N(\pi)$ over the unimodal and symmetric unimodal classes may be attained at elements of Γ_2 and Γ_1, respectively. Following these results, we have

Proposition 4

$$\begin{aligned}\sup_{\pi\in\Gamma_{US}(\theta_0)} N(\pi) &= \sup_{\pi\in\Gamma_1} N(\pi) = \sup_{z>0} \frac{1}{2z}\int_{\theta_0-z}^{\theta_0+z} h(\theta)d\theta,\\ \sup_{\pi\in\Gamma_U(\theta_0)} N(\pi) &= \sup_{\pi\in\Gamma_2} N(\pi) = \sup_{z>0} \frac{1}{2z}\int_{r}^{t} h(\theta)d\theta,\end{aligned}$$

with $(r,t) = (\theta_0, \theta_0 + z)$ *or* $(r,t) = (\theta_0 - z, \theta_0)$.

In cases in which a primitive for h is available, we may appeal to nonlinear programming. Otherwise, we may easily devise an approximation method.

If we consider a quantile class with unimodality constraints, slight modifications of Dall'Aglio and Salinetti (1994), algorithm allow us to approximate the supremum.

9.2.6 *Total variation neighborhoods*

Consider now total variation neighborhoods:

$$\Gamma_{TV}^{\varepsilon}(\pi_0) = \{\pi : \ ||\pi - \pi_0||_{TV} \leq \varepsilon\},$$

where $||\pi||_{TV} = \sup_{A\in\beta} |\pi(A)|$ is the total variation norm.

Assume $\Theta = \{\theta_1, \ldots, \theta_n\}$. The computation of $\sup_{\pi\in\Gamma_{TV}^{\varepsilon}(\pi_0)} N(\pi)$ leads to the linear programming problem:

$$\begin{aligned}&\max \sum_{i=1}^{n} p_i h(\theta_i)\\ &s.t.\\ &\sum_{i\in I} \pi_0(\{\theta_i\}) - \varepsilon \leq \sum_{i\in I} p_i \leq \sum_{i\in I} \pi_0(\{\theta_i\}) + \varepsilon, \quad \forall I \subset \{1, \ldots, n\}\\ &\sum_{i=1}^{n} p_i = 1\\ &p_i \geq 0,\ i = 1, \ldots, n.\end{aligned}$$

The number of constraints is $2^n + 1$, nonnegativity apart. For big n, we could solve the dual linear programming problem. Alternatively, we might use the following algorithm:

Step 1 $\varepsilon_1 = \varepsilon$, $\theta^* = \arg\max_{\theta \in \Theta} h(\theta)$, $q(\theta^*) = \min\{1, \pi_0(\theta^*) + \varepsilon\}$ and $i = 1$.

Step 2 If $q(\theta^*) = 1$, then $\sup_{\pi \in \Gamma^{\varepsilon}_{TV}(\pi_0)} N(\pi) = h(\theta^*)$. Stop. (I)
Else, let $\theta^i_* : \arg\min_{\theta \in \Theta}, h(\theta)$, $q(\theta^i_*) = \max\{0, \pi_0(\theta^i_*) - \varepsilon_1\}$ and $\forall \theta \in \Theta$, $q(\theta) = \pi_0(\theta)$, $\theta \neq \theta^*, \theta^i_*$.

Step 3 If $q(\theta^i_*) = \pi_0(\theta^i_*) - \varepsilon_1$, then $\sup_{\pi \in \Gamma^{\varepsilon}_{TV}(\pi_0)} N(\pi) = N(q)$. Stop. (II)
Else, $\varepsilon_1 = \varepsilon_1 - \pi_0(\theta^i_*)$ $\Theta = \Theta \backslash \{\theta^i_*\}$, $i = i + 1$. Go to **Step 2**. (III)

Let us check the correctness of the algorithm. The result is obvious in (I). Just consider the point mass at θ^*, which belongs to $\Gamma^{\varepsilon}_{TV}(\pi_0)$. In case (II), we need to check:

1. $q \in \Gamma^{\varepsilon}_{TV}(\pi_0)$
2. $\sup_{\pi \in \Gamma^{\varepsilon}_{TV}(\pi_0)} N(\pi) = N(q)$.

1 is immediate. 2 follows by contradiction. Since $\Gamma^{\varepsilon}_{TV}(\pi_0)$ is closed, the supremum is attained. Suppose $\pi \neq q$ is a maximum, and $N(\pi) > N(q)$. Clearly, $\pi(\theta_*) > \pi_0(\theta_*) - \varepsilon$ and $\pi(\theta^*) < \pi_0(\theta^*) + \varepsilon$ since, otherwise, $\pi = q$. Moreover, $N(\pi) > N(q) \geq N(\pi_0)$. Hence, there is θ_i such that $h(\theta_i) < h(\theta^*)$ and $\pi(\theta_i) > \pi_0(\theta_i)$. Let $\pi(\theta_i) + p = \pi_0(\theta_i)$, with $p > 0$. Define π' as follows: $\pi'(\theta^*) = \pi(\theta^*) + p$, $\pi'(\theta_i) = \pi_0(\theta_i)$ and $\pi'(\theta) = \pi(\theta)$, $\forall \theta \neq \theta_i, \theta^*$. Then $N(\pi') > N(\pi)$, a contradiction.

For (III), the proof is reduced to that of (II), taking into account the values $\theta_*, \theta^1_*, \ldots, \theta^m_*$ obtained in the m iterations of the algorithm before the end.

In the countable case, the algorithm is valid but may not stop if case (III) appears. We could solve this using a stopping threshold: when ε_1 is small enough, consider $q(\theta_*) = \pi_0(\theta_*) - \varepsilon_1$ and go to (II).

9.3 Loss robustness

When addressing decision theoretic issues, we immediately think of considering classes of loss functions and extend sensitivity to the prior procedures to this new setting in a straightforward manner. For example, we could compute ranges of expected losses when the loss ranges in a class. These extensions demand some care, however. Suppose, for example, that the class of losses includes functions of the type $L(a, \theta) + k$, for $k \in [k_0, k_1]$.

This happens, for example, in the universal class (see Hwang, 1985), which is the class of functions $L(|\theta - a|)$, where L is an arbitrary nondecreasing function. Then, the range of posterior expected losses is greater than $k_1 - k_0$, which, if "large," may suggest that we do not have robustness. Clearly, this is not true since all these loss functions are strategically equivalent, as they lead to the same optimal decision. See Martín et al. (1998) and Dey et al. (1998) for some issues about loss robustness.

Some of the classes will be parallel to those used in sensitivity to the prior studies described in Section 9.2. Others will be associated with specific methods of loss elicitation. Finally, another source of classes of loss functions is the stochastic dominance literature (see Levy, 1992) for a review, which concentrates mainly on loss functions whose derivatives of increasing order alternate in sign. We shall assume monotonicity constraints in all our classes. Note that, in this case, we have one prior only and we consider that we can obtain the posterior distribution. So, we use operator T instead of N. If we lack posterior distributions, the results are useful after little modifications.

9.3.1 ε-contaminated class

As we mention in Section 9.2.1, *ε-contaminated classes* are popular in robust Bayesian analysis. We may use the same principle to define a neighborhood around a loss function L_0 as follows:

$$\mathcal{L}^{\varepsilon}(L_0, \mathcal{W}) = \{L : \ L(c) = (1-\varepsilon)L_0(c) + \varepsilon M(c) \ : \ M \in \mathcal{W}\},$$

where ε represents imprecision over L_0 and $\mathcal{W}$ is a class of loss functions, typically including L_0.

The computation of the supremum of $T(L)$ when $L \in \mathcal{L}^{\varepsilon}(L_0, \ \mathcal{W})$ leads to the computation of the supremum of $T(M)$ when $M \in \mathcal{W}$, as in Section 9.2.1:

$$\sup_{\mathcal{L}^{\varepsilon}(L_0, \ \mathcal{W})} T(L) \leq 0 \iff \sup_{\mathcal{W}} T(M) \leq -\frac{1-\varepsilon}{\varepsilon} T(L_0).$$

9.3.2 Parametric class

In many applications, we shall attempt to model preferences with a parametric class and try to get information for that parameter. Some examples may be seen in Bell (1995). We consider the *general parametric class*

$$\mathcal{L}_{\omega} = \{L_{\omega}, \ \omega \in \Omega\}.$$

Note that if we have an explicit expression for the posterior expected loss $T(L_{\omega})$, the computation of the supremum is relatively simple, since we have a nonlinear programming problem. If we lack such an expression, we may use a Monte Carlo scheme similar to that of Section 9.2.2, which we illustrate in Section 9.4.1. In this case, we may sample from $\pi(\cdot|x)$, perhaps using importance sampling or a Markov chain Monte Carlo method. Alternatively, we could use the operator N instead of T and sample from the prior.

9.3.3 Partially known losses

Another realistic case is that in which we have assessed bounds on the loss of some consequences. These may arise naturally if we use the *probability equivalent* elicitation method see, e.g, French and Ríos Insua (2000).

Assume that consequences are of the type $e(|a-\theta|) \in \mathbb{R}$, with e nondecreasing. Let $\{C_1, \dots, C_n\}$ be a finite partition of the space of consequences $\mathcal{C}$. C_i will be intervals $\langle c_{i-1}, c_i \rangle$[1] of $\mathbb{R}$. Assume that C_i are to the left of 0, for $i = 1, \dots, m-1$, the C_i are to the right of 0, for $i = m+1, \dots, n$, and $0 \in C_m$. For a given alternative, we shall use the notation $\Theta_a(c) = \{\theta \in \Theta : c(a,\theta) = e(|a-\theta|) = c\}$.

We define a class of *partially known loss functions* as

$$\begin{aligned}\mathcal{L}_K &= \{L : \ell_{i-1} \geq L(c) \geq \ell_i,\ \forall c \in C_i,\ i = 1, \dots, m-1; \\ &\qquad \ell_{i-1} \leq L(c) \leq \ell_i,\ \forall c \in C_i,\ i = m+1, \dots, n,\ 0 \in C_m\}.\end{aligned}$$

Some of the sets C_i could be empty. We add to $\mathcal{L}_K$ the condition that $L \in \mathcal{L}_K$ must be nonincreasing in $(-\infty, 0)$ and nondecreasing in $(0, \infty)$. Vickson (1977) has used related classes in stochastic dominance analysis.

Consider first the case in which $\mathcal{C}$ is finite. This may happen, for instance, in hypothesis testing problems. In this case, we have a linear programming problem. To see this, suppose $c_1 < c_2 < \cdots < c_m < 0 < c_{m+1} < \cdots < c_N$. Then, we would have to solve, in $L(c_i)$,

$$\begin{aligned}T(L) &= \sum_{i=1}^{N} L(c_i)\left[\pi(\Theta_a(c_i)|x) - \pi(\Theta_b(c_i)|x)\right] \\ s.t. & \\ & L(c_j) \geq L(0), \quad j \in \{m, m+1\}, \\ & L(c_i) \geq L(c_{i+1}), \quad i = 1, \dots, m-1, \\ & L(c_i) \leq L(c_{i+1}), \quad i = m+1, \dots, N-1, \\ & \forall c_i \in C_j,\ \ell_{j-1} \geq L(c_i) \geq \ell_j, \quad j = 1, \dots, m-1, \\ & \forall c_i \in C_j,\ \ell_{j-1} \leq L(c_i) \leq \ell_j, \quad j = m+1, \dots, n.\end{aligned}$$

Consider now the continuous case. We shall assume that the sets C_i will be intervals $\langle c_{i-1}, c_i \rangle$ of $\mathbb{R}$, with $\ell_{m-1} = \ell_{m+1}$, to simplify the notation.

[1] $\langle c_{i-1}, c_i \rangle$ denotes one of the following intervals: (c_{i-1}, c_i), $(c_{i-1}, c_i]$, $[c_{i-1}, c_i)$, $[c_{i-1}, c_i]$.

Proposition 5

$$\sup_{L\in\mathcal{L}_K} T(L) = \sum_{i=1}^{n} \Big[\ell_{i-1}\Big(\pi(\Theta_a(C_i^*)|x) - \pi(\Theta_b(C_i^*)|x)\Big)$$
$$+\ell_i\Big(\pi(\Theta_a(C_i\setminus C_i^*)|x) - \pi(\Theta_b(C_i\setminus C_i^*)|x)\Big)\Big],$$

with $C_i^* \subset C_i$ defined as follows:

1. *If $0\in C_i$, then $C_i^* = \langle c_{i-1}^*, c_i^*\rangle$ such that*

$$\pi(\Theta_a(C_i^*)|x) - \pi(\Theta_b(C_i^*)|x) = \inf_{C\subset C_i}(\pi(\Theta_a(C)|x) - \pi(\Theta_b(C)|x))$$

for all $C = \langle c_{i-1}, c_i\rangle \subset C_i$.

2. *If $0 < c, \forall c\in C_i$, then $C_i^* = \langle c_{i-1}, c_i^*]$ such that*

$$\pi(\Theta_a(C_i^*)|x) - \pi(\Theta_b(C_i^*)|x) = \inf_{C\subset C_i}(\pi(\Theta_a(C)|x) - \pi(\Theta_b(C)|x))$$

for all $C = \langle c_{i-1}, c] \subset C_i$.

3. *If $0 > c, \forall c\in C_i$, then $C_i^* = \langle c_{i-1}, c_i^*]$ such that*

$$\pi(\Theta_a(C_i^*)|x) - \pi(\Theta_b(C_i^*)|x) = \inf_{C\subset C_i}(\pi(\Theta_a(C)|x) - \pi(\Theta_b(C)|x))$$

for all $C = \langle c_{i-1}, c] \subset C_i$.

Proof. We have

$$\int (L(a,\theta)-L(b,\theta))d\pi(\theta|x) = \sum_{i=1}^{n}\Big[\int_{\Theta_a(C_i)} L(a,\theta)d\pi(\theta|x) - \int_{\Theta_b(C_i)} L(b,\theta)d\pi(\theta|x)\Big]$$

$$= \sum_{i=1}^{n}\Big[\int_{C_i} L(c)d\pi^a(c) - \int_{C_i} L(c)d\pi^b(c)\Big] = \sum_{i=1}^{n}\int_{C_i} L(c)\big(d\pi^a - d\pi^b\big)(c),$$

with π^a and π^b defined by a change of variable.

We prove the result for case 2, the reasoning being similar for 1 and 3. Let $\delta = \pi^a - \pi^b$. Let C_i^* be as above. Then, we have to prove

$$\int_{C_i} L(c)d\delta(c) \le \ell_{i-1}\delta(C_i^*) + \ell_i\delta(C_i\setminus C_i^*),\ \ \forall L\in\mathcal{L}_K.$$

Fix $L\in\mathcal{L}_K$, and define d_1, d_2 as follows:

$$\begin{aligned} d_1 &= \inf\{c\in C_i : L(c) > \ell_{i-1}\},\\ d_2 &= \inf\{c\in C_i : L(c) < \ell_i\}. \end{aligned}$$

We have

$$\int_{C_i} L(c)d\delta(c) = \int_{c_{i-1}}^{d_1} \ell_{i-1}d\delta(c) + \int_{d_1}^{c_i^*} L(c)d\delta(c) + \int_{c_i^*}^{d_2} L(c)d\delta(c) + \int_{d_2}^{c_i} \ell_i d\delta(c).$$

We now prove that

$$(I) \int_{c_i^*}^{d_2} L(c)d\delta(c) \leq \ell_i\, \delta(c_i^*, d_2),$$

$$(II) \int_{d_1}^{c_i^*} L(c)d\delta(c) \leq \ell_{i-1}\, \delta(d_1, c_i^*).$$

(I) Since $L(c)$ is nondecreasing in (c_i^*, d_2), there is a sequence of step functions I_n such that $I_n \uparrow L$ in that interval. Since δ is finite,

$$\lim_{n\to\infty} \int_{c_i^*}^{d_2} I_n(c)d\delta(c) = \int_{c_i^*}^{d_2} L(c)d\delta(c).$$

Let $E_1^n, \ldots, E_{m_n}^n$ be a partition of $(c_i^*, d_2]$ defined as follows:

$$I_n(c) = \omega_i^n,\ c \in E_i^n \text{ with } \omega_{m_n}^n \leq \omega_{m_n-1}^n \leq \cdots \leq \omega_1^n \leq \ell_i.$$

We have

- $\delta(E_{m_n}^n) \geq 0$, by definition of c_i^*. Then

$$\int_{E_{m_n}^n} I_n(c)d\delta(c) = \omega_{m_n}\, \delta(E_{m_n}^n) \leq \ell_i\, \delta(E_{m_n}^n).$$

- $\delta(E_{m_n}^n) + \delta(E_{m_n-1}^n) \geq 0$, by definition of c_i^*. Then

$$\begin{aligned} \int_{E_{m_n}^n \cup E_{m_n-1}^n} I_n(c)d\delta(c) &= \omega_{m_n}\, \delta(E_{m_n}^n) + \omega_{m_n-1}\, \delta(E_{m_n-1}^n) \\ &\leq \omega_{m_n-1}(\delta(E_{m_n-1}^n) + \delta(E_{m_n}^n)) \\ &\leq \ell_i\, \delta(E_{m_n}^n \cup E_{m_n-1}^n) \\ &\ \ \vdots \end{aligned}$$

- $\delta(E_{m_n}^n) + \cdots + \delta(E_1^n) \geq 0$. Then

$$\int_{c_i^*}^{d_2} I_n(c)d\delta(c) \leq \ell_i\, \delta(c_i^*, d_2).$$

Hence (I) follows.

(II) By definition of c_i^*, $\delta(d_1, c_i^*) \leq 0$, then

$$\int_{d_1}^{c_i^*} V(c)d\delta(c) \leq \Big(\inf_{c\in(d_1,c_i^*)} V(c)\Big)\delta(d_1, c_i^*) \leq \ell_{i-1}\, \delta(d_1, c_i^*).$$

□

9.3.4 Other classes

The previous results should give a flavor of our approach to this computational problem in Bayesian robustness. Results for other classes like bands of convex loss functions, mixtures of convex loss functions or finite classes may be seen in Martín et al. (1998).

9.4 Prior-loss robustness

We consider now the case in which there is imprecision in both beliefs and preferences. We illustrate the problem with two classes. The problem is more complicated than previous ones because now the operator N is bilinear instead of linear.

9.4.1 Parametric classes

We study now the class Γ_λ defined in Section 9.2.2 and the class $\mathcal{L}_\omega$ defined in Section 9.3.2. In this case, to check dominance, we have to compute

$$\sup_{\Lambda,\Omega} \int_\Theta (L_\omega(a,\theta) - L_\omega(b,\theta)) f(x|\theta) d\pi_\lambda(\theta) = S. \tag{9}$$

When it is posible to find a primitive, the operator T^{ab} is an explicit function of the parameters λ and ω; so, the calculation of the supremum is a mathematical programming problem, as shown in the example.

Example 1 We consider parametric classes of loss functions and prior distributions widely used:

- LINEX class of loss functions (Varian, 1975):

$$\mathcal{L}_{(\lambda,\varphi,\alpha)} = \left\{ L_{(\lambda,\varphi,\alpha)}(c) = \lambda c + \varphi e^{\alpha c}, \text{ con } \lambda, \varphi, \alpha > 0 \right\}.$$

- Class of normal probability distributions with mean and standard deviation belonging to bounded intervals:

$$\Gamma_{(\nu,\varsigma)} = \{\pi \sim N(\nu,\varsigma) : \nu \in (\nu_1,\nu_2),\ \varsigma \in (\varsigma_1,\varsigma_2)\}.$$

Suppose that actions a and b are defined by $(a,\theta) = a_1 + a_2\theta$, $(b,\theta) = b_1 + b_2\theta$. Suppose also that the likelihood is normal, so that the class of posterior distributions is normal with mean μ and standard deviation σ, where $\mu \in (\mu_*, \mu^*)$, $\sigma \in (\sigma_*, \sigma^*)$, $\sigma_* > 0$.

We have that a dominates b if and only if

1. $|b_2| \geq |a_2|$ and

2. $a_1 - b_1 + (a_2 - b_2)\mu_* \leq 0$, if $a_2 \leq b_2$, or $a_1 - b_1 + (a_2 - b_2)\mu^* \leq 0$, if $b_2 < a_2$

It is easy to prove that

$$\begin{aligned} T(L_{(\lambda,\varphi,\alpha)}, \pi_{N(\mu,\sigma)}; a, b) &= \lambda(a_1 - b_1 + (a_2 - b_2)\mu) \\ &\quad + \varphi(e^{\alpha(a_1 + a_2\mu + \frac{1}{2}\alpha a_2^2\sigma^2)} - e^{\alpha(b_1 + b_2\mu + \frac{1}{2}\alpha b_2^2\sigma^2)}). \end{aligned}$$

So, as λ, φ take values in $(0, \infty)$, $T(L_{(\lambda,\varphi,\alpha)}, \pi_{N(\mu,\sigma)}; a, b) \leq 0 \; \forall L \in \mathcal{L}_{(\lambda,\varphi,\alpha)}$, $\forall \pi \in \Gamma_{(\nu,\varsigma)}$ iff, $\forall \mu \in (\mu_*, \mu^*)$, $\sigma \in (\sigma_*, \sigma^*)$, $\alpha > 0$;

$$a_1 - b_1 + (a_2 - b_2)\mu \leq 0, \text{ and} \tag{10}$$

$$e^{\alpha(a_1 + a_2\mu + \frac{1}{2}\alpha a_2^2\sigma^2)} - e^{\alpha(b_1 + b_2\mu + \frac{1}{2}\alpha b_2^2\sigma^2)} \leq 0. \tag{11}$$

To prove this, note that, if (10) and (11) are true, and as $\varphi, \lambda > 0$, then $T(L_{(\lambda,\varphi,\alpha)}, \pi_{N(\mu,\sigma)}; a, b) \leq 0$, and, therefore $b \preceq a$. On the other hand, if (10) doesn't hold, then for $\lambda \to \infty$ and φ finite, we can get $T(L_{(\lambda,\varphi,\alpha)}, \pi_{N(\mu,\sigma)}; a, b) > 0$ for some $L_{(\lambda,\varphi,\alpha)}$ and $\pi_{N(\mu,\sigma)}$. This is similar if (11) is not true.

Due to the fact that the exponential function is strictly increasing Inequality (11) becomes

$$a_1 - b_1 + (a_2 - b_2)\mu \leq \frac{1}{2}\alpha\sigma^2(b_2^2 - a_2^2). \tag{12}$$

Fix μ and σ; then (12) will hold only if $a_2^2 \leq b_2^2$; otherwise, we can take α big enough that $a_1 - b_1 + (a_2 - b_2)\mu^* > \frac{1}{2}\alpha\sigma^2(b_2^2 - a_2^2)$.

Moreover, if $a_2^2 \leq b_2^2$, then (10) implies (12). In this way, we obtain condition 1. Condition 2 is easily obtained from condition 1 and Inequality (10). △

If we lack a primitive, we may appeal to a simulation approach similar to that of Section 9.2.2, where g is an importance sampling distribution:

1. Generate n i.i.d. vectors $\theta_1, \ldots, \theta_n \sim g$.

2. Approximate $N(\pi_\lambda, L_\omega)$ by

$$N_n(\lambda, \omega) = \frac{1}{n} \sum_{i=1}^{n} (L_\omega(a, \theta_i) - L_\omega(b, \theta_i)) \pi_\lambda(\theta_i) \frac{f(x|\theta_i)}{g(\theta_i)}.$$

3. Approximate $\sup_{\lambda,\omega} N(\pi_\lambda, L_\omega)$ by

$$S_n = \sup_{\lambda,\omega} N_n(\pi_\lambda, L_\omega).$$

Proposition 6 *Suppose that Λ and Ω are compact and, for each θ, $L_\omega(a,\theta)$, $L_\omega(b,\theta)$ and $\pi_\lambda(\theta)$ are continuous functions of ω and λ, respectively, and measurable functions of θ for (λ,ω). Assume also that, for a certain density $g(\theta)$,*

$$\left|(L_\omega(a,\theta)-L_\omega(b,\theta))\frac{f(x|\theta)\pi_\lambda(\theta)}{g(\theta)}\right|\leq q(\theta)$$

and q is integrable with respect to the probability measure associated with g. Then we have $S_n \rightarrow S$.

Proof. We are under the conditions of a uniform version of the strong law of large numbers (Jennrich, 1969), and, for almost every sequence $(\theta_1,\theta_2,\ldots)\sim g$,

$$\frac{1}{n}\sum_{i=1}^{n}(L_\omega(a,\theta_i)-L_\omega(b,\theta_i))\frac{f(x|\theta_i)\pi_\lambda(\theta)}{g(\theta_i)} \stackrel{n}{\longrightarrow}$$

$$\int(L_\omega(a,\theta)-L_\omega(b,\theta))f(x|\theta)\pi_\lambda(\theta)d\theta,$$

uniformly $\forall(\lambda,\omega)\in\Lambda\times\Omega$. Hence, for any ϵ, there is n such that, for almost all $(\theta_1,\theta_2,\ldots)\sim g$,

$$\sup\left(\left|\frac{1}{n}\sum_{i=1}^{n}(L_\omega(a,\theta_i)-L_\omega(b,\theta_i))\frac{f(x|\theta_i)\pi_\lambda(\theta_i)}{g(\theta_i)}-S\right|\right)\leq\epsilon.$$

□

Recall that we are actually interested in the sign of the supremum. Hence, we may use an interval estimate of S to estimate an appropriate sample size n. For example, if λ_n^1, ω_n^1 designates the optimum of (3) and $SE(N_n(\lambda_n^1,\omega_n^1))$ is the standard error,

$$\frac{1}{\sqrt{n}}\sqrt{\frac{\sum_{i=1}^{n}((L_\omega(a,\theta_i)-L_\omega(b,\theta_i))(f(x|\theta_i)\pi_\lambda(\theta_i)/g(\theta_i))-S_n)^2}{n-1}}$$

1. Compute $S_n = T_n(\lambda_n^1,\omega_n^1)$.

2. Estimate $\widehat{EE_n} = SE(N_n(\lambda_n^1,\omega_n^1))$.

3. If $S_n + 2\widehat{EE_n} < 0$, then $b \prec a$. If $S_n - 2\widehat{EE_n} > 0$, stop. Otherwise, if $\widehat{EE_n}$ is small enough, stop. Otherwise, resample (unless a certain maximum sample size is attained).

Example 2 Suppose we have a problem with $\mathcal{A} = \{a \in [0,1]\}$ and $c(a,\theta) = 1.1 + (\theta - .1)a$. Moreover, we have parametric classes for preferences and beliefs:

$$\begin{aligned}
\mathcal{L}_\omega &= \{L_\omega(c) = e^{-\omega c}, \text{ with } \omega \in [1,\ 2.5]\}, \\
\Gamma &= \{\pi(\theta|\mu) \sim \mathcal{N}(\mu, .1), \text{ with } \mu \in [.12, .15]\}.
\end{aligned}$$

For two alternatives $a, b \in \mathcal{A}$, we have,

$$b \preceq a \iff \sup_{\omega,\mu} \int \left(e^{-\omega(1.1+(\theta-.1)a} - e^{-\omega(1.1+(\theta-.1)b)}) \right) \pi(\theta|\mu) d\theta \leq 0. \tag{13}$$

To illustrate our method, we consider a normal distribution with mean 0 and standard deviation 0.1 as the importance sampling distribution. Then,

$$\begin{aligned}
\pi(\theta|\mu) &= (\sqrt{2\pi}0.1)^{-1} \exp\left(-\frac{1}{2} \left(\frac{\theta-\mu}{0.1} \right)^2 \right), \\
g(\theta) &= (\sqrt{2\pi}0.1)^{-1} \exp\left(-\frac{1}{2} \left(\frac{\theta}{0.1} \right)^2 \right), \text{ and} \\
\pi(\theta|\mu)/g(\theta) &= \exp(-50(\mu^2 - 2\mu\theta)).
\end{aligned}$$

Fix a and b. We have to compute, as an approximation to the supremum,

$$\sup_{\omega,\mu} \left(\frac{1}{n} \sum_{i=1}^{n} \left(e^{-\omega(\theta_i-.1)a} - e^{-\omega(\theta_i-.1)b} \right) e^{-50(\mu^2-2\mu\theta_i)} \right),$$

where $\theta_1, \ldots, \theta_n$ is i.i.d. from $\mathcal{N}(0, .1)$. The table shows the infima for $b = 0$ and $a = 1$.

n	S_n	$\widehat{EE}_n$	$S_n + 2\widehat{EE}_n$
500	0.0008	0.00112	0.0028
1000	0.0010	0.00083	0.0028
5000	−0.0003	0.00041	0.0004
10000	−0.0002	0.00025	0.0003
20000	−0.0004	0.00015	−0.0001

Then a dominates b, since $S_n + 2\widehat{EE}_n < 0$. △

9.4.2 Quantile and partially known classes

In decision analytic contexts, it is natural to consider quantile classes for probability distributions and partially known classes for loss functions, since they are associated with standard elicitation methods based on calibration methods using, for example, fortune wheels. We already defined $\Gamma_{\vec{p}}$ quantile classes in Section 9.2.3 and $\mathcal{L}_K$ in Section 9.3.3.

When $\mathcal{C}$ and Θ are discrete, dominance is checked by solving a nonlinear programming problem. Therefore, we shall describe only the more complex continuous case. Note that we need to pay attention only to discrete distributions.

Lemma 1 *Let $\Gamma_D \subset \Gamma_{\vec{p}}$ be such that*

$$\Gamma_D = \{\pi \in \Gamma_{\vec{p}} : \exists\, \theta_i \in A_i \;\; \text{such that } \pi(\theta_i) = p_i,\ i = 1, \ldots, n\}.$$

Then

$$\inf_{\mathcal{L}_K \times \Gamma_{\vec{p}}} N(L, \pi) = \inf_{\mathcal{L}_K \times \Gamma_D} N(L, \pi).$$

We then have:

Proposition 7 *For C_i, $i = 1, \ldots, m$, define the subsets of Θ*

$$\begin{aligned} A_{aij} &= \{\theta \in A_j : (a, \theta) \in C_i\},\ j = 1, \ldots, n, \\ A_{bij} &= \{\theta \in A_j : (b, \theta) \in C_i\},\ j = 1, \ldots, n, \\ A_j^{ik} &= A_{aij} \cap A_{bkj}. \end{aligned}$$

If

- *$\mathcal{C}$ and Θ are continuous,*
- *(a, θ) is continuous in θ,*
- *$(a, \theta) \neq (a, \theta')$, $\forall \theta \neq \theta'$ almost surely, $\forall a \in \mathcal{A}$,*

then

$$\sup_{\mathcal{L}_K \times \Gamma_{\vec{p}}} N^{ab}(L, \pi) = \sum_{j=1}^{n} \left[\sup_{i,k} \{ \sup_{\theta \in A_j^{ik}} \{(l_i - l_{k-1}) f(x|\theta)\}\} \right] p_j. \tag{14}$$

9.5 Efficient set

We consider now the problem of computing the set of all nondominated alternatives. This problem is one of the main challenges in multiobjective programming and stochastic dominance. When there is a finite number of alternatives we can appeal to a pairwise comparison scheme like that of Section 9.5.2. But, in general, the problem is a difficult task (see e.g. the CONDOR report, 1988). We study this problem in two cases; first, when we can compute exactly the efficient set, and second, with the help of an approximation scheme.

9.5.1 *Specific computation of the efficient set*

In some situations, due to the analytic form of the loss functions and prior distributions, we can compute the efficient set. We illustrate the idea with two examples. In Example. 3 the efficient set contains only one alternative. In Example. 4 the efficient set is an interval.

Example 3 (Martín et al. (1998)) Let $\pi(\theta|x)$ be a symmetric, unimodal distribution. Without loss of generality, assume that the mode is at 0. Let $L(a,\theta) = L(\theta - a)$ be any convex, nonconstant, symmetric loss function. Let $\mathcal{L}$ be any class of such loss functions. It follows that the mode 0 is the Bayes action for any $L \in \mathcal{L}$ and the unique nondominated action.

For any $a \in \mathcal{A}(= \Theta)$ and any $L \in \mathcal{L}$, it follows that

$$\int [L(0,\theta) - L(a,\theta)]d\pi(\theta|x)$$

$$\begin{aligned} &= \int_0^\infty [L(\theta) - L(\theta - a)]d\pi(\theta|x) + \int_0^\infty [L(-\theta) - L(-\theta - a)]d\pi(-\theta|x) \\ &= \int_0^\infty [2L(\theta) - L(\theta - a) - L(\theta + a)]d\pi(\theta|x) < 0, \end{aligned}$$

the inequality following from the convexity of L, which is nonconstant on a subset with positive probability. The result follows immediately. $\triangle$

Example 4 (Cont. Example 2) For two alternatives $a, b \in \mathcal{A}$, we have

$$b \preceq a \iff \sup_{\omega,\mu} \int \left(e^{-\omega(1.1+(\theta-.1)a} - e^{-\omega(1.1+(\theta-.1)b)}) \right) \pi(\theta|\mu)d\theta \leq 0. \tag{15}$$

So,

$$b \preceq a \iff (.1 - \mu)(b - a) \geq \frac{1}{2}\omega(.1)^2(a^2 - b^2), \quad \forall \mu, \omega. \tag{16}$$

If $b > a$, then (15) never holds. If $b < a$, (15) becomes

$$(.1 - \mu) \geq -\frac{1}{2}\omega(.1)^2(a + b). \tag{17}$$

Then, as $\mu \in [.12, .15]$ and $\omega \in [.12, .15]$, $a = .8$ dominates $b \in [0, .8)$ and it is easy to prove that $[.8, 1]$ is the efficient set. $\triangle$

9.5.2 *Approximation scheme*

In most cases, however, we shall not be able to compute exactly the nondominated set, and we shall need a scheme to approximate it. The following algorithm (Bielza et al., 2000) provides an approximation. Basically, we sample a subset A_n of alternatives, and compute the set N_n of nondominated alternatives within that subset, suggesting it as an approximation to the nondominated set.

1. Sample $A_n = \{a_1, \ldots, a_n\} \subset \mathcal{A}$.
2. Choose $\pi \in \Gamma, L \in \mathcal{L}$ and relabel the a's as $(e_1, \ldots, e_n)$ so that $N(\pi, L; e_i, e_{i+1}) \leq 0, \forall i = 1, \ldots, n-1$.
3. Let $d(i) = 0$, $\forall i$.
4. For $i = 1$ to $n - 1$

```
    Do
    If d(i) = 0, for j = i + 1 to n
      Do
      If d(j) = 0
        If e_j ⪯ e_i, let d(j) = 1
                    Else, if (N(π, L; e_i, e_j) = 0 and e_i ⪯ e_j)
                          then d(i) = 1 and next i
      End Do
    End Do
```

5. Let $N_n = \{e_i : d(i) = 0\}$.

As shown in Bielza et al. (2000), under certain conditions, as the sample size increases, N_n converges to the true nondominated set in an appropriate sense. Step 2 permits some reduction in the computational burden.

Note that the basic problem is the provision of procedures to check whether an alternative a dominates an alternative b for step 4. We have illustrated this problem in Sections 9.2, 9.3 and 9.4 for several classes of loss functions and prior distributions.

Example 5 (Cont. Example 2) We apply the discretisation scheme considering only a sample size of 500. First, we apply step 2 with $\mu = .15$ and $\omega = 1$. The expected loss of alternatives is represented in Figure 1a. So we have ordered them in decreasing order. Then, we apply step 4, using formula (17). The values of d are represented in Figure 1b. As $N_n = \{a : d(a) = 0\}$, we can conclude that the nondominated set is approximately [.8,1]. $\triangle$

9.6 Screening methods

In some cases, nondominance is a very powerful concept leading to a unique or a finite number of nondominated alternatives, from which it is easy to choose one of them. However, in most cases the nondominated set will be too big to reach a final decision. Note that as a byproduct of the procedures to compute or approximate the nondominated set, we obtain estimates on the differences in posterior expected losses among nondominated alternatives. If these were not too large, we would conclude that these do not

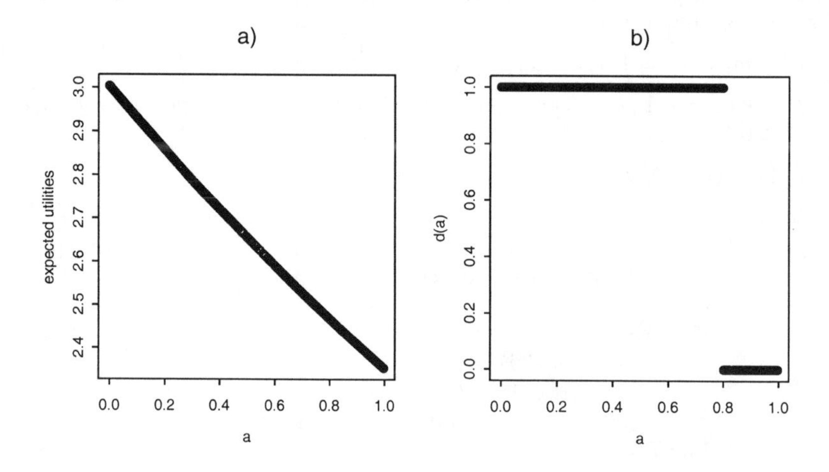

FIGURE 1. a) expected loss, b) nondominated ($d = 0$) points

perform too differently and basically we could recommend any of those alternatives, that is, if robustness holds, any of the nondominated solutions is acceptable.

On the contrary, if these were large there would be notorious differences among nondominated alternatives and we should proceed with care. One possibility would be to elicit additional information from the decision maker and further constrain the classes of probability distributions and loss functions. Clearly, in this case the set of nondominated alternatives will be smaller and we would hope that this iterative process would converge until the nondominated set is small enough to reach a final decision. Martín and Ríos Insua (1996) provides ideas for aiding in eliciting additional information when robustness lacks.

It is conceivable that, in such a context, at some stage we might not be able to gather additional information, yet there are several nondominated alternatives with very different expected utilities. In this case, we need a method to reduce the efficient set. For example, minimax solutions may be useful as a way to select a single robust solution. We associate with each alternative its biggest expected loss; we then suggest the alternative with minimum greatest expected loss. Specifically, an alternative $a_M \in \mathcal{A}$ is $\mathcal{L} \times \Gamma$-minimax if $\max_{L \in \mathcal{L}, \pi \in \Gamma} T(L, \pi, a_M) = \min_{a \in \mathcal{A}} \max_{L \in \mathcal{L}, \pi \in \Gamma} T(L, \pi, a)$. Minimax (and related concepts such as minimax regret solutions) are usually introduced as ad hoc concepts. We view them as heuristics that allow us to choose a nondominated alternative in cases where we are not able to gather additional information, as results like the following show:

Proposition 8 *If the set of $\mathcal{L} \times \Gamma$-minimax alternatives is finite, one of them is nondominated.*

As a corollary, if there is a unique $\mathcal{L} \times \Gamma$-minimax alternative, it is nondominated. Interestingly enough, some of these minimax concepts have axiomatic foundations compatible with those described for robust Bayesian analysis.

Computation of minimax alternatives is not simple in general cases, but, again, we may appeal to discretisation schemes. Once with a discrete sample from the set of alternatives, we compute the corresponding nondominated set and, then, the maximum posterior expected loss of each alternative (this is a conventional global sensitivity analysis problem), picking such with the minimum maximum.

Other methods to select a decision from an efficient set may be seen in Ríos Insua (1990) and Ríos Insua et al. (1997).

9.7 Conclusions

We have analyzed the case in which beliefs are modelled with a class of probability distributions, the typical context in robust Bayesian analysis, and preferences modelled with a class of loss functions. However, rather than studying the local or global behavior of a posterior probability or a posterior expectation, we have concentrated on the global behavior of differences in expected losses. We believe that this should be the main computational issue for robust Bayesians, since it allows us to check dominance among alternatives.

This problem is essential in looking for nondominated alternatives. If the set of alternatives is finite, we may apply the approach of Section 9.5.2 based on pairwise comparisons. In the continuous case, we propose an approximation method. Also, we have described the case in which imprecision in preferences is modelled with a class of loss functions; and the case of imprecision in both preferences and beliefs.

Finally, for problems with a big efficient set, we have mentioned some screening methods.

For future work, there are many open problems like exploring new screening methods, checking dominance with imprecision in both beliefs and preferences, and incorporating imprecision in likelihood to the robustness analysis. Also, it is important to produce results to compute the efficient set for fixed classes (like quantile class, partially known class) without approximations.

Acknowledgments

We are very grateful to David Ríos and a referee for discussions and helpful comments. This work has been partially supported by a grant from CICYT.

References

BAZARAA, M. and SHETTY, S. (1979). *Nonlinear Programming.* New York: Wiley.

BELL, D. (1995). Risk, return and utility. *Management Science,* **41**, 23–30.

BERGER, J. (1990). Robust Bayesian analysis: sensitivity to the prior. *Journal of Statistical Planning and Inference,* **25**, 303–328.

BERGER, J. (1994). An overview of robust Bayesian analysis (with discussion). *Test,* **3**, 5–124.

BERGER, J. and BERLINER, L.M. (1986). Robust Bayes and empirical Bayes analysis with ε-contaminated priors. *Annals of Statistics,* **14**, 461–486.

BERLINER, L.M. and GOEL, P. (1986). Incorporating partial prior information: ranges of posterior probabilities. *Technical Report,* **357**, Dept. of Statistics, Ohio State University.

BIELZA, C., MARTÍN, J. and RÍOS INSUA, D. (2000). Approximating nondominated sets in continuous problems. *Technical Report,* Technical University of Madrid.

CONDOR (1988). Operations Research: the next decade. *Operations Research,* **36**, 619–637.

DALL'AGLIO, M. and SALINETTI, G. (1994). Bayesian robustness, moment problem and semi-infinite linear programming. *Technical Report,* Universitá "La Sapienza," Roma.

DEY, D., LOU, K. and BOSE, S. (1998). A Bayesian approach to loss robustness. *Statistics and Decisions,* **16**, 65–87.

FRENCH, S. and RÍOS INSUA, D. (2000). *Statistical Decision Theory.* London: Arnold.

HWANG, J.T. (1985). Universal domination and stochastic domination: estimation simultaneously under a broad class of loss functions. *Annals of Statistics,* **13**, 295–314.

JENNRICH, R. (1969). Asymptotic properties of non-linear least squares estimators. *Annals of Mathematical Statistics,* **40**, 639–643.

LEVY, H. (1992). Stochastic dominance and expected utility analysis: survey and analysis. *Management Science,* **38**, 555–593.

MARTÍN, J. and RÍOS INSUA, D. (1996). Local sensitivity analysis in Bayesian decision theory. In *Bayesian Robustness* (J.O. Berger et al., eds.) *Lectures Notes Institute of Mathematical Statistics*, **29**, 119–135. Hayward: IMS.

MARTÍN, J., RÍOS INSUA D. and RUGGERI, F. (1998). Issues in Bayesian loss robustness. *Sankhya: The Indian Journal of Statistics, Series A,* **60**, 405–417.

MORENO, E. and PERICCHI, L. (1992). Subjetivismo sin dogmatismo: Inferencia bayesiana robusta. *Estadística Española,* **34**, 5–60.

NAU, R., RÍOS INSUA, D. and MARTÍN, J. (1996). The shape of incomplete preferences and the foundation of robust Bayesian statistics. *Technical Report*, Madrid Technical University.

RÍOS INSUA, D. (1990). *Sensitivity Analysis in Multiobjective Decision Making.* New York: Springer Verlag.

RÍOS INSUA, D. and FRENCH, S. (1991). A framework for sensitivity analysis in discrete multiobjective decision making. *European Journal of Operational Research,* **54**, 176–190.

RÍOS INSUA, D. and MARTÍN, J. (1994a). On the foundations of robust decision making. In *Decision Theory and Decision Analysis: Trends and Challenges* (Ríos, ed.). Amsterdam: Kluwer.

RÍOS INSUA, D. and MARTÍN, J. (1994b). Robustness issues under imprecise beliefs and preferences. *Journal of Statistical Planning and Inference,* **48**, 383–389.

RÍOS INSUA, D., MARTÍN, J., PROLL, L., FRENCH, S. and SALHI, A. (1997). Sensitivity analysis in statistical decision theory: A decision Analytic view. *Journal of Statistical Computation and Simulation,* **57**, 197–218.

RÍOS INSUA, D., RUGGERI, F. and MARTÍN, J. (1999). Bayesian sensitivity analysis: a review. In *Mathematical and Statistical Methods for Sensitivity Analysis* (A. Saltelli et al., eds.). New York: Wiley.

RUGGERI, F. (1992). Bounds on the prior probability of a set and robust Bayesian analysis. *Theory of Probability and Its Applications,* **37**, 358–359.

SHAO, J. (1989). Monte Carlo approximations in Bayesian decision theory. *Journal American Statistical Association,* **84**, 727–732.

SIVAGANESAN, S. and BERGER, J. (1989). Ranges of posterior measures for priors with unimodal contaminations. *Annals of Statistics,* **17**, 868–889.

VARIAN, H.R. (1975). A Bayesian approach to real state assessment. In *Studies in Bayesian Econometrics and Statistics in Honor of Leonard J. Savage* (S. Feinberg and A. Zellner, eds.). Amsterdam: North-Holland.

VICKSON, R. (1977). Stochastic orderings from partially known utility functions. *Mathematics of Operations Research,* **2**, 244–252.

WASSERMAN, L., LAVINE, M. and WOLPERT, R. (1993). Linearization of Bayesian robustness problems. *Journal of Statistical Planning and Inference,* **37**, 307–316.

10

Stability of Bayes Decisions and Applications

Joseph Kadane, Gabriella Salinetti and Cidambi Srinivasan

ABSTRACT This article reviews the recent developments on the stability of Bayes decision problems. Following a brief account of the earlier works of Kadane and Chuang, who gave the initial formulation of the problem, the article focuses on a major contribution due to Salinetti and the subsequent developments. The discussion also includes applications of stability to the local and global robustness issues relating to the prior distribution and the loss function of a Bayes decision problem.

Key words: elicitation, topology, Gâteaux derivative, locally equi-lower bounded growth.

10.1 Introduction

Stability of Bayes decision problems is the qualitative study of the sensitivity of the Bayesian inferences to uncertainty in the inputs. Typically, these inputs are the likelihood, the prior distribution and the loss function. In practical situations these inputs are often chosen to be approximations to the truth in some sense with the hope that the errors in the approximations will not lead to inferences and decisions with loss substantially greater than would have been obtained if the true inputs could be written down precisely. So it is important to know not only the exact answer for an approximately specified initial position, but also what happens if the inputs are changed in a neighborhood of the assumed initial position. In particular, it would be of interest to know if the inference based on approximate input would converge to the "true" inference as the approximation improves. Informal studies of stability of this nature have long been a part of applied Bayesian analysis (Edwards et al., 1963, and Box, 1980).

The reasons for the study of stability are based on foundational as well as practical considerations. The foundational motivation is guided by the widely accepted fact that subjective Bayesian analysis is the only coherent method of inference. This, of course, assumes that one can make arbitrarily fine discriminations in the judgment of the inputs and the inferences are relatively stable (qualitatively robust) as the judgments become more accurate.

The practical motivation for stability lies in quantitative Bayesian robustness analysis. The theory of stability is the foundation of the development of tools (Kadane and Srinivasan, 1996) needed for quantitative analysis of global and local Bayesian robustness issues. Indeed, stability leads to a general linearizing algorithm for global Bayesian robustness analysis. Also, quantitative tools analogous to the classical influence functions for Bayesian robustness analysis can be derived for stable decision problems.

This article is an incomplete review of stability. This focuses only on the developments in stability since the work of Kadane and Chuang (1978). The literature on stability contains several articles preceding the work of the above mentioned authors. Stone (1963) studied decision procedures with respect to the use of wrong prior distributions and emphasized the usefulness of non-optimal decisions that do not require full specification of the prior probability distribution. Fishburn et al. (1967) and Pierce and Folks (1969) also discussed decision making under uncertainty in the prior distribution. Britney and Winkler (1974) have investigated stability of Bayesian point estimates with respect to the uncertainty in the loss.

10.2 Notation

The entire data will be denoted by X, which is assumed to arise from a density $f(x|\theta)$ (with respect to a fixed dominating measure), with θ denoting the finite-dimensional unknown parameter of f. A prior probability measure will be denoted $\pi(\theta)$. In addition to the prior probability and the likelihood, another important input in a decision theoretic problem is the loss function $L(\theta, d)$, where d denotes a decision. The set of all feasible decisions is denoted by D.

Key Bayesian decision theoretic quantities are the posterior distribution

$$\pi(\theta|x) = f(x|\theta)\pi(\theta)/m(x|\pi)$$

with the assumption that the marginal distribution $m(x|\pi)$ is positive, and the posterior risk

$$\psi\left(\pi, d(x)\right) = \int L\left(\theta, d(x)\right) d\pi(\theta|x)\,.$$

Since the data X is fixed in Bayesian analysis, we will suppress x wherever possible. In particular, $d(x)$ will be denoted by d and the likelihood $f(x|\theta)$ with x fixed will be denoted by $l(\theta)$.

10.3 Stability

The theory of stability of Bayes decision problems, as formulated by Kadane and Chuang (1978), is an approach to formalize the robustness of the elicited inputs, the prior and the loss, of the decision problem. If stability obtains, close enough elicitation leads to decisions with posterior expected loss, evaluated correctly, nearly as small as achievable. If, on the other hand, the problem is unstable, a very small elicitation error may lead to a disproportionately large expected loss. This motivation led to the following framework.

To formulate the definitions of stability of a decision problem, suppose that the parameter space is $\Theta \subset R^m$, the decision space is $D \subset R^m$ and the likelihood is a function $\ell(\theta)$. Let $L_0(\theta, d)$ be a loss function and π_0 be a prior distribution on Θ. Also, let L_n, $n = 1, 2, \ldots$ denote a sequence of loss functions converging (in some topology) to L_0 and $\pi_n, n = 1, 2 \ldots$ a sequence of priors converging weakly to π_0. We will denote the weak convergence by "$\pi_n \Rightarrow \pi_0$."

Definition I: The decision problem (L_0, ℓ, π_0) is strongly stable I (SSI) if for every sequence $\pi_n \Rightarrow \pi_0$ and $L_n \rightarrow L_0$

$$\lim_{\epsilon \downarrow 0} \limsup_{n \to \infty} \left[\int L_n(\theta, d_0(\epsilon))\ell(\theta) d\pi_n(\theta) - \inf_{d \epsilon D} \int L_n(\theta, d)\ell(\theta) d\pi_n(\theta) \right] = 0 \tag{1}$$

for every $d_0(\epsilon)$ such that

$$\int L_0(\theta, d_0(\epsilon))\ell(\theta) d\pi_0(\theta) \leq \inf_{d \in D} \int L_0(\theta, d)\ell(\theta) d\pi_0(\theta) + \epsilon. \tag{2}$$

The decision problem (L_0, ℓ, π_0) is said to be weakly stable I (WSI) if (1) holds for a particular choice $d_0(\epsilon)$.

Viewing (L_0, ℓ, π_0) as the approximate specification by the statistician and assuming the "truth" lies along the sequence (L_n, ℓ, π_n), the strong stability I of (L_0, ℓ, π_0) implies that "small" errors in the specification of the loss and the prior will not result in substantially worse decisions in terms of the risk. This essentially motivates the above definition.

A more general and stringent definition of stability is possible and is as follows.

Definition II: The decision problem (L_0, ℓ, π_0) is strongly stable II if for all sequences $\pi_n \Rightarrow \pi_0$, $Q_n \Rightarrow \pi_0$, $L_n \rightarrow L_0$ and $W_n \rightarrow L_0$,

$$\lim_{\epsilon \downarrow 0} \limsup_{n \to \infty} \left[\int L_n(\theta, d_{Q_n}(\epsilon))\ell(\theta) d\pi_n(\theta) - \inf_{d \in D} \int L_n(\theta, d)\ell(\theta) d\pi_n(\theta) \right] = 0 \tag{3}$$

for every $d_{Q_n}(\epsilon)$ satisfying

$$\int W_n(\theta, d_{Q_n}(\epsilon))\ell(\theta)dQ_n(\theta) \leq \inf_{d \in D} \int W_n(\theta, d)\ell(\theta)dQ_n(\theta) + \epsilon. \quad (4)$$

The problem (L_0, ℓ, π_0) is weakly stable II if (3) holds for a particular choice of $d_{Q_n}(\epsilon)$.

The reader is referred to Kadane and Chuang (1978) for a motivation of the second definition of stability. It permits the following interpretation: for any two sequences, (W_n, ℓ, Q_n) and (L_n, ℓ, π_n), of approximations to (L_0, ℓ, π_0), the optimal decisions d_{Q_n} of the first sequence tend to be nearly optimal when evaluated at (L_n, ℓ, π_n). In other words, stability in the sense of Definition I holds uniformly in a small neighborhood around the initial specification (L_0, ℓ, π_0).

In the same paper, Kadane and Chuang studied the stability of a decision problem under the topology of uniform convergence (i.e., $L_n(\theta, d)$ converges to $L_0(\theta, d)$ uniformly in θ and d) for the losses and obtained sufficient conditions for stability. In particular, they noted that under uniform convergence (1) is equivalent to

$$\lim_{\epsilon \downarrow 0} \limsup_{n \to \infty} \left[\int L_0(\theta, d_0(\epsilon))\ell(\theta)d\pi_n(\theta) - \inf_d \int L_0(\theta, d)\ell(\theta)d\pi_n(\theta) \right] = 0 \quad (5)$$

and (3) and (4) are equivalent to

$$\lim_{\epsilon \downarrow 0} \limsup_{n \to \infty} \left[\int L_0(\theta, d_{Q_n}(\epsilon))\ell(\theta)d\pi_n(\theta) - \inf_d \int L_0(\theta, d)\ell(\theta)d\pi_n(\theta) \right] = 0 \quad (6)$$

and

$$\int L_0(\theta, d_{Q_n}(\epsilon))\ell(\theta)dQ_n(\theta) \leq \inf_d \int L_0(\theta, d)\ell(\theta)dQ_n(\theta) + \epsilon, \quad (7)$$

respectively. This fact considerably simplifies the task of verifying the stability.

It is straightforward to show that SSII implies SSI. However, the converse is not in general true even under the uniform convergence of losses.

Example 10.3.1 Let $\Theta = D = [-1, 1]$ and the loss L_0 be given by

$$L_0(\theta, d) = \begin{cases} 1 & \text{if } d \neq 0, \ \theta d \leq 0, \\ (\theta - d)^2 & \text{otherwise.} \end{cases}$$

△

It is easy to show that the problem is strongly stable I under the uniform convergence of the loss. However, strong stability II does not hold if $\pi_n = \delta_{\{1/n\}}$ and $Q_n = \delta_{\{1/n\}}$. See Kadane and Srinivasan (1998) for details.

10.4 Examples of unstable decision problems

Stability I is not always guaranteed, even in standard statistical problems like tests of hypotheses and estimation. The following two examples are illustrative of this phenomenon.

The first example illustrates instability in a testing of a hypothesis problem with a simple null hypothesis versus a composite alternative hypothesis.

Example 10.4.1 *Let θ be a real-valued parameter, and consider testing $H_0 : \theta = \theta_0$ versus $H_1 : \theta \neq \theta_0$, where θ_0 is specified. Then the decision space is finite, say, $D = \{1, 2\}$, and the canonical loss function is*

$$L(\theta, 1) = 0 \text{ and } L(\theta, 2) = 1 \text{ if } \theta = \theta_0,$$

$$L(\theta, 1) = 1 \text{ and } L(\theta, 2) = 0 \text{ if } \theta \neq \theta_0.$$

Let π be a prior distribution on θ. Then the testing problem is unstable I if the posterior CDF $\Pi(\theta|x)$ has a significant jump at θ_0. More precisely, the testing problem is unstable in the sense of Definition I if $\Pi(\theta_0|x) - \Pi(\theta_0^-|x) > 1/2$.

A similar example can be obtained for composite hypotheses. Also, the example holds even when the likelihood function is very smooth.

The next example shows that even estimation (or prediction) problems with smooth continuous loss functions may be unstable.

Example 10.4.2 *Consider estimating the parameter θ of a likelihood $l(\theta) \equiv 1$. Let the loss function be the familiar squared error loss, namely, $L(\theta, d) = (\theta - d)^2$. If $\Theta = D = R$, then the estimation problem is unstable for any prior probability distribution π on Θ.*

The above example on estimation is very important in that the familiar estimation problems are generally unstable in the sense of Kadane and Chuang (1978). The instability in the example occurs because weak convergence does not imply convergence in expectation. This raises two important issues:

1. What are the necessary and sufficient conditions for stability?
2. Can stability be obtained by suitably restricting the approximating priors to suitable subclasses of prior distributions? If so, characterize the subclasses.

These two issues will be the focus of the subsequent sections.

10.5 Sufficient conditions for stability I

The examples in the preceding section clearly indicate that the notion of stability as formulated by Kadane and Chuang (1978) is fairly stringent.

Stability depends very much on the smoothness of the true prior distribution π_0, the loss function L_0 and the likelihood $l(\theta)$. In addition, a decision problem with smooth inputs may be unstable along particular sequences of approximating priors $\{\pi_n\}$ and loss functions $\{L_n\}$. Therefore, it is of interest to develop sufficient conditions for stability as well as characterize classes of approximating prior distributions and loss functions that guarantee stability. In the following discussion of these issues we assume the topology of uniform convergence for the losses.

There are several sets of necessary and sufficient conditions for strong stability I developed by Chuang (1984), Salinetti (1994) and Kadane and Srinivasan (1998). One of the most general results in this direction is due to Salinetti (1994) which subsumes the results of Chuang (1984). The following definition is needed to summarize Salinetti's theorem.

Let d_0 be the optimal Bayes decision with respect to the true prior π_0.

Definition III: The loss $L_0(\theta, d)$ has a *locally equi-lower bounded* growth if for every $d \in D$ there exists a neighborhood $N(d)$ of d and a constant $b(d)$ such that for every $d' \in N(d)$,

$$L_0(\theta, d') - L_0(\theta, d_0) > b(d)$$

for all $\theta \in \Theta$.

The following theorem holds under the assumptions:

1. $L_0(\theta, d)$ is jointly lower semi-continuous;
2. $L_0(\theta, d_0)$ is continuous in θ;
3. $L_0(\theta, d)$ has a locally equi-lower bounded growth.

Theorem 10.5.1 (Salinetti) *The decision problem (L_0, l, π_0) is strongly stable I if and only if for any sequence of approximating priors $\pi_n \Rightarrow \pi_0$ and every $\epsilon > 0$, the sequence (L_0, l, π_n) of decision problems has a bounded sequence of ϵ-optimal Bayes decisions.*

The above theorem is fairly general, though the verification of the conditions can be tedious. The existence of a bounded sequence of ϵ-optimal decisions often holds for unbounded bowl-shaped loss functions, examples of which include the familiar squared error and absolute deviation losses. General sufficient conditions for the existence of a bounded sequence of ϵ-optimal decisions are available in Kadane and Srinivasan (1998). The condition of equi-lower boundedness of L_0 is too strong and often not satisfied by unbounded loss functions. In particular, the squared error loss function fails to have equi-lower boundedness. However, this is not a serious issue, as observed by Kadane and Srinivasan (1998). The authors have shown that the equi-lower boundedness can be replaced by the weaker condition "$L_0(\theta, d)l(\theta)$ is bounded continuous in θ for each d." This latter condition is often satisfied if the likelihood is smooth and vanishes at infinity at a sufficiently fast rate. For details see Theorem 4.4 in Kadane and Srinivasan (1998).

Salinetti's theorem is not applicable to finite decision problems because the loss functions in those problems are often discontinuous. Kadane and Srinivasan (1998) have developed easily verifiable sufficient conditions for the stability of finite decision problems. Their results are applicable, in particular, to multiple decision problems and set estimation problems.

Strong stability often holds if a slightly weaker notion is adopted wherein the approximating prior distributions are restricted to suitable subclasses of all prior distributions. The choice of the subclasses may be determined by the statistical problem under consideration. In Bayesian robustness analysis reviewed in Berger (1994) it is a common practice to restrict consideration to special classes of prior distributions: ϵ-contamination class, density ratio class, as treated by DeRobertis and Hartigan (1981); unimodal quantile class of Berger and O'Hagan (1988); and Gustafson class (1994). Strong stability is easily established (see Kadane and Srinivasan, 1996) when the approximating prior distributions are restricted to these classes. In a similar vein, Salinetti (1994) showed that strong stability holds if the approximating prior distributions π_n are $L_0(\bullet, d_0)$ tight in the sense that the posterior risk of d_0 under π_n converges to the posterior risk under π_0.

Finally, a few comments are in order regarding strong stability II. As observed earlier, SSII is far more stringent than SSI. However, sufficient conditions for SSII and structural results establishing the equivalence of the two notions of stability are discussed at some length in Kadane and Chuang (1978), Chuang (1984) and Kadane and Srinivasan (1998).

10.6 Bayesian robustness and stability

The notion of stability was introduced by Kadane and Chuang (1978) with the goal of laying foundation to the study of Bayesian robustness. The Bayesian decision theoretic framework involves assumptions regarding the likelihood, the loss and the prior distribution of the relevant unknown parameters. These specifications are often based on convenient rationalizations of somewhat imprecise knowledge of the underlying statistical problem and justified on the grounds of a vague notion of continuity. The concept of strong stability makes the vague notion of continuity precise. In addition, strong stability facilitates the study of robustness by enabling the development of tools for the quantitative robustness analysis. An important tool in local as well as global Bayesian robustness analysis is the derivative of the relevant posterior quantity with respect to the prior distribution (Ruggeri and Wasserman, 1993, Srinivasan and Trusczynska, 1992, Gustafson et al., 1996). Of particular interest are the derivatives of the optimal Bayes decisions and the posterior expected losses of Bayes decisions.

An important consequence of strong stability is the Gâteaux differentiability (Kadane and Srinivasan, 1996) of the optimal Bayes decision and smooth functions of it with respect to the prior distribution.This is stated below in the context of estimation where the loss $L_0(\theta, d)$ is a function of only the difference $(\theta - d)$, which is denoted by $L_0(\theta - d)$.

More precisely, suppose that the loss function $L_0(\theta - d)$ is twice differentiable, strictly convex and the likelihood is bounded continuous. Let d_{π_0} denote the optimal Bayes decision with respect to π_0. Then we have the following result.

Theorem 10.6.1 (Kadane and Srinivasan) *Suppose* (L_0, l, π_0) *is strongly stable I. Then* d_{π_0} *is Gâteaux differentiable and the derivative in the direction of* $Q \in \mathcal{P}$ *is*

$$d'_{\pi_0}(Q) = \frac{-2 \int L'_0(\theta - d_{\pi_0}) l(\theta|x) dQ(\theta)}{\int L''_0(\theta - d_{\pi_0}) l(\theta|x) d\pi_0(\theta)},$$

where $\mathcal{P}$ is a convex set of prior distributions containing π_0.

Similar results establishing the derivative for other functions of the decision problem can be found in Srinivasan and Trusczynska (1992) and Ruggeri and Wasserman (1993).

The derivative $d'_{\pi_0}(Q)$ (or the derivatives of other smooth functions of d_{π_0}) plays a crucial role in global robustness analysis. Global analysis involves the computation of the ranges of relevant functions of d_π as π varies in a specified class Λ of prior distributions and making judgments about the robustness of the posterior quantity. The derivative can be used to develop an algorithm to compute the range. For details see Kadane and Srinivasan (1996) and Perone Pacifico et al. (1998).

The derivatives can also be used for local robustness analysis. A local sensitivity measure analogous to the frequently used "influence function" can be developed using the derivative for studying the sensitivity of the Bayes decision or a functional of it. See Ruggeri and Wasserman (1993), Kadane and Srinivasan (1996) and Conigliani et al. (1995) for the relevant results and examples.

10.7 Future work

Many extensions of the ideas of stability have not been explored. There are other topologies specifying closeness of priors, likelihoods and losses to be considered, and the extension of these ideas to infinite-dimensional parameter spaces. There are also more refined results that may be available on rates of convergence, when suitable stability holds. Finally, there always remains the fundamental question with which work on stability began: what advice can these results offer to aid the improvement of methods of elicitation?

Acknowledgments

J. Kadane's research is supported in part by the National Science Foundation under Grant DMS 9801401. G. Salinetti's research is supported in part by CNR under Grant "Decisioni Statistiche: Teoria e Applicazioni" and C. Srinivasan's research is supported in part by the National Institute of Health under Grant CA-77114.

References

BERGER, J. (1994). An overview of robust Bayesian analysis. *Test*, **3**, 5–59.

BERGER, J.O. and O'HAGAN, A. (1988). Ranges of posterior probabilities for unimodal priors with specified quantiles. In *Bayesian Statistics 3* (J.M. Bernardo et al., eds.). Oxford: Oxford University Press.

BOX, G.E.P (1980) Sampling and Bayes' inference in sicentific modelling and robustness (with discussion). *J. Roy. Statist. Soc. Ser. A*, **143**, 383–430.

BRITNEY, R. and WINKLER, R. (1974). Bayesian point estimation and prediction. *Annals of the Institute of Statistical Mathematics*, **26**, 15–34.

CHUANG, D.T. (1984). Further theory of stable decisions. In *Robustness of Bayesian Analysis* (J. Kadane, ed.), 165–228. New York: North Holland.

CONIGLIANI, C., DALL'AGLIO, M., PERONE PACIFICO, M. and SALINETTI, G. (1995). Robust statistics: from classical to Bayesian analysis. *Metron*, **LII, 3–4**, 89–109.

DEROBERTIS, L. and HARTIGAN, J. (1981). Bayesian inference using intervals of measures. *Annals of Statistics*, **9**, 235–244.

EDWARDS, W., LINDEMAN, H. and SAVAGE, L.J. (1963). Bayesian statistical inference for psychological research. *Psychological Review*, **70**, 193–242.

FISHBURN, P., MURPHY, A. and ISAACS, H. (1967). Sensitivity of decision to probability estimation errors: a reexamination. *Operations Research*, **15**, 254–267.

GUSTAFSON, P. (1994). Local Sensitivity of Posterior Expectations. *Ph.D. Dissertation*, Dept. Statistics, Carnegie-Mellon University.

GUSTAFSON, P., SRINIVASAN, C. and WASSERMAN, L. (1996). Local sensitivity. In *Bayesian Statistics 5* (J.M. Bernardo, J.O. Berger, A.P. Dawid and A.F.M. Smith, eds.). Oxford: Oxford University Press.

KADANE, J. and CHUANG, D.T. (1978). Stable decision problems. *Annals of Statistics*, **6**, 1095–1110.

KADANE, J. and SRINIVASAN, C. (1994). Comment on "An Overview of Robust Bayesian Analysis" by J.O. Berger. *Test*, **3**, 93–95.

KADANE, J. and SRINIVASAN, C. (1996). Bayesian Robustness and stability. In *Bayesian Robustness*, (J.O. Berger, B. Betrò, E. Moreno, L.R. Pericchi, F. Ruggeri, G. Salinetti and L. Wasserman, eds.), 81–100. IMS Lecture Notes - Monograph Series Volume 29, Hayward, CA: Institute of Mathematical Statistics.

KADANE, J. and SRINIVASAN, C. (1998). Bayes decision Problems and stability. *Sankhya*, **60**, 383–404.

O'HAGAN, A. and BERGER, J. (1988). Ranges of posterior probabilities for quasi-unimodal priors with specified quantiles. *Journal of the American Statistical Association*, **83**, 503–508.

PERONE PACIFICO, M., SALINETTI, G. and TARDELLA, L. (1998). A note on the geometry of Bayesian global and local robustness. *Journal of Statistical Planning and Inference*, **69**, 51–64.

PIERCE, D. and FOLKS, J.L. (1969). Sensitivity of Bayes procedures to the prior distribution. *Operations Research*, **17**, 344–350.

RUGGERI, F. and WASSERMAN, L. (1993). Infinitesimal sensitivity of posterior distributions. *Canadian Journal of Statistics*, **21**, 195–203.

SALINETTI, G. (1994). Stability of Bayesian decisions. *Journal of Statistical Planning and Inference*, **40**, 313–320.

SRINIVASAN, C. and TRUSCZYNSKA, H. (1992). On the ranges of posterior quantities. *Technical Report*, **294**, Department of Statistics, University of Kentucky.

STONE, M. (1963). Robustness of nonideal decision procedures. *Journal of the American Statistical Association*, **58**, 480–486.

11

Robustness Issues in Bayesian Model Selection

Brunero Liseo

ABSTRACT One of the most prominent goals of Bayesian robustness is to study the sensitivity of final answers to the various inputs of a statistical analysis. Also, since the use of extra-experimental information (typically in the form of a prior distribution over the unknown "elements" of a model) is perceived as the most discriminating feature of the Bayesian approach to inference, sensitivity to the prior has become perhaps the most relevant topic of Bayesian robustness.

On the other hand, the influence of the prior on the final answers of a statistical inference does not always have the same importance. In estimation problems, for example, priors are often picked for convenience, for, if the sample size is fairly large, the effect of the prior gets smaller. But, this is not the case in model selection and hypothesis testing. Both Bayes factors and other alternative Bayesian tools are quite sensitive to the choice of the prior distribution, and this phenomenon does not vanish as the sample size gets larger. This makes the role of Bayesian robustness crucial in the theory and practice of Bayesian model selection.

In this chapter we briefly outline the formal problem of Bayes model selection and indicate the ways robust Bayesian techniques can be a valuable tool in the model choice process.

Due to the prominent role of Bayes factors in model selection problems, we will discuss in detail sensitivity of Bayes factors (and their recent ramifications) to prior distribution. Examples of theoretical interest (precise hypothesis testing, default analysis of mixture models) will also be presented.

Key words: Bayes factor, classes of prior distributions, default priors for testing, nested models.

11.1 Introduction

Suppose that a researcher has collected some data x and he wants to use them to draw some conclusions about his research. The researcher knows there are a number of existing statistical models he can use, say $M_1, M_2, \ldots, M_k$, and many others that can be created. Among the many possible questions the researcher can ask himself are the following:

- Among $M_1, M_2, \ldots, M_k$, which model best explains the data at hand?
- Among $M_1, M_2, \ldots, M_k$, which model is best suited to predict a future observation?
- Among $M_1, M_2, \ldots, M_k$, which is, in some sense, the *closest* to the true mechanism that generated the data?

Some other times $k = 1$, that is, for several reasons, he would like to make use of a prespecified model: in this case the obvious question is

- Is M_1 a sufficiently good approximation to the *true* model?

Bayesian model selection (BMS, hereafter) is the area of Bayesian statistics which helps to answer these (and other) questions. Needless to say, the entire topic is rather complex and a unified description cannot be done without a decision theoretic viewpoint, as in Bernardo and Smith (1994), although some Bayesian statisticians, more radically, view BMS as an exploratory task (Bayarri, 1999). However, we shall not give an exhaustive account of BMS literature; excellent reviews are given by Kass and Raftery (1995), De Santis and Spezzaferri (1999), Berger (1999) and George (1999). Rather, we want to describe the reason why BMS techniques are very sensitive to the various inputs of a Bayesian analysis. Also, we aim at illustrating how robust Bayesian tools can be valuable in the art of selecting a model.

In BMS and hypothesis testing as well, the Bayes factor plays a central role, although in itself it is not a fully coherent Bayesian measure of support (see Lindley, 1997, and Lavine and Schervish, 1999). This paper will deal mainly with the Bayes factor and its ramifications; other Bayesian approaches to model selection will be briefly outlined. Also, we will concentrate on sensitivity to the prior: likelihood robustness is addressed in Shyamalkumar (2000). The chapter is organised as follows: in §2 we describe, somewhat subjectively, what we believe BMS is, and what ingredients are needed to perform it: we will explain why the Bayes factor (as well as other BMS tools) is typically very sensitive to the prior distribution over the parameters of the competing models. There are three ways in which Bayesian analysis tries to face the problem of sensitivity to the prior:

- using classes of prior distributions,
- using *intrinsically robust* procedures,
- robustifying procedures.

In §3 to §5 we illustrate how these ideas can be usefully applied to BMS. We will use simple examples to highlight the general issues. In §6, we consider other aspects of sensitivity in model selection; §7 deals with the problem of the validation of a single model, when there are no competitors. In the last section we draw some partial conclusions and provide some side comments.

11.2 Bayesian model selection

Before going into the theory of BMS, it is necessary to agree on a reasonable definition of what a model is and, also, on what is the event *"Model M is true"*. Very loosely speaking, a model M (in a structural sense) is a class of probability measures over the sampling space $\mathcal{X}$; the event *Model M holds* is true if and only if the empirical distribution function converges, in some sense, to a member of the class M as $n \to \infty$ (see Piccinato, 1998). Hence, a statistical model is a concept that Bayesianism shares with frequentism: there is no role for prior probabilities in the definition of a model (we are not considering here a predictive approach). Also, events such as *Model M holds* are events only in an asymptotic sense: they are not conceivably observable.

For each model M_j, $j = 1, \ldots, k$, let $f_j(\cdot|\theta_j)$ be the sampling distribution of observables, where θ_j is the d_j-dimensional vector of parameters for model M_j. Also, $\pi_j(\theta_j)$ will denote the prior distribution of θ_j.

It is quite difficult to formalise the general problem of BMS. The main difficulty is the postulated existence (and/or availability) of a *true* model (see Bernardo and Smith, 1994). In a rather simplified way, we can distinguish two cases:

A The "true" model is within $M_1, M_2, \ldots, M_k$.

B There is *not* a true model within $M_1, M_2, \ldots, M_k$.

At a first glance, case A seems rather artificial. But we must consider this as an acceptable surrogate for all those situations where we can act *as if* one of the models were the correct one.

Example 1. *Testing precise hypothesis.* Suppose we observe an i.i.d. sample $X_1, X_2, \ldots, X_n$ from a distribution $f(\cdot|\theta)$ and wish to test $H_0 : \theta = \theta_0$ versus $H_1 : \theta \neq \theta_0$. Though one rarely believes that θ is exactly equal to θ_0, it can be very convenient to behave this way, by considering deviations from the null hypothesis irrelevant for practical purposes (see Berger and Delampady, 1987). $\triangle$

Trying to quantify the above irrelevance in terms of the events we defined above is quite an interesting problem. In case A, things are logically simple: we give prior probabilities to each model, namely $(p_1, p_2, \ldots, p_k)$, where $p_j = \text{Prob}(M_j)$, $j = 1, \ldots, k$.

After observing the data x, the Bayes theorem gives the posterior probability of the models,

$$\text{Prob}(M_j|x) = \frac{p_j m_j(x)}{\sum_{i=1}^{k} p_i m_i(x)}, \quad j = 1, \ldots, k, \tag{1}$$

where $m_j(x)$ denotes the prior predictive distribution of observables under model M_j, that is, assuming for sake of simplicity that π_j is absolutely continuous with respect to the Lebesgue measure,

$$m_j(x) = \int_{\Theta_j} f_j(x|\theta_j)\pi_j(\theta_j)d\theta_j.$$

The comparison between two models, say M_j and M_h, can be based on the ratio

$$\frac{\text{Prob}(M_j|x)}{\text{Prob}(M_h|x)} = \frac{p_j}{p_h}\frac{m_j(x)}{m_h(x)}. \tag{2}$$

The last factor in the right-hand side of (2),

$$\frac{m_j(x)}{m_h(x)}, \tag{3}$$

is called the Bayes factor for M_j against M_h and is denoted $B_{jh}(x)$. It is given by the ratio of posterior odds to prior odds and can be interpreted as the change from the latter to the former induced by the observed data. The Bayes factor is not a measure in a strictly probabilistic sense. Lavine and Schervish (1999) have shown a simple testing example where the Bayes factor gives enough evidence that the parameter being tested, say θ, belongs to a set A and, at the same time, it does not provide enough evidence that $\theta \in C$, with A strictly included in C. Note also that, when prior probabilities of the models are equal, then

$$\text{Prob}(M_j|x) = \left(\sum_{i=1}^{k} B_{ij}\right)^{-1}, \qquad j = 1, \ldots, k.$$

As mentioned at the beginning of this section, case A is not well suited for all situations: there are cases where a single true model cannot even be imagined and the available statistical models represent, at best, approximations to the real mechanism (is there one?) of data generation. As a result, in case B, attaching prior probabilities to single models is not a simple task: in a certain sense, we cannot state events like M_j *is true*. In these situations, a weaker Bayesian approach is typically adopted: prior probabilities of the models are left unspecified and comparisons between models are performed via the Bayes factor only. Although this approach is not exempt from criticism (see, for example, Lavine and Schervish, 1999, and Lindley, 1997), it is perhaps the most widely accepted and it is also justified from a decision-theoretic viewpoint (Bernardo and Smith, 1994). Therefore, given k models $M_1, M_2, \ldots, M_k$, the comparisons between M_h and M_j are performed via B_{hj}, $h, j = 1, \ldots, k$.

Example 2. *Selection of covariates in multiple linear regression.* Suppose y is an n-dimensional vector of observations which is assumed to depend linearly on some subset of a set of p covariates $X_1, X_2, \ldots, X_p$. Here $k = 2^p$ and it is clear that none of the models can be thought of as true and the aim is finding the "best" subset of covariates, conditionally on the assumed linear relation. $\triangle$

11.3 Robustness issues

Bayes factors are very sensitive to the choice of prior distributions of the parameters for the various models. This is a well-documented fact as the next example shows.

Example 1 *(continued) Jeffreys–Lindley's Paradox.* Assume that the i.i.d. sample in Example 1 comes from a normal distribution $N(\theta, \sigma^2)$, with σ^2 known. Suppose, moreover, that the prior distribution for θ under M_2 is $N(\theta_0, \tau^2)$, with τ^2 known. In testing precise hypotheses, it is quite natural, although not mandatory, to assume the null value θ_0 as the prior mean under the alternative hypothesis; without loss of generality, we can set $\tau^2 = c\sigma^2$. The Bayes factor B_{12} of M_1 to M_2 is easy to calculate,

$$B_{12} = \sqrt{1+nc}\exp\left\{-\frac{nc}{2(1+nc)}z^2\right\},$$

where $z = \sqrt{n}(\bar{x} - \theta_0)/\sigma$ is the usual standardised statistic and $\bar{x}$ denotes the sample mean. One can note that, as n goes to infinity, for a fixed z, B_{12} goes to infinity as well. The same happens for fixed n, as c goes to infinity. This happens because, under M_2, the variance of the prior predictive distribution of the data, $m_2(x)$, goes to infinity with c: then, for fixed z, the prior variance can be chosen in such a way that $m_2(x)$ can be made arbitrarily small. This example has been seen as a paradox since, no matter which hypothesis is true, as n goes to infinity the Bayes factor seems to select the sharp hypothesis. Many authors have commented on this apparent paradox. References include Bartlett (1957), Lindley (1957), Jeffreys (1961), Shafer (1982) and Bernardo (1999). Berger and Delampady (1987) argue that the bad behaviour for large n is actually not a problem, since for large n a precise hypothesis ceases to be a valid approximation of a more reasonable small interval hypothesis, say $\Omega : \{\theta : |\theta - \theta_0| < \epsilon\}$, for some small ϵ. Also, under M_2, it is not reasonable to let n go to infinity and take the sample observation $Z = z$ as fixed, since, under M_2, Z is a non-central χ^2 with a noncentrality parameter which increases with n. However, the problem remains unsolved as c goes to infinity, and this is not confined to conjugate priors. Lavine and Wolpert (1995) have considered the case when $\sigma^2 = 1$, and a uniform prior over $|\theta| < R$, for a given constant R, is adopted for θ under M_2. If a single observation is taken and $x = 5$, then B_{21} will depend on R. If we consider all R's greater than 10, we realise that B_{21} will range from 0 to 10^{10}! This is particularly astonishing, because the posterior distribution is practically the same for all $R > 10$. $\triangle$

This simple example shows that Bayes model selection procedures are inherently nonrobust and care should be taken in assessing prior opinion. The technical explanation for the extreme sensitivity of Bayes factors lies in that they are ratios of densities: as such, they are quite sensitive to small variations in one or both the components. Similar problems do not arise, for example, with posterior calculations, where, in general, small variations of the prior are irrelevant, due to the effect of the normalising constant (O'Hagan, 1995). Also, in estimation problems, one can often provide robust answers simply by using appropriate noninformative improper priors. This solution cannot be adopted in testing hypotheses, at least in general: improper priors are specified only up to a constant and, if they are used to obtain one prior predictive distribution of the data, the resulting Bayes factor will also contain undefined constants. Bayesian statisticians have tried to overcome these difficulties in three different ways, namely using classes of proper prior distributions, using intrinsically robust procedures stemming from noninformative priors, and robustifying procedures. In the next sections we will discuss in some detail these alternative approaches.

11.4 Classes of priors

A way to deal with prior sensitivity is to consider reasonably large classes of priors for the parameters in M_1 and/or M_2 and analyse the range of possible values for the resulting Bayes factors. We mainly consider here the case when M_1 is nested into M_2, as in testing a precise hypothesis. We say that M_1 is nested into M_2 if the following conditions are met:

i) $\theta_2 = (\xi, \phi), \quad \xi \in \Theta_1, \ \phi \in \Phi;$

ii) $f_1(x|\theta_1) = f_2(x|\xi, \phi = \phi_0)$, for a fixed ϕ_0 and suitable θ_1 and ξ.

In these cases, the Bayes factor is the Bayesian alternative to classical significance tests. There is a vast literature showing dramatic differences between the two approaches: classical p-values tend to give answers which are biased against the null hypothesis (see Edwards et al., 1963, Berger and Delampady, 1987, Berger and Sellke, 1987, Moreno and Cano, 1989, Delampady and Berger, 1990, and many others).

Example 1. *(continued) Jeffreys–Lindley's Paradox.* For this example, putting together and reinterpreting results from different papers (Berger and Sellke, 1987, Berger and Delampady, 1987, De Santis and Spezzaferri, 1997), we consider several classes of priors for the parameter θ under M_2, namely

- $\Gamma = \{\text{All prior distributions}\}$,
- $\Gamma_{US} = \{\text{All symmetric prior distributions unimodal at } \theta_0\}$,
- $\Gamma_N = \{\text{All normal prior distributions } N(\theta_0, c\sigma^2), c > 0\}$.

These classes aim at modelling our degree of prior uncertainty about the parameter of the model. Berger and Sellke (1987) show in a series of theorems how to compute the exact range of $B_{12}(x)$ and $P(M_1|x)$ when the prior varies in the three classes above. As is easily seen, the supremum of $B_{12}(x)$ over the three classes is always infinite and the supremum of $P(M_1|x)$ is always 1. Table 1 shows the results for some representative p-values one would obtain with the usual frequentist test based on $z = \sqrt{n}|\bar{x} - \theta_0|/\sigma$, and assuming $\text{Prob}(M_1) = .5$.△

	Γ		Γ_{US}		Γ_N	
p-value	$\underline{B}_{12}$	$\underline{P}(M_1\|z)$	$\underline{B}_{12}$	$\underline{P}(M_1\|z)$	$\underline{B}_{12}$	$\underline{P}(M_1\|z)$
.62	.883	.469	1	.500	1	.500
.32	.607	.377	1	.500	1	.500
.10	.258	.205	.701	.412	.637	.389
.05	.147	.128	.473	.321	.409	.290
.01	.036	.035	.154	.133	.123	.109
.001	.004	.004	.024	.023	.018	.018

TABLE 1. Lower bounds of $B_{12}(x)$ and $P(M_1|x)$ for Γ, Γ_{US} and Γ_N.

Some comments are important at this point.

1. Since $\Gamma_N \subset \Gamma_{US} \subset \Gamma$, one obviously obtains that, for all $x, \underline{B}_{12}(\Gamma) \leq \underline{B}_{12}(\Gamma_{US}) \leq \underline{B}_{12}(\Gamma_N)$; a similar relation holds for the posterior probabilities of M_1. The important fact here is that the $\underline{B}_{12}(\Gamma)$'s are significantly larger than the corresponding p-values. Although p-values and Bayes factor are logically different entities, they are both perceived as measures of evidence in favour of the null hypothesis. A robust Bayesian analysis clearly shows that the evidence provided by the p-value can be misleading, at least in the sense it is usually interpreted by practitioners.

2. Whereas, for a fixed prior, the Bayes factor obviously depends on the sample size, the lower bounds do not.

3. The (nonreported) corresponding upper bounds for Bayes factors and posterior probabilities are always infinite and 1, respectively. In other words, unless we subjectively restrict the class of priors, we can always find a prior which gives overwhelming evidence in favour of M_1, no matter what the data are. This fact seems unavoidable and it provides an easy example of inherent sensitivity of model choice to prior opinion. The use of a prior with large variance, in an effort to be "noninformative", will force the Bayes factor to prefer the null model. This unpleasant fact convinced many Bayesians, Jeffreys in primis, that improper priors should be avoided in hypothesis testing and model selection. We will pursue discussion of these issues in the next section.

The use of bounds of Bayes factors over large classes of priors has been proposed in several papers with a twofold objective: criticise the classical measures of evidence and perform a robust selection of a model. Delampady and Berger (1990) compared lower bounds of Bayes factors with the classical χ^2 test for goodness of fit; Bayarri and Berger (1994) proposed a robust testing procedure for detection of outliers; Berger and Mortera (1994) compared frequentist, objective and robust Bayesian methods for eliminating nuisance parameters in testing problems. The Bayesian robustness literature has proposed several other classes of priors (Berger, 1994; Moreno, 2000). Their impact in model selection and testing has not been fully analysed.

11.5 Inherently robust procedures

Sensitivity analysis based on classes of priors is undoubtedly the most reasonable one. However, in real data analysis, when more than two models are competing, elicitation of (classes of) priors for the parameters of each single model is a very difficult task.

Example 2 *(continued).* Here we have to select, in a "robust" way, a subset of a set of p explanatory variables $(X_1, X_2, \ldots, X_p)$: there are, of course, 2^p possible choices. A naive solution would be to elicit a prior (or a class of priors) for the larger model (the one including all the X_i's) and then derive the other priors via marginalisation. This solution is clearly unacceptable because parameters have a meaning, which depends on the model they belong to. Berger and Pericchi (1998) reported the case when there are two competing nested linear models,

$$M_1 : Y = \beta_0 + \beta_1 X_1 + \varepsilon_1 \tag{4}$$

and

$$M_2 : Y = \beta_0 + \beta_1 X_1 + \beta_2 X_2 + \varepsilon_2. \tag{5}$$

The actual meaning of β_1 is absolutely different in the two models, especially when X_1 and X_2 show some degree of correlation. $\triangle$

Example 3. *Default testing for the number of components in a mixture.* Suppose that $X_1, \ldots, X_n$, are distributed independently as

$$X_i \sim \sum_{j=1}^{k} w_j N(\mu_j, \sigma_j^2), \qquad i = 1, \ldots, n.$$

Here, k is the unknown number of components of the mixture, and it represents the parameter of main interest. Conditional on k, each X_i is a mixture of normal random variables with unknown parameters $\theta_j = (\mu_j, \sigma_j)$,

$j = 1, \ldots, k$, and $= (w_1, \ldots, w_k)$ denotes the (usually unknown) classification probabilities vector. In this context, model M_k equals to assume a k-components mixture, and the goal is to select a model among $M_1, \ldots, M_{K^*}$, where K^* is a fixed integer. Notice that the K^* models are nested. In the mixture models problem, the main difficulty to perform a robust Bayes analysis via classes of priors is the large number of parameters, and a default approach based on improper priors would be highly desirable. $\triangle$

The general issue here is that there are several difficulties in specifying any reasonable joint prior for the parameters of each model and, a fortiori, in performing a careful sensitivity analysis. A possible way out is to use inherent robust priors. The ideal would be to have priors whose influence on the model selection procedure tends to be modest. On the other hand, we have already seen that improper priors cannot be safely used in BMS. Moreover, Wasserman (1996) highlights the conflict between robustness and improper priors: in words, either classes of neighbourhoods of improper priors are not sufficiently rich to safely check robustness or they provide trivial bounds for posterior quantities of interest.

A possible alternative would consist in trying to convert improper priors into proper, data-driven priors, using a small part of the data. In the next subsections we will discuss some of these approaches.

11.5.1 Conventional priors

Jeffreys (1961) developed some "desiderata" to construct conventional proper priors for model comparison. His procedure unavoidably generates ad hoc priors, tailored for the particular model at hand. Though sensible, the Jeffreys' approach lacks the "automatism" often required for a statistical procedure supposed to be routinely used. Among other possible choices of conventional priors are the *intrinsic* prior (Berger and Pericchi, 1996a), the *fractional* prior (Moreno, 1997) and the *expected posterior* prior (Pérez and Berger, 1999). These will be further analysed in §5.4.

11.5.2 Approximations

Schwarz (1978) has shown that, as the sample size gets larger, the contribution of the priors to the Bayes factor is bounded (that is, it is neither zero nor infinity). This suggests ignoring, at a first approximation, the prior contribution, leading to the well-known Bayesian information criterion (BIC). Schwarz's approximation to B_{12} is given by

$$B_{12}^S = \frac{f_1(\cdot|\hat{\theta}_1)}{f_2(\cdot|\hat{\theta}_2)} n^{\frac{d_2 - d_1}{2}}. \tag{6}$$

Under general regularity conditions, as $n \to \infty$ and both under M_1 and M_2,

$$\frac{\log B_{12}^S - \log B_{12}}{\log B_{12}} \to 0. \tag{7}$$

Refinements of this approach have been proposed in the last years (Kass and Vaidyanathan, 1992; Kass and Wasserman, 1995; Verdinelli and Wasserman, 1995; Pauler, 1998). Although B_{12}^S works extremely well in many practical applications, problems arise when the sample size is not large. Moreover, the bounded contribution of the prior is simply ignored, which, sometimes, can be crucial.

Example 1 *(continued).* In the simple normal testing example, $B_{12}^S = \sqrt{n}$ $\exp\{-z^2/2\}$. For a conjugate normal prior $N(\theta_0, c\sigma^2)$, the actual Bayes factor is approximately $B_{12} \approx \sqrt{1+nc}\exp\{-z^2/2\}$ and $B_{12}/B_{12}^S \approx \sqrt{c}$. The Schwarz criterion "ignores" the prior standard deviation. A way to "hide" this discrepancy is to adopt, as a conventional prior, a normal unit-information prior (Kass and Wasserman, 1995). △

BIC is also not applicable with irregular models like Example 3, where the usual regularity conditions underlying Schwarz approximation are not fulfilled (Roeder and Wasserman, 1997).

Another way of approximating Bayes factor makes use of Laplace's method to calculate prior predictive distributions of the data for model M_j, namely (Kass and Raftery, 1995),

$$m_j(x) \approx (2\pi)^{d_j/2}|\hat{\Sigma}|^{1/2} f_j(x|\hat{\theta}_j)\pi_j(\hat{\theta}_j), \tag{8}$$

where $\hat{\Sigma}$ is the observed precision matrix. Provided that the approximation is sufficiently good, its relative error is of order $O(n^{-1})$, the contribution of the priors is only through their values at the MLE's under the two models: this enables us to check sensitivity to priors varying in given classes in a computationally fast and easy way, simply by changing the values of $\pi_i(\hat{\theta}_i)$ and $\pi_j(\hat{\theta}_j)$ in (8).

Prior and model robustness via posterior simulation is developed in Geweke and Petrella (1998) and Geweke (1999). In particular, when M_i is nested into M_j, Geweke (1999) proposes a calculation of B_{ji} which is based both on analytic expansion and on posterior simulation: but the simulation is needed only for the simplest model M_i. This can be of considerable advantage in complex problems when a posterior sample from the more complex model would be costly (see, however, Pericchi's discussion of Geweke's paper).

11.5.3 Partial Bayes factors

The idea behind the partial Bayes factor of M_i against M_j is to split the sample into two components $x = (x_T, x_D)$; x_T is called the *training* sample and is used to update the initial (almost always improper) prior distributions of the two competing models. It is common to use a *proper minimal* training sample, that is, a subsample x_T such that its prior predictive distributions (under M_i and M_j) exist and no subsets of x_T are proper. The two obtained posteriors are then used as priors to compute the Bayes factor using the likelihoods provided by x_D, the portion of the data which is used to discriminate between M_i and M_j. In formulae, if $B_{ij}^N(x)$ denotes, up to arbitrary constants, the Bayes factor obtained via the improper initial priors, then the partial Bayes factor is given by

$$B_{ij}^p(x_T; x_D) = B_{ij}^N(x) B_{ji}^N(x_T). \tag{9}$$

The indeterminacy due to the use of improper priors is then removed, but now $B_{ij}^p(x_T; x_D)$ depends on the particular splitting of the sample. A way to overcome this new indeterminacy is to consider an average over all the possible splitting of the sample, thus producing an *intrinsic* Bayes factor (*IBF*, hereafter). This idea has been put forward by Berger and Pericchi in a series of papers (1996a, 1996b, 1997, 1998). There are many ways of taking an average, each producing a different *IBF*. For example, the arithmetic *IBF* is given by

$$AIBF(x) = B_{ij}^N(x) \frac{1}{H} \sum_{h=1}^{H} B_{ji}^N(x_T^{(h)}), \tag{10}$$

where $x_T^{(h)}$ is the generic training sample ($h = 1, \ldots, H$) and H is the number of all possible training samples. Berger and Pericchi (1998) suggest that all different *IBF*s should be part of the *objective* Bayesian statistician's toolkit, and the choice among them should be based on the particular concern (robustness, stability, simplicity,...). A careful review of each of them would require a much deeper discussion than is possible here. From a robust Bayesian perspective, it is important to recall the case of nested models where $AIBF$ is particularly indicated. Even there, however, if the sample size is too small, Berger and Pericchi (1996a) suggest substituting each partial Bayes factor in the second factor of (10) with its expected value, where the expectation is taken over the sampling space, under the larger model with the parameters set equal to their maximum likelihood estimates. Thus, with exchangeable observations, and if M_j is nested into M_i, one obtains the expected *AIBF* (EAIBF),

$$EAIBF(x) = B_{ij}^N(x) E_{\hat{\theta}_i}^{M_i} \left(B_{ji}^N(x_T^{(h)}) \right). \tag{11}$$

One of the main reasons for the success of the *IBF* methodology is that an *IBF*, under particular circumstances (nested models being the most important), is often equivalent, at least asymptotically, to an actual Bayes factor, that is, one obtained with the adoption of a suitable pair of prior distributions (one for each competing model), which are called intrinsic (see §5.4).

The fractional Bayes factor (*FBF*, hereafter) was introduced by O'Hagan (1995) and may be considered an alternative way to compute Bayes factors using improper priors. Here, a fraction b $(0 < b < 1)$ of each likelihood function is "transferred" to the priors in order to make them proper. The rest of the two likelihoods are then used to compute the Bayes factor. It is easy to see that, for a given fraction b, the *FBF* of M_i against M_j is given by

$$FBF(x;b) = B_{ij}^{N}(x)B_{ji}^{b}(x), \tag{12}$$

where $B_{ji}^{b}(x) = m_j^b(x)/m_i^b(x)$, and $m_h^b(x) = \int_{\Theta_h} f_h^b(x|\theta_h)\pi_h^N(\theta_h)d\theta_h$. The *FBF* approach delays the problem of the arbitrariness of x_T until the choice of a particular b. There are strong indications, at least from a consistency perspective, that one should take $b = m/n$, where m is the minimal training sample size. O'Hagan (1995) and De Santis and Spezzaferri (1999) argue that different choices of b can lead to more robust conclusions: we analyse these issues in §6.

Theoretical development and practical use of default Bayes factors for comparing two models have increased enormously in the last few years. The problem of comparing two nested models when prior information is lacking seems to have found several possible solutions. Much work is still to be done in the nonnested case. The only method which does not require any particular adjustment for this more difficult problem is the *FBF* approach, although it is now much harder to justify the *FBF* in terms of correspondence to a *fractional* prior (see §5.4). The *IBF* approach attempts to transform the nonnested problem into a nested one by encapsulating M_i and M_j into a larger model M_* (Berger and Pericchi, 1996a); the two competing models are then compared indirectly, that is,

$$BF_{ij}^{I}(x) = BF_{*j}^{I}(x)/BF_{*i}^{I}(x), \tag{13}$$

BF^I being any type of *IBF*.

But the main issue here is: *how to deal with multiple models in a robust way?* Is there any practical and fairly robust way to select a model among k (possibly nonnested) competitors?

Let us assume, at least formally, that $P(M_j) = k^{-1}$, for all j. Then the posterior probability of M_j is given by

$$P(M_j|x) = \frac{m_j(x)}{\sum_{i=1}^{k} m_i(x)}. \tag{14}$$

Default Bayes factors do not provide us with the $m_i(x)$'s. However, as it is easily understood, both *IBF* and *FBF* methodologies merely require the computation of k different Bayes factors in order to produce the posterior probabilities in (14). Indeed, in the *FBF* case, after choosing a fixed model M_h, one can obtain

$$P(M_j|x) = \frac{(B_{hj}(x))^{-1}}{\sum_{i=1}^{k}(B_{hi}(x))^{-1}}, \tag{15}$$

where the result does not depend on M_h. The same is true for the *IBF* if we take the encompassing model M_* as M_h. These procedures give us, at least, a common ground to compare multiple models. The above solution, however, is not completely satisfactory. In fact, the posterior probabilities in (15) are only "pseudo"-probabilities since they are based on Bayes factors which are not necessarily coherent among each other. When multiple models are to be compared, it would be much more satisfactory to know the prior predictive distributions of the data under each single model. This goal is achieved via the use of *intrinsic* priors (when they do exist) and/or the *expected posterior* priors.

Example 3 *(continued).* The selection of the number of components of a mixture is a challenging problem for default Bayesian analysis. The above mentioned methods are simply not applicable (see, for example, Diebolt and Robert, 1994, Berger and Pericchi, 1998). The main problem is that, under M_k, if we adopt the usual location-scale improper prior for each θ_j, namely $\pi_j^N(\mu_j, \sigma_j) \propto \sigma_j^{-1}$, and assume that the θ_j are independent of each other, then the posterior will be improper, no matter what the sample size is and, consequently, no prior predictive distribution is available. As a result, we can neither use *FBF* nor compute any kind of *IBF*, since there are no training samples. $\triangle$

In such a framework it would be desirable to know the prior predictive distribution of the data, under each single model. This goal can be achieved via a related, but different approach: the derivation of *intrinsic*, *fractional*, and *expected posterior* priors. We discuss these topics in the next section.

11.5.4 Default priors for testing

One of the most appealing features of *IBF* and *FBF* is that they often produce answers that are asymptotically equivalent to those which would be obtained by direct use of some prior distributions on the parameters of the two models. These priors are generally defined as *intrinsic* and *fractional* priors. This issue is important for several reasons:

1. It is somehow reassuring that "default" procedures correspond, at least asymptotically, to a formal Bayesian reasoning, although the resulting priors are generally weakly data-dependent.

2. Since computation of *IBF* and *FBF* requires the least amount of subjective input, one can argue that the resulting priors are, in a vague sense, "noninformative" and, hopefully, robust.

If a pair of intrinsic (or fractional) priors can be easily derived for a specific comparison, then one can directly use them, without passing through the *IBF* (*FBF*) computations. This would avoid the occurrence of a series of disturbing features of IBFs, for example, that they are not naturally coherent in multiple model comparisons.

The definition of the intrinsic prior is based on the following approximation formula for the Bayes factor, which can be derived from a Laplace approximation:

$$B_{ji} = B_{ji}^N \frac{\pi_j(\hat{\theta}_j)\pi_i^N(\hat{\theta}_i)}{\pi_j^N(\hat{\theta}_j)\pi_i(\hat{\theta}_i)}(1 + o(1)), \tag{16}$$

where π_i and π_j are the priors used to compute the actual Bayes factor and π_i^N and π_j^N are the noninformative priors used to compute B_{ji}^N. Then one equates any of the various IBFs to (16) and tries to solve the ensuing functional equation. For simplicity consider here exchangeable observations and the arithmetic IBF; suppose also that the MLEs under the wrong models behave properly, that is (Berger and Pericchi, 1996a): as $n \to \infty$, under M_j,

$$\hat{\theta}_j \to \theta_j, \qquad \hat{\theta}_i \to \psi_i(\theta_j), \qquad \frac{1}{H}\sum_{h=1}^{H} B_{ij}^N(x_T^{(h)}) \to E_{\theta_j}^{M_j}\left(B_{ij}^N(X_T^{(h)})\right);$$

also, under M_i,

$$\hat{\theta}_i \to \theta_i, \qquad \hat{\theta}_j \to \psi_j(\theta_i), \qquad \frac{1}{H}\sum_{h=1}^{H} B_{ij}^N(x_T^{(h)}) \to E_{\theta_i}^{M_i}\left(B_{ij}^N(X_T^{(h)})\right),$$

where ψ_i and ψ_j are suitable limits of the sequences of MLE's under the wrong model.

Then, Berger and Pericchi (1996a) show that the resulting *intrinsic priors* are the solution of the following equations:

$$\frac{\pi_j(\theta_j)\pi_i^N(\psi_i(\theta_j))}{\pi_j^N(\theta_j)\pi_i(\psi_i(\theta_j))} = E_{\theta_j}^{M_j} B_{ij}^N(X_T^{(h)}), \tag{17}$$

$$\frac{\pi_i^N(\theta_i)\pi_j(\psi_j(\theta_i))}{\pi_i(\theta_i)\pi_j^N(\psi_j(\theta_i))} = E_{\theta_i}^{M_i} B_{ij}^N(X_T^{(h)}). \tag{18}$$

Solutions to these equations are not necessarily proper nor unique. It can actually be shown (Dmochowsky, 1996; Sansó et al., 1996; Moreno et al., 1998) that, when M_i is nested into M_j, one of the above equations is redundant and the general solution to the *intrinsic* equation is given by

$$\begin{cases} \pi_i^I(\theta_i) = \pi_i^N(\theta_i)u(\theta_i) \\ \pi_j^I(\theta_j) = \pi_j^N(\theta_j)E_{\theta_j}^{M_j} B_{ij}^N(X_T^{(h)})u(\psi_i(\theta_j)), \end{cases} \tag{19}$$

where $u(\cdot)$ can be any continuous function. The class of intrinsic priors is then extremely huge. An important exception is when the nested hypothesis (or model) is simple: in this case the pair of intrinsic priors is univocally determined. A robustness study within the class of all possible intrinsic (and fractional) priors has been performed by Moreno (1997).

Similar arguments (Moreno, 1997; De Santis and Spezzaferri, 1999) are needed to derive *fractional* priors. It is important to note that, in the asymptotic construction of the fractional prior, the sequence $\{b_n\}$ of fractions of the likelihoods must go to zero. Also, $\{b_n\}$ must be fixed in such a way that the quantity

$$\frac{\int_{\Theta_j} f_j(x|\theta_j)^{b_n} \pi_j^N(\theta_j) d\theta_j}{\int_{\Theta_i} f_i(x|\theta_i)^{b_n} \pi_i^N(\theta_i) d\theta_i} \tag{20}$$

converges in probability (under the nesting model M_j) to a degenerate random variable $F(\theta_j)$ in order to obtain a solution.

A new approach to the selection of default priors for model comparison is developed in Pérez and Berger (1999). Suppose one wants to compare k different models and (at least) one of them, say M_i, has an improper prior $\pi_i^N(\theta_i)$ for θ_i. Pérez and Berger (1999) suggest adjusting the improper prior by computing the *expected posterior* (EP for short) prior

$$\pi_i^*(\theta_i) = \int_{\mathcal{Y}^*} \pi_i^N(\theta_i|y^*) m^*(y^*) dy^*, \tag{21}$$

where m^* is a suitable (not necessarily proper) prior predictive distribution on imaginary training samples $y^* \in \mathcal{Y}^*$. The size of y^* should be chosen as small as possible provided that priors (21) exist for all M_i's. Of course, the resulting procedure strongly depends on the choice of m^*. In the nested case, when one of the models, say M_1, is nested in every other model, the use of the prior predictive distribution (under M_1) of the training sample as the measure m^* is asymptotically equivalent to the use of the arithmetic *IBF* and the resulting EP priors coincide with the intrinsic prior obtained via the expected arithmetic *IBF*.

Example 3 (*continued*). Moreno and Liseo (1998) derived, in a closed form, intrinsic and fractional priors for testing k, the number of components of the mixture. Their approach is based on comparing each single model M_k with a simpler model (nested into all the competitors) M_*, namely

$$M_*: \quad X \sim N(\mu_0, \sigma_0^2), \qquad \text{for some fixed } (\mu_0, \sigma_0).$$

The priors they obtained have many relevant properties: for example, they are consistent across models, that is, the intrinsic prior (under M_j) for the comparison of M_j against M_* is equal to the restriction on $(\theta_1, \ldots, \theta_j)$ of the intrinsic prior (under M_k) for the comparison of M_k against M_*, with $j < k$.

Quite similar results are obtained by Pérez (1998) via the use of expected posterior priors whenever $m^*(x)$ is taken to be the marginal from the model with a single component. $\triangle$

11.5.5 Robustifying procedures

Although the *FBF* and the various *IBF*s have been proposed when dealing with improper priors, from a strict mathematical viewpoint, nothing prevents us from using them also with proper priors. O'Hagan (1995) claims that *"even when proper prior distributions are specified, and the usual Bayes factor is available for use, the FBF may be preferred because of its greater robustness to misspecification of the prior."*

Example 1 *(continued).* As a prior under the nesting model M_2, O'Hagan (1995) considers a conjugate prior not centred at θ_0, namely $N(\theta_1, c\sigma^2)$. The resulting *FBF* is

$$FBF(x;b) = \left(\frac{1+nc}{1+nbc}\right)^{1/2} \exp\left\{-\frac{n}{2\sigma^2}\left[(1-b)(\bar{x}-\theta_0)^2 - \lambda(\bar{x}-\theta_1)^2\right]\right\},$$

where $\lambda = (1-b)/[(1+nbc)(1+nc)]$; λ can be regarded as a factor of sensitivity to changes in the prior mean. This sensitivity decreases with b. Conigliani and O'Hagan (2000) put forward these ideas by investigating more robust choices of b. De Santis and Spezzaferri (1997, 1999) have proposed the use of *FBF* in the presence of (classes of) proper priors for a more robust model choice. In a series of examples, including our Example 1, they compare the behaviour of *IBF*s and the *FBF* (with b always equal to the ratio of minimal training sample size to sample size) with the usual Bayes factor. Here we report the results for the conjugate class Γ_N, with sample size $n = 10$ (note that lower bounds for *IBF* and *FBF* actually depend on n).

p-value	$\underline{B}_{12}$	$\overline{B}_{12}$	$\underline{B}^{EAI}_{12}$	$\overline{B}^{EAI}_{12}$	$\underline{FBF}_{12}$	$\overline{FBF}_{12}$
.62	1	∞	1	3.972	1	2.826
.32	1	∞	1	2.781	1	2.016
.10	.701	∞	.705	1.236	.751	1
.05	.473	∞	.478	1	.521	1
.01	.154	∞	.158	1	.158	1
.001	.024	∞	.026	1	.024	1

TABLE 2. Ranges of Bayes factor, expected arithmetic *IBF* and *FBF*, when $n = 10$, $b = 1/n$, and $\pi_2 \in \Gamma_N$

Table 2 shows that the use of *IBF* and/or *FBF* can lead to a robust model choice more often than the usual Bayes factor. For example, for data with a p-value as small as .05, the upper bounds of *IBF* and *FBF* are equal to 1, indicating a robust choice of the nesting model. This kind of behaviour can be emphasised by choosing a larger fraction b. In a sense, the real purpose is to find a reasonable calibration of b in order to attain both robustness and discriminatory power. Wasserman (1996) noticed that the use of *FBF* actually corresponds to assuming, for model M_j, a prior distribution proportional to $\pi_j^N(\theta_j)(f_j(x|\theta_j))^b$ and a likelihood $(f_j(x|\theta_j))^{1-b}$. This non-standard use of the data "transforms" the prior and typically provides a more concentrated (data-driven) prior. De Santis and Spezzaferri (1997) show that when the nested models M_1 and M_2 are compared, whenever the prior on the common parameters of the two models is the same and the prior for the extra parameters in M_2 is sufficiently diffuse, then $FBF_{12}(x;b) < B_{12}(x)$, for all $0 < b < 1$. This can be of some relevance when the range of the Bayes factor is computed over large classes which include diffuse priors: in fact, the *FBF* seems to compensate the bias in favour of the smaller model induced by the regular Bayes factor. For example, turning to Example 1, Table 2 shows that for $n = 10$, $x = .196$, and $b = .1$, the range of FBF_{12} is $[.521, 1]$, which approximately corresponds to the range of the regular Bayes factor with prior variance varying in the two disjoint intervals $(0, 0.12\,\sigma^2)$ and $(.67\sigma^2, 4.16\,\sigma^2)$. The *FBF* acts in the sense of shrinking the prior variance, and the choice of b modulates the extent of such a shrinkage. Sansó et al. (1996) provide a similar analysis using the expected *IBF*. $\triangle$

11.6 Other aspects of sensitivity

The Bayes factor is certainly the most popular Bayesian tool for global model criticism. However, many authors have argued that Bayes factors fail to detect "local" aspects of sensitivity (e.g., the presence of outliers). Also, the Bayes factor approach is *predicated on the truth of one of the models being compared, and if all the models are wrong, then selecting the best according to the Bayes factor can be practically disastrous* (Rubin, 1995); for a different viewpoint, see Berger and Pericchi (1998).

Several alternative Bayesian approaches to model checking are based on a broader definition of predictive density. Gelfand and Dey (1994) consider the following framework: starting with the original sample $x = (x_1, \ldots, x_n)$, assume that, under model M_j, the x_i's are i.i.d. with density $f_j(x_i|\theta_j)$; also, θ_j is distributed according to some density $\pi_j(\theta_j)$, possibly improper. For each subdivision of the n-vector x into subsets x_A and x_B, one can compute the conditional predictive density of x_A given x_B, namely

$$f_j(x_A|x_B) = \frac{\int_{\Theta_j} f_j(x_A|\theta_j) f_j(x_B|\theta_j)\pi_j(\theta_j)}{\int_{\Theta_j} f_j(x_B|\theta_j)\pi_j(\theta_j)}, \qquad j = 1, \ldots, k. \tag{22}$$

Different choices of x_A and x_B lead to different predictive densities. When $x_A = x$, then (22) is the usual prior predictive density: this is the case of regular Bayes factors and in this case priors must be proper. When x_B is a minimal training sample, then (22) is used to calculate partial Bayes factors, as in the *IBF* approach. Note that in the two cases above the size of x_B does not change with n.

Geisser (1975) has used the cross-validation density

$$\prod_{i=1}^{n} f_j(x_{(i)}|x_{(-i)}), \qquad j = 1, \ldots, k,$$

where $x_{(i)}$ and $x_{(-i)}$ represent the ith observation and the entire sample with the ith observation omitted, respectively. Each term in the above product is, of course, a special case of (22): this approach has often been proposed to compare the actual observed data with prediction based on the rest of the data (conditional predictive ordinate: see Pettitt, 1990).

The main objective of these methods is not a global model checking: rather, they aim to detect particular discrepancies, like the presence of outliers.

A more general approach to sensitivity with respect to the components of a model is developed in Weiss (1996). Here the idea is that the usual Bayes formula, under the base model M_0, $\pi_0(\theta|x) \propto \pi_0(\theta) \prod f_0(x_i|\theta)$, is perturbed by a multiplicative factor $h(\theta, x)$. Then the perturbed posterior can be written as $\pi^*(\theta|x) = \pi(\theta|x)h^*(\theta, x)/E(h^*(\theta, x)|x)$.

According to the desired sensitivity check, one can choose a suitable h^*. For example, sensitivity to changes in the prior (from π_0 to, say, π_1) can be analysed via the function $h^*(\theta) = \pi_1(\theta)/\pi_0(\theta)$. The influence of the perturbation can then be measured in terms of some divergence measure. See Weiss (1996) for details. In §2 we mentioned the case when none of the models under comparison can be considered as the "true" model. In this case, interesting and promising relations between a cross validation approach and the geometric *IBF* are discussed by Key et al. (1999) and their discussants.

11.7 Robustness and measures of surprise

Sometimes there is no more than one available model. Strictly speaking, this is not a model comparison problem: rather, we want to verify whether a given model is sufficiently compatible with the data at hand. Suppose, first, that the model under consideration is simple, that is, M_0: $X \sim f(x)$, and there are no unknown parameters in f. The classical approach uses p-values (based on a suitable statistic $T(X)$) to validate or refute the null model M_0. The Bayesian criticism to the use of p-values is well known but a general, alternative approach is still lacking. Box (1980) proposed a significance test based on the prior predictive distribution of the data, under the null hypothesis. Of course, one cannot invoke the Bayes factor unless an alternative model M is proposed. Recently, Sellke et al. (1999) and Bayarri and Berger (1999) have proposed a Bayesian *calibration* of p-values. Instead of the observed $p = p(x)$, they suggest using the quantity

$$\tilde{B}(p) = -e \ p \log p. \tag{23}$$

The rationale behind (23) is both Bayesian and robust. It is well known that, under M_0, $p(X)$ is a uniform random variable in $(0, 1)$. Consequently, $Z = -\log p(X)$ has an exponential distribution (with constant failure rate) with mean equal to 1. As a possible class of alternative models one can imagine the set of all the distributions (for $Z = -\log p(X)$) with non-increasing failure rate (NIFR). The NIFR property for Z is equivalent to stating that

$$\frac{p(X)}{p(x_0)} \mid p(X) < p(x_0)$$

is stochastically decreasing in $p(x_0)$. Intuitively, it means that, conditionally on the observed $p(x_0)$, the r.v. $p(X)$ (under the alternative hypotheses) tends to be "smaller in law" if compared with the uniform distribution. In this framework, Sellke et al. (1999) proved the following theorem:

Theorem 1. *Consider testing*

$$M_0 : p(X) \sim \mathcal{U}(0,1) \qquad versus \qquad M_1 : p(X) \sim g(p),$$

with g such that $Z = -\log p(X)$ has the NIFR property. Then, if $p(x_0) < 1/e$, the (attainable) lower bound of the Bayes factor of M_0 against M_1 is given by $\underline{\mathrm{B}}(p(x_0)) = \tilde{B}(p(x_0)) = -e \ p(x_0) \log p(x_0)$.

Then, the Bayesian calibration of the observed p-value assumes a precise robust Bayes meaning as the minimum support the data give to M_0 when it is compared with a relatively large (but reasonable) class of alternatives. As such, it should be interpreted as an index of surprise, rather than as an actual Bayes factor. This result indicates that robust Bayesian analysis can have a valuable role in the process of model validation and/or in exploring plausible alternatives.

More complex (and fascinating) is the case of a composite null hypothesis, say M_0: $X \sim f(x;\theta)$, $\theta \in \Theta$, where θ is a vector of unknown parameters. In this case, before introducing the calibration (23), one has to "define" a p-value, via elimination of θ. Bayarri and Berger (1999, 2000) introduce two new proposals.

11.8 Conclusions

In §2 we underlined the importance of an unambiguous definition of a statistical model. We have already argued that, in general, one must assume the existence of a *true* value for θ. Consequently, the likelihood function is the only ingredient in the specification of the statistical model; the prior distribution is *what we know about* θ. Exceptions to this rule lead us to more sophisticated scenarios, like hierarchical models in Bayesian analysis or superpopulation models from a classical viewpoint. In this framework, however, the meaning of the parameters and the way in which we make model comparison should be reconsidered.

Example 1 *(continued).* This example is paradigmatic and can be used to describe several different real problems. Suppose, for example, that we are interested in measuring the distance of a star from the Earth. Assume, for the moment, that, under some conditions, this distance can be regarded as a fixed constant we need to estimate and that θ_0, the value to be tested, originated from a given theory. In this case the analysis described in §3 makes perfectly sense. But it could also be the case that the distance of the star cannot be regarded as a fixed constant, for it possibly depends on the time of the day and/or on the day of the year. Had our observations been taken under different condition would force the role of our statistical ingredients to change radically: there is no room at all for a true θ_0 now unless we want to use it as an approximate model; the prior distribution over θ would have a different meaning, for it should now describe a second stage variability (across time) and as such it would probably be accepted (as a part of the model) by non-Bayesians too. In this second case, the null hypothesis would correspond to the use of $\theta = \theta_0$ as a possible simplified model designed to describe the (more complex) physical reality. If the distribution of θ under the alternative model is taken to be, as in §2, $N(\theta_0, \tau^2)$, the two competing models become

$$M_0 : X \sim N(\theta_0, \sigma^2) \qquad \text{vs.} \qquad M_1 : X \sim N(\theta_0, \sigma^2 + \tau^2),$$

and the real issue here is not about θ; rather, it is about τ: can it safely be set equal to zero or not? $\triangle$

These considerations are important from a robust Bayes perspective: the above scenario can be seen as the simplest hierarchical model we can think of. From the arguments outlined in §3, it is easy to see how important it is to check sensitivity with respect to higher-level components of a statistical model (see Cano Sanchez, 1993, and De Santis and Spezzaferri, 1996).

Acknowledgments

I am grateful to Susie Bayarri and Ludovico Piccinato for useful discussions. The research has been supported by *C.N.R.* and *M.U.R.S.T.* (Italy) grants.

References

BARTLETT, M.S. (1957). Comment on "A statistical paradox", by D.V. Lindley. *Biometrika*, **44**, 533-534.

BAYARRI, M.J. (1999). Discussion of "Bayesian model choice: what and why?" by J.T. Key, L-R. Pericchi and A.F.M. Smith. In *Bayesian Statistics 6* (J.M. Bernardo, J.O. Berger, A.P. Dawid and A.F.M. Smith, eds.), 357-359. Oxford: Clarendon Press.

BAYARRI, M.J. and BERGER, J.O. (1994). Applications and limitations of robust Bayesian bounds and Type II MLE. In *Statistical Decision Theory and Related Topics V* (S.S. Gupta and J.O. Berger, eds.), 121–134. New York: Springer Verlag.

BAYARRI, M.J. and BERGER, J.O. (1999). Quantifying surprise in the data and model verification (with discussion). In *Bayesian Statistics 6* (J.M. Bernardo, J.O. Berger, A.P. Dawid and A.F.M. Smith, eds.), 53–82. Oxford: Clarendon Press.

BAYARRI, M.J. and BERGER, J.O. (2000). P-values for composite null models. To appear in *Journal of the American Statistical Association.*

BERGER, J.O. (1994). An overview of robust Bayesian analysis (with discussion). *Test*, **3**, 5–59.

BERGER, J.O. (1999). Bayes factors. In *Encyclopedia of Statistical Sciences* (S. Kotz et al., eds.), Update Volume 3, 20–29. New York: Wiley.

BERGER, J.O. and DELAMPADY, M. (1987). Testing precise hypotheses (with discussion). *Statistical Science*, **2**, 317–352.

BERGER, J.O. and MORTERA, J. (1994). Robust Bayesian hypothesis testing in the presence of nuisance parameters. *Journal of Statistical Planning and Inference*, **40**, 357–373.

BERGER, J.O. and PERICCHI, L.R. (1996a). The intrinsic Bayes factor for model selection and prediction. *Journal of American Statistical Association*, **91**, 109–122.

BERGER, J.O. and PERICCHI, L.R. (1996b). The intrinsic Bayes factor for linear models. In *Bayesian Statistics 5* (J.M. Bernardo, J.O. Berger, A.P. Dawid and A.F.M. Smith, eds.), 23–42. Oxford: Clarendon Press.

BERGER, J.O. and PERICCHI, L.R. (1997). On the justification of default and intrinsic Bayes factors. In *Modelling and Prediction* (J.C. Lee et al., eds.), 276–293. New York: Springer-Verlag.

BERGER, J.O. and PERICCHI, L.R. (1998). On criticisms and comparisons of default Bayes factors for model selection and hypothesis testing. In *Proceedings of the Workshop on Model Selection* (W. Racugno, ed.), 1–50. Bologna: Pitagora.

BERGER, J.O. and SELLKE, T. (1987). Testing a Null Hypothesis: The irreconcilability of P-values and evidence. *Journal of American Statistical Association*, **82**, 112–122.

BERNARDO, J.M. (1999). Nested hypotheses testing: the Bayesian reference criterion (with discussion). In *Bayesian Statistics 6* (J.M. Bernardo, J.O. Berger, A.P. Dawid and A.F.M. Smith, eds.), 101–130. Oxford: Clarendon Press.

BERNARDO, J.M. and SMITH, A.F.M. (1994). *Bayesian Theory.* New York: Wiley.

BOX, G.E.P. (1980). Sampling and Bayes' inference in scientific modelling and robustness. *Journal of Royal Statistical Society*, **A**, 383–430.

CANO SANCHEZ, J.A. (1993). Robustness of the posterior mean in normal hierarchical models. *Communications in Statistics: Theory and Methods*, **22**, 1999–2014.

CONIGLIANI, C. and O'HAGAN, A. (2000). Sensitivity of the fractional Bayes factor to prior distributions. To appear in *Canadian Journal of Statistics.*

DELAMPADY, M. and BERGER, J.O. (1990). Lower Bounds on Bayes factor for multinomial distributions with applications to Chi-squared tests of fit. *Annals of Statistics*, **18**, 1295–1316.

DE SANTIS, F. and SPEZZAFERRI, F. (1996). Comparing hierarchical models using Bayes factor and fractional Bayes factor. In *Bayesian Robustness* (J.O. Berger et al., eds.), IMS Lecture Notes, Vol. 29, 305–314. Hayward: IMS.

DE SANTIS, F. and SPEZZAFERRI, F. (1997). Alternative Bayes factors for model selection. *Canadian Journal of Statistics*, **25**, 503–515.

DE SANTIS, F. and SPEZZAFERRI, F. (1999). Methods for default and robust Bayesian model comparison: the fractional Bayes factor approach. *International Statistical Review*, **67**, 1–20.

DIEBOLT, J. and ROBERT, C.P. (1995). Estimation of finite mixture distributions through Bayesian sampling. *Journal of Royal Statistical Society, B*, **56**, 363–375.

DMOCHOWSKY, J. (1996). Intrinsic priors via Kullback-Leibler geometry. In *Bayesian Statistics 5* (J.M. Bernardo, J.O. Berger, A.P. Dawid and A.F.M. Smith, eds.), 543–549. Oxford: Clarendon Press.

EDWARDS, W., LINDMAN, H. and SAVAGE, L.J. (1963). Bayesian statistical inference for psychological research. *Psychological Review*, **70**, 193–242.

GEISSER, S. (1975). The predictive sample reuse with applications. *Journal of American Statistical Association* **70**, 320–328.

GELFAND, A.E. and DEY, D.K. (1994). Bayesian model choice: asymptotic and exact calculations. *Journal of Royal Statistical Society, B*, **56**, 501–514.

GEORGE, E.I. (1999). Bayesian model selection. In *Encyclopedia of Statistical Sciences* (S. Kotz et al., eds.), Update Volume 3, 39–46. New York: Wiley.

GEORGE, E.I. and MCCULLOGH, R.E. (1993). Variable selection via Gibbs sampling. *Journal of American Statistical Association*, **88**, 881–889.

GEWEKE, J. (1999). Simulation methods for model criticism and robustness analysis (with discussion). In *Bayesian Statistics 6* (J.M. Bernardo, J.O. Berger, A.P. Dawid and A.F.M. Smith, eds.), 275–299. Oxford: Clarendon Press.

GEWEKE, J. and PETRELLA, L. (1998). Prior density ratio class in econometrics. *Journal of Business and Economics Statistics*, **16**, 469–478.

JEFFREYS, H. (1961). *Theory of Probability.* London: Oxford University Press.

KASS, R. and RAFTERY, A. (1995). Bayes factors. *Journal of American Statistical Association*, **90**, 773–795.

KASS, R. and VAIDYANATHAN, S. (1992). Approximate Bayes factors and orthogonal parameters, with application to testing equality of two binomial proportions. *Journal of Royal Statistical Society, B*, **54**, 129–144.

KASS, R. and WASSERMAN, L. (1995). A reference Bayesian test for nested hypotheses and its relationship to the Schwarz criterion. *Journal of American Statistical Association*, **90**, 928–934.

KEY, J.T., PERICCHI, L.R. and SMITH, A.F.M. (1999). Bayesian model choice: what and why? (with discussion). In *Bayesian Statistics 6* (J.M. Bernardo, J.O. Berger, A.P. Dawid and A.F.M. Smith, eds.), 343–370. Oxford: Clarendon Press.

LAVINE, M. and SCHERVISH, M. (1999). Bayes factors: what they are and what they are not. *The American Statistician*, **53**, 119–122.

LAVINE, M. and WOLPERT, R.L. (1995). Discussion of "Fractional Bayes factors for model comparison", by A. O'Hagan. *Journal of Royal Statistical Society*, B, **57**, 132.

LINDLEY, D.V. (1957). A statistical paradox. *Biometrika*, **44**, 187–192.

LINDLEY, D.V. (1997). Some comments on Bayes factors. *Journal of Statistical Planning and Inference*, **61**, 181–189.

MORENO, E. (1997). Bayes factors for intrinsic and fractional priors in nested models. In *Bayesian Robustness* (J.O. Berger et al., eds.), IMS Lecture Notes, Vol. 29. Hayward: IMS.

MORENO, E. (2000). Global Bayesian Robustness for some classes of prior distributions. In *Robust Bayesian Analysis* (D. Ríos Insua and F. Ruggeri, eds.). New York: Springer-Verlag.

MORENO, E., BERTOLINO, F. and RACUGNO, W. (1998). An intrinsic limiting procedure for model selection and hypotheses testing. *Journal of American Statistical Association*, **93**, 1451–1460.

MORENO, E. and CANO SANCHEZ, J.A. (1989). Testing a point null hypothesis: asymptotic robust Bayesian analysis with respect to priors given on a sub-sigma field. *International Statistical Review*, **57**, 221–232.

MORENO, E. and LISEO, B. (1998). Default Bayes factors for testing the number of components of a mixture. *Technical Report*, Dept de Estadistica y I.O., Universidad de Granada, Spain.

O'HAGAN, A. (1995). Fractional Bayes factors for model comparison (with discussion). *Journal of Royal Statistical Society*, B, **57**, 99–138.

PAULER, D.K. (1998). The Schwarz criterion and related methods for normal linear models. *Biometrika*, **85**, 13–27.

PÉREZ, J. (1998). Development of Expected Posterior Prior Distributions for Model Comparisons. Ph.D. Thesis, Dept. of Statistics, Purdue University, USA.

PÉREZ, J. and BERGER, J.O. (1999). Expected posterior prior distributions for model selection. *Technical Report*, ISDS, Duke University, USA.

PERICCHI, L.R. (1999). Discussion of "Simulation methods for model criticism and robustness analysis" by J. Geweke. In *Bayesian Statistics 6* (J.M. Bernardo, J.O. Berger, A.P. Dawid and A.F.M. Smith, eds.), 293–295. Oxford: Clarendon Press.

PETTITT, L.I. (1990). The conditional predictive ordinate for the normal distribution. *Journal of Royal Statistical Society*, B, **52**, 175–184.

PICCINATO, L. (1998). Discussion of "Choosing among models when none of them is true", by J.T. Key, L.R. Pericchi, A.F.M. Smith. In *Proceedings of the Workshop on Model Selection* W. Racugno, ed., 350–354, Bologna: Pitagora.

ROEDER, K. and WASSERMAN, L. (1997). Practical Bayesian density estimation using mixtures of normals. *Journal of American Statistical Association*, **92**, 894–902.

RUBIN, D. (1995). Discussion of "Fractional Bayes factors for model comparison" by A. O'Hagan. *Journal of Royal Statistical Society, B*, **57**, 133.

SANSÓ, B., MORENO, E. and PERICCHI, L.R. (1996). On the robustness of the intrinsic Bayes factor for nested models. In *Bayesian Robustness* (J.O. Berger et al., eds.), IMS Lecture Notes, Vol. 29, 155–174. Hayward: IMS.

SCHWARZ, G. (1978). Estimating the dimension of a model. *Annals of Statistics*, **6**, 461–464.

SELLKE, T., BAYARRI, M.J. and BERGER, J.O. (1999). Calibration of p-values for testing precise null hypotheses. *Technical Report*, ISDS, Duke University.

SHAFER, G. (1982). Lindley's paradox (with discussion). *Journal of American Statistical Association*, **77**, 325–351.

SHYAMALKUMAR, N.D. (2000). Likelihood robustness. In *Robust Bayesian Analysis* (D. Ríos Insua and F. Ruggeri, eds.). New York: Springer-Verlag.

VERDINELLI, I. and WASSERMAN, L. (1995). Computing Bayes factors using a generalisation of the Savage-Dickey density ratio. *Journal of American Statistical Association*, **90**, 614–618.

WASSERMAN, L. (1996). The conflict between improper priors and robustness. *Journal of Statistical Planning and Inference*, **52**, 1–15.

WEISS, R. (1996). An approach to Bayesian sensitivity analysis. *Journal of Royal Statistical Society, B*, **58**, 739–750.

ZELLNER, A. and SIOW, A. (1980). Posterior odds for selected regression hypotheses. In *Bayesian Statistics* (J.M. Bernardo, M.H. DeGroot, D.V. Lindley and A.F.M. Smith, eds.), 585–603. Valencia: Valencia University Press.

12

Bayesian Robustness and Bayesian Nonparametrics

Sanjib Basu

ABSTRACT Bayesian robustness studies the sensitivity of Bayesian answers to user inputs, especially to the specification of the prior. Nonparametric Bayesian models, on the other hand, refrain from specifying a specific prior functional form P, but instead assume a second-level hyperprior on P with support on a suitable space of probability measures. Nonparametric Bayes thus appears to have the same goals as robust Bayes, and nonparametric Bayes models are typically presumed to be robust. We investigate this presumed robustness and prior flexibility of nonparametric Bayes models. In this context, we focus specifically on Dirichlet process mixture models. We argue that robustness should be an issue of concern in Dirichlet process mixture models and show how robustness measures can actually be evaluated in these models in a computationally feasible way.

Key words: Dirichlet process, Gâteaux derivative, local sensitivity.

12.1 Introduction

Robust statistics is concerned with the fact that assumptions and inputs commonly made in statistics are, at most, approximations to reality. In a Bayesian framework, the inputs are the functional form of the model and the observed data. The model consists of the sampling model and the prior model though the exact separation line between the two is not always clear cut. A decision-theoretic framework adds a third item, namely, the form of the loss function.

Robust Bayesian analysis studies the sensitivity of Bayesian answers to imprecision of these inputs, namely, the prior, the sampling model, the loss function, and the data. Most of the work on Bayesian robustness has concentrated on prior imprecision as the choice of the prior has typically been an issue of controversy. Excellent reviews on prior robustness can be found in Berger (1990, 1994), Wasserman (1992), Berger et al. (2000), and many other papers in this volume. However, the functional forms of the sampling distribution and the loss function are also inputs in the model; the sensitivity to their imprecision is studied in Basu (1999), Shyamalkumar (2000) and the loss robustness papers in this volume. Clarke and Gustafson

(1998) study the overall sensitivity to the prior, the sampling distribution and the data.

Our focus in this article will be on imprecision of the prior specification. This specification (or some part of the specification) is deferred to a higher level in hierarchical models. Nonparametric (or semiparametric) Bayesian models (NPBMs) go one step further, where in the beginning level, the functional form for the prior (and sometimes the sampling model) is left arbitrary. A prior probability model is then assumed on this functional space. The most well-known among the NPBMs is the Dirichlet process (DP) model (Ferguson, 1973) where the prior at the first level is an arbitrary probability measure P and a DP prior is assumed on P at the next level. Other processes that are used in this context are the Gaussian process (Verdinelli and Wasserman, 1998), the general class of Neutral to the right processes (Doksum, 1974) and Polya tree processes (Lavine, 1992). Similar ideas are used in Bayesian nonparametric function estimation (Shively et al., 1999, Dennison et al., 1998, Vidakovic, 1998, Rios Insua and Müller, 1998), density estimation (MacEachern, 1994, Escobar and West, 1995), modeling link functions (Basu and Mukhopadhyay, 1998) and many other contexts. Sinha and Dey (1997) describe other prior processes, such as the Lévy process, and the martingale process, which are used in modeling hazard and survival functions.

At the onset, nonparametric Bayes models (NPBM) refrain from specifying a specific functional form for the prior. Instead, an NPBM assumes a prior process on the functional form and hence appears to have the same goals as robust Bayes. However, this similarity between nonparametric and robust Bayes is only apparent. As Hampel et al. (1986) write, "robust statistics consider the effects of only approximate fulfillment of assumptions, while nonparametric statistics make rather weak but nevertheless strict assumptions." In the Bayesian context, the contrast is even stronger as the semiparametric or nonparametric Bayesian models are, in some sense, "superparametric." To illustrate this point, suppose we consider the popular NPBM of the Dirichlet process mixture model (DPM), where we independently observe $Y_i \sim p(y_i|\theta_i)$, $i = 1, \ldots, n$; $\theta_1, \ldots, \theta_n$ are i.i.d. $\sim P(\theta)$ and P has a Dirichlet process $\mathrm{DP}(n_0, P_0)$ prior. It is well known from Ferguson (1973) or Antoniak (1974) that this model implies a specific and strictly structured marginal joint prior distribution on the $(\theta_1, \ldots, \theta_n)$ space which assigns masses to each of the subspaces where some of the θ_i are equal. Moreover, the distribution of the probability mass among these different subspaces and the distribution within each subspace are completely determined by the hyperparameters n_0 and P_0, respectively. When one thinks about the DPM model in this marginalized form, robustness with respect to choices of n_0 and P_0 even in this nonparametric model becomes an issue of crucial importance.

In this article, we investigate the (prior) robustness of Bayesian nonparametric models. We will specifically focus on the DPM model, as it is the most popular among the NPBMs. The methodology can be used in other NPBMs, though we do not pursue it here.

12.2 Dirichlet process: preliminaries

The Dirichlet process (DP) defines a probability measure on the space of probability measure. Let $(\Omega, \mathcal{A})$ denote a generic space and its associated σ-field. Let P denote a random probability measure and $\mathcal{P}$ denote the space of all probability measures on $(\Omega, \mathcal{A})$. The Dirichlet process (DP) is a probability measure defined on $\mathcal{P}$. The random probability $P \sim DP(n_0, P_0)$ if for any measurable partition $B_1, \ldots, B_k$ of Ω, the finite dimensional distribution of $(P(B_1), \ldots, P(B_k))$ is Dirichlet$(n_0 P_0(B_1), \ldots, n_0 P_0(B_k))$. Here $n_0 > 0$ (n_0 can be non-integer) is known as the concentration parameter. The location parameter $P_0 \in \mathcal{P}$, that is, P_0 is a probability measure on Ω. Ferguson (1973) showed that the above system of finite-dimensional distributions satisfies the Kolmogorov consistency condition and hence uniquely characterizes the law of the process generating P.

In the Bayesian context, the most important property of a DP prior is probably its conjugacy. If $\theta_1, \ldots, \theta_n$ are i.i.d. observations from P where $P \sim DP(n_0, P_0)$, then the posterior distribution of P is again a Dirichlet process (n_*, P_*), where $n_* = n_0 + n$, $P_* = \{n_0 P_0 + nP_n\}/\{n_0 + n\}$ and $P_n = (1/n) \sum_{i=1}^{n} \delta_{\{\theta_i\}}$ is the empirical distribution. Note that P_* provides a bridge between the completely nonparametric estimate (the empirical distribution P_n) and a fixed parametric specification P_0 (see Doss, 1994).

If $P \sim DP(n_0, P_0)$, then the prior mean of P is P_0. If $\theta_1 \ldots, \theta_n$ are i.i.d. $\sim P$ where $P \sim DP(n_0, P_0)$, then marginally, $\theta_1 \sim P_0$. However, since $P\,|\theta_1, \ldots, \theta_i \sim DP(n_0 + i, (n_0/((n_0 + i))P_0 + (1/(n_0 + i)) \sum_{j=1}^{i} \delta_{\{\theta_j\}})$, we have

$$p(\theta_{i+1}|\theta_1, \ldots, \theta_i) = \frac{n_0}{n_0 + i} P_0 + \frac{1}{n_0 + i} \sum_{j=1}^{i} \delta_{\{\theta_j\}}. \tag{1}$$

The above illustrates that the DP puts mass on each of the subspaces where some of the θ_i's are equal.

Sethuraman (1994) provides a constructive definition of $P \sim DP(n_0, P_0)$ as follows. Consider the two infinite sequences $\phi_1, \phi_2, \ldots$, i.i.d. $\sim P_0$ and $q_1, q_2, \ldots$, i.i.d. $\sim$ Beta$(1, n_0)$. Define $p_1 = q_1$ and $p_i = q_i \prod_{j=1}^{i-1}(1 - q_j), i > 1$. Then the random probability measure P defined as $\sum_{i=1}^{\infty} p_i \delta_{\{\phi_i\}}$ follows a DP(n_0, P_0) distribution. It is obvious from this construction that P is almost surely discrete. This fact was noted by Ferguson (1973) and Blackwell and MacQueen (1974). Ruggeri (1994) studied Bayesian robustness when the sampling distribution of the Bayes model comes from a Dirichlet process $DP(n_0, P_0)$. Carota (1996) investigated the local sensitivity of the Bayes factor for comparing a fixed parametric sampling model against a Dirichlet process-based nonparametric alternative.

In many statistical applications, the underlying distribution is assumed to be continuous. Continuous distributions can be nonparametrically modeled by a Dirichlet process mixture (DPM) model (also called mixture of Dirichlet process models), which goes back to Antoniak (1974) and Ferguson (1983). If Y is an observation from such a distribution, then its distribution is specified in two stages. In the first stage, $Y|\theta \sim F(y|\theta)$, where F is a known parametric distribution indexed by θ. In the second stage, one specifies $\theta \sim P$ where $P \sim DP(n_0, P_0)$. It follows that the marginal distribution of Y is a mixture, $p(y|P) = \int F(y(\theta)dP(\theta)$, and if F is continuous then so is $p(y|P)$. More generally, F may involve covariates x and other hyperparameters β. For example, in a DP scale-mixed normal linear regression, we have $Y \sim N(x^T\beta, \sigma^2)$; $\sigma^2 \sim P$; $P \sim DP(n_0, P_0)$ and the regression parameter β follows a prior $p(\beta)$.

In the general formulation of the DPM model, we observe independent observations Y_i (which could be multivariate, with possibly categorical components) with covariates x_i, $i = 1, \ldots, n$. The DPM model is then hierarchically constructed as follows:

(*i*) $Y_i \sim F(y_i|\theta_i, x_i, \beta_i)$, $i = 1, \ldots, n$.

(*ii*) $\theta_i \sim P(\theta_i), i = 1, \ldots, n$.

(*iii*) $P \sim DP(n_0, P_0)$.

(*iv*) The hyperparameter $\beta = (\beta_1, \ldots, \beta_n)$ has prior $p(\beta)$.

This DPM model has been extensively used in recent NPBM literature. Examples include Escobar (1994) and MacEachern (1994) who use it in the normal means problem, Bush and MacEachern (1996) who use it in design of experiments, Escobar and West (1995) who use it in density estimation, Basu and Mukhopadhyay (1998) who apply it in binary regression, Müller et al. (1996) who use it in nonparametric regression and the recent paper by Gopalan and Berry (1998) on multiple comparisons. This list is by no means exhaustive.

Use of DPM models has become computationally feasible with the advent of Markov chain methods for sampling from the posterior. Before the explosion of Markov chain methods in the 1990s, Kuo (1986) devised a Monte Carlo-based method can be used for small n. Escobar (1994) and MacEachern (1994) simultaneously developed Gibbs sampling algorithms for DPM models. These and almost all other Markov chain methods for DPM models that were later developed use the marginal distribution of $(\theta_1, \ldots, \theta_n)$; that is, the random probability P is not generated in the sampler. From (1), the conditional prior distribution of θ_i is $p(\theta_i|\theta_{-i}) = (n_0/((n_0 + n - 1))P_0 + (1/((n_0 + n - 1)) \sum_{j \neq i} \delta_{\{\theta_j\}}$, where $\theta_{-i} = (\theta_1, \ldots, \theta_{i-1}, \theta_{i+1}, \ldots, \theta_n)$. It follows that the conditional posterior distribution of θ_i is

$$p(\theta_i|\theta_{-i}, Y) = \frac{c\, n_0}{n_0 + n - 1} P_0(\theta)\, F(y_i|\theta_i, x_i, \beta_i) + \frac{c}{n_0 + n - 1} \sum_{j \neq k} F(y_i|\theta_j, x_i, \beta_i)\delta_{\{\theta_j\}}. \tag{2}$$

If P_0 and F are conjugate, the normalizing constant c can easily be obtained.

Instead of sequentially generating θ_i, $i = 1, \ldots, n$ as in Escobar (1994) Gibbs sampler, one can run the sampler on the space of partitions resulting in better mixing. Such samplers are developed by MacEachern (1994) and by Bush and MacEachern (1996). Finally, when P_0 and F are not conjugate, the computational methods are more involved. MacEachern and Müller (1998) present the "no gaps" algorithm for the non-conjugate case. Neal (1998) and MacEachern and Müller (2000) provide excellent reviews of all these computational methods as well as some other methods for the non-conjugate case.

As we mentioned above, all of these computational methods use the marginal distribution of $(\theta_1, \ldots, \theta_n)$ induced by the DP structure. In the following, we describe this marginal distribution. Let $B_1, \ldots, B_k$ denote a partition of $\{1, 2, \ldots, , n\}$. Each B_l denotes a cluster of equal θ values; if $i, j \in B_l$, then $\theta_i = \theta_j$ and vice versa. Let ϕ_l and n_l denote the distinct value of θ_i and the frequency of θ_i in cluster l, respectively. Given this partition, the distribution of $(\theta_1, \ldots, \theta_n)$ is supported only on this particular k-dimensional subspace with joint distribution $\prod_{l=1}^{k} P_0(\phi_l)$. The unconditional (prior) distribution of $(\theta_1, \ldots, \theta_n)$ over all such partitions can then be written as a mixture of these conditional distributions as

$$p(\theta_1, \ldots, \theta_n) = \sum \frac{\Gamma(n_0) n_0^k}{\Gamma(n + n_0)} \prod_{l=1}^{k} \Gamma(n_l) P_0(\phi_l), \tag{3}$$

where the sum is over all partitions $B_1, \ldots, B_k$ of $\{1, 2, \ldots, n\}$; $k = 1, \ldots, n$. For (3), we see that the prior distribution of $(\theta_1, \ldots, \theta_n)$ in the DPM model is, in fact, a single distribution that is completely and rigidly specified by the hyperparameters n_0 and P_0. Moreover, all computational methods for the DPM model directly use this marginalized prior $p(\theta_1, \ldots, \theta_n)$ of (3) instead of the random probability P described earlier.

To reiterate, we argue that a DPM model actually specifies a single joint prior distribution $p(\theta_1, \ldots, \theta_n)$ indexed by the hyperparameters n_0 and P_0. Thus, sensitivity of results to prior imprecision is an issue of utmost importance. This issue is pursued in the following sections.

12.3 Robustness and functional derivatives

Robustness of Bayesian answers to the choice of the prior (and the sampling model) goes back to Good (1959, 1961). Recent interest in this area can be attributed to Berger (1984). Berger (1984, 1990, 1994), Wasserman (1992) and papers in this volume provide excellent reviews of this area. Three different approaches seem to have emerged on how prior robustness should be analyzed. The informal approach simply considers a few different prior distributions and the resulting inferences. The second is called the global robustness approach, where prior uncertainty is modeled by a neighborhood or a class of priors and the range of inferences for this class is determined; see Berger (1990), Berger et al. (2000) and Moreno (2000). The third is the local sensitivity approach where the effect of infinitesimal perturbations of the prior is considered and the rate of change of inference with respect to change in the prior is measured. Gustafson (2000) is an excellent review of the local approach.

In this article, we pursue robustness of NPBMs via the local approach. Suppose we focus on a scalar posterior quantity $T(p)$ for a prior $p(\cdot)$. We measure the rate of change of the functional $T(p)$ by the Gâteaux derivative.

Definition 1 *The Gâteaux derivative of the functional $T(p)$ at prior $p(\ )$ in the direction of another prior $p'(\ )$ is a scalar-valued linear functional $T'(p'-p)$ such that $\frac{1}{\epsilon}|T((1-\epsilon)p+\epsilon p') - T(p) - \epsilon T'(p'-p)| \to 0$ as $\epsilon \to 0$.*

This definition is from Huber (1981) and Basu (1996) and is slightly different from what is generally found in standard mathematics texts. Conceptually, the Gâteaux derivative $T'(p'-p)$ roughly measures the rate of change in $T(p)$ when we move from prior $p(\ )$ to $p'(\ )$ along the path $(1-\epsilon)p(\ )+\epsilon p'(\ )$, $0 \le \epsilon \le 1$.

One often starts from a mathematically convenient prior $p_0(\ )$ and wants to measure the rate of change of $T(p_0)$ towards other priors $p'(\ )$. However, these other prior choices are often not completely known. Instead, one can model the prior uncertainty by a class of priors Γ as is done in the global approach. For example, Γ can be the class of all possible priors on the relevant parameter space. Sensitivity can then be measured by the fastest rate of changes over all $p' \in \Gamma$ or $\sup_{p'\in\Gamma} |T'(p_0 - p')|$.

Here, the linearity of the Gâteaux derivative turns out to be extremely useful. For a large number of convex prior classes Γ, Basu (1996) has shown that the above supremum can actually be computed. This computation has two steps. First, moment theory and Krein–Milman type results can be used to show that

$$\sup_{p'\in\Gamma} |T'(p_0 - p')| = \sup_{p'\in\mathcal{E}} |T'(p_0 - p')|, \tag{4}$$

where $\mathcal{E}$ consists of only the extreme points of Γ. Second, the extreme points class $\mathcal{E}$ is generally driven by one or two parameters, and hence the supremum over $\mathcal{E}$ can often be obtained numerically. Basu (1996) used this approach to measure sensitivity of posterior median, posterior mode and posterior variance. Basu (1999) combined this technique with the steepest-descent method to measure likelihood robustness.

12.4 Robustness in NPBM

The question of robustness can be broadly subdivided into parametric and nonparametric robustness. These two approaches are most easily explained in an example. Consider a Bayesian problem where an $N(\mu_0, \tau_0^{-1})$ prior is used. Parametric robustness studies the sensitivity of the hyperparameter (μ_0, τ_0) specification within the normal functional form; this sensitivity may be measured by either the global or the local approach. Nonparametric robustness questions the normal functional form itself and measures sensitivity over a (nonparametric) class of other prior functional forms.

Here we review results from Basu (2000), who investigates the parametric and nonparametric robustness of NPBMs. Consider the hierarchical DPM model described earlier in Section 12.2, which we restate here briefly: $Y_i \sim F(y|\theta_i, x_i, \beta_i)$, $\theta_i \sim P(\theta_i)$, $i = 1, \ldots, n$; $P \sim DP(n_0, P_0)$ and the hyperparameter $\beta = (\beta_1, \ldots, \beta_n)$ has prior $p(\beta)$. The DP specification has two hyperparameters, n_0 and P_0. Let $T(n_0, P_0)$ be a posterior quantity based on the above DPM model.

Parametric robustness here would study the effect of different hyperparameter specifications, for example, changes in the functional $T(\cdot, \cdot)$ if (n_0, P_0) is changed to (n_0', P_0'). Before we state what is nonparametric robustness of NPBMs, we need a few notational clarifications. Recall that $DP(n_0, P_0)$ defines a probability measure on $\mathcal{P}$ where $\mathcal{P}$ is the space of all probability measures on Θ. We use μ to denote a generic probability measure on $\mathcal{P}$. $DP(n_0, P_0)$ is one such measure, which we denote by μ_{DP}. Let $T(\mu_{DP})$, as above, denote a posterior quantity for the DPM model. Nonparametric robustness in this context then questions the DP structure itself and studies the effect on the posterior function $T(\cdot)$ if μ_{DP} is changed to μ'. Here μ' can be another DP or any other probability on the space of probability measures $\mathcal{P}$.

Here we review results from Basu (2000) on the parametric robustness of DPM models. Recall that parametric robustness studies the effect of the imprecision of (n_0, P_0). However, since P_0 itself is a probability measure and hence infinite-dimensional, one needs to use nonparametric classes and functional derivatives even to study parametric robustness in DPM models.

We mention here that if a specific functional form such as Gamma (α, β) is assumed for P_0, Doss and Narasimhan (1998) have developed a dynamic graphics environment based on reweighting and importance sampling where one can dynamically see the effect of changing the hyperparameters α and

β. We have broader goals in this paper: we want to nonparametrically study the effect of changing the functional form of P_0 as well.

In this context, we will focus on posterior quantities $T(n_0, P_0)$ that can be obtained as posterior expectations of some function of the θ_i's, namely, there is a function $h(\theta, \beta) = h(\theta_1, \ldots, \theta_n, \beta)$ such that

$$T(n_0, P_0) = \int h(\theta, \beta) dp(\theta|Y) = \frac{\int h(\theta, \beta)\, p(Y|\theta, \beta)\, p(\beta)p(\theta)d\beta d\theta}{\int p(Y|\theta, \beta)p(\beta)p(\theta)\, d\beta d\theta} = \frac{H(n_0, P_0)}{m(n_0, P_0)}, \tag{5}$$

where $Y = (Y_1, \ldots, Y_n)$ denote the observed data, the likelihood fucntion is $p(Y|\theta, \beta) = \prod_{i=1}^n F(Y_i|\theta_i, x_i, \beta_i)$ and $p(\theta)$ is the marginal prior for $\theta = (\theta_1, \ldots, \theta_n)$. The effect of the hyperparameters n_0, P_0 enters only through the prior $p(\theta)$. From (1), it follows that

$$p(\theta) = \left\{1/\prod_{i=1}^n (n_0 + i - 1)\right\} \prod_{i=1}^n \left\{ n_0 P_0(\theta_i) + \sum_{j=1}^{i-1} \delta_{\{\theta_j\}}(\theta_i) \right\}. \tag{6}$$

12.4.1 Concentration parameter n_0

Let us first consider the sensitivity of $T(n_0, P_0)$ with respect to n_0. The concentration parameter n_0 controls the number of ties among $\theta_1, \ldots, \theta_n$. From Section 12.2, if we have $\theta_i \sim P$, $i = 1, \ldots, n$; $P \sim DP(n_0, P_0)$, then the conditional mean $E[P|\theta]$ is $P_* = (n_0 P_0 + n P_n)/(n_0 + n)$, where $P_n = (1/n) \sum_{i=1}^n \delta_{\{\theta_i\}}$ is the empirical distribution. Thus, if n_0 is close to 0, the posterior mean P_* will be close to the completely nonparametric estimate P_n, whereas when n_0 is large compared to n, P_* will be close to the prior guess P_0. It is thus clear that posterior results will change with changes in n_0. The effect of change in n_0 on posterior results has been illustrated in Escobar and West (1998). Escobar and West (1995, 1998) further show how one can add one more hierarchy to the DPM model by putting a prior on n_0 and learn about n_0.

We measure robustness by measuring the rate of change of $T(n_0, P_0)$ with respect to changes in n_0. If this rate of change is significantly smaller at n_0' compared to n_0'', it would imply that n_0' is a more robust choice as perturbations have much smaller effects at this value.

Since n_0 is a scalar parameter, the rate of change is simply the derivative with respect to n_0. From (5), we have

$$\frac{\partial}{\partial n_0} T(n_0, P_0) = \frac{\frac{\partial}{\partial n_0} H(n_0, P_0)}{m(n_0, P_0)} - T(n_0, P_0) \frac{\frac{\partial}{\partial n_0} m(n_0, P_0)}{m(n_0, P_0)}. \tag{7}$$

Typically, $\partial H(n_0; P_0)/\partial n_0$ and $\partial m(n_0, P_0)/\partial n_0$ are individually evaluated, and finally (7) is used to evaluate $\partial T(n_0, P_0)/\partial n_0$. In the DPM model, however, all computations are generally done via Markov chain Monte Carlo (MCMC). $\partial H(n_0, P_0)/\partial n_0$ and $\partial m(n_0, P_0)/\partial n_0$ are not functionals of the posterior and hence they cannot be directly estimated from the posterior draws obtained by MCMC methods. What is needed is to show that the entries in the right hand side of (7) can actually be written as expectations of the posterior from a DPM model.

Note that n_0 enters $H(n_0, P_0)$ and $m(n_0, P_0)$ only through the prior $p(\theta)$ given in (6). A straightforward calculation and careful rearrangement show that

$$\frac{\frac{\partial}{\partial n_0} H(n_0, P_0)}{m(n_0, P_0)} = -g_2(n_0)\, T(n_0, P_0) + \frac{\int h(\theta, \beta)\, g_1(n_0, \theta)\, p(Y|\theta, \beta)\, p(\beta)\, p(\theta)\, d\beta d\theta}{m(n_0, P_0)} \tag{8}$$

and

$$\frac{\frac{\partial}{\partial n_0} m(n_0, P_0)}{m(n_0, P_0)} = -g_2(n_0) + \frac{\int g_1(n_0, \theta)\, p(Y|\theta, \beta)\, p(\beta)\, p(\theta) d\beta d\theta}{m(n_0, P_0)}, \tag{9}$$

where

$$g_1(n_0, \theta) = \sum_{i=1}^{n} \frac{P_0(\theta_i)}{n_0 P_0(\theta_i) + \sum_{j=1}^{i-1} \delta_{\{\theta_j\}}(\theta_i)}$$

and $g_2(n_0) = \sum_{i=1}^{n} 1/(n_0 + i - 1)$. From (8) and (9) and recalling that $T(n_0, P_0)$ is simply the posterior expectation of $h(\theta, \beta)$, we have the following:

(*i*) $T(n_0, P_0) = \int h(\theta, \beta) dp(\theta, \beta|\text{data})$;

(*ii*) $\frac{\frac{\partial}{\partial n_0} H(n_0, P_0)}{m(n_0, P_0)} = -g_2(n_0) T(n_0, P_0) + \int h(\theta, \beta) g_1(n_0, \theta) dp(\theta, \beta|\text{data})$; and

(*iii*) $\frac{\frac{\partial}{\partial n_0} m(n_0, P_0)}{m(n_0, P_0)} = -g_2(n_0) + \int g_1(n_0, \theta) dp(\theta, \beta|\text{data})$.

It is now obvious how one can estimate these three terms. Given posterior draws $\{(\theta^{(l)}, \beta^{(l)}), l = 1, \ldots, N\}$ from $p(\theta, \beta|$ data), one can estimate each of the above integrals by their Monte Carlo averages. Moreover, when one plugs in (8) and (9) into (7), the term involving $g_2(n_0)$ cancels out. We illustrate this computation in Section 12.5.

12.4.2 Location parameter P_0

Here we review the method developed in Basu (2000) to measure the sensitivity of the posterior quantity $T(n_0, P_0)$ with respect to P_0 in a DPM model. Since P_0 itself is a probability measure, we use Gâteaux derivatives to measure the rate of change of $T(n_0, P_0)$ with respect to changes in P_0. Recall that the Gâteaux derivative of $T(n_0, P_0)$ in the direction of Q_0 is $T'(Q_0 - P_0) = \partial T(n_0, (1-\epsilon)P_0 + \epsilon Q_0)/\partial \epsilon|_{\epsilon=0}$. It easily follows that

$$T'(Q_0 - P_0) = \frac{H'(Q_0 - P_0)}{m(n_0, P_0)} - T(n_0, P_0)\, \frac{m'(Q_0 - P_0)}{m(n_0, P_0)}, \tag{10}$$

where $H'(Q_0 - P_0)$ and $m'(Q_0 - P_0)$ are Gâteaux derivatives of $H(n_0, P_0)$ and $m(n_0, P_0)$, respectively. After careful rearrangement it follows that

$$
\begin{aligned}
\frac{H'(Q_0 - P_0)}{m(n_0, P_0)} &= \\
& -n_0 \sum_{i=1}^{n} \int h(\theta,\beta)\{\frac{P_0(\theta_i)}{n_0 P_0(\theta_i) + \sum_{j=1}^{i-1} \delta_{\{\theta_j\}}(\theta_i)}\} \frac{p(Y|\theta,\beta)p(\beta)p(\theta)d\beta d\theta}{m(n_0, P_0)} \\
& + n_0 \sum_{i=1}^{n} \int h(\theta,\beta)\{\frac{Q_0(\theta_i)}{n_0 P_0(\theta_i) + \sum_{j=1}^{i-1} \delta_{\{\theta_j\}}(\theta_i)}\} \frac{p(Y|\theta,\beta)\cdot p(\beta)p(\theta)d\beta d\theta}{m(n_0, P_0)} \\
= & \quad -n_0 \int h(\theta,\beta) g_1(n_0,\theta) dp(\theta,\beta|\text{ data}) \\
& + n_0 \int h(\theta,\beta) f(Q_0, P_0, \theta) dp(\theta,\beta|\text{ data}), \qquad (11)
\end{aligned}
$$

where $f(Q_0, P_0, \theta) = \sum_{i=1}^{n} Q_0(\theta_i)/(n_0 P_0(\theta_i) + \sum_{j=1}^{i-1} \delta_{\{\theta_j\}}(\theta_i))$. Further, an expression for $m'(Q_0 - P_0)/(m(n_0, P_0))$ can be easily obtained from above simply by replacing $h(\theta, \beta)$ by 1 in (11).

From (11), it appears that we can evaluate $(H'(Q_0 - P_0))/(m(n_0, P_0))$ by using importance sampling ideas and Monte Carlo averaging as we did in Section 12.4.1. While this is true for the first term in (11), the second term is problematic. We obtain the above expression for the second term because, instead of integration with respect to $Q_0(\theta_i)$, we are using $n_0/(n_0 + i - 1)P_0(\theta_i) + 1/(n_0 + i - 1)\sum_{j=1}^{i-1} \partial_{\{\theta_j\}}(\theta_i)$ as an importance sampling distribution and hence multiplying the integrand by the ratio of these two distributions. This will work as long as Q_0 and P_0 have similar support and P_0 is heavier tailed than Q_0. If Q_0 and P_0 have very different tail-behavior, this importance sampling will be inefficient. These issues and alternatives methods are discussed in Basu (2000).

Once $(H'(Q_0 - P_0))/(m(n_0, P_0))$ and $(m'(Q_0 - P_0))/(m(n_0, P_0))$ are evaluated using (11) or other methods, the Gâteaux derivative $T'(Q_0 - P_0)$ is evaluated from (10). However, $T'(Q_0 - P_0)$ is the rate of change of the posterior quantity $T(n_0, P_0)$ in the direction of specific alternative measure Q_0. Instead of a single Q_0, one typically wants to measure $\sup_{Q_0 \in \Gamma} |T'(Q_0 - P_0)|$ for a class Γ of alternative choices. If $\Gamma = \Gamma_1 = \{\text{all distributions}\}$, this supremum will measure the fastest rate of change of $T(n_0, P_0)$ when the base measure P_0 of the DP can be changed to any arbitrary distribution. Another common choice for Γ is the contamination class $\Gamma_2 = \{Q_0 = (1-\epsilon)P_0 + \epsilon \cdot Q_0', Q_0' \in \Gamma_1\}$. This models the case where instead of changing P_0 completely, only $100\epsilon\%$ of P_0 is changed to another distribution Q_0'. Finally, an alternative attractive choice of Γ is the mixture class. To illustrate this idea, consider the case where the base measure P_0 of the DP is indexed by another hyperparameter η_0, such as $P_0(\cdot|\eta_0) = N(\mu_0, \tau_0^{-1})$, where $\eta = (\mu, \tau)$. One can then model imprecision of P_0 as $\Gamma_3 = \{Q_0 = \int_E P_0(\cdot|\eta)dG(\eta),\ G$ is an arbitrary distribution on $E\}$. For the normal $N(\mu_0, \tau_0^{-1})$ example, one can choose $E = (\mu_L, \mu_U) \otimes (\tau_L, \tau_U)$ and choose the bounds to reflect the degree of allowed variation.

Each of the Γ_l, $l = 1, 2, 3$, classes described above is convex and (4) holds, that is, the evaluation of $\sup_{Q_0 \in \Gamma_l} |T'(Q_0 - P_0)|$ can be reduced to the supremum over only the class of extreme points. These extreme point classes are $\mathcal{E}_1 = \{$all degenerate distributions$\}$, $\mathcal{E}_2 = \{Q_0 = (1-\epsilon)P_0 + \epsilon Q_0',\ Q_0'$ a degenerate distribution$\}$ and $\mathcal{E}_3 = \{Q_0 = P_0(\cdot|\eta), \eta \in E\}$, respectively. The supremum of $|T'(Q_0 - P_0)|$ over these extreme point classes can be obtained numerically. For this optimization, one needs to evaluate $T'(Q_0 - P_0)$ repeatedly for many different $Q_0 \in \mathcal{E}$. The importance sampling and related ideas discussed in (11) for evaluation of $T'(Q_0 - P_0)$ turns out to be really useful at this point. One needs to obtain a large MCMC sample only once, and once this is obtained, it can be recycled repeatedly to evaluate $T'(Q_0 - P_0)$ at many different Q_0.

12.5 Application: O-ring failure

Dalal et al. (1989) present data on o-ring failures in the 23 pre-Challenger Space Shuttle launches. The Rogers Commission concluded that the Challenger accident was caused by a gas leak through the six o-ring joints of the shuttle. Dalal et al. looked at the number of distressed o-rings (among the six) versus launch temperature (Temp.) for 23 previous shuttle flights. The previous shuttles were launched at temperatures between 53° F and 81° F. Dalal et al. (1989) report that a maximum likelihood analysis with temperature as a covariate (they also included pressure as a covariate but its effect is insignificant) predicts a strong probability ($\approx 82\%$) of distress in o-rings at temp. = 31°F, the actual launch conditions of the Challenger explosion. However, since 31° is clearly outside the observed data range of 53° − 81°, such an extrapolated prediction is sensitive to the model assumptions, as was pointed out by Lavine (1991).

We consider the binary regression model with DP mixed normal scale mixture link that was proposed in Basu and Mukhopadhyay (1998). Let $y_{ij} = 1$ if the jth o-ring showed distress in the ith launch, $j = 1, \ldots, 6$, $i = 1, \ldots, 23$. We then consider the following model:

(*i*) y_{ij}'s are independent Bernoulli(p_{ij}).

(*ii*) Moreover, $p_{ij} = \int \Phi(\sqrt{\lambda_{ij}}\ x_i^T \beta) dP(\lambda_{ij})$ a normal scale mixture. Here $x_i = (1, \text{temp}_i)^T$, $\beta = (\beta_0, \beta_1)^T$ and $\Phi(\ )$ is the standard normal cdf.

(*iii*) The random mixing distribution $P \sim \text{DP}(n_0, P_0)$. We choose $P_0 = \text{Gamma}(\nu/2, \nu/2)$ with density proportional to $\lambda^{\nu/2-1}\exp(-\lambda\nu/2)$.

(*iv*) The regression parameter β has prior $p(\beta)$, which is $\mathrm{N}(\beta_*, A)$.

If P was simply P_0, then the resulting link is $\int \Phi(\sqrt{\lambda}\, x_i^T \beta) dP_0(\lambda)$, which, for $P_0 = \mathrm{Gamma}(\nu/2, \nu/2)$, is a t-link with ν degrees of freedom. Addition of the DP structure in (*iii*) adds more flexibility than a simple t-link. (*ii*) can be written hierarchically in two stages as

(*iia*) $p_{ij} = \Phi(\sqrt{\lambda_{ij}}\, x_i^T \beta)$, and

(*iib*) λ_{ij} are i.i.d. $\sim P$.

When $n_0 \to 0$, then the DP structure implies that all the λ_{ij}'s are equal, in which case the above is simply a probit link model. On the other hand, when $n_0 \to \infty$, the λ_{ij}'s are i.i.d. $\sim P_0$, and then the proposed model is a t-link model with ν degrees of freedom. The DP structure thus provides a bridge between the probit and the t-link.

The Markov chain-based posterior analysis for this model is described in Basu and Mukhopadhyay (1998). The analysis is done via data augmentation as proposed in Albert and Chib (1993) where we introduce independent latent variables $Z_{ij} \sim N(x_i^T \beta, 1/\lambda_{ij})$ and $Z_{ij} > 0$ if observed $y_{ij} = 1$, $Z_{ij} < 0$ if observed $y_{ij} = 0$. See Basu and Mukhopadhyay (1998) for further details.

Let $T(n_0, P_0)$ be the predictive mean of the probability of an o-ring failure at 31° (the launch temperature of the fatal shuttle). We obtain a Rao-Blackwell-ized estimate of this predictive mean using the predictive distribution of a new λ based on the DP model. See MacEachern and Müller (2000) for predictive inference from DPM models.

We expect the predicted probability $T(n_0, P_0)$ to be sensitive to the model specification since we are predicting well outside the observed range of previous launch temperatures. We study the robustness of $T(n_0, P_0)$ to the specification of the hyperparameters n_0 and P_0.

Recall that $P_0 = \mathrm{Gamma}(\nu/2, \nu/2)$. We first compute $T(n_0, P_0)$ for a range of n_0 and ν values, namely, $n_0 = .14, .37, 1, 7.39, 20.09, 54.60, 148.41,$ 403.43 and 1096.63 ($\log n_0 = -2,\ -1,\ 0,\ 1,\ 2,\ 3,\ 4,\ 5,\ 6$ and 7, respectively) and $\nu = 1,\ 1.5,\ 2,\ 4,\ 6$ and 10. In the top plot of Figure 1, we see that $T(n_0,\ \mathrm{Gamma}(\nu/2, \nu/2))$ shows a general upward trend as n_0 (or $\log n_0$) increases. Moreover, when n_0 is small, $T(n_0,\ \mathrm{Gamma}(\nu/2, \nu/2))$ changes very little with ν. But, as n_0 increases, $T(n_0,\ \mathrm{Gamma}(\nu/2, \nu/2))$ changes rapidly when ν is small but changes slowly for large ν values. Overall, the values of $T(n_0,\ \mathrm{Gamma}(\nu/2, \nu/2))$ cover a wide range from 0.66 to 0.90. This illustrates the sensitivity of the predictive inference to the underlying model specification.

We measure sensitivity of $T(n_0, P_0)$ with respect to n_0 by evaluating the derivative $\partial T(n_0, P_0)/\partial n_0$ as described in (7). The derivative is estimated via (8) and (9) based on MCMC samples drawn from the posterior. The results are shown in the bottom plot of Figure 1. We see that the rate of change of $T(n_0, P_0)$ is substantially higher when n_0 is small. This rate can be as high as ≈ 0.30 (at $\log n_0 = -2$, $\nu = 1$). When we consider that $T(n_0, P_0)$ is a probability, the rate of 0.30 indicates that the result is highly sensitive to the model specifications. We do want to point out that there is a danger of misinterpreting Figure 1 since the derivative is taken with respect to n_0 and not in terms of $\log n_0$.

We measure robustness with respect to the location parameter $P_0 = \text{Gamma}(\nu/2, \nu/2)$ by considering the class of Gamma mixtures, $\Gamma = \{Q_0 = \int_E \text{Gamma}(\eta/2, \eta/2) dG(\eta), G$ is any arbitrary mixing distribution on the space $E\}$. To avoid very flat or very concentrated distributions, we take $E = \{0.02 \leq \eta \leq 16\}$. Robustness of the predictive probability $T(n_0, P_0)$ is measured by the supremum of its Gâteaux derivative $\sup_{Q_0 \in \Gamma} |T'(Q_0 - P_0)|$, which is evaluated following the method outlined in Section 12.4.2. The value of this supremum Gâteaux derivative is evaluated at different $P_0 = \text{Gamma}(\nu/2, \nu/2)$ and is plotted against ν in Figure 2 for different n_0 (or $\log n_0$) values. We only used $\nu = 1, 1.5$ and 2 to avoid inefficiency in the importance sampling based method (see Section 12.4.2). We see from Figure 2 that the fastest rate of change is often of the order of 0.2. Given that $0 \leq T(n_0, P_0) \leq 1$ (being a probability), this indicates that $T(n_0, P_0)$ is sensitive to specification of P_0. Moreover, Figure 2 shows that the rate of change can be as high as 0.6 or even 0.8.

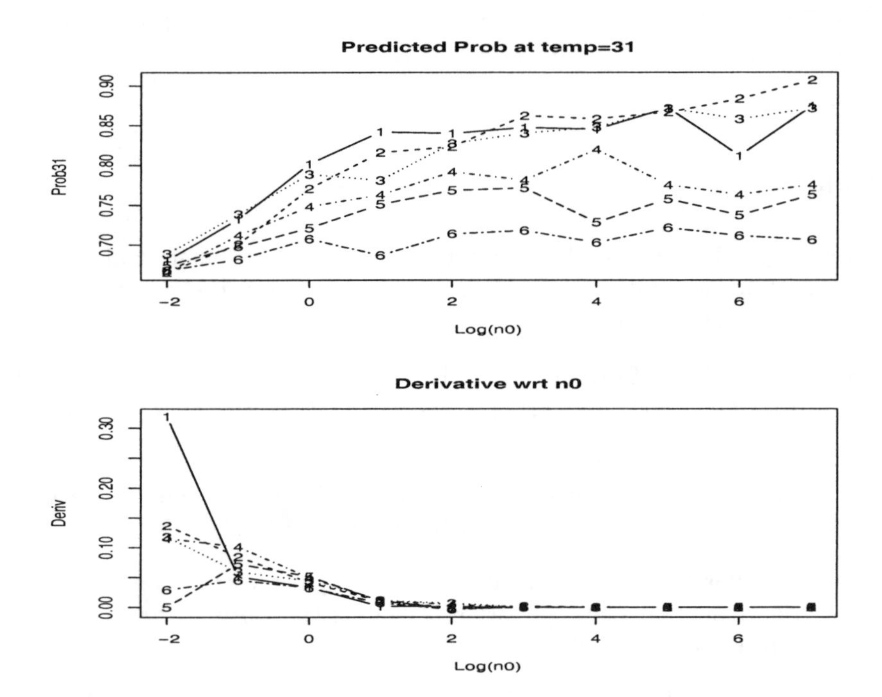

FIGURE 1. (1) $\nu = 1$, (2) $\nu = 1.5$, (3) $\nu = 2$, (4) $\nu = 4$, (5) $\nu = 6$, (6) $\nu = 10$

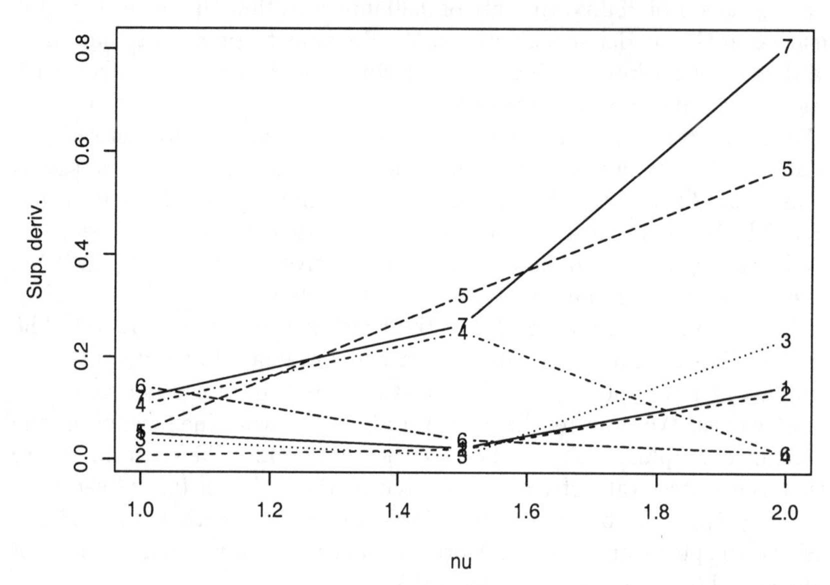

FIGURE 2. (1) $\log n_0 = -1$, (2) $\log n_0 = 0$, (3) $\log n_0 = 1$, (4) $\log n_0 = 2$, (5) $\log n_0 = 3$, (6) $\log n_0 = 4$, (7) $\log n_0 = 5$

References

ALBERT, J.H. and CHIB, S. (1993). Bayesian analysis of binary and polychotomous response data. *Journal of American Statistical Association*, **88**, 669–679.

ANTONIAK, C.E. (1974). Mixture of Dirichlet processes with applications to Bayesian nonparametric problems. *Annals of Statistics*, **2**, 1152–1174.

BASU, S. (1996). Local sensitivity, functional derivatives and nonlinear posterior quantities. *Statistics and Decisions*, **14**, 405–418.

BASU, S. (1999). Posterior sensitivity to the sampling distribution and the prior: more than one observation. *Annals of the Institute of Statistical Mathematics*, **51**, 499–513.

BASU, S. (2000). Robustness of Dirichlet process mixture models. (Manuscript under preparation).

BASU, S. and MUKHOPADHYAY, S. (1998). Binary response regression with normal scale mixture links. In *Generalized Linear Models: A Bayesian Perspective* (D.K. Dey et al. eds.). To appear.

BLACKWELL, D. and MACQUEEN, J.B. (1973). Ferguson distributions via Pólya urn schemes. *Annals of Statistics* **1**, 353- 355.

BERGER, J.O. (1984). The robust Bayesian viewpoint (with discussion). In *Robustness of Bayesian Analysis.* (J.B. Kadane ed.). Amsterdam: North-Holland.

BERGER, J.O. (1990). Robust Bayesian analysis: sensitivity to the prior. *Journal of Statistical Planning and Inference*, **25**, 303–328.

BERGER, J.O. (1994). An overview of robust Bayesian analysis. *TEST*, **2**, 5–58.

BERGER, J.O., RÍOS INSUA, D. and RUGGERI, F. (2000). Bayesian robustness. In *Robust Bayesian Analysis* (D. Ríos Insua and F. Ruggeri eds.). New York: Springer-Verlag.

BUSH, C.A. and MACEACHERN, S.N. (1996). A semiparametric Bayesian model for randomised block designs. *Biometrika*, **83**, 275–285.

CAROTA, C. (1996). Local robustness of Bayes factors for nonparametric alternatives. In *Bayesian Robustness* (J.O. Berger, B. Betrò, E. Moreno, L.R. Pericchi, F. Ruggeri, G. Salinetti and L. Wasserman, eds.), 283–291. IMS Lecture Notes - Monograph Series, Volume 29. Hayward, CA: Institute of Mathematical Statistics.

CLARKE, B. and GUSTAFSON, P. (1998). On the overall sensitivity of the posterior distribution to its inputs. *Journal of Statistical Planning and Inference*, **71**, 137–150.

DALAL, S.R., FOWLKES, E.B. and HOADLEY, B. (1989). Risk analysis of Space Shuttle:pre-Challenger prediction of failure. *Journal of American Statistical Association*, **84**, 945–957.

DENNISON, D.G.T., MALLICK, B.T. and SMITH, A.F.M. (1998). Automatic Bayesian curve fitting. *Journal of the Royal Statistical Society, Ser. B*, **80**, 331–350.

DOKSUM, K. (1974) Tail-free and neutral random probabilities and their posterior distributions. *Annals of Probability*, **2**, 183–201.

DOSS, H. (1994). Bayesian nonparametric estimation for incomplete data via substitution sampling. *Annals of Statistics*, **22**, 1763–1786.

DOSS, H. and NARASIMHAN, B. (1998). Dynamic display of changing posterior in Bayesian survival analysis. In *Practical Nonparametric and Semiparametric Bayesian Statistics* (D. Dey, P. Müller and D. Sinha, eds.), 63–87. New York: Springer-Verlag.

ESCOBAR, M.D. (1994). Estimating normal means with a Dirichlet process prior. *Journal of American Statistical Association*, **89**, 268–277.

ESCOBAR, M.D. and WEST, M. (1995). Bayesian density estimation and inference using mixtures. *Journal of American Statistical Association*, **90**, 577–588.

ESCOBAR, M.D. and WEST, M. (1998). Computing nonparametric hierarchical models. In *Practical Nonparametric and Semiparametric Bayesian Statistics* (D. Dey, P. Müller and D. Sinha, eds.), 1–22. New York: Springer-Verlag.

FERGUSON, T. (1973). A Bayesian analysis of some nonparametric problems. *Annals of Statistics*, **1**, 209–230.

FERGUSON, T. (1983). Bayesian density estimation by mixtures of normal distributions. In *Recent Advances in Statistics: Papers in Honor of Herman Chernoff* (Rizvi et al., eds.), 287–302. New York: Academic Press.

GOOD, I.J. (1959). Could a machine make probability judgments? *Computers and Automation*, **8**, 14–16 and 24–26.

GOOD, I.J. (1961). Discussion of C.A.B. Smith: Consistency in statistical inference and decision. *Journal of the Royal Statistical Society, Ser. B*, **23**, 28–29.

GOPALAN, R. and BERRY, D.A. (1998). Bayesian multiple comparisons using Dirichlet process priors. *Journal of American Statistical Association*, **93**, 1130–1139.

GUSTAFSON, P. (2000). Local robustness of posterior quantities. In *Robust Bayesian Analysis* (D. Ríos Insua and F. Ruggeri, eds.). New York: Springer-Verlag.

HAMPEL, F.R., RONCHETTI, E.M., ROUSSEEUW, P.J. and STAHEL, W.A. (1986). *Robust Statistics: The Approach Based on Influence Functions.* New York: John Wiley.

HUBER, P.J. (1981). *Robust Statistics.* New York: John Wiley.

KUO, L. (1986). Computations of mixtures of Dirichlet processes. *SIAM Journal of Scientific and Statistical Computation*, **7**, 60–71.

LAVINE, M. (1991). Problems in extrapolation illustrated with Space Shuttle o-ring data. *Journal of American Statistical Association*, **86**, 919–922.

LAVINE, M. (1992). Some aspects of Polya tree distributions for statistical modeling. *Annals of Statistics*, **20**, 1203–1221.

MACEACHERN, S.N. (1994). Estimating normal means with a conjugate style Dirichlet process prior. *Communications in Statistics: Simulation and Computation*, **23**, 727–741.

MACEACHERN, S.N. and MÜLLER, P. (1998). Estimating mixture of Dirichlet process models. *Journal of Computational and Graphical Statistics*, **7**, 223–238.

MACEACHERN, S.N. and MÜLLER, P. (2000). Sensitivity analysis by MCMC in encompassing Dirichlet process mixture models. In *Robust Bayesian Analysis* (D. Ríos Insua and F. Ruggeri eds.). New York: Springer-Verlag.

MORENO, E. (2000). Global Bayesian robustness for some classes of prior distributions. In *Robust Bayesian Analysis* (D. Ríos Insua and F. Ruggeri eds.). New York: Springer-Verlag.

MÜLLER, P. ERKANLI, A. and WEST, M. (1996). Bayesian curve fitting using multivariate normal mixtures. *Biometrika*, **83**, 67–79.

NEAL, R.M. (1998). Markov chain sampling methods for Dirichlet process mixture models. *Technical Report*, University of Toronto.

RÍOS INSUA D. and MÜLLER, P. (1998). Feedforward neural networks for nonparametric regression. In *Practical Nonparametric and Semiparametric Bayesian Statistics* (D. Dey, P. Müller and D. Sinha eds.), 181–191. New York: Springer-Verlag.

RUGGERI, F. (1994). Nonparametric Bayesian robustness. *Technical Report*, **94.8**, CNR-IAMI.

SETHURAMAN, J. (1994). A constructive definition of Dirichlet priors. *Statistica Sinica*, **4**, 639–650.

SHIVELY, T.S., KOHN, R. and WOOD, S. (1999). Variable selection and function estimation in additive nonparametric regression using a data-based prior (with discussions), *Journal of American Statistical Association*, **94**, 777–806.

SHYAMALKUMAR, N.D. (2000). Likelihood robustness. In *Robust Bayesian Analysis* (D. Ríos Insua and F. Ruggeri eds.). New York: Springer-Verlag.

SINHA, D. and DEY, D.K. (1997). Semiparametric Bayesian analysis of survival data. *Journal of American Statistical Association*, **92**, 1195–1212.

VERDINELLI, I. and WASSERMAN, L. (1998). Bayesian goodness-of-fit testing using infinite-dimensional exponential families. *Annals of Statistics*, **26**, 1215–1241.

VIDAKOVIC, B. (1998). Wavelet-based nonparametric Bayes methods. In *Practical Nonparametric and Semiparametric Bayesian Statistics* (D. Dey, P. Müller and D. Sinha eds.), 133–150. New York: Springer-Verlag.

WASSERMAN, L. (1992). Recent methodological advances in robust Bayesian inference. In *Bayesian Statistics 4* (J.M. Bernardo et al,. eds.), 483–502. Oxford: Oxford University Press.

13

Γ-Minimax: A Paradigm for Conservative Robust Bayesians

Brani Vidakovic

ABSTRACT In this chapter a tutorial overview of Gamma minimaxity (Γ-minimaxity) is provided. One of the assumptions of the robust Bayesian analysis is that prior distributions can seldom be quantified or elicited exactly. Instead, a family of priors, Γ, reflecting prior beliefs is elicited. The Γ-minimax decision-theoretic approach to statistical inference favors an action/rule which incorporates information specified via Γ and guards against the least favorable prior in Γ. This paradigm falls between Bayesian and minimax paradigms; it coincides with the former when prior information can be summarized in a single prior and with the latter when no prior information is available (or equivalently, possible priors belong to the class of *all* distributions).

Key words: Conditional Gamma minimax, Gamma minimax regret.

13.1 What is Γ-minimax?

The Gamma minimax (Γ-minimax) approach was originally proposed by Robbins (1951). Under this approach the statistician (decision maker) is unable (or unwilling) to specify a single prior distribution reflecting his or her prior knowledge about model parameters of interest. Rather, the statistician is able to elicit a class Γ of plausible prior distributions. This idea is vividly expressed by Efron and Morris (1971):

> ... We have referred to the "true prior distribution" ... but in realistic situations there is seldom any one population or corresponding prior distribution that is "true" in an absolute sense. There are only more or less relevant priors, and the Bayesian statistician chooses among those as best he can, compromising between his limited knowledge of subpopulation distributions and what is usually an embarrassingly large number of identifying labels attached to the particular problem.

Optimality of a decision rule is generally judged by some form of cost for the inaccuracy of the rule. In Γ-minimax, the rule or action that minimizes the supremum of the cost functional over distributions in Γ is selected.

If prior information is scarce, the class Γ of priors under consideration is large and the decision maker's actions are close to the minimax actions. In the extreme case when no information is available, the Γ-minimax setup is equivalent to the usual minimax setup.

If, on the other hand, the statistician has substantial prior information, then the class Γ can be narrow. An extreme case is a class Γ that contains a single prior. In this case, the Γ-minimax framework becomes the usual Bayes framework.

When the model is regular, the Γ-minimax decision coincides with the Bayes decision with respect to the least favorable prior. Thus, establishing the regularity of the model is fundamental in solving minimax-type problems.

Depending on the cost functional, the spirit of the Γ-minimax paradigm can be (i) classical, if the payoff for the decision maker is measured by the Bayes risk [integrated frequentist risk] and (ii) Bayesian, if the payoff is measured by the posterior expected loss. For such payoffs, Γ-minimax is often called *conditional* Γ*-minimax;* see Section 4.

Next we give a simple example to illustrate some of the notions.

Example 1. Suppose we observe $X|\theta \sim \mathcal{N}(\theta, 1)$ where $\theta \in \{-1, 1\}$. Let

$$\theta \sim \pi_p(\theta) \in \Gamma = \left\{ \begin{pmatrix} -1 & 1 \\ 1-p & p \end{pmatrix}, \ 0 \leq p \leq 1 \right\},$$

and let the class of decision rules under consideration $\mathcal{D}_a$ be indexed by a,

$$\delta_a = \begin{cases} -1, & x < a \\ 1, & x \geq a. \end{cases}$$

What is the Γ-minimax rule in the class $\mathcal{D}_a$ when the loss is squared error?

The frequentist risk of δ_a is

$$R(\theta, \delta_a) = E^{X|\theta}(\theta - \delta_a)^2 = (\theta+1)^2\Phi(a-\theta) + (\theta-1)^2[1 - \Phi(a-\theta)].$$

The integrated frequentist risk (or payoff function) is

$$r(\pi_p, \delta_a) = 4p\Phi(a-1) + 4(1-p)[1 - \Phi(a+1)]. \tag{1}$$

The problem is regular. That is, the payoff function as a function of two variables p and a has a saddle point at $a = 0$ and $p = 1/2$. This can be demonstrated by simple calculus arguments (by equating partial derivatives to 0 and verifying analytic requirements for saddle point).

Since δ_0 is Bayes with respect to $\pi_{1/2}$, δ_0 is Γ-minimax in $\mathcal{D}_a$ and $\pi_{1/2}$ is the least favorable prior. One can demonstrate that δ_0 is overall Γ-minimax, that is, Γ-minimax with respect to all (measurable) decision rules.

When one of Γ, $\mathcal{D}_a$ or the payoff function is more complicated, the above approach may not be possible. Results described in the next section help solve more general Γ-minimax problems. $\triangle$

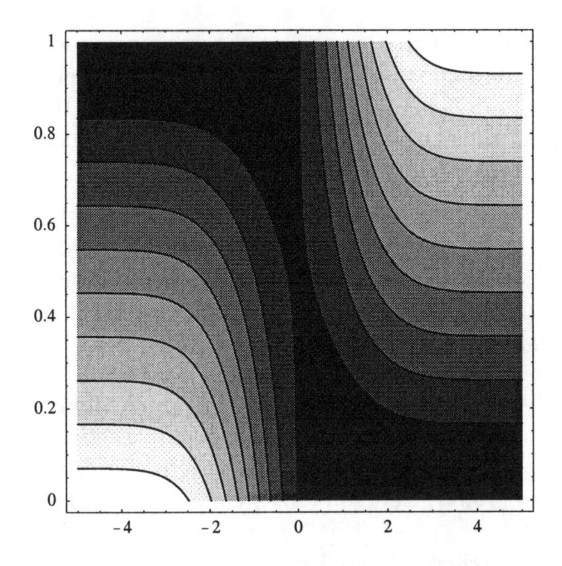

FIGURE 1. Contour plot of the integrated frequentist risk in (1). The horizontal axis is the value of a in $\mathcal{D}_a$ and the vertical axis is p in π_p. Note the saddlepoint at $(0, 1/2)$.

The Γ-minimax paradigm has been criticized on several grounds. Some Bayesians object that belief in the Γ-minimax principle may produce "demonstrable incoherence" since there are examples in which the Γ-minimax rule is not Bayes.

Other complaints concern the use of frequentist measures as cost (payoff) measures, as well as the fact that the Γ-minimax rules often guard against priors from Γ deemed "unreasonable." While in principle agreeing with such concerns, we emphasize that the cost function may not be the decision maker's choice, especially when the inference is interpreted as a statistical game. Also, least favorable priors, against which Γ-minimax rules are guarding, can be made "more reasonable" by careful elicitation of Γ.

There is also widespread justification for Γ-minimax; for a detailed discussion we direct the reader to Berger (1984) and Rios (1990). Experiments and observations showed that in the presence of uncertainty, the decision maker often takes conservative actions even though they might be suboptimal. We give an example that is based on *Ellsberg's paradox* (Ellsberg, 1954) representing a behavioral defense of Γ-minimax.

> Suppose box I contains 50 white and 50 black balls while box II also contains 100 white and black balls, although their proportion is unknown (in absence of information all proportions might be considered equally likely).

> A subject in the experiment first chooses a box and then draws a ball at random from the selected box. If the selected ball is white, the subject receives a prize of, say, $1,000. If a black ball is selected, the subject receives no prize. Which box should the subject choose?
>
> Although preferences for selecting either of the boxes should be equal, most subjects in the experiment prefer the box I. An explanation for such decisions is that subjects act in Γ-minimax fashion, protecting themselves against unfavorable proportions, or *unreliable probabilities*, as Ellsberg calls them.

There is substantial research published on applications of Γ-minimax to statistical estimation, testing, and ranking and selection. For the foundations and the philosophy of Γ-minimaxity, the reader is referred to, among others, Robbins (1951), George (1969), Good (1952), Berger (1984, 1985), and Rios (1990).

A group of researchers affiliated with the Technische Hochschule at Darmstadt (W. Bischoff, L. Chen, J. Eichenauer-Herrmann, K. Ickstadt, J. Lehn, E. Weiß, and others) have a substantial body of research on Γ-minimax estimation, foundations, and certain testing problems.

Γ-minimax research in problems of ranking and selection has been studied by the Purdue decision theory group (S.S. Gupta, D-Y. Huang, R. Berger, W.-C. Kim, K.J. Miescke, and P. Hsiao).

Relevant references on Γ-minimaxity can be found in Berger (1994).

13.2 The mathematical formulation and some standard results

In this section we formulate the Γ-minimax problem in the estimation setup and provide the basic results and definitions.

Let X be a random variable whose distribution is in $\{P_\theta,\ \theta \in \Theta\}$, a family which is indexed by a parameter (random variable) θ. The goal is to make an inference about the parameter θ, given an observation X. A solution is a *decision procedure (decision rule)* $\delta(x)$, which identifies a particular inference for each value of x that can be observed. Let $\mathcal{A}$ be the class of all *actions*; that is, all possible realizations of $\delta(x)$. The *loss function* $L(\theta, a)$ maps $\Theta \times \mathcal{A}$ into the set of real numbers and defines the cost to the statistician when action a is taken and the true value of the parameter is θ. A *risk function* $R(\theta, \delta)$ characterizes the performance of the rule δ for each value of parameter $\theta \in \Theta$. The risk is usually defined in terms of the underlying loss function $L(\theta, a)$ as

$$R(\theta, \delta) = E^{X|\theta} L(\theta, \delta(X)), \tag{2}$$

where $E^{X|\theta}$ is the expectation with respect to P_θ. Since the risk function is defined as an average loss with respect to a sample space, it is called the *frequentist risk.* Let $\mathcal{D}$ be the collection of all measurable decision rules. There are several principles for determining preferences among the rules in $\mathcal{D}$. The three most relevant here are the Bayes principle, the minimax principle, and the Γ-minimax principle.

Under the Bayes principle, the prior distribution π is specified on the parameter space Θ. Any rule δ is characterized by its *Bayes risk*

$$r(\pi, \delta) = \int R(\theta, \delta)\pi(d\theta) = E^\pi R(\theta, \delta). \tag{3}$$

The rule δ_π that minimizes Bayes risk is called the *Bayes rule* and is given by

$$\delta_\pi = \arg \inf_{\delta \in \mathcal{D}} r(\pi, \delta). \tag{4}$$

The *Bayes risk of the prior distribution* π (*Bayes' envelope function*) is

$$r(\pi) = r(\pi, \delta_\pi). \tag{5}$$

The distribution π^* for which

$$r(\pi^*) \geq r(\pi), \text{ for any } \pi, \tag{6}$$

is called the *least favorable prior.*

Under the minimax principle, the optimal rule δ^* which minimizes the maximum of the frequentist risk $R(\theta, \delta)$ and is given by

$$\delta^* = \arg \inf_{\delta \in \mathcal{D}} (\sup_{\theta \in \Theta} R(\theta, \delta)) \tag{7}$$

is called the *minimax rule.*

Now suppose that instead of a single prior on θ, the statistician elicits a family of priors, Γ. Under the Γ-minimax principle, the rule δ_0 is optimal if it minimizes $\sup_{\pi \in \Gamma} r(\pi, \delta)$; specifically,

$$\delta_0 = \arg \inf_{\delta \in \mathcal{D}} (\sup_{\pi \in \Gamma} r(\pi, \delta)). \tag{8}$$

Such a rule δ_0 is called Γ-*minimax rule.* The corresponding Γ-*minimax risk* is given by

$$\bar{r}_\Gamma = \inf_{\delta \in \mathcal{D}} \sup_{\pi \in \Gamma} r(\pi, \delta). \tag{9}$$

Let $\underline{r}_\Gamma$ be the supremum of $r(\pi)$ over the class Γ,

$$\underline{r}_\Gamma = \sup_{\pi \in \Gamma} \inf_{\delta \in \mathcal{D}} r(\pi, \delta) = \sup_{\pi \in \Gamma} r(\pi) = r(\pi_0). \tag{10}$$

Then

$$\underline{r}_\Gamma \leq \bar{r}_\Gamma.$$

This fact has an interpretation as a lower-upper value inequality in the theory of statistical games. For instance, an intelligent player chooses a prior distribution on θ, and the statistician responds by choosing the rule $\delta \in \mathcal{D}$. The payoff function for the statistician is $r(\pi, \delta)$. The rule δ_0 is the minimax strategy for the statistician and $\bar{r}_\Gamma$ is an upper value of the game. It is of interest that a statistical game has a value since in this case finding the Γ-minimax rules is straightforward. The following theorem gives conditions under which the statistical game has a value.

Theorem 13.2.1 *If δ_0 is the Bayes rule with respect to the prior $\pi_0 \in \Gamma$, and for all $\pi \in \Gamma$*

$$r(\pi, \delta_0) \leq r(\pi_0, \delta_0), \tag{11}$$

then $\underline{r}_\Gamma = \bar{r}_\Gamma$, δ_0 is Γ-minimax, and π_0 is the least favorable prior.

Example 2. Let $X \sim \mathcal{N}(\theta, 1)$ and let $\Gamma = \{\pi \mid E^\pi\theta = \mu, E^\pi(\theta-\mu)^2 = \tau^2,\ \mu, \tau \text{ fixed}\}$. Let $\delta_0(x) = (\tau^2/(1+\tau^2))x + (1/(1+\tau^2))\mu$. Then, for any $\pi \in \Gamma$, $r(\pi, \delta_0) = (\tau^2/(1+\tau^2))$. In addition, δ_0 is Bayes with respect to $\pi_0 = \mathcal{N}(\mu, \tau^2)$ from Γ and $r(\pi, \delta_0) = r(\pi_0, \delta_0) = (\tau^2/(1+\tau^2))$. By Theorem 13.2.1, δ_0 is the Γ-minimax rule. For more details about extensions of this example, consult Berger (1985) and Jackson et al. (1970). $\triangle$

The following example discusses a case in which the form of the Γ-minimax rule is known in principle. However, effective calculations are possible only for limited parameter spaces.

Example 3. Assume that $X|\theta \sim \mathcal{N}(\theta, 1)$ and the prior for θ belongs to the class of all symmetric and unimodal distributions supported on $[-m, m]$. The form of the Γ-minimax rule in estimating θ is given in Vidakovic (1992) and Vidakovic and DasGupta (1996). The least favorable prior π_0 depends on m. In general, π_0 is a finite linear combination of uniform distributions and the point mass at zero,

$$\pi_0(\theta) = \alpha_0 \mathbf{1}(\theta = 0) + \sum_{i=1}^{n} \frac{\alpha_i}{2m_i} \mathbf{1}(-m_i \leq \theta \leq m_i),$$

$$0 < m_1 < \ldots < m_n = m, \quad \alpha_i \geq 0, \quad \sum\nolimits_{i=0}^{n} \alpha_i = 1.$$

The marginal density of X is

$$m(x) = \alpha_0 \phi(x) + \sum_{i=1}^{n} \frac{\alpha_i}{2m_i} (\Phi(x + m_i) - \Phi(x - m_i)),$$

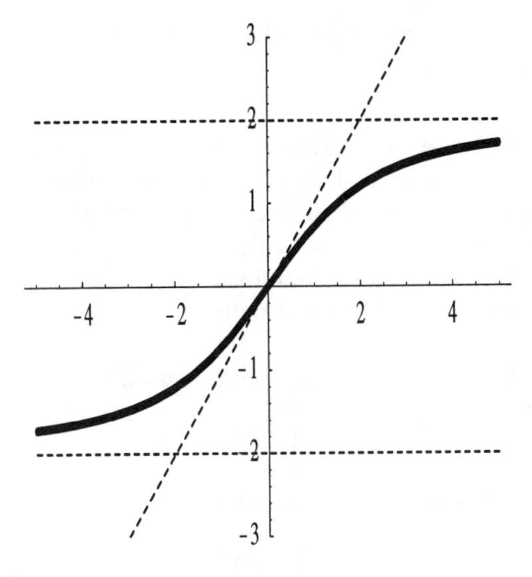

FIGURE 2. Γ-minimax rule for $m = 2$.

and the Bayes rule $\delta_0(x) = x + m'(x)/m(x)$ has the form

$$\delta_0(x) = x - \frac{\alpha_0 x \phi(x) - \sum_{i=1}^{n} \frac{\alpha_i}{2m_i}(\phi(x+m_i) - \phi(x-m_i))}{\alpha_0 \phi(x) + \sum_{i=1}^{n} \frac{\alpha_i}{2m_i}(\Phi(x+m_i) - \Phi(x-m_i))}$$
$$(= \delta(x; 0, \alpha_0, m_1, \alpha_1, \dots, m_n, \alpha_n)).$$

For instance, for $m \leq m_0 = 2.532258$,[1] the least favorable prior is uniform $\mathcal{U}[-m, m]$, the Γ-minimax rule is

$$\delta_0(x) = x + \frac{\phi(x+m) - \phi(x-m)}{\Phi(x+m) - \Phi(x-m)}, \tag{12}$$

and Γ-minimax risk is given by

$$r_\Gamma = 1 - \frac{1}{m}\int_0^\infty \frac{(\phi(x+m) - \phi(x-m))^2}{\Phi(x+m) - \Phi(x-m)} dx.$$

Figure 2 depicts the Γ-minimax rule for $m = 2$, for which the uniform prior $\mathcal{U}[-2, 2]$ is the least favorable. △

---

[1]This constant has been obtained independently by Eichenauer-Herrmann and Ickstadt (1992).

13.3 Constrained rules in Gamma minimax estimation

It is often beneficial to the statistician to restrict the class of all decision rules to a particular subset of all decision rules. Why would the statistician restrict his choices in the inference process? There are two reasons. Although unrestricted Γ-minimax rules may be difficult or even impossible to find, finding restricted Γ-minimax rules is usually straightforward. Furthermore, restricted rules are usually simple, which means that their implementation is easy.

Let $\mathcal{D}_L \subset \mathcal{D}$ be a subclass of all decision rules, let

$$\delta_L^* = \arg \inf_{\delta \in \mathcal{D}_L} (\sup_{\pi \in \Gamma} r(\pi, \delta)) \tag{13}$$

be the *restricted* Γ*-minimax rule*, and let

$$\bar{r}_L = \inf_{\delta \in \mathcal{D}_L} \sup_{\pi \in \Gamma} r(\pi, \delta) \tag{14}$$

be the Γ*-minimax risk of the restricted rule.*

The most common subclass $\mathcal{D}_L$ is the class of all linear decision rules; that is, rules of the form $\delta_L(x) = ax + b$. In the multivariate case the analogy is the class of all affine rules.

There are several measures for comparing restricted and unrestricted rules. The measure defined as a ratio of Γ-minimax risks of restricted and unrestricted rules,

$$\rho = \frac{r_L}{r_\Gamma}, \tag{15}$$

was investigated by Donoho et al. (1990) in the context of estimating a normal mean. The restricted rule δ_L is "good" and can be used instead of an unrestricted one if ρ is "close" to 1.

For example, Vidakovic and DasGupta (1996) show that, under the model discussed in Example 3, the loss of efficiency due to the use of linear rules is at most 7.4% uniformly over m ($\rho \leq 1.074$). This efficiency improves even more if one considers n-degree polynomial rules. In the example that follows, we give the form of a Γ-minimax polynomial rule and work out details for the cubic case.

Example 4. Let $\mathcal{D}_n$ denote the class of all polynomial rules of the form

$$\delta_n(x) = \sum_{i=0}^{n} a_i x^i, \quad n \in \mathbf{N}.$$

Exact Γ-minimax rules can be approximated arbitrarily well by Γ-minimax polynomial rules. It is straightforward to show that the polynomial Γ-minimax rules are skewed symmetric; that is, $a_0 = a_2 = \cdots = a_{2k} = 0$ for

$n = 2k+1$. Define $\mathbf{a} = (a_1, a_3, \ldots, a_{2k+1})'$ and $\mathbf{y} = (x, x^3, x^5, \ldots, x^{2k+1})'$. The frequentist risk of $\delta_n(x) = \mathbf{a}'\mathbf{y}$ is

$$R(\theta, \delta_n(x)) = (\theta - \mathbf{a}'E\mathbf{y})^2 + \mathbf{a}'\Sigma\mathbf{a}, \tag{16}$$

where $\Sigma = \text{Cov}(\mathbf{y}, \mathbf{y})$. The quantities $E\mathbf{y}$ and Σ can be expressed through Chebyshev–Hermite-like polynomials of θ. Let $\phi(x) = (1/\sqrt{2\pi})e^{(-x^2/2)}$ and $D = d/dx$. The polynomials defined as

$$H_n(x) = \frac{(-D)^n \phi(x)}{\phi(x)}$$

are the standard Chebyshev–Hermite polynomials. Let $Q_n(x) = (1/i^n)H_n(ix)$ and $t_k = Q_k(\theta)$. Then,

$$E\mathbf{y} = (t_1, t_3, \ldots, t_{2k+1})'$$

and

$$\Sigma = \begin{pmatrix} t_2 - t_1^2 & t_4 - t_1 t_3 & \cdots & t_{2k+2} - t_1 t_{2k+1} \\ t_4 - t_3 t_1 & t_6 - t_3^2 & \cdots & t_{2k+4} - t_3 t_{2k+1} \\ \vdots & \vdots & \ddots & \vdots \\ t_{2k+2} - t_{2k+1} t_1 & t_{2k+4} - t_{2k+1} t_3 & \cdots & t_{4k+2} - t_{2k+1}^2 \end{pmatrix}.$$

The frequentist risk needed for finding the Γ-minimax solution is obtained by simplifying (16):

$$R(\theta, \mathbf{a}'\mathbf{y}) = \sum_{i \in \mathbf{O}_k} a_i^2 t_{2i} + 2\sum\sum\nolimits_{i,j \in \mathbf{O}_k, i<j} a_i a_j t_{i+j} - 2\theta \sum_{i \in \mathbf{O}_k} a_i t_i + \theta^2, \tag{17}$$

where $\mathbf{O}_k = \{1, 3, 5, \ldots, 2k+1\}$.

We elaborate on the case $n = 3$. Indeed, larger values of odd n differ from the case $n = 3$ only by the complexity of calculation.

In finding the minimax solution we can interchange the *sup* and *inf*, because the corresponding statistical game can be formulated as a finite S-game and therefore has a value; see Berger (1985). Thus, to find the polynomial Γ-minimax rule, we first minimize $E^\pi R(\theta, \mathbf{a}'\mathbf{y})$ with respect to $\mathbf{a}$, for fixed π, and then maximize this minimum with respect to $\pi \in \Gamma$. By standard moment theory, for a fixed k, the least favorable distributions are linear combinations of at most $k+2$ uniform distributions, if we consider a point mass at zero as a degenerate uniform distribution.

Theorem 13.3.1

$$\inf_{\delta \in \mathcal{D}_n} \sup_{\pi \in \Gamma} r(\pi, \delta) = \sup_{0 \le p_1, p_2, p_3 \le 1} 2BDE - CD^2 - AE^2 + F, \tag{18}$$

where

$$\begin{aligned}
A &= \frac{1}{3}m^2\nu_1 + 1, \\
B &= \frac{1}{5}m^4\nu_2 + 2m^2\nu_1 + 3, \\
C &= \frac{1}{7}m^6\nu_3 + 3m^4\nu_2 + 15m^2\nu_1 + 15, \\
D &= \frac{1}{3}m^2\nu_1, \\
E &= \frac{1}{5}m^4\nu_2 + m^2\nu_1, \qquad \textit{and} \\
F &= \frac{1}{3}m^2\nu_1,
\end{aligned}$$

and $\nu_i = \nu_i(p_1, p_2, p_3)$ *are first three canonical moments:* $\nu_1 = p_1$, $\nu_2 = p_1(p_1 + q_1 p_2)$, *and* $\nu_2 = p_1(p_1(p_1 + q_1 p_2) + q_1 p_2(p_1 + q_1 p_2 + q_2 p_3))$, $q_i = 1 - p_i$.

Proof: Let

$$\delta_3(x) = (a_1, a_3)\begin{pmatrix} x \\ x^3 \end{pmatrix} = a_1 x + a_3 x^3.$$

Then

$$\begin{aligned}
R(\theta, \delta_3(x)) &= \left(\theta - \mathbf{a}'\begin{pmatrix} \theta \\ \theta^3 + 3\theta \end{pmatrix}\right)^2 + \mathbf{a}'\begin{pmatrix} 1 & 3\theta^2 + 3 \\ 3\theta^2 + 3 & 9\theta^4 + 36\theta^2 + 15 \end{pmatrix}\mathbf{a} \\
&= a_1^2(\theta^2 + 1) + 2a_1 a_3(\theta^4 + 6\theta^2 + 3) + a_3^2(\theta^6 + 15\theta^4 + 45\theta^2 + 15) \\
&\qquad -2\, a_1\theta^2 - 2a_3(\theta^4 + 3\theta^2) + \theta^2.
\end{aligned}$$

If we take the expectation of $R(\theta, \delta_3(x))$ with respect to θ, and use the representation[2] $\theta = U \cdot Z$ to replace $E\theta^n$ with $(1/(n+1))EZ^n$, we get (18).

If we minimize $r(\pi, \delta_3)$ with respect to a_1 and a_3 first, then by standard calculus arguments the minimum

$$r(\pi, \delta_3^\star) = 2BDE - CD^2 - AE^2 + F$$

is achieved for the rule $\delta_3^\star = a_1^\star x + a_3^\star x^3$, where

$$a_1^\star = \frac{DC - BE}{AC - B^2} \quad \text{and} \quad a_3^\star = \frac{AE - BD}{AC - B^2}.$$

To maximize $r(\pi, \delta_3^\star)$ with respect to the moments ν_1, ν_2, and ν_3, the *canonical moments* (see Skibinsky, 1968) are employed. After expressing

[2] Any symmetric and unimodal random variable θ supported on $[-m, m]$ is equal in distribution to $U \cdot Z$, where U is uniform on $[-1, 1]$ and Z is the corresponding random variable supported on $[0, m]$.

the ν_i's through the canonical moments p_1, p_2, and p_3, the original extremal moment problem with complex boundary conditions is transformed to an equivalent problem where the boundary conditions are independent and simple. In fact, the maximization over the unit cube $[0,1] \times [0,1] \times [0,1]$ is performed.

The numerical maximization (Fortran IMSL routine for constrained maximization DBCONF) used suggests that only two types of distributions can be least favorable in the cubic Γ-minimax problem. For $m < 2.7599$, the maximizing p_1 is equal to 1, which corresponds (regardless of values for p_2 and p_3) to the uniform $\mathcal{U}[-m,m]$ distribution on θ. Some selected values of $m < 2.7599$, and the corresponding values of $a_1^\star, a_3^\star, r_C, r_C/r_\Gamma$ (where r_C is the Γ-minimax risk of the cubic rule $\delta_3^\star$), are given in Table 1.

m	$a_1^\star$	$a_3^\star$	r_C	r_C/r_Γ
0.3	0.02962	−0.00016	0.02912	1
0.5	0.08023	−0.00102	0.07691	1
1	0.28032	−0.00777	0.24922	1.0001
1.5	0.50624	−0.01597	0.42241	1.0014
2	0.69311	−0.02000	0.55315	1.0035
2.5	0.82672	−0.02000	0.64193	1.0047
2.7	0.86704	0.01928	0.66862	1.0038

TABLE 1. Values of coefficients for the cubic rule and corresponding ρ

If $m \geq 2.7599$, then the maximizing p_1 is strictly less than 1, p_2 is equal to 1, and p_3 is arbitrary. This corresponds to the least favorable distribution on θ, which is a linear combination of uniforms $\mathcal{U}[-m,m]$ and a point mass at zero, namely $\pi_0(\theta) = \alpha\delta(\{0\}) + (1-\alpha)(1/2m)\mathbf{1}(-m \leq \theta \leq m)$. Table 2 gives the risks and the efficiency in this case. $\triangle$

13.4 Conditional Gamma minimax and Gamma minimax regret

Conditional Γ-minimax had been at hinted by Watson (1974) but was first explored in a particular context by DasGupta and Studden (1989). As the name suggests, conditional on observations, the statistician is interested in devising an action that minimizes the payoff expressed in terms of posterior expected loss. For an action a, the posterior expected loss is $\rho(\pi, a) = E^{\pi^*} L(\theta, a)$. For any action a, let π_a be a density such that $\rho(\pi_a, a) = \sup_\pi \rho(\pi, a)$. Such π_a (not necessarily unique) is called a *least favorable prior.* An action a^* that minimizes the supremum of posterior expected loss is called a *conditional* Γ*-minimax action.* The existence of conditional

m	$a_1^\star$	$a_3^\star$	$\alpha = 1 - p_1$	r_C	r_C/r_Γ
2.8	0.87906	−0.01845	0.01011	0.68054	1.0027
3	0.88499	−0.01581	0.05447	0.70300	1.0015
3.5	0.89928	−0.01080	0.13326	0.75217	1.0043
4	0.91228	−0.00748	0.18379	0.79215	1.0087
m	$a_1^\star$	$a_3^\star$	$\alpha = 1 - p_1$	r_C	$r_C/r_\Gamma <$
5	0.93347	−0.00379	0.24253	0.85037	1.0151
6	0.94885	−0.00207	0.27412	0.88861	1.0192
8	0.96793	−0.00075	0.30535	0.93256	1.0228
10	0.97835	−0.00033	0.31976	0.95526	1.0177
12	0.98451	−0.00016	0.32758	0.96831	1.0147
15	0.98984	−0.00007	0.33398	0.97938	1.0116
20	0.99417	−0.00002	0.33895	0.98825	1.0094
50	0.99905	$-5.82\ 10^{-7}$	0.34428	0.99809	1.0020
100	0.99976	$-3.65\ 10^{-8}$	0.34513	0.99952	~ 1

TABLE 2. Continuation of the previous table when $m \geq 2.7599$

Γ-minimax actions for general models/losses was explored in Betrò and Ruggeri (1992).

Although this approach is more in the Bayesian spirit, there is an abundance of examples in which the conditional Γ-minimax actions fail to be Bayes actions with respect to any single prior in the class Γ. We give an example by Watson (1974). For more interesting theory and examples about conditional Γ-minimax, we direct the reader to Betrò and Ruggeri (1992) and DasGupta and Studden (1989). The stability of conditional Γ-minimax actions is explored in the work of Boratyńska (1997), Męczarski (1993, 1998), and Męczarski and Zieliński (1991).

Example 5. Consider estimating the parameter θ of a Poisson process, and suppose that prior is known to belong to the family Γ of gamma distributions,

$$\pi(\theta) \in \left\{ \frac{m^r \theta^{r-1}}{(r-1)!} e^{-\theta m}, r \text{ known}, m \in [m_L, m^U] \right\}.$$

Assume that the loss is weighted squared error, $L(\theta, a) = (\theta - a)^2/\theta$. If the process is observed up to time T and n events have been recorded, the Γ-minimax action in estimating θ is

$$a_{GM} = \sqrt{\frac{(n+r-1)(n+r)}{(m_L+T)(m_U+T)}}.$$

However, for a particular prior in Γ (indexed by m), the Bayes action is

$$a_B = \frac{n+r-1}{m+T},$$

and clearly no such value of m exists for which $a_{GM} = a_B$ for all values of n. $\triangle$

DasGupta and Studden (1989) discuss conditional Γ-minimax actions for the squared error loss, multivariate normal model, and multivariate normal prior. We discuss one of their interesting results. Let $\mathbf{y} \sim \mathcal{MVN}_p(\theta, \Sigma_0)$ be a multivariate normal distribution where θ is unknown and Σ_0 is a known positive definite matrix.

Let the distribution for θ belong to the class

$$\Gamma = \{\pi|\ \pi \sim \mathcal{N}(\mu, \Sigma_0), \mu \text{ fixed}, \Sigma_1 \leq \Sigma \leq \Sigma_2\},$$

where the matrix inequality $A \leq B$ means that $B - A$ is a nonnegative definite matrix. For a given c, suppose that $c'\theta$ is the parameter of interest. Let $S(c)$ be the set of two-dimensional vectors consisting of the posterior mean and posterior variance,

$$S(c) = \{(E(c'\theta|\mathbf{y}), Var(c'\theta|\mathbf{y})|\pi \in \Gamma\}.$$

It is curious that the set $S(c)$ forms an ellipse

$$\{u|\ (u-u_0)'D^{-1}(u-u_0) \leq 1\},$$

where $u_0 = (c'\mu + c'\bar{\Lambda}v, c'\bar{\Lambda}c)'$ and

$$D = A^2 \begin{pmatrix} v'(\Lambda_2 - \Lambda_1)v & c'(\Lambda_2 - \Lambda_1)v \\ & c'(\Lambda_2 - \Lambda_1)c \end{pmatrix},$$

with $\Lambda_i = (\Sigma_0^{-1} + \Sigma_i^{-1})$, $i = 1, 2$; $\bar{\Lambda} = (\Lambda_1 + \Lambda_2)/2$, $v = \Sigma_0^{-1}(\mathbf{y} - \mu)$, and $A^2 = (1/4)c'(\Lambda_2 - \Lambda_1)c$.

The conditional Γ-minimax action in estimating $c'\theta$ is

$$a^* = \begin{cases} c'(\Lambda_2 v + \mu) & \text{if } v'(\Lambda_2 - \Lambda_1)v \leq 1 \\ c'(\Lambda^* v + \mu) & \text{if } v'(\Lambda_2 - \Lambda_1)v > 1, \end{cases} \tag{19}$$

where $\Lambda^* = \bar{\Lambda} + ((\Lambda_2 - \Lambda_1)/(2v'(\Lambda_2 - \Lambda_1)v))$.

It is interesting that

$$\theta^* = \begin{cases} (\Lambda_2 v + \mu) & \text{if } v'(\Lambda_2 - \Lambda_1)v \leq 1 \\ (\Lambda^* v + \mu) & \text{if } v'(\Lambda_2 - \Lambda_1)v > 1 \end{cases}$$

is *not* a conditional Γ-minimax action for θ, although for each c, (19) is a conditional Γ-minimax estimator of $c'\theta$.

However, DasGupta and Studden show that θ^* is a conditional Γ-minimax action for θ if $\Lambda_2 - \Lambda_1$ is a multiple of the identity matrix, I. For instance, if $\mu = 0$, $\Sigma_0 = I$, $\Sigma_1 = \alpha I$, and $\Sigma_2 = \infty I$ $(\Sigma \geq \alpha I)$, the conditional Γ-minimax action is

$$\theta^* = \begin{cases} \mathbf{y} & \text{if } \mathbf{y}'\mathbf{y} \leq \alpha + 1 \\ y - \frac{1}{2(\alpha+1)}\left(1 - \frac{\alpha+1}{\mathbf{y}'\mathbf{y}}\right)\mathbf{y} & \text{if } \mathbf{y}'\mathbf{y} > \alpha + 1. \end{cases}$$

This estimator resembles a shrinkage James-Stein–type estimator as well as some restricted-risk Bayesian estimates; see Berger (1982) and DasGupta and Rubin (1987).

There is another interesting example from Betrò and Ruggeri (1992).

Example 6. Assume $X = 1$ was observed from the model $f(x|\theta) = \theta^x(1-\theta)^{1-x}$, and Γ has two distributions only, $\pi_1(\theta) = \mathbf{1}(0 \leq \theta \leq 1)$ and $\pi_2(\theta) = (3/2)\mathbf{1}(0 \leq \theta 1/2) + (1/2)\mathbf{1}(1/2 < \theta \leq 1)$. The posterior expected losses for an action a are $\rho(\pi_1, a) = a^2 - 4/3a + 1/2$ and $\rho(\pi_2, a) = a^2 - 10/9a + 3/8$. Thus $a^\star = 9/16$ is the conditional Γ-minimax action.

It is curious that the Γ-minimax rule $\delta^*(x) = (x+1)/3$ does not produce $9/16$ for $X = 1$. $\triangle$

13.4.1 Conditional Gamma minimax regret

Regret-type rules in decision theory have been time-honored; see, for example, the discussion in Berger (1984, 1985). In the conditional context, Γ-minimax regret rules were by Zen and DasGupta (1993).

As before, let X be an observation from a distribution P_θ with density $p_\theta(x)$, indexed by the parameter $\theta \in \Theta$. Suppose that θ has a prior distribution with density $\pi(\theta)$. Let $\mathcal{A}$ be the action space, $L(\theta, a)$ the loss if the action $a \in \mathcal{A}$ is adopted, and θ the state of nature. Let $\pi_x(\theta)$ be the posterior density when x is observed, and $\rho(\pi_x, a)$ the posterior expected loss of a.

The *posterior regret* of an action a is

$$d(\pi_x, a) = \rho(\pi_x, a) - \rho(\pi_x, a_{\pi_x}),$$

where a_{π_x} is an action minimizing $\rho(\pi_x, a)$. Informally, d measures the loss of optimality due to choosing a instead of the optimal action a_{π_x}.

Definition 13.4.1 *$a_M \in \mathcal{A}$ is the posterior regret Γ-minimax (PRGM) action if*

$$\inf_{a \in \mathcal{A}} \sup_{\pi \in \Gamma} d(\pi_x, a) = \sup_{\pi \in \Gamma} d(\pi_x, a_M). \tag{20}$$

Suppose that $\Theta = \mathcal{A} \subset \mathbf{R}$ is an interval, and the loss function is $L(\theta, a) = (\theta - a)^2$. It is easy to see that

$$r(\pi_x, a) = (a - a_{\pi_x})^2,$$

where a_{π_x} is the posterior expected value of π_x. The following proposition provides a guide for finding the posterior regret action.

Proposition 1 *Let* $\underline{a} = \inf_{\pi \in \Gamma} a_{\pi_x}$ *and* $\bar{a} = \sup_{\pi \in \Gamma} a_{\pi_x}$ *be finite. Then*

$$a_M = \frac{1}{2}(\underline{a} + \bar{a}). \tag{21}$$

Thus, besides its heuristic appeal, computing a_M is simple provided that we have procedures to compute the range of posterior expectations; see Berliner and Goel (1990), DasGupta and Bose (1988), Rios et al. (1995), Sivaganesan and Berger (1989), and Zen and DasGupta (1993) for details.

In the following example of Zen and DasGupta (1993), it is demonstrated how the PRGM action changes with enrichment of the class Γ.

Example 7. Let $X|\theta \sim \text{Bin}(n, p)$ and suppose the prior on θ belongs to
(i) $\Gamma_1 = \{\text{Beta}(\alpha, \alpha),\ \alpha_1 \le \alpha \le \alpha_2\}$,
(ii) $\Gamma_2 = \Gamma_{SU}[0, 1]$ (symmetric, unimodal, supported on [0,1]), or
(iii) $\Gamma_3 = \Gamma_S[0, 1]$ (symmetric, supported on [0,1]).
In case (i), the PRGM action is

$$\delta_1(X) = \frac{X + \frac{n(\alpha_1+\alpha_2)+4\alpha_1\alpha_2}{2(n+\alpha_1+\alpha_2)}}{n + \frac{n(\alpha_1+\alpha_2)+4\alpha_1\alpha_2}{n+\alpha_1+\alpha_2}}.$$

Hence, δ_1 is Bayes with respect to $\pi^*(\theta) = \text{Beta}(\alpha^*, \alpha^*)$, where $\alpha^* = (n(\alpha_1 + \alpha_2) + 4\alpha_1\alpha_2)/(n + \alpha_1 + \alpha_2)$. Note that $\pi^*(\theta) \in \Gamma_1$ and $\alpha^* \to (\alpha_1 + \alpha_2)/2$, when $n \to \infty$.

If $\theta \sim \pi \in \Gamma_2$, the PRGM action is

$$\delta_2(X) = \frac{X + \frac{n}{2} + 2}{2n + 4},$$

which is Bayes with respect to $\pi^*(\theta) = \text{Beta}(n/2 + 2, n/2 + 2) \in \Gamma$.

When $\theta \sim \pi \in \Gamma_3$, the PRGM rule is rather unattractive,

$$\delta_3(X) = \begin{cases} 1/4, & 0 \le X < [\frac{n}{2}] \\ 1/2, & X = [\frac{n}{2}] \\ 3/4, & [\frac{n}{2}] < X \le n. \end{cases}$$

$\triangle$

Example 8. Let $X_1, \ldots, X_n \sim \mathcal{N}(\theta, 1)$. We seek the PRGM action for θ. Let Γ_S and Γ_{SU} be the families of all symmetric distributions and all symmetric, unimodal distributions on $[-m, m]$, respectively. In deriving the PRGM action, we use the well-known fact that any distribution $\pi \in \Gamma_S$ can be represented as a mixture of symmetric two-point priors, and

$$\sup_{\pi \in \Gamma_S} a_{\pi_x} = \sup_{\pi \in \Gamma_S'} a_{\pi_x},$$

$$\inf_{\pi \in \Gamma_S} a_{\pi_x} = \inf_{\pi \in \Gamma_S'} a_{\pi_x},$$

where

$$\Gamma'_S = \left\{ \pi \in \Gamma_S \mid \pi(\{-z\}) = \pi(\{z\}) = \frac{1}{2},\ z \in [0, m] \right\}.$$

Then, for $\theta \sim \pi \in \Gamma_S$, the PRGM action is

$$a_M = \frac{m}{2} \tanh mn\bar{X}.$$

Consider now $\pi \in \Gamma_{SU}[-m, m]$. Here,

$$\sup_{\pi \in \Gamma_{SU}} a_{\pi_x} = \sup_{0 \le z \le m} f(z, x),$$
$$\inf_{\pi \in \Gamma_{SU}} a_{\pi_x} = \inf_{0 \le z \le m} f(z, x),$$

where

$$f(z, x) = \frac{\int_{-z}^{z} \theta \phi_{1/n}(x - \theta) d\theta}{\int_{-z}^{z} \phi_{1/n}(x - \theta) d\theta},$$

and $\Phi_{1/n}(x)$ and $\phi_{1/n}(x)$ denote the cdf and pdf of the normal $\mathcal{N}(0, \frac{1}{n})$ law.

For $\theta \sim \pi \in \Gamma_{SU}$, the PRGM action is

$$a_M = \frac{\bar{X}}{2} + \frac{1}{2n} \frac{\phi_{1/n}(\bar{X} + m) - \phi_{1/n}(\bar{X} - m)}{\Phi_{1/n}(\bar{X} + m) - \Phi_{1/n}(\bar{X} - m)}.$$

$\triangle$

13.5 Conclusions

In this chapter we gave an overview of the Γ-minimax inference standpoint as a way to select a robust action/rule from a multitude of possible rules. When the problem is "regular", that is, when the sup and inf can interchange places in the game-theoretic formulation of the problem, the selected action becomes Bayes with respect to a *least favorable prior*. We also provided a list of examples in connection with the different flavors of the Γ-minimax formulation, namely standard, conditional, regret, and restricted.

Acknowledgments

The author thanks the Editors for their kind invitation and for providing some recent references. Constructive comments of Anirban DasGupta, Michael Kozdron, and an anonymous referee improved the presentation.

References

BERGER, J. (1982). Estimation in continuous exponential families: Bayesian estimation subject to risk restrictions and inadmissibility results. In *Statistical Decision Theory and Related Topics III,* **1**, 109–141. New York: Academic Press.

BERGER, J. (1984). The robust Bayesian viewpoint. In *Robustness of Bayesian Analyses* (J. Kadane, ed.), 63–124. Amsterdam: Elsevier Science Publishers.

BERGER, J. (1985). *Statistical Decision Theory and Bayesian Analysis, Second Edition.* New York: Springer-Verlag.

BERGER, J. (1990). Robust Bayesian analysis: sensitivity to the prior. *Journal of Statistical Planning and Inference*, **25**, 303–328.

BERGER, J. (1994). An overview of robust Bayesian analysis (with discussion). *Test,* **1**, 5–124.

BERGER, R. (1979). Gamma minimax robustness of Bayes rules. *Communications in Statistics, Part A - Theory and Methods*, **8**, 543–560.

BERLINER, M. and GOEL, P. (1990). Incorporating partial prior information: ranges of posterior probabilities. In *Bayesian and Likelihood Methods in Statistics and Econometrics: Essays in Honor of George A. Barnard* (S. Geisser, J. Hodges, F. J. Press and A. Zellner, eds.), 397–406. Amsterdam: North-Holland.

BETRÒ, B. and RUGGERI, F. (1992). Conditional Γ-minimax actions under convex losses. *Communications in Statistics, Part A - Theory and Methods*, **21**, 1051–1066.

BORATYŃSKA, A. (1997). Stability of Bayesian inference in exponential families. *Statistics & Probability Letters*, **36**, 173–178.

DASGUPTA, A. and BOSE, A. (1988). Γ-minimax and restricted-risk Bayes estimation of multiple Poisson means under ϵ-contaminations of the subjective prior. *Statistics & Decisions*, **6**, 311–341.

DASGUPTA, A. and RUBIN, H. (1987). Bayesian estimation subject to minimaxity of the mean of a multivariate normal distribution in the case of a common unknown variance. In *Statistical Decision Theory and Related Topics IV* (S.S. Gupta and J. O. Berger eds.) Vol.1, 325–345.

DASGUPTA, A. and STUDDEN, W. (1989). Frequentist behavior of robust Bayes estimates of normal means. *Statistics and Decisions,* **7**, 333–361.

DONOHO, D., LIU, R. and MACGIBBON, B. (1990). Minimax risk over hyperrectangles, and implications. *Annals of Statistics*, **18**, 1416–1437.

EICHENAUER-HERRMANN, J. and ICKSTADT, K. (1992). Minimax estimators for a bounded location parameter. *Metrika*, **39**, 227–237.

EFRON, B. and MORRIS, C. (1971). Limiting the risk of Bayes and empirical Bayes estimators – Part I: The Bayes case. *Journal of the American Statistical Association*, **66,** 807–815.

ELLSBERG, D. (1954). Classic and current notions of "measurable utility." *Economic Journal,* **LXIV**, 528–556.

GEORGE, S. (1969). Partial prior information: some empirical Bayes and G-minimax decision functions. *Ph.D. Thesis*, Southern Methodist University.

GOOD, I.J. (1952). Rational decisions. *Journal of the Royal Statistical Society* (Ser. B), **14,** 107–114.

JACKSON, D., O'DONOVAN, T., ZIMMER, W. and DEELY, J. (1970). $\mathcal{G}_2$-minimax estimators in the exponential family. *Biometrika,* **57**, 439–443.

MĘCZARSKI, M. (1993). Stability and conditional gamma-minimaxity in Bayesian inference. *Applicationes Mathematicae*, **22**, 117–122.

MĘCZARSKI, M. (1998). *Robustness Problems in Bayesian Statistical Analysis* (in Polish). Monograph Series No. 446. Warszawa: Publishing House of Warsaw School of Economics.

MĘCZARSKI, M. and ZIELIŃSKI, R. (1991). Stability of the Bayesian estimator of the Poisson mean under the inexactly specified Gamma prior. *Statistics & Probability Letters,* **12**, 329–333.

RIOS INSUA, D. (1990). *Sensitivity Analysis in Multiobjective Decision Making.* Lecture Notes in Economics and Mathematical Systems, **347**. New York: Springer-Verlag.

RIOS INSUA, D., RUGGERI, F. and VIDAKOVIC, B. (1995). Some results on posterior regret Γ-minimax estimation. *Statistics & Decisions,* **13**, 315–331.

ROBBINS, H. (1951). Asymptotically sub-minimax solutions to compound statistical decision problems. In *Proceedings of the Second Berkeley Symposium on Mathematical Statistics and Probability,* **1**. Berkeley: University of California Press.

SKIBINSKY, M. (1968). Extreme nth moments for distributions on $[0, 1]$ and the inverse of a moment space map. *Journal of Applied Probability*, **5**, 693–701.

SIVAGANESAN, S. and BERGER, J. (1989). Range of posterior measures for priors with unimodal contaminations. *Annals of Statistics,* **17**, 868–889.

VIDAKOVIC, B. (1992). A study of the properties of computationally simple rules in estimation problems. *Ph.D. Thesis*, Department of Statistics, Purdue University.

VIDAKOVIC, B. and DASGUPTA, A. (1996). Efficiency of linear rules for estimating a bounded normal mean. *Sankhyā, Series A, Indian Journal of Statistics*, **58**, 81–100.

WATSON, S.R. (1974). On Bayesian inference with incompletely specified prior distributions. *Biometrika*, **61**, 193–196.

ZEN, M. and DASGUPTA, A. (1993). Estimating a binomial parameter: is robust Bayes real Bayes? *Statistics & Decisions*, **11**, 37–60.

ZIELIŃSKI, R. (1994). Comment on "Robust Bayesian methods in simple ANOVA models" by W. Polasek. *Journal of Statistical Planning and Inference*, **40**, 308–310.

14

Linearization Techniques in Bayesian Robustness

Michael Lavine, Marco Perone Pacifico, Gabriella Salinetti and Luca Tardella

ABSTRACT This paper deals with techniques which permit one to obtain the range of a posterior expectation through a sequence of linear optimizations. In the context of Bayesian robustness, the linearization algorithm plays a fundamental role. Its mathematical aspects and its connections with fractional programming procedures are reviewed and a few instances of its broad applicability are listed. At the end, some alternative approaches are briefly discussed.

Key words: fractional programming, prior distributions, sensitivity.

14.1 Introduction

Let $\mathcal{F}$ be a class of densities with respect to a common measure, $f(x|\theta_f) \in \mathcal{F}$ a density for the data with θ_f denoting unknown parameters of f, Γ_f a class of prior distributions for θ_f and π a member of Γ_f. Berger (1994, p. 21) says:

> If $\psi(\pi, f)$ is the posterior functional of interest (e.g., the posterior mean), global robustness is concerned with computation of
>
> $$\underline{\psi} = \inf_{f \in \mathcal{F}} \inf_{\pi \in \Gamma_f} \psi(\pi, f), \qquad \overline{\psi} = \sup_{f \in \mathcal{F}} \sup_{\pi \in \Gamma_f} \psi(\pi, f). \tag{1}$$

Typically one is interested in a function $h(\theta_f)$, and $\psi(\pi, f)$ is the posterior expectation of $h(\theta_f)$.

This paper is concerned mainly with a particular method for computing $(\underline{\psi}, \overline{\psi})$ that has come to be known as the *linearization algorithm*. The linearization algorithm became popular in robust Bayesian statistics after the articles by Lavine (1991) and Lavine et al. (1993), although it dates back at least to DeRobertis and Hartigan (1981) in statistics and much farther elsewhere, as noted in Perone Pacifico et al. (1994). As we show in Section 14.2, the linearization algorithm is a special case of more general techniques, widely known in the optimization literature. In this paper we concentrate on the computation of $\overline{\psi}$.

In contrast to finding $\overline{\psi}$ exactly, a seemingly easier problem is, for a fixed number q, to determine whether $\overline{\psi}$ is greater than q. But

$$\overline{\psi} \equiv \sup_{f,\pi} \frac{\int h(\theta_f) f(x|\theta_f)\pi(d\theta_f)}{\int f(x|\theta_f)\pi(d\theta_f)} > q \tag{2}$$

if and only if

$$\sup_{f,\pi} \int (h(\theta_f) - q) f(x|\theta_f)\pi(d\theta_f) > 0. \tag{3}$$

The point is that the nonlinear functional of the likelihood f and the prior distribution π in (2) has become a linear function of f and π in (3) so the supremum may be much easier to find. And if we can find the supremum for a fixed q in (3), then we can find the suprema for a sequence of q's and thereby determine $\overline{\psi}$ as accurately as desired.

Section 14.2 gives a wider mathematical background and more details; Section 14.3 shows how to determine the supremum in (3) in a variety of examples and Section 14.4 discusses alternative approaches.

14.2 Mathematical aspects

Consider the optimization problem $\overline{\psi}$ in (1) in the more general context, namely, that of finding

$$\overline{\psi} \equiv \sup_{x \in X} \frac{N(x)}{D(x)}, \tag{4}$$

where X is a nonempty subset of some space S, N and D are real-valued functions defined on X and it is assumed that $D(x) > 0$ for each x in X (in (1) $x = (f, \pi)$).

In the mathematical programming literature (4) is known as a *fractional optimization problem.* Pioneered by Isbell and Marlow (1956) in the particular case $S = \Re^n$, a number of solution procedures for (4) have been proposed; for a survey see Ibaraki (1981) and Schaible (1981). Besides the direct approach based on nonlinear programming methods, the solution procedures essentially fall into two categories: the *parametrization* approach (Isbell and Marlow, 1956; Jagannathan, 1966; Dinkelbach, 1967) and the *variable transformation* approach (Charnes and Cooper, 1962; Schaible, 1976; Bradley and Frey, 1974).

In this section we present an almost elementary "transposition" in an abstract setting of the parametrization approach. Connections between Bayesian robustness computations and the variable transformation approach will be discussed in Section 14.4.

For each $q \in \Re$ define the auxiliary problem

$$G(q) = \sup_{x \in X} [N(x) - qD(x)]. \tag{5}$$

It will be shown how $\overline{\psi}$ can be solved via a family of $G(q)$.

We can start by observing that $q \mapsto G(q)$ is nonincreasing and convex; if, in addition, $\inf_{x \in X} D(x) > 0$, then $q \mapsto G(q)$ is strictly decreasing. >From these properties follows

Proposition 1 *For every real q we have:*
(i) $G(q) > 0$ *if and only if* $\sup_{x \in X} N(x)/D(x) > q$;
if, in addition, $\inf_{x \in X} D(x) > 0$ *and* $\sup_{x \in X} D(x) < +\infty$, *then*
(ii) $G(q) = 0$ *if and only if* $q = \sup_{x \in X} N(x)/D(x)$.

Proof. To suppose $G(q) > 0$ is equivalent to stating that $N(x') - qD(x') > 0$ for some x' in X. Thus there exists $x' \in X$ such that $N(x')/D(x') > q$, which is true if and only if $\sup_{x \in X} N(x)/D(x) > q$.

To prove (ii) suppose $\sup_{x \in X} N(x)/D(x) = q < +\infty$. For each $\varepsilon > 0$ there exists $x_\varepsilon \in X$ such that $N(x_\varepsilon)/D(x_\varepsilon) > q - \varepsilon$ and then $N(x_\varepsilon) - qD(x_\varepsilon) > -\varepsilon D(x_\varepsilon) \geq -\varepsilon \sup_{x \in X} D(x)$. It follows that $G(q) \geq 0$ and, from (i), that $G(q) = 0$. Conversely, suppose $G(q) = 0$. From (i) it follows that $q \geq \sup_{x \in X} N(x)/D(x)$. Arguing by contradiction, if $q > \sup_{x \in X} N(x)/D(x)$ then, from the strict monotonicity of $q \mapsto G(q)$, $G(\sup_{x \in X} N(x)/D(x)) > 0$; this contradicts (i) and the result follows. □

In the rest of the paper we will assume that $\overline{\psi}$ is finite; otherwise, as a consequence of (i), the function G would be positive everywhere. The condition $\sup_{x \in X} D(x) < +\infty$ gives the *if* part of (ii) that makes the optimal value $\overline{\psi}$ a solution of $G(q) = 0$; the condition $\inf_{x \in X} D(x) > 0$ ensures the *only if* part of (ii) and makes that solution unique. Actually, the condition on the $\sup_{x \in X} D(x)$ can be avoided if it is known that there exists $x^\star \in X$ where the supremum $\overline{\psi}$ is attained. In fact, if there exists $x^\star \in X$ such that $\overline{\psi} = N(x^\star)/D(x^\star)$, then $q^\star = N(x^\star)/D(x^\star)$ is the solution of the equation $G(q) = 0$. This solution is unique if $\inf_{x \in X} D(x) > 0$.

It is now clear that every optimizing solution of problem (4) is the solution for problem (5) with $q = q^\star$ and every $x^\star$ such that $G(q^\star) = N(x^\star) - q^\star D(x^\star)$ maximizes the ratio N/D.

It follows that, under the conditions specified above, the fractional optimization problem (4) reduces to finding the unique root of the equation

$$G(q) = \sup_{x \in X} [N(x) - qD(x)] = 0. \tag{6}$$

Obviously, the possibility to solve (6) depends on the expression $q \mapsto G(q)$, and specifically on the supremum problem which defines it. The problem is certainly simplified when, as usual in Bayesian robustness problems such as (1), the functionals N and D are linear on a vector space.

The parametrization approach described in Proposition 1 represents a general version of the *linearization algorithm* as introduced in Lavine (1991) and formalized in the linearization theorem of Lavine et al. (1993). Observe that the condition $\sup_{x \in X} D(x) < +\infty$ for problem (1) is always satisfied when, as in Lavine et al. (1993), the likelihood is bounded.

An easy alternative way of visualizing geometrically how the linearization works is presented in Perone Pacifico et al. (1998). We will not go into details here, but just give the main idea: one can map the abstract space X into the subset of the real plane

$$E(X) = \{(D(x), N(x)) : x \in X\}$$

and realize that the geometric objective then becomes finding the maximum slope of the line which departs from the origin and supports $E(X)$. Many other geometric analogues are explained starting from this representation.

Root finding in (6) can be pursued by an easy bisection algorithm once two starting values q_1 and q_2 are provided such that $q_1 \leq q^\star \leq q_2$. If the computational burden of the optimization defining $G(q)$ is heavy, a great amount of computing time can be saved by an efficient root finding that exploits all the convexity properties of the function G. Namely, it can be shown (see Proposition 1 in Perone Pacifico et al., 1998), that if there exists an optimizing x_q such that $G(q) = N(x_q) - qD(x_q) > 0$, then a more efficient approximation of $q^\star$ is determined by

$$\frac{N(x_q)}{D(x_q)} \leq q^\star \leq q + \frac{G(q)}{D_{\min}},$$

where $D_{\min} = \inf_{x \in X} D(x)$. Sharper bounds are obtained in case two optimizing points x_q and x_r are available such that

$$G(q) = N(x_q) - qD(x_q) > 0 \qquad \text{and} \qquad G(r) = N(x_r) - rD(x_r) < 0.$$

In that case, the following holds:

$$\max\left\{\frac{N(x_q)}{D(x_q)}, \frac{N(x_r)}{D(x_r)}\right\} < q^\star \leq \frac{r\,G(q) - q\,G(r)}{G(q) - G(r)}.$$

14.3 Examples

14.3.1 Robustness with respect to the prior

Here we examine (3) in some specific examples and show how to compute the supremum. Unless otherwise noted, we concentrate on finding bounds on the posterior mean ($h(\theta_f) = \theta_f$), and where there is only one class of distributions, that is, $\mathcal{F}$ is a singleton and $\theta_f = \theta$, so (3) reduces to finding

$$\sup_{\pi \in \Gamma} \int (\theta - q) f(x|\theta) \pi(d\theta). \tag{3'}$$

The point of most of the examples is that the supremum over a class of probability measures can be simplified to the supremum over a finite-dimensional space, which can then be solved by standard methods. Many of the examples below can also be found in Lavine (1991) or Lavine et al. (1991, 1993). We refer the reader to those papers for further details and to Moreno (2000) for a recent overview.

All distributions: As an easy immediate example consider the class $\mathcal{P}(\Theta)$ of all distributions over Θ. When π varies in $\mathcal{P}(\Theta)$ the distribution maximizing (3′) puts all its mass at the θ that maximizes $(\theta - q)f(x|\theta)$, and the maximization over a class of distributions reduces to a maximization over Θ.

Density bounded class: A density *band* is the set of measures π (not necessarily satisfying $\pi(\Theta) = 1$) such that $L \leq \pi \leq U$ for two prespecified measures L and U. When π varies in a density band, the optimizing measure for (3′) is set equal to U on the θs where $(\theta - q)f(x|\theta)$ is nonnegative and equal to L elsewhere; thus the maximization over the density band is reduced to solving an inequality over Θ. This is the class of priors used by DeRobertis and Hartigan (1981) and for which they introduced the linearization algorithm to statistics. The *constant odds-ratio* class of Walley (1991) is a special case.

Lavine (1991) introduced the density *bounded* class considering only the probability measures in a density band and used the linearization algorithm to find bounds.

Fréchet class: Let $\Theta = (\theta_1, \theta_2)$ be two-dimensional and suppose that the marginal priors π_1 for θ_1 and π_2 for θ_2 are given. Let Γ be the set of all priors π having the given marginals. In this case, (3′) can be solved using standard linear programming techniques. See Lavine et al. (1991) for details.

ϵ-contamination class: Let π_0 be a prior distribution and G be a class of prior distributions. The class

$$\Gamma_\epsilon = \{\pi : \pi = (1 - \epsilon)\pi_0 + \epsilon g, g \in G\}$$

is called an ϵ-contamination class of priors. See Berger (1985) for a more thorough discussion. When π varies in an ϵ-contamination class,

(3′) becomes

$$\sup_{\pi \in \Gamma_\epsilon} \int (\theta - q) f(x|\theta) \pi(d\theta)$$
$$= (1-\epsilon) \int (\theta - q) f(x|\theta) \pi_0(d\theta) + \epsilon \sup_{g \in G} \int (\theta - q) f(x|\theta) g(d\theta),$$

and the supremum can be found when the class G is tractable, like any of the classes listed here.

Quantile class: Berliner and Goel (1990) and Betrò and Guglielmi (2000) consider the following class of priors: for a given partition of Θ into sets $I_1, \ldots, I_m$ and given nonnegative numbers $p_1, \ldots, p_m$ such that $\sum p_j = 1$, define the class

$$\Gamma_Q = \{\pi : \pi(I_j) = p_j \ \forall j \in \{1, \ldots, m\}\}$$

which can be represented as

$$\Gamma_Q = \{\pi : \pi = \sum_{j=1}^{m} p_j \pi_j, \pi_j \in \mathcal{P}(I_j) \ \forall j \in \{1, \ldots, m\}\}.$$

Using this representation and the linearity of (3′), we have

$$\sup_{\pi \in \Gamma_Q} \int (\theta - q) f(x|\theta) \pi(d\theta)$$
$$= \sum_j p_j \sup_{\pi_j \in \mathcal{P}(I_j)} \int_{I_j} (\theta - q) f(x|\theta) \pi_j(d\theta)$$
$$= \sum_j p_j \sup_{\theta \in I_j} (\theta - q) f(x|\theta),$$

and the problem has been reduced to m separate maximizations over subsets of Θ.

Shape constraints: Berger and O'Hagan (1988) and O'Hagan and Berger (1988) compute bounds on the posterior probabilities of sets when Γ is roughly the class of unimodal priors. Lavine et al. (1993) show how to achieve the same results more easily using the linearization algorithm; different types of symmetry and unimodality in multidimensional parametric spaces are studied in Bose (1994) and Liseo *et al.* (1993). All those results are based on the representation of the class as a *class of mixtures*

$$\Gamma_S = \{\pi : \pi(\cdot) = \int_Y k(y, \cdot) \nu(dy), \nu \in \mathcal{P}(Y)\},$$

where Y is finite-dimensional and k is an appropriate Markov kernel. Applying Fubini's theorem to (3′) yields

$$\begin{aligned}\sup_{\pi\in\Gamma_S}\int(\theta-q)f(x|\theta)\pi(d\theta)\\ &= \sup_{\nu\in\mathcal{P}(Y)}\int_Y\left(\int_\Theta(\theta-q)f(x|\theta)k(y,d\theta)\right)\nu(dy)\\ &= \sup_{y\in Y}\int_\Theta(\theta-q)f(x|\theta)k(y,d\theta),\end{aligned}$$

which is again an optimization in finite dimension.

Generalized moment class: Consider some class G of distributions, m measurable functions $h_1,\dots,h_m$ and two m-dimensional vectors α and β such that $\alpha_j\le\beta_j$ for all $j\in\{1,\dots,m\}$. The generalized moment class

$$\Gamma_M=\{\pi\in G:\alpha_j\le\int h_j(\theta)\pi(d\theta)\le\beta_j\ \forall j\in\{1,\dots,m\}\}$$

has, as particular cases, the quantile class, the density bounded class and many others. For this unifying view of Bayesian robustness, see Salinetti (1994).

The solution of problem (3′) is based on the *generalized moment problem theory,* mainly developed by Kemperman (see the overview in Kemperman, 1987); for numerical procedures, specifically related to Bayesian robustness, see Dall'Aglio (1995) and Betrò and Guglielmi (1996, 2000).

The most relevant aspect of this kind of class is that it can be constructed by simultaneously considering constraints of different nature: for instance, exploiting the moment problem theory; Perone Pacifico et al. (1996) consider density bounded classes with fixed quantiles or prior correlation, while Liseo et al. (1996) treat ϵ-contamination classes with fixed marginals.

14.3.2 Robustness with respect to the sampling distribution

Let Ω be the set of all probability distributions on $\mathcal{X}$. For each $\theta\in\Theta$, let $P_\theta\in\Omega$ be the probability measure corresponding to θ. Let π_0 be a prior on θ. Viewed as a prior on Ω, π_0 gives probability one to the set $\{P_\theta:\theta\in\Theta\}$. Let $N(\theta)\subset\Omega$ be a neighborhood of P_θ. One can explore sensitivity with respect to sampling distributions by considering the class of priors

$$\Gamma=\{\pi:\forall B\subset\Theta,\pi(\cup_{\theta\in B}N(\theta))\ge\pi_0(B)\}.$$

The idea is to begin with the base measure π_0 and allow the "mass" on θ to spread throughout the neighborhood $N(\theta)$. Lavine (1991) considers this class simultaneously with perturbations to π_0 and uses the linearization algorithm to bound posterior expectations. Equation (3) takes a form for which finding the supremum is easy. One interesting result is that if $N(\theta)$ is a Levy, Prohorov, Kolmogorov, total variation or ϵ-contamination neighborhood, then for every q the supremum in (3) is greater than 0. Consequently, the bounds on the predictive cumulative distribution function are 0 and 1. To achieve tighter bounds one must use neighborhoods in which densities are bounded away from 0 and ∞.

14.3.3 Robustness with respect to the regression function

In parametric regressions, the parameter β is mapped to a single regression function $r_\beta(x) = E[Y|x, \beta]$. One may explore sensitivity by considering a neighborhood N_β of regression functions and allowing the prior "mass" on β to spread throughout the neighborhood. Lavine (1994) uses the linearization algorithm to explore such sensitivity when the neighborhood is defined by

$$N_\beta = \{r : r(x) \in [\ell_\beta(x), u_\beta(x)]\}$$

for some specified functions ℓ_β and u_β. The optimization in (3) again simplifies to a maximization over a Euclidean space.

14.4 Alternative computational techniques

Even though the linearization algorithm is often used for computations in Bayesian robustness problems, it is not the only approach available. In fact, as mentioned in Section 14.2, the *variable transformation* approach can also be adapted to more abstract spaces and fruitfully employed in Bayesian robustness.

In particular, the algorithmic procedures used in Betrò and Guglielmi (1996, 2000), Perone Pacifico et al. (1996) and Cozman (1999) can be viewed as different variable transformations.

In general, the variable transformation approach to problem (4) proceeds by dealing with the denominator as a real variable, appropriately chosen to simplify the structure of the problem. In this spirit one natural variable transformation is

$$t = \frac{1}{D(x)}, \quad y = xt,$$

so that the domain becomes

$$Z = \{(t, y) : D(y/t)t = 1,\ y/t \in X,\ t > 0\}.$$

It is easy to check that, in general,

$$\overline{\psi} = \sup_{(t,y)\in Z} N(y/t)t \tag{7}$$

and that $(t^\star, y^\star) \in Z$ is an optimal solution for (7) if and only if the corresponding $x^\star = y^\star/t^\star$ is an optimal solution for (4) (see Charnes and Cooper, 1962; Schaible, 1976; Bradley and Frey, 1974, for details and Perone Pacifico et al., 1994, for the use of this approach in Bayesian robustness).

Solving problem (7) instead of (4) is particularly helpful when the functionals N and D are linear since, in such a situation,

$$\overline{\psi} = \sup_{y\in Y} N(y)$$

with $Y = \{y \in cone(X) : D(y) = 1\}$.

Moreover, if X is a class of measures defined through a finite number m of linear constraints, the corresponding class Y is still of the same nature, with $m+1$ linear constraints. This approach has been adopted in Betrò and Guglielmi (1996, 2000) and Cozman (1999).

With respect to the same problem, Perone Pacifico et al. (1996) introduced a different variable transformation constraining the marginal value and obtaining

$$\sup_{x\in X} \frac{N(x)}{D(x)} = \sup_{w\in W} \frac{1}{w} \sup_{x\in X} \{N(x) : D(x) = w\} = \sup_{w\in W} \frac{n(w)}{w},$$

where $W \subset \Re_+$ is the range of $D(x)$ and, for each w, the constrained supremum $n(w)$ can be computed through the moment problem theory. Furthermore, since $w \mapsto n(w)$ is concave, the above formulation makes the problem a standard fractional programming problem on $\Re$ with concave numerator and linear denominator that can be solved making use of any suitable algorithm. Following this approach, Perone Pacifico et al. (1996) got as an additional bonus the possibility of dealing with a nonlinear numerator as in the case of computing the range of the posterior variance.

References

BERGER, J.O. (1985). *Statistical Decision Theory and Bayesian Analysis, 2nd ed.* New York: Springer-Verlag.

BERGER, J.O. (1994). An overview of robust Bayesian analysis. *Test*, **3**, 5–59.

BERGER, J.O. and O'HAGAN, A. (1988). Ranges of posterior probabilites for unimodal priors with specified quantiles. In *Bayesian Statistics 3* (J.M. Bernardo, M.H. DeGroot, D.V. Lindley, and A.V.M. Smith, eds.), 45–65. Oxford: Oxford University Press.

BERLINER, L. M. and GOEL, P. (1990). Incorporating partial prior information: ranges of posterior probabilities. In *Bayesian and Likelihood Methods in Statistics and Econometrics: Essays in Honor of George A. Barnard* (S. Geisser, J. Hodges, F. J. Press and A. Zellner, eds.), 397–406. Amsterdam: North-Holland.

BETRÒ, B. and GUGLIELMI, A. (1996). Numerical robust Bayesian analysis under generalized moment conditions. In *Bayesian Robustness, IMS Lecture Notes - Monograph Series*, (J.O. Berger, B. Betrò, E. Moreno, L.R. Pericchi, F. Ruggeri, G. Salinetti, and L. Wasserman, eds.), 3–20. Hayward: IMS.

BETRÒ, B. and GUGLIELMI, A. (2000). Methods for global prior robustness under generalized moment conditions. In *Robust Bayesian Analysis*, (D. Ríos Insua and F. Ruggeri, eds.). New York: Springer-Verlag.

BOSE, S. (1994). Bayesian robustness with mixture classes of priors. *Annals of Statistics*, **22**, 652–667.

BRADLEY, S.P. and FREY, S.C. (1974). Fractional programming with homogeneous functions. *Operations Research*, **22**, 350–357.

CHARNES, A. and COOPER, W.W. (1962). Programming with linear fractional functionals. *Naval Research Logistic Quarterly*, **9**, 181–186.

COZMAN, F. (1999). Calculation of posterior bounds given convex sets of prior probability measures and likelihood functions. To appear in *Journal of Computational and Graphical Statistics.*

DALL'AGLIO, M. (1995). Problema dei momenti e programmazione lineare semi infinita nella robustezza bayesiana. *Ph.D. Thesis*, Dipartimento di Statistica, Probabilità e Statistiche Applicate dell'Università di Roma "La Sapienza."

DEROBERTIS, L. and HARTIGAN, J. (1981). Bayesian inference using intervals of measures. *Annals of Statistics*, **9**, 235–244.

DINKELBACH, W. (1967). On nonlinear fractional programming. *Management Science*, **13**, 492–498.

IBARAKI T. (1981). Solving mathematical programming problems with fractional objective functions. In *Generalized Concavity in Optimization and Economics* (S. Schaible and W.T. Ziemba, eds.), 441–472.

ISBELL, J.R. and MARLOW, W.H. (1956). Attrition games. *Naval Research Logistic Quarterly*, **3**, 71–93.

JAGANNATHAN, R. (1966). On some properties of programming problems in parametric form pertaining to fractional programming. *Management Science*, **12**, 609–615.

KEMPERMAN, J.H.B. (1987). Geometry of the moment problem. In *Moments in mathematics*, Proceedings of Symposia in Applied Mathematics, **37**, 16–53.

LAVINE, M. (1991). Sensitivity in Bayesian statistics: the prior and the likelihood. *Journal of the American Statistical Association*, **86**, 396–399.

LAVINE, M. (1994). An approach to evaluating sensitivity in Bayesian regression analyses. *Journal of Statistical Planning and Inference,* **40**, 242–244.

LAVINE, M., WASSERMAN, L. and WOLPERT, R.L. (1991). Bayesian inference with specified prior marginals. *Journal of the American Statistical Association*, **86**, 964–971.

LAVINE, M., WASSERMAN, L. and WOLPERT, R.L. (1993). Linearization of Bayesian robustness problems. *Journal of Statistical Planning and Inference*, **37**, 307–316.

LISEO, B., MORENO, E. and SALINETTI, G. (1996). Bayesian robustness for classes of bidimensional priors with given marginals. In *Bayesian Robustness, IMS Lecture Notes - Monograph Series*, (J.O. Berger, B. Betrò, E. Moreno, L.R. Pericchi, F. Ruggeri, G. Salinetti, and L. Wasserman, eds.), 101–115. Hayward: IMS.

LISEO, B., PETRELLA, L. and SALINETTI, G. (1993). Block unimodality for multivariate Bayesian robustness. *Journal of the Italian Statistical Society*, **2**, 55–71.

MORENO, E. (2000). Global Bayesian robustness for some classes of prior distributions. In *Robust Bayesian Analysis* (D. Ríos Insua and F. Ruggeri, eds.). New York: Springer-Verlag.

O'HAGAN, A. and BERGER, J.O. (1988). Ranges of posterior probabilities for quasiunimodal priors with specified quantiles. *Journal of the American Statistical Association,* **83**, 503–508.

PERONE PACIFICO, M., SALINETTI, G. and TARDELLA L. (1994). Fractional optimization in Bayesian robustness. *Technical Report*, **A 23**, Dipartimento di Statistica, Probabilità e Statistiche Applicate, Università di Roma "La Sapienza."

PERONE PACIFICO, M., SALINETTI, G. and TARDELLA, L. (1996). Bayesian robustness on constrained density band classes. *Test*, **5**, 395–409.

PERONE PACIFICO, M., SALINETTI, G. and TARDELLA, L. (1998). A note on the geometry of Bayesian global and local robustness. *Journal of Statistical Planning and Inference*, **69**, 51–64.

SALINETTI, G. (1994). Discussion to "An Overview of Robust Bayesian Analysis" by J.O. Berger. *Test*, **3**, 109–115.

SCHAIBLE, S. (1976). Fractional programming I, Duality. *Management Science*, **22**, 858–867.

SCHAIBLE, S. (1981). A survey of fractional programming, in *Generalized Concavity in Optimization and Economics* (S. Schaible and W.T. Ziemba, eds.), 417–440.

WALLEY, P. (1991). *Statistical Reasoning with Imprecise Probabilities.* London: Chapman and Hall.

15

Methods for Global Prior Robustness under Generalized Moment Conditions

Bruno Betrò and Alessandra Guglielmi

ABSTRACT Generalized moments have the form $\int_{\mathcal{X}} h d\mu$ (where h is a real function and μ is a positive measure on a measurable space $\mathcal{X}$), which in the case that $h(x) = x^n$ and μ is a probability measure reduces to the ordinary definition of moment. Bounds on generalized moments of distribution functions have rcently been considered in the context of globally robust Bayesian analysis as a way for defining classes of priors (that we shall call generalized moment classes) covering a great variety of situations. An interesting issue in Bayesian robustness, not thoroughly investigated, is the problem of optimizing posterior functionals over a generalized moment class. The problem can be rephrased as a linear optimization problem over the set of finite measures still constrained by generalized moment conditions, so that it is possible to deal with it in the framework of the comprehensive theory developed for linear optimization under generalized moment conditions.
The first part of this paper is devoted to a review of this theory. In the second part, its application to robust Bayesian analysis is considered, giving sufficient conditions for the existence and the identification of the required extremum values. Finally, algorithms which have been proposed for actual computations are reviewed in the third part.

Key words: Bayesian inference, moment theory, linear semi-infinite programming, algorithms for robustness analysis.

15.1 Introduction

Robustness is an important issue in Bayesian analysis and it has been explored to some extent in the last decade. Literature on the subject up to 1994 is reviewed in Berger (1994), which is still the basic reference for terminology and classes of problems. Not much work has been done about the computational aspects, and most of the test cases examined in the literature are rather academic ones.

The aim of this paper is to review the state of the art for what concerns global prior robustness (in short, robustness henceforth) under generalized moment conditions. Classes of priors determined by bounds on generalized moments have been considered in connection with robustness first by Betrò et al. (1994) and then by Betrò and Guglielmi (1994), Goutis (1994), Dall'Aglio (1995), Smith (1995), Betrò et al. (1996), and Betrò and Guglielmi (1997).

Generalized moment classes incorporate a number of interesting situations – the most common is the fixing of bounds on quantiles of the prior distribution – and have been widely studied in other contexts, so that a rather comprehensive theory exists for optimization of linear functionals defined over them, mainly due to Kemperman (1971, 1983, 1987), which we review in Section 2. From an algorithmic point of view, there is an interesting connection between optimization of a linear functional on a generalized moment class and linear semi-infinite programming, so that methods developed in this latter framework can be fruitfully exploited. In robustness analysis, the functional to be optimized is not linear but, as it is shown in Section 3, it is possible to obtain linearity by a suitable transformation. In this way, it is possible to solve prior robustness problems by means of linear semi-infinite programming; algorithms proposed so far in the literature are reviewed in Section 4.

15.2 Some theory on the moment problem

In this section we review to some extent the theory of generalized moment problems, essentially from Kemperman (1971, 1983, 1987). In subsections 2.1 and 2.2 we investigate generalized moment problems with inequality and equality constraints, respectively. Of course, the second case is included in the first one, but for equality constraints it is possible to state the relevant results under relaxed conditions.

15.2.1 Inequality constraints

Let S be a Polish space, that is, there exists some separable complete metrization d of S, consistent with $\mathcal{B}(S)$, the Borel σ-algebra generated by all open subsets in S. Let $\mathcal{M}$ be the set of all finite Borel measures on S, with the topology of the weak convergence ($\mu_n \xrightarrow{w} \mu$ in $\mathcal{M}$ if, and only if, $\int_S f d\mu_n \to \int_S f d\mu \; \forall f \in \mathcal{C}_b(S)$, i.e. for all bounded continuous f defined on S). Moreover, $\mathcal{M}$ is Polish in this topology (Prohorov, 1956, p. 167).

Let $\mathcal{M}_0 = \mathcal{M}_0(S)$ be a convex set of finite measures on S, J be an index set, not necessarily finite, and $h_j : S \to \mathbb{R}$, $j \in J$, $u : S \to \mathbb{R}$ be measurable functions. If $\mathcal{M}_1 = \mathcal{M}_1(S) := \{\mu \in \mathcal{M}_0 : \int_S h_j^+ d\mu < +\infty, \int_S h_j d\mu \leq \eta_j, j \in J\}$, where η_j are real constants, we are interested in determining the upper bound

$$U(u) := \sup_{\mu \in \mathcal{M}_1} \int_S u d\mu, \tag{1}$$

where we assume the existence of μ^* in $\mathcal{M}_1$ such that $\int u^- d\mu^* > -\infty$, so that $U(u) > -\infty$. Problems of type (1) are referred to as generalized moment problems. It is easy to show that

$$U(u) \leq U^*(u) := \inf\{\sum_{j \in J} \beta_j \eta_j + \sup_{\mu \in \mathcal{M}_0} \int_S (u - \sum_j \beta_j h_j) d\mu, \beta_j \geq 0 \forall j\} \tag{2}$$

with all but finitely many (a.b.f.m.) β_js equal to 0. If $\mathcal{M}_0 = \mathcal{M}$, then

$$U^*(u) = \inf\{\sum_{j \in J} \beta_j \eta_j : \beta_j \geq 0,\ j \in J, \sum_j \beta_j h_j(s) \geq u(s)\ s \in S\}. \tag{3}$$

If J is finite, the problem of calculating $U^*(u)$ as in (3) is usually called a *linear semi-infinite programming* (LSIP) problem.

We state the following conditions that will be useful later:

(i) h_j is lower semicontinuous (l.s.c.), that is, $\{s \in S : h_j(s) > b\}$ is an open set for any $b \in \mathbb{R}$, for all $j \in J$;

(ii) u is upper semicontinuous (u.s.c.), that is, $-u$ is l.s.c.;

(iii) $h_j \in H_0$ for all $j \in J$, where H_0 is the convex cone of $\{h_k, k \in J\}$, namely, the set of all functions f such that there exist $\{\rho_k \geq 0, k \in J\}$, ρ_ks a.b.f.m. equal to 0, $\sum_J \rho_k h_k(s) \geq 0$ on S, and $\forall \varepsilon > 0$ the function $f(s) + \varepsilon \sum_J \rho_k h_k(s)$ is bounded below on S;

(iv) $-u \in H_0$.

Remark 1 If h_j is bounded from below, then $h_j \in H_0$. This is obviously true when S is compact, since h_j is l.s.c.

The following results hold.

Theorem 1 (Kemperman, 1983) *Under (i), if $\mathcal{M}_0$ is compact, then $\mathcal{M}_1$ is compact too.*

Theorem 2 (Kemperman, 1983) *Under conditions (i)–(iv), if $\mathcal{M}_0$ is compact and $\mathcal{M}_1 \neq \emptyset$, then (2) holds with the equality sign. Moreover, if $U(u)$ is finite, then the supremum in (1) is assumed.*

15.2.2 Equality constraints

In this section S is merely a topological space and $J = \{1, 2, \ldots, n\}$ is a finite set. As before, let $h_j : S \to \mathbb{R}$, $j = 1, \ldots, n$, $u : S \to \mathbb{R}$ be measurable functions. Let A_n be the convex cone generated by the points $(h_1(s), \ldots, h_n(s))$, $s \in S$, namely, $A_n = \{y \in \mathbb{R}^n : y_j = \sum_{i=1}^{q} \lambda_i h_j(s_i), j = 1, \ldots, n, \lambda_i \geq 0, s_i \in S, i = 1, \ldots, q, q < +\infty\}$; A_n is equal to the moment cone, that is, the set $\{\int h_j d\mu, j = 1, \ldots, n, \mu \in \mathcal{M}\}$. For any $y \in A_n$, let $\mathcal{M}(y) := \{\mu$ measure on S such that $\int_S h_j d\mu = y_j, j = 1, \ldots, n\}$, and $\mathcal{M}_b(y)$ be the set of $\mu \in \mathcal{M}(y)$ whose support consists in not more than n points; by Caratheodory's theorem (see Lemma 1 in Kemperman, 1983), $\mathcal{M}_b(y) \neq \emptyset$. As before, we are interested in determining

$$U_1(y) := \sup_{\mu \in \mathcal{M}(y)} \int_S u d\mu,$$

where $|u|$ is integrable for all $\mu \in \mathcal{M}(y)$. It can be shown that

$$U_1(y) = \sup_{\mu \in \mathcal{M}_b(y)} \int u d\mu \quad \text{and} \quad U_1(y) = \sup\{y_{n+1} : (y, y_{n+1}) \in A_{n+1}\},$$

where A_{n+1} is the convex cone generated by the points $(h_1(s), \ldots, h_n(s), u(s))$, $s \in S$. We will further assume that the functions $h_1, \ldots, h_n$ are linearly independent, which implies $\text{int}(A_n) \neq \emptyset$ ($\text{int}(A)$ denotes the interior of the set A). Now, if $C := \{\underline{c} = (c_1, \ldots, c_n) \in \mathbb{R}^n : \sum_1^n c_i h_i(s) \geq u(s)\ s \in S\}$, then

$$U_1(y) \leq \inf_C \sum_1^n c_i y_i.$$

The following result holds.

Theorem 3 (Kemperman, 1983) *If $y \in int(A_n)$, then*

$$U_1(y) = \inf_C \sum_1^n c_i y_i. \tag{4}$$

In particular, $U_1(y) = +\infty$ when $C = \emptyset$. Moreover, if $C \neq \emptyset$, then $U_1(y)$ is finite and the infimum in (4) is assumed.

A characterization of the optimal solutions in (4) is given by the following:

Complementary Slackness Condition (see Smith, 1995). *If $\overline{\mu}$ and $\underline{c}$ are optimal solutions in the left and right hand sides of (4), respectively, then $\overline{\mu}$ has mass only at those points s such that $\sum_0^n \underline{c}_i h_i(s) = u(s)$.*

Observe that Theorem 3 does not require the semicontinuity of the constraint or objective functions, but only that $y \in \text{int}(A_n)$. Anyhow, this condition may not be easy to check, so that other sufficient conditions can be considered. For instance, if $U_1(y)$ is u.s.c. on A_n, then (4) holds for all $y \in A_n$ (if A_{n+1} is closed, then $U_1(y)$ is u.s.c. on A_n), or we can use the following:

Theorem 4 (Kemperman, 1983) *If S is compact, h_j is continuous for all j, u is u.s.c. and there exist $d_1, \ldots, d_n \in \mathbb{R}$, $q \geq 0$ such that $\sum_1^n d_i h_i(s) - qu(s) \geq 1$ for all s, then $C \neq \emptyset$, $\inf_C \sum_1^n c_i y_i$ is finite, and (4) holds for all $y \in A_n$.*

15.3 Robustness analysis under generalized moment conditions

Let X be a random variable on a dominated statistical space $(\mathcal{X}, \mathcal{F}_{\mathcal{X}}, \{P_\theta, \theta \in (\Theta, \mathcal{F})\})$, with density $f(x|\theta)$ with respect to the dominant measure λ, where the parameter space $(\Theta, \mathcal{F})$ is such that Θ is a subset of a finite-dimensional Euclidean space; denote by $l_x(\theta)$ (or simply by $l(\theta)$) the likelihood function $f(x|\theta)$, which we assume $\mathcal{F}_{\mathcal{X}} \otimes \mathcal{F}$-measurable. Let $\mathcal{P}$ be the space of all probability measures on $(\Theta, \mathcal{F})$ and define

$$\Gamma = \{\pi \in \mathcal{P} : \int_\Theta H_i(\theta)\pi(d\theta) \leq \alpha_i,\ i = 1, \ldots, m\},$$

where the H_i are integrable with respect to any $\pi \in \mathcal{P}$ and the α_i are fixed real constants, $i = 1, \ldots, m$. Suppose that Γ is nonempty. We are interested in determining

$$\sup_{\pi \in \Gamma} \frac{\int_\Theta g(\theta) l(\theta) \pi(d\theta)}{\int_\Theta l(\theta)\pi(d\theta)}, \tag{5}$$

where $g : \Theta \to \mathbb{R}$ is a given function such that $\int g(\theta)\pi(d\theta|x)$ exists for all $\pi \in \Gamma$, $\pi(d\theta|x)$ being the posterior distribution of θ corresponding to the prior π.

The functional to be optimized in (5) is not linear in π. However, it is possible to build up a problem equivalent to (5) in which the objective functional is linear. As pointed out in Perone Pacifico et al. (1994) (see also Cozman, 1999), usual approaches of fractional programming give guidelines for the linearization.

The parametrization approach: Define the auxiliary problem

$$G(\lambda) = \sup_{\pi \in \Gamma} \left\{ \int_\Theta [g(\theta) - \lambda] l(\theta) \pi(d\theta) \right\} \tag{6}$$

and look for a root of $G(\lambda) = 0$. A unique root exists if and only if $\inf_{\pi \in \Gamma} \int_\Theta l(\theta)\pi(d\theta) > 0$ and $\sup_{\pi \in \Gamma} \int_\Theta g(\theta) l(\theta)\pi(d\theta) < \infty$; such a root coincides with the sup in (5). This approach is known in robustness analysis literature as Lavine's algorithm (Lavine, 1991; Lavine et al., 1993; Lavine et al., 2000), but it is well known in fractional programming as Dinkelbach or Jagannathan algorithm since the sixties (Dinkelbach, 1967; Jagannathan, 1966). Notice that, in general, solving $G(\lambda) = 0$ requires solving a sequence of optimization problems like (6).

The variable transformation approach: Consider the following map from $\mathcal{P}$ into $\mathcal{M}$ (here $S = \Theta$):

$$\nu(A) = \frac{\int_A \pi(d\theta)}{\int_\Theta l(\theta)\pi(d\theta)}, \quad A \in \mathcal{F}, \tag{7}$$

assuming that $0 < \int_\Theta l(\theta)\pi(d\theta) < +\infty$. From (7) we have

$$\int_\Theta l(\theta)\nu(d\theta) = 1. \tag{8}$$

Moreover, (7) is invertible, that is,

$$\pi(A) = \frac{\int_A \nu(d\theta)}{\int_\Theta \nu(d\theta)}, \quad A \in \mathcal{F}.$$

Therefore, (5) turns into

$$U(gl) = \sup_{\nu \in \mathcal{M}_1} \int_\Theta g(\theta)l(\theta)\nu(d\theta), \tag{9}$$

where $\mathcal{M}_1 = \mathcal{M}_1(\Theta) = \{\nu \in \mathcal{M} : \int_\Theta f_i(\theta)\nu(d\theta) \leq 0, i = 1, \ldots, m, \int_\Theta l(\theta)\nu(d\theta) = 1\}$ and $f_i(\theta) := H_i(\theta) - \alpha_i$, $i = 1, \ldots, m$.

The variable transformation approach was first proposed for Bayesian analyis by Betrò and Guglielmi (1994), but again in fractional programming it has been known since the sixties as the Charnes-Cooper method (Charnes and Cooper, 1962) or, in a variant of this latter, as the White-Snow (White, 1986; Snow, 1991) algorithm (in our context equivalent to considering instead of ν as in (7) the posterior measure $\nu(A) = \int_A l(\theta)\pi(d\theta)/\int_\Theta l(\theta)\pi(d\theta)$, which has the drawback of introducing division by zero where l vanishes). Observe that (8) is obviously equivalent to two inequality constraints, so that problem (9) corresponds to (1) when the objective function is gl, the total number of (inequality) constraints is $m+2$, $h_j = f_j$, $j = 1, \ldots, m$, $h_{m+1} = l$, $h_{m+2} = -l$, and $\eta_j = 0, j = 1, \ldots, m$, $\eta_{m+1} = 1$, $\eta_{m+2} = -1$.

Another variable transformation has been proposed in Perone Pacifico et al. (1996):

$$U(gl) = \sup_y \frac{1}{y} \sup_{\pi \in \Gamma} \left\{ \int_\Theta g(\theta)l(\theta)\pi(d\theta) : \int_\Theta l(\theta)\pi(d\theta) = y \right\} = \sup_y \frac{n(y)}{y};$$

here $n(y)$ is a concave function so that standard fractional programming algorithms can be adopted; however, $n(y)$ has a complicated structure, even if in some circumstances specific properties of it can be exploited (Perone Pacifico et al., 1996).

In the rest of the paper we consider the transformed problem (9). In order to apply the theory of generalized moments, we consider separately the case of compact Θ and the case of noncompact Θ.

15.3.1 Θ is compact

We state first a set of conditions which will be useful later:

H0: Θ is a compact subset;

H1: f_j is l.s.c., $j = 1, \ldots, m$, and l is continuous;

H2: g is u.s.c.;

H3: (Slater condition) there exist $d_0 \in \mathbb{R}$, $d_1, \ldots, d_m \geq 0$, $q \geq 0$ such that

$$d_0 l(\theta) + \sum_1^m d_j f_j(\theta) - q g(\theta) l(\theta) \geq 1, \ \theta \in \Theta;$$

H4: there exist $\varepsilon > 0$, $\bar{\beta}_1, \ldots, \bar{\beta}_m \geq 0$: $\sum_1^m \bar{\beta}_i f_i(\theta) > 1$ for all θ such that $l(\theta) \leq \varepsilon$.

Referring to conditions (i)–(iv) in Subsection 2.1, observe that H1–H2 and (i)–(ii) coincide and that H0–H2 yield (iii)–(iv). In H3, we can always assume that $d_0 > 0$, since $l(\theta) \geq 0$, as we will do from now on. Besides, it can be easily seen that H4 yields H3 with $q = 0$ when the constraint functions f_js are bounded from below.

Lemma 1 *Under H0–H1, H4 holds if and only if*

H4′: *there exist $\beta_1^*, \ldots, \beta_m^* \geq 0$: $\sum_1^m \beta_i^* f_i(\theta) > 0$ for all θ such that $l(\theta) = 0$*

holds.

Proof. The "if" part is obvious. To prove the "only if" part, observe that, since $B := \{\theta : l(\theta) = 0\}$ is compact and $\sum_1^m \beta_i^* f_i(\theta)$ is l.s.c., this function attains its minimum over B, namely, strictly positive by hypothesis. Therefore, there exist $\bar{\beta}_1, \ldots, \bar{\beta}_m \geq 0$ such that if $\theta \in B$, $\theta \in A := \{\theta : \sum_1^m \bar{\beta}_i f_i(\theta) > 1\}$. Obviously, A^c is compact and $A^c \subset B^c = \cup_{\varepsilon > 0}\{\theta : l(\theta) > \varepsilon\}$ ($\{\theta : l(\theta) > \varepsilon\}$ is open $\forall \varepsilon > 0$). By compactness, it follows that there exist $0 < \varepsilon_1 \leq \varepsilon_2 \leq \ldots \leq \varepsilon_n$ such that $A^c \subset \cup_1^n\{\theta : l(\theta) >$

$\varepsilon_i\} \subset \{\theta : l(\theta) > \varepsilon_1\}$. Therefore, $\{\theta : \sum_1^m \bar{\beta}_i f_i(\theta) > 1\} \supset \{\theta : l(\theta) \leq \varepsilon_1\}$, that is, H4 holds. □

Moreover:

Lemma 2 *If $f_j \geq C_j$, $j = 1, \ldots, m$, where the C_js are real constants, then H4 yields*

$$\mathcal{M}_1 \subset \mathcal{M}_C := \{\nu \in \mathcal{M} : \nu(\Theta) \leq C\},$$

for some suitable $C > 0$.

Proof. Let $\mathcal{M}_1 \neq \emptyset$. If $\nu \in \mathcal{M}_1$, since $1 = \int_\Theta l(\theta)\,\nu(d\theta) \geq \int_{\{\theta:l(\theta)>\varepsilon\}} l(\theta)\nu(d\theta) > \varepsilon\nu(\{\theta : l(\theta) > \varepsilon\})$, then $\nu(\{\theta : l(\theta) > \varepsilon\}) < 1/\varepsilon$. Moreover, by H4, $0 \geq \int_\Theta(\sum_1^m \bar{\beta}_i f_i(\theta))\nu(d\theta) > \int_{\{\theta:l(\theta)>\varepsilon\}} (\sum_1^m \bar{\beta}_i C_i)d\nu + \int_{\{\theta:l(\theta)\leq\varepsilon\}} d\nu = (\sum_1^m \bar{\beta}_i C_i)\nu(\{\theta : l(\theta) > \varepsilon\}) + \nu(\{\theta : l(\theta) \leq \varepsilon\})$. Note that $C_j \leq 0\ \forall j$, so that $(\sum_1^m \bar{\beta}_i C_i) \leq 0$. Therefore, $\nu(\{\theta : l(\theta) \leq \varepsilon\}) < -(\sum_1^m \bar{\beta}_i C_i)\nu(\{\theta : l(\theta) > \varepsilon\}) < -(\sum_1^m \bar{\beta}_i C_i)/\varepsilon$, and $\nu(\Theta) < 1/\varepsilon - \sum_1^m \bar{\beta}_i C_i/\varepsilon = C > 0$. □

Remark 2 Since H3 yields H4′, by Lemma 1 we conclude that, under H0–H1, H3 and H4 are equivalent.

Remark 3 By 2 and Remark 2, under H0–H1 both H4 and H3 imply that the feasible measures are uniformly bounded.

Observe now that, by definition, for any $C > 0$, $\mathcal{M}_C$ is closed; moreover, under H0, $\mathcal{M}_C$ is compact. Indeed, this follows by the relative compactness of $\mathcal{M}_C$; that is straightforward (a subset $M \in \mathcal{M}$ is *relatively compact* in the weak topology if $\sup_{\nu\in\mathcal{M}} \nu(\Theta) < +\infty$ and the infimum over any bounded sets B of $\sup_{\nu\in\mathcal{M}} \nu(B^c)$ is 0 – see Kallenberg, 1986, p. 170).

The main result of this section is the following:

Proposition 1 *Under H0–H2 and H3 or H4, if $\mathcal{M}_1 \neq \emptyset$, then $\mathcal{M}_1$ is compact,*

$$U(gl) := \sup_{\nu\in\mathcal{M}_1} \int_\Theta g(\theta)l(\theta)\nu(d\theta) = \inf\{\beta_0 + \sup_{\nu\in\mathcal{M}_C} \int_\Theta (g(\theta)l(\theta) - \beta_0 l(\theta) - \sum_1^n \beta_i f_i(\theta))d\nu : \beta_0 \in \mathbb{R}, \beta_1, \ldots, \beta_m \geq 0\} =: U^*(gl), \tag{10}$$

for some suitable $C > 0$, and the sup *in the left hand side of (10) is assumed.*

Proof. As observed in Remark 3, there exists $C > 0$ such that $\mathcal{M}_1 \subset \mathcal{M}_C$, so that we may apply Theorem 2 with $\mathcal{M}_0(S) \subset \mathcal{M}_C(\Theta)$. □

Corollary 1 *Under the same hypotheses as in Proposition 1,*

$$\sup_{\nu \in \mathcal{M}_1} \int gld\nu = \inf\{\beta_0 : \beta_0 \in \mathbb{R}, \beta_1, \dots, \beta_m \geq 0,$$
$$\beta_0 l(\theta) + \sum_1^m \beta_i f_i(\theta) \geq g(\theta)l(\theta)\ \theta \in \Theta\}.$$

Proof. If C is as in Proposition 1, we have $\mathcal{M}_1 \subset \mathcal{M}_C \subset \mathcal{M}_{C_1}$, for any $C_1 \geq C$, so that $U^*(gl)$ can be written as

$$\inf\{\beta_0 + \sup_{\nu \in \mathcal{M}_{C_1}} \int (gl - \beta_0 l - \sum_1^n \beta_i f_i) d\nu : \beta_0 \in \mathbb{R}, \beta_1, \dots, \beta_m \geq 0\}.$$

If $B_1 := \{(\beta_0, \dots, \beta_m) : \beta_0 \in \mathbb{R}, \beta_1, \dots, \beta_m \geq 0, g(\theta)l(\theta) - \beta_0 l(\theta) - \sum_1^n \beta_i f_i(\theta) \leq 0 \forall \theta \in \Theta\}$ and $B_2 := \{(\beta_0, \dots, \beta_m) : \beta_0 \in \mathbb{R}, \beta_1, \dots, \beta_m \geq 0, \exists \theta^* \in \Theta : g(\theta^*)l(\theta^*) - \beta_0 l(\theta^*) - \sum_1^n \beta_i f_i(\theta^*) > 0\}$, then $U^*(gl)$ is

$$\min_{j=1,2} \inf_{B_j}\{\beta_0 + \sup_{\nu \in \mathcal{M}_{C_1}} \int (gl - \beta_0 l - \sum_1^n \beta_i f_i) d\nu : \beta_0 \in \mathbb{R}, \beta_1, \dots, \beta_m \geq 0\}.$$

The first infimum is equal to $\inf_{B_1}\{\beta_0\}$, while the second one is equal to $\inf_{B_2}\{\beta_0 + C_1 \sup_\Theta (gl - \beta_0 l - \sum_1^n \beta_j f_j)\}$. Since $U^*(gl)$ and $U(gl)$ coincide by Proposition 1, and $U(gl)$ does not depend on C_1, $U^*(gl)$ does not depend on C_1 either, yielding $U^*(gl) = \inf_{B_1}\{\beta_0\} = U(gl)$. □

15.3.2 Θ is not compact

Let us state some further conditions.

H5: $f_j(\theta) \geq C_j$, $j = 1, \dots, m$, where the C_js are real constants;

H6: l is bounded;

H7: gl is bounded from above;

H8: $\forall \varepsilon > 0$ there exists a compact K_ε s.t. at least one of the following conditions holds:

i) there exist $\beta_1^\varepsilon, \dots, \beta_m^\varepsilon \geq 0$, $\sum_1^m \beta_i^\varepsilon \leq B$ (a positive constant), such that $\sum_1^m \beta_i^\varepsilon f_i(\theta) > 1/\varepsilon\ \forall \theta \in K_\varepsilon^c$;

ii) $g(\theta)l(\theta) < -1/\varepsilon\ \forall \theta \in K_\varepsilon^c$.

Let us denote by $\mathcal{N}_D$ the set $\{\nu \in \mathcal{M} : \int f_j d\nu \leq 0, \int l d\nu \leq 1 + D\}$, for $D > 0$. Obviously, $\mathcal{M}_1 \subset \mathcal{N}_D$. The following result holds.

Lemma 3 *Under H1–H2 and H5, $\mathcal{M}_1$ and $\mathcal{N}_D$, for any $D > 0$, are closed.*

Proof. The result trivially holds if $\mathcal{M}_1 = \emptyset$. If $\mathcal{M}_1 \neq \emptyset$, let $\nu_n \in \mathcal{M}_1$, $n = 1, 2, \ldots$, and $\nu_n \xrightarrow{w} \nu$. Since f_j, $j = 1, \ldots, m$ is l.s.c., then $\int f_j d\nu_n = \sup_{\{h \in \mathcal{C}_b(\Theta), h \leq f_j\}} \int h d\nu_n$ for all n (see Lemma 3 in Kemperman, 1983). Thus, if $h \in \mathcal{C}_b(\Theta)$, $h \leq f_j$, $\int h d\nu = \lim_{n \to +\infty} \int h d\nu_n \leq \lim_{n \to +\infty} \int f_j d\nu_n \leq 0$, yielding $\int f_j d\nu = \sup_{\{h \in \mathcal{C}_b(\Theta), h \leq f_j\}} \int h d\nu \leq 0$; in the same way we prove that $\int l d\nu = \lim_{n \to +\infty} \int l d\nu_n = 1$ holds, so that $\nu \in \mathcal{M}_1$. The proof for $\mathcal{N}_D$ is similar. □

We can say something more:

Lemma 4 *Assume that H1–H2, H4–H5 hold and that $\mathcal{M}_1 \neq \emptyset$. Then, under H8(i), $\mathcal{N}_D$ is relatively compact; under H8(ii) and H7, $\mathcal{N}_D(\gamma) := \{\nu \in \mathcal{N}_D : \int (g - \gamma) l d\nu \geq 0\}$ is relatively compact for any $\gamma \leq \sup_{\nu \in \mathcal{M}_1} \int g l d\nu$.*

Proof. First of all, we prove that $\mathcal{N}_D$ is contained in $\mathcal{M}_C$ for some $C > 0$. Indeed, if $\nu \in \mathcal{N}_D$, $1 + D \geq \int_{\{\theta : l(\theta) > \varepsilon\}} l(\theta) \nu(d\theta) > \varepsilon \nu(\{\theta : l(\theta) > \varepsilon\})$, so that $\nu(\{\theta : l(\theta) > \varepsilon\}) < (1 + D)/\varepsilon$; moreover, by H4, $\nu(\{\theta : l(\theta) \leq \varepsilon\}) \leq -(1 + D)(\sum_1^m \bar{\beta}_i C_i)/\varepsilon$, so that $\nu(\Theta) < ((1 + D)/\varepsilon)(1 - \sum_1^m \bar{\beta}_i C_i) =: C(D)$. Observe that $C(D)$ increases with D. Therefore, for $\nu \in \mathcal{N}_D$, by H8(i) we have $0 \geq \int \sum_i \beta_i^\varepsilon f_i d\nu > \int_{K_\varepsilon} \sum_i \beta_i^\varepsilon f_i d\nu + (1/\varepsilon)\nu(K_\varepsilon^c) \geq \nu(K_\varepsilon) B C_{\min} + (1/\varepsilon)\nu(K_\varepsilon^c)$, where $C_{min} := \min_j C_j$. This yields $(1/\varepsilon - BC_{\min})\, \nu(K_\varepsilon^c) + BC_{\min}\nu(\Theta) \leq 0$, so that $\sup_{\nu \in \mathcal{N}_D} \nu(K_\varepsilon^c) \leq ((-BC_{\min})/(1/\varepsilon - BC_{\min}))C(D)$. We conclude that $\mathcal{N}_D$ is relatively compact. When H8(ii) holds, the proof is similar. □

The key results of the section are the following proposition and its corollary.

Proposition 2 *Under H1–H2, H4–H7, if $\mathcal{M}_1 \neq \emptyset$ is relatively compact, then*

$$\sup_{\mu \in \mathcal{M}_1} \int g l d\nu$$

$$= \inf\{\beta_0 + \sup_{\mathcal{N}_D} \int (gl - \beta_0 l - \sum_1^m \beta_i f_i) d\nu, \beta_0 \in \mathbb{R}, \beta_1, \ldots, \beta_m \geq 0\}, \quad (11)$$

$$= \inf\{\beta_0 : \beta_0 \in \mathbb{R}, \beta_1, \ldots, \beta_m \geq 0, \beta_0 l(\theta) + \sum_1^m \beta_i f_i(\theta) \geq g(\theta) l(\theta)\ \theta \in \Theta\},$$

and the supremum over $\mathcal{M}_1$ is assumed.

Proof. Conditions H1–H2, H4–H5 imply that $\mathcal{M}_1 \subset \mathcal{N}_D$ and that there exist $C = C(D)$ such that $\mathcal{N}_D \subset \mathcal{M}_{C(D)}$ for all $D > 0$. By Lemmas 3 and 4, $\mathcal{N}_D$ is compact. The first equality in (11) holds, applying Theorem 2 with $\mathcal{M}_0(S) = \mathcal{N}_D(\Theta)$. The proof of the second equality in (11) is identical to that of Corollary 1, substituting $\mathcal{M}_{D_1}$ to $\mathcal{M}_{C_1}$. □

Corollary 2 *Under H1–H2, H4–H7 and H8(i) or (ii), if $\mathcal{M}_1 \neq \emptyset$, then (11) holds and the supremum in its left-hand side is assumed.*

Proof. If H8(i) holds, Lemma 4 implies that $\mathcal{M}_1$ is relatively compact. Then Proposition 2 gives the result. If H8(ii) holds, let $\hat{\nu} \in \mathcal{M}_1$ and $\gamma \leq \int gld\hat{\nu}$; then $\mathcal{N}_D(\gamma) := \{\nu \in \mathcal{N}_D : \int gld\nu \geq \gamma\}$ is relatively compact by Lemma 4. Since $\mathcal{M}_1(\gamma) = \{\nu \in \mathcal{M}_1 : \int gld\nu \geq \gamma\} \subset \mathcal{N}_D(\gamma)$, $\mathcal{M}_1(\gamma)$ is relatively compact too. Therefore, by Proposition 2,

$$\sup_{\nu \in \mathcal{M}_1} \int gld\nu = \sup_{\nu \in \mathcal{M}_1(\gamma)} \int gld\nu$$

$$= \inf_{\beta_0 \in \mathbb{R}, \beta_1, \ldots, \beta_m \geq 0} \{\beta_0 + \sup_{\mathcal{N}_D(\gamma)} \int (gl - \beta_0 l + \beta_{m+1}(g-\gamma)l - \sum_1^m \beta_i f_i) d\nu\}$$

$$= \inf_{\beta_0 \in \mathbb{R}, \beta_1, \ldots, \beta_m \geq 0} \{\beta_0 : \beta_0 l(\theta) + \sum_1^m \beta_i f_i(\theta) - \beta_{m+1}(g-\gamma)l(\theta) \geq g(\theta)l(\theta)\, \forall\theta\}$$

$$= \inf_{\beta_0 \in \mathbb{R}, \beta_1, \ldots, \beta_m \geq 0} \{\beta_0 : \beta_0 l(\theta) + \sum_1^m \beta_i f_i(\theta) + \beta_{m+1}\gamma l(\theta) \geq g(\theta)l(\theta)\, \forall\theta\}.$$

Now, if we choose $\gamma < 0$, $\beta_0 l(\theta) + \sum_1^m \beta_i f_i(\theta) + \beta_{m+1}\gamma l(\theta) \geq gl(\theta)$ for all $\theta \in \Theta$ yields $\beta_0 l(\theta) + \sum_1^m \beta_i f_i(\theta) \geq gl(\theta)$ for all $\theta \in \Theta$, so that the result holds. □

Remark 4 Condition H8(i) or (ii) in Corollary 2 could be replaced by

$$\forall \varepsilon > 0 \text{ there exists a compact set } K_\varepsilon \subset S :$$

$$\exists \gamma_1, \ldots, \gamma_m \geq 0 \text{ with } \gamma_0 l(\theta) + \sum_1^m \gamma_j f_j(\theta) \geq I_{K_\varepsilon^c}(\theta), \quad \gamma_0 \leq \varepsilon,$$

since this condition also yields the relative compactness of $\mathcal{M}_1$. It has been assumed as a hypothesis in Theorem 6 in Kemperman (1983), but it appears rather difficult to check.

15.3.3 Discreteness of the solutions

Once it is ensured that the supremum in (9) is reached by some measure ν^*, then ν^* can be assumed to have a finite support of at most $m+1$ points. Indeed, setting

$$z_0 = \int gld\nu^*, \quad z_i = \int f_i d\nu^*, \quad i = 1 \ldots m,$$

and recalling that $1 = \int ld\nu^*$, Theorem 1 in Rogosinsky (1958) (see also Lemma 1 in Kemperman, 1983) states that there exists a measure ν satisfying the above conditions and having a finite support of at most $m+2$

points. We observe that such a measure is still optimal, so that we can assume that ν^* coincides with it. Consequently, considering the set $\widehat{\mathcal{M}}$ of the measures in $\mathcal{M}_1$ which have the same support as ν^*, it turns out that this latter also solves the problem

$$\sup_{\nu \in \widehat{\mathcal{M}}} \int gl d\nu. \tag{12}$$

Problem (12) is now an ordinary (finite) linear programming problem in $m+2$ variables and $m+1$ constraints for which it is well known that, if a solution exists, then it can be assumed to have at most $m+1$ nonnull coordinates.

15.4 Algorithms for prior robustness under generalized moment conditions

From now on we assume the hypotheses of Corollary 1, in the case of Θ compact, and of Corollary 2, when Θ is not compact.

Corollaries 1 and 2 provide a fundamental link between prior robustness under generalized moment conditions and linear semi-infinite programming (LSIP). This enables the numerical treatment of robustness analysis by means of algorithms developed in this latter context which has been widely considered in the optimization literature in recent years (see, e.g., Goberna and López, 1998, for a recent and comprehensive survey on LSIP). Indeed, various authors have followed this path, as mentioned in this section.

Before we consider algorithms, a preliminary matter needs some attention. In the fundamental equalities stated by the above mentioned corollaries, while the sup in the left hand side is guaranteed to be attained by some measure $\nu \in \mathcal{M}_1$, so that the dual problem on the right hand side has a feasible solution and the corresponding β_0 is bounded from below, an optimal solution of the dual is not guaranteed to exist. Therefore, it is worthy to have conditions ensuring that the infimum is attained, which is obviously useful also from an algorithmic viewpoint. Corollary 9.3.1 in Goberna and López (1998) gives necessary and sufficient conditions for the set of optimal solutions of an LSIP problem being not empty. We restate here the part of the corollary which is relevant to our purposes in the following terms:

Corollary 3 *Let (P) denote the problem*

$$\beta_0^* = \inf\{\beta_0 : \beta_0 \in \mathbb{R}, \beta_1, \ldots, \beta_m \geq 0, \beta_0 l(\theta) + \sum_1^m \beta_i f_i(\theta) \geq g(\theta) l(\theta), \theta \in \Theta\};$$

then the following conditions are equivalent:

(i) F^*, *the optimal solution set of* (P), *is a non-empty bounded set;*

(ii) (P) *is bounded and the non-empty level sets of* (P) *are bounded;*

(iii) *there exists a finite subproblem of* (P) *whose non-empty level sets are bounded;*

(iv) *the vector* $\underline{c} := (1, 0, \ldots, 0)^T \in int\ (M_{m+1})$, *where* M_{m+1} *is the moment cone* $\{y_0, y_1, \ldots, y_m : \int l d\nu = y_0, \int f_i d\nu \leq y_i, i = 1, \ldots, m,$ ν *is a measure with finite support* $\}$.

We observe that condition (iv) is true if there exists a finite measure ν (with finite support) such that $\int l d\nu = 1$ and $\int (f_i + \epsilon_i) d\nu \leq 0, \epsilon_i > 0$; indeed, it is easily seen that in this case, given $0 < \epsilon_0 < 1$, then for any $|y_0 - 1| < \epsilon_0, y_i \geq -(1-\epsilon_0)\epsilon_i \nu(\Theta)$, $\nu' := y_0 \nu$ gives $\int l d\nu' = y_0, \int f_i d\nu' \leq y_i$. This is never true when $f_i = -f_j$ for some i and j, as it occurs if we write an equality constraint as a couple of inequality constraints. More generally, the interior of M_{m+1} is empty if the functions $l, f_i, i = 1, \ldots, m$ are not linearly independent, as already noticed in Section 2.2. In this case the validity of (iv) requires the preliminary aggregation of inequality constraints into a single equality constraint (if $f_i = -f_j$, then in the definition of M_{m+1} the equality $\int f_i d\nu = y_i$ takes the place of $\int f_i d\nu \leq y_i$ and $\int f_j d\nu \leq y_j$) and the exclusion of redundant constraints.

Coming to algorithms for solving LSIP problems, and hence for determining the bounds required by robustness analysis via Corollary 1 or 2, we first recall the classification into five categories stated by Goberna and López (1998, p.253):

(A) *discretization methods* (*by grids and by cutting planes*);
(B) *local reduction methods;*
(C) *exchange methods;*
(D) *simplex-like methods;*
(E) *descent methods.*

Goberna and López (1998) also report that the ranking of each family in the list corresponds, by some means, to its reputation in terms of computational efficiency, according to most experts' opinions. Taking into account this classification and the fact that local reduction methods require regularity conditions on the functions which may not hold in robustness analysis (e.g., when indicator functions are considered), in the following two subsections we review in some details algorithms belonging to category (A), which are the most popular and easy to implement. In subsection 4.3 we briefly review the application of LSIP algorithms to robustness analysis in the Bayesian literature, including also algorithms not falling into category (A).

15.4.1 Grid discretization methods

We assume that the set of optimal solutions F^* is a non-empty bounded set, so that a discretization of problem (P) exists, with Θ replaced by a finite grid S (by *grid* we mean a finite or countable set of points belonging to Θ, not necessarily regularly displaced), having non-empty bounded level sets (see Corollary 3 and the remarks following it).

Let Θ_r be a sequence of finite grids such that $\Theta_r \subset \Theta_{r+1}$ (expansive grids), $\Theta_0 := S$ and $\overline{\cup_{r=0}^{\infty}\Theta_r} = \Theta$ (that is, Θ is closed and $\cup_{r=0}^{\infty}\Theta_r$ is dense in Θ). Then the basic grid discretization algorithm is as follows:

Grid discretization algorithm. Given a tolerance $\epsilon \geq 0$, initialize the iteration index $r = 0$ and cycle through the following steps:

Step 1 Solve the LP subproblem

$$(P_r)\ \inf\{\beta_0 :\ \beta_0 \in \mathbb{R}, \beta_1, \ldots, \beta_m \geq 0, \beta_0 l(\theta) + \sum_1^m \beta_i f_i(\theta) \geq g(\theta) l(\theta), \qquad \theta \in \Theta_r\},$$

and let β^r be its optimal solution; then go to Step 2.

Step 2 Minimize the function $s(\theta, \beta^r) := \beta_0^r l(\theta) + \sum_1^m \beta_i^r f_i(\theta) - g(\theta)l(\theta)$ (slack function) over Θ and set $s_r = \inf_{\theta \in \Theta} s(\theta, \beta^r)$.
If $s_r \geq -\epsilon$, then stop and accept β_r as the optimal solution of (P).
Otherwise, replace r by $r+1$ and loop to Step 1.

Observe that the final β^r is not guaranteed to be feasible for (P) (in this case it would be also optimal), so that the algorithm produces in general an underestimate β_0^r of $U(gl)$. In actual robustness analysis, this means that we might compute an underestimate of the posterior range of the quantity of interest, coming to an erroneous conclusion about the presence of robustness. However, applying Theorem 16 of Chapter V in Glashoff and Gustafson (1983) (see also Dall'Aglio, 1995, p. 60), if $d_0 > 0$ is like in H3 when $q = 0$ (remember that such a d_0 exists under H4), we obtain $\beta_0^* \leq \beta_0^r + \epsilon d_0$ and hence a measure of the achieved accuracy. Obviously, this result is useful only if it is easy to determine a d_0 such that ϵd_0 is small enough.

Convergence of the grid discretization algorithm is obtained under the hypothesis that all the functions l, f_i, g are continuous (see Theorem 11.1 in Goberna and López, 1998), in the sense that either finite termination occurs (this is always the case when $\epsilon > 0$) or β^r has at least a cluster point in F^*. However, the continuity assumption rules out the possibility that some of the f_is are indicator functions.

15.4.2 Cutting-plane discretization methods

The basic cutting-plane (CP) algorithm. Let a sequence $\{\epsilon_r\}$ be given such that $\epsilon_r > 0$ and $\lim_r \epsilon_r = 0$. Let $\epsilon \geq 0$ be a given tolerance. Initialize the iteration index $r = 0$ and take $\Theta_0 = S$.

Step 1 Solve the LP subproblem

$$(P_r)\ \inf\{\beta_0 :\ \beta_0 \in \mathbb{R}, \beta_1, \ldots, \beta_m \geq 0, \beta_0 l(\theta) + \sum_1^m \beta_i f_i(\theta) \geq g(\theta) l(\theta),\ \theta \in \Theta_r\},$$

and let β^r be its optimal solution; then go to Step 2.

Step 2 Compute $s_r := \inf_{\theta \in \Theta} s(\theta, \beta^r)$.
If $s_r \geq -\epsilon$, then stop and accept β_r as optimal solution of (P).
If $s_r < \epsilon$, then find a non-empty finite set S_r such that $s(\theta, \beta^r) \leq s_r + \epsilon_r$ for each $\theta \in S_r$ (that is, S_r is a set of ϵ_r-minimizer); set $\Theta_{r+1} = \Theta_r \cup S_r$, replace r by $r + 1$ and go to Step 1.

The algorithm proceeds by adding to the current constraints a finite number of constraints which are violated, or almost violated, at β^r (local cutting planes). Convergence can be proved under general conditions, as illustrated by the following result:

Theorem 5 (see Theorem 11.2 in Goberna and López, 1998) *Assume that the set $\{(l(\theta), f_1(\theta), \ldots, f_m(\theta)), \theta \in \Theta\}$ is bounded. Then the CP algorithm generates either a finite sequence or an infinite sequence having cluster points which are optimal solutions of (P). In particular, finite termination occurs if $\epsilon > 0$.*

The condition assumed in Theorem 5 is not a restriction as, for any $\theta \in \Theta$, we can normalize the vector $(l(\theta), f_1(\theta), \ldots, f_m(\theta))$ without altering problem (P).

The alternating algorithm by Gustafson and Kortanek (1973) is the first example of a CP method presented in the literature for LSIP problems, and it coincides with the basic CP algorithm presented above when $\epsilon = \epsilon_r = 0$ and the set S_r is restricted to a single point for each r. Several variants of the alternating algorithm have been proposed by various authors, in order to increase its effectiveness (notice that the original alternating algorithm requires the exact global minimization of the slack function). Hu (1990) replaces the condition $s(\theta, \beta^r) \leq s_r + \epsilon_r$ by $s(\theta, \beta^r) \leq \alpha \max\{-1, s_r\}$ for any fixed $0 < \alpha < 1$, so that the size of the cut is regulated in an automatic way.

Within the same scheme of things, but not included in the general setting of the basic cutting-plane algorithm, is the central cutting-plane (CCP) method by Elzinga and Moore (1975), originally proposed for general convex programming problems; the method is aimed at providing at each iteration a feasible point (unlike both grid discretization methods and ordinary CP methods) and at preventing the numerical instability of CP algorithms due to the fact that cuts tend to be parallel near optimal points.

The CCP algorithm can be described in our context as follows:

The Central Cutting-Plane Algorithm. Let $\alpha_0 \in \mathbb{R}$ be available such that $\alpha_0 \geq \beta_0^*$. Let a tolerance $\epsilon \geq 0$ and a scalar $\gamma \in (0,1)$ be given. Set $\Theta_0 = S$.

Step 1 Find an optimal solution (β^r, ρ_r) of the problem

$$\begin{aligned}(Q_r)\ \sup\{\rho :\ & \rho \geq 0, \beta_0 \in \mathbb{R}, \beta_1, \dots, \beta_m \geq 0,\\ & \beta_0 + \rho \leq \alpha_r,\\ & \beta_0 l(\theta) + \sum_1^m \beta_i f_i(\theta) - \rho \parallel (l(\theta), f_1(\theta), \dots, f_m(\theta)) \parallel\\ & \geq \qquad g(\theta) l(\theta),\ \theta \in \Theta_r\}\end{aligned}$$

If $\rho_r \leq \epsilon$, then stop.
Otherwise, go to Step 2.

Step 2 Compute $s_r := \inf_{\theta \in \Theta} s(\theta, \beta^r)$.
If $s_r \geq 0$, set $\alpha_{r+1} = \beta_0^r$ and $\Theta_{r+1} = \Theta_r$ and replace r by $r+1$. Go to Step 1.

If $s_r < 0$, find θ_r such that $s(\theta_r, \beta^r) < 0$, set $\alpha_{r+1} = \alpha_r$ and $\Theta'_{r+1} = \Theta_r \cup \{\theta_r\}$; then go to Step 3.

Step 3 Let S_r be the set $\{\theta \in \Theta'_{r+1} : s(\theta, \beta^p) < 0,\ p < r, \rho_r \leq \gamma \rho_p$, $\beta_0^r l(\theta) + \sum_1^m \beta_i^r f_i(\theta) - \rho_r \parallel (l(\theta), f_1(\theta), \dots, f_m(\theta)) \parallel > g(\theta) l(\theta)$. Then take $\Theta_{r+1} = \Theta'_{r+1} \backslash S_r$ and go to Step 1.

The solution (β^r, ρ_r) of (Q_r) defines the center and the radius of the maximal sphere inscribed into $F_r := \{\beta \in \mathbb{R}^{m+1}, \beta_1, \dots, \beta_m \geq 0 : \beta_0 \leq \alpha_r, \beta_0^r l(\theta) + \sum_1^m \beta_i^r f_i(\theta) \geq g(\theta) l(\theta), \theta \in \Theta_r\}$. Step 1 is well defined as $F^* \subset F_r \subset F_0$, so that (Q_r) has an optimal solution. In Step 2 either an "objective cut" or a "feasibility cut" of F_r is obtained, reducing in both cases the polytope containing F^*. It is easily seen that $\{\alpha_r\}$ is a non-increasing sequence bounded from below by β^*.

The following result holds.

Theorem 6 (Lemma 11.4 and Theorem 11.5 in Goberna and López, 1998) *The CCP algorithm either terminates at the r–th iteration, for some $r \geq 0$, and it is $\alpha_r = \beta^*$ if $\epsilon = 0$, or $\lim_r \alpha_r = \beta^*$. Moreover, if z^0 is a feasible solution of (P), then the sequence $\{z^{r+1}\}$, such that $z^{r+1} = \beta^r$ if $s_r \geq 0$, $z^{r+1} = z^r$ otherwise, either is finite (and the final z^r is an optimal solution if $\epsilon =$) or has cluster points, all of them being optimal solutions.*

Looking at the sequence z^{r+1}, the CCP algorithm can be interpreted as an interior-point algorithm.

15.4.3 LSIP algorithms for robustness analysis in the Bayesian literature

The cutting-plane algorithm by Hu (1990) was incorporated into the procedure described in Betrò and Guglielmi (1997), which, developing the approach first introduced in Betrò and Guglielmi (1994), considers at each step an upper iteration and a lower iteration, bracketing β_0^* from above and from below, respectively. In the upper iteration, the primal problem in (10) is substituted by one in which the functions involved are piecewise constant over a regular grid (assuming that Θ is a rectangular region) and are defined through inclusion functions provided by interval analysis (see, e.g., Ratschek and Rokne, 1998); then the grid is refined splitting intervals to which the solution of the latter problem assigns positive mass. The lower iteration is essentially Hu's algorithm with s_r replaced by a lower bound obtained by interval arithmetics on inclusion functions too. Application of this method to robustness analyis in reliability has been considered in Betrò et al. (1996).

In Dall'Aglio (1995) a modification of Hu's algorithm is proposed, in which accuracy with respect to the optimal value of (P) is controlled by the upper bound $\beta_0^* \leq \beta_0^r + \epsilon d_0$, and examples of its application are given. For the case of equality constraints, a simplex-like method adapted from Glashoff and Gustafson (1983) is also presented, in which at each step points of a grid are substituted by new points so that the cardinality of the grid remains constant, say r, and the optimal value of (P_r) on the new grid is not smaller than that on the old grid; however, convergence to β^* is not guaranteed.

Fandom Noubiap and Seidel (1998) introduce an algorithm for calculating a Γ-minimax test, where Γ is given by a finite set of generalized moment conditions. The inner maximization problem, which is required to solve, is a generalized moment problem in the form (9) with $l \equiv 1$, that is, ν is a probability measure. The solution method considered is the basic cutting-plane presented above.

The case when ν is a probability measure is considered also in Smith (1995), where the constraints in $\mathcal{M}_1$ are equality constraints. The algorithm proposed there, without proving its convergence, is based on (4) and on the complementary slackness condition introduced in Section 2.2. The algorithm is similar to the "three phase algorithm" for LSIP, described in Glashoff and Gustafson (1983) (see also Algorithm 11.5.1 in Goberna and López, 1998), and combines the grid discretization approach with the local reduction one (see Goberna and López, 1998, Sect. 11.2 for this latter), so that we present it here with some details to illustrate the idea behind local reduction. Smith's algorithm consists essentially of two steps:

I. Approximate Θ by a finite grid and solve the corresponding finite linear program using the simplex method. The solution to the approximate problem provides a lower bound on the optimal value of the original problem.

II. Using the solution in Step I as a guide, construct a feasible and approximately optimal solution to the dual problem. The main point in this step is to approximately locate the points $\theta_0, \ldots, \theta_q$ $(q \leq m)$ at which the optimal solution $\overline{\mu}$ is concentrated. If these points were the points of the exact solution, by the complementary slackness condition there would be a polynomial $\beta_0 l(\theta) + \sum_1^m \beta_i f_i(\theta)$ such that $\beta_0 l(\theta_j) + \sum_1^m \beta_i f_i(\theta_j) = g(\theta_j) l(\theta_j)$, $j = 0, \ldots, q$, and $\beta_0 l(\theta) + \sum_1^m \beta_i f_i(\theta) \geq g(\theta) l(\theta)$ $\forall \theta$. In this case $\beta_0 l + \sum_1^m \beta_i f_i$ and gl would be tangent at those points where $l, f_1, \ldots, f_m$ and gl are all continuous. But, if $\theta_0, \ldots, \theta_q$ are not necessarily optimal, we can select $m+1$ conditions, q equations of the form $\beta_0 l(\theta_j) + \sum_1^m \beta_i f_i(\theta_j) = g(\theta_j) l(\theta_j)$ and $m - q + 1$ of the form $\beta_0 l(\theta_j) + \sum_1^m \beta_i \nabla f_i(\theta_j) = \nabla(gl)(\theta_j)$, and solve for β. If this β is such that $\beta_0 l(\theta) + \sum_1^m \beta_i f_i(\theta) \geq g(\theta) l(\theta)$ $\forall \theta$, then β_0 is an upper bound on the exact optimal solution.

Observe that the algorithm, as the other ones based on local reduction, requires differentiability properties of the functions defining the problem.

Betrò (1999a) recently introduced a modification of the CCP algorithm, called ACCP (accelerated CCP) algorithm, aimed at accelerating its convergence. In problem (Q_r), α_r is tentatively replaced by a lower value α_r'; if (Q_r) has still a solution, then the algorithm proceeds to the other steps with $\alpha_r = \alpha_r'$; otherwise, α_r' is retained as a lower bound to the optimal value and a new α_r', intermediate between the previous α_r' and α_r, is tried until (Q_r) becomes solvable. The global optimization problem required in Step 2 is solved by means of interval analysis. Some examples taken from the LSIP literature show that much faster convergence may be achieved in this way.

Betrò (1999b) discusses application of ACCP to robustness analysis with some examples where a solution is quickly found.

15.5 Conclusions

We have reviewed to some extent both the theoretical and the algorithmic aspects of robustness analysis under generalized moment conditions. While the theory is well established and leads to sound results, none of the algorithms proposed so far has been convincing enough to be adopted for routine robustness analysis, which has actually prevented the wide diffusion of the Bayesian robust viewpoint among statisticians, even Bayesian ones. The ACCP algorithm seems to be a candidate for filling the gap because of its wide applicability and speed of convergence, but more testing is needed, investigating in particular the response to the increase in the dimension of the parameter space, together with the achievement of reliable and user-friendly software.

References

BERGER, J.O. (1994). An overview of robust Bayesian analysis. *Test*, **3**, 5–59.

BETRÒ, B. (1999a). An accelerated central cutting plane algorithm for semi-infinite linear programming. *Quaderno IAMI*, **99.16**, CNR-IAMI.

BETRÒ, B. (1999b). The accelerated central cutting plane algorithm in the numerical treatment of Bayesian global prior robustness problems. *Quaderno IAMI*, **99.17**, CNR-IAMI.

BETRÒ, B. and GUGLIELMI, A. (1994). An algorithm for robust Bayesian analysis under generalized moment conditions. *Quaderno IAMI*, **94.6**, CNR-IAMI.

BETRÒ, B. and GUGLIELMI, A. (1997). Numerical robust Bayesian analysis under generalized moment conditions. In *Bayesian Robustness, IMS Lecture Notes - Monograph Series* (J.O. Berger, B. Betrò, E. Moreno, L.R. Pericchi, F. Ruggeri, G. Salinetti, and L. Wasserman, eds.), 3–20. Hayward: IMS.

BETRÒ B., GUGLIELMI, A. and ROSSI, F. (1996). Robust Bayesian Analysis for the power law process. *ASA 1996 Proceedings of the Section on Bayesian Statistical Science*, 288–291.

BETRÒ, B., MĘCZARSKI, M. and RUGGERI, F. (1994). Robust Bayesian analysis under generalized moments conditions. *Journal of Statistical Planning and Inference*, **41**, 257–266.

CHARNES, A. and COOPER, W.W. (1962). Programming with linear fractional functionals. *Naval Research Logistic Quarterly*, **9**, 181–186.

COZMAN, F. (1999). Calculation of posterior bounds given convex sets of prior probability measures and likelihood functions. To appear in *Journal of Computational and Graphical Statistics.*

DALL'AGLIO, M. (1995). Problema dei Momenti e Programmazione Lineare semi-infinita nella Robustezza Bayesiana. *Ph.D. Thesis*, Dip. Statistica, Probabilità e Statistiche Applicate, Università "La Sapienza," Roma.

DINKELBACH, D. (1967). On nonlinear fractional programming. *Management Science*, **13**, 492–498.

ELZINGA, J. and MOORE, TH. (1975). A central cutting plane algorithm for the convex programming problem. *Mathematical Programming*, **8**, 134–145.

FANDOM NOUBIAP, R. and SEIDEL, W. (1998). A minimax algorithm for calculating optimal tests under generalized moment conditions. *DP in Statistics and Quantitative Economics*, **85**, Universitat der Bundeswehr, Hamburg.

GLASHOFF, K. and GUSTAFSON, S. (1983). *Linear Optimization and Approximation.* New York: Springer-Verlag.

GOBERNA, M.A. and LÓPEZ, M.A. (1998). *Linear semi-Infinite Optimization.* New York: Wiley.

GOUTIS, C. (1994). Ranges of posterior measures for some classes of priors with specified moments. *International Statistical Review*, **62**, 245–357.

GUSTAFSON, S. and KORTANEK, K.O. (1973). Numerical treatment of a class of semi-infinite programming problems. *Naval Research Logistic Quarterly*, **20**, 477–504.

HU, H. (1990). A one-phase algorithm for semi-infinite linear programming. *Mathematical Programming*, **46**, 85–103.

JAGANNATHAN, R. (1966). On some properties of programming problems in parametric form pertaining to fractional programming. *Management Science*, **12**, 609–615.

KALLENBERG, O. (1986). *Random Measures.* Berlin: Akademie-Verlag.

KEMPERMAN, J.H.B. (1971). On a class of moment problems. In *Proceedings Sixth Berkeley Symposium on Mathematical Statistics and Probability*, **2**, 101–106.

KEMPERMAN, J.H.B. (1983). On the role of duality in the theory of moments. In *Lecture Notes in Economics and Mathematical Systems*, **215**, 63–92. Berlin: Springer-Verlag.

KEMPERMAN, J.H.B. (1987). Geometry of the moment problem. In *Moments in mathematics, Proceedings of Symposia in Applied Mathematics* (H.J. Landau ed.), 16–53. Providence: AMS.

LAVINE, M. (1991). Sensitivity in Bayesian statistics: the prior and the likelihood. *Journal of the American Statistical Association*, **86**, 396–399.

LAVINE, M., PERONE PACIFICO, M., SALINETTI, G. and TARDELLA, L. (2000). Linearization techniques in Bayesian robustness. In *Robust Bayesian Analysis* (D. Ríos Insua and F. Ruggeri, eds.). New York: Springer-Verlag.

LAVINE, M., WASSERMAN, L. and WOLPERT, R.L. (1993). Linearization of Bayesian robustness problems. *Journal of Statistical Planning and Inference*, **37**, 307–316.

PERONE PACIFICO, M., SALINETTI, G. and TARDELLA, L. (1994). Fractional optimization in Bayesian robustness. *Technical report*, **A 23**, Dipartimento di Statistica, Probabilità e Statistiche Applicate dell'Università di Roma 'La Sapienza'.

PERONE PACIFICO, M., SALINETTI, G. and TARDELLA, L. (1996). Bayesian robustness on constrained density band classes. *Test*, **5**, 395–409.

PROHOROV, Y.V. (1956). Convergence of random processes and limit theorems in probability theory. *Theory of Probability and its Applications*, **1**, 157–174.

RATSCHEK, H. and ROKNE, J. (1998). *New Computer Methods for Global Optimization*. Chichester: Ellis Horwood.

ROGOSINSKY, W.W. (1958). Moments of non-negative mass. *Proceedings of the Royal Society of London*, Ser. A, **245**, 1–27.

SMITH, J.E. (1995). Generalized Chebychev inequalities: theory and applications in decision analysis. *Operations Research*, **43**, 807–825.

SNOW, P. (1991). Improved posterior probability estimates from prior and conditional linear constraint systems. *IEEE Transactions on Systems, Man and Cybernetics*, **21**, 464–469.

WHITE, C.C., III (1986). A posteriori representations based on linear inequalities descriptions of a priori and conditional probabilities. *IEEE Transactions on Systems, Man and Cybernetics*, **16**, 570–573.

16

Efficient MCMC Schemes for Robust Model Extensions Using Encompassing Dirichlet Process Mixture Models

Steven MacEachern and Peter Müller

ABSTRACT We propose that one consider sensitivity analysis by embedding standard parametric models in model extensions defined by replacing a parametric probability model with a nonparametric extension. The nonparametric model could replace the entire probability model, or some level of a hierarchical model. Specifically, we define nonparametric extensions of a parametric probability model using Dirichlet process (DP) priors. Similar approaches have been used in the literature to implement formal model fit diagnostics (Carota et al., 1996).
In this paper we discuss at an operational level how such extensions can be implemented. Assuming that inference in the original parametric model is implemented by Markov chain Monte Carlo (MCMC) simulation, we show how minimal additional code can turn the same program into an implementation of MCMC in the larger encompassing model, providing an alternative to traditional sensitivity analysis. If the base measure of the DP is assumed conjugate to the appropriate component of the original probability model, then implementation is straightforward. The main focus of this paper is to discuss general strategies allowing implementation of models without this conjugacy.

Key words: Dirichlet process, MCMC, mixtures.

16.1 Introduction

We propose that one consider sensitivity analysis by embedding standard parametric models in nonparametric extensions. We use random measures with DP priors to define these encompassing nonparametric extensions. We present a framework which makes the implementation of posterior inference in such extensions always possible with minimum additional effort, essentially requiring only one additional multinomial sampling step in a Markov chain Monte Carlo (MCMC) posterior simulation. This is straightforward for models which are conjugate (conjugate in a sense which we shall make formal). In models without such conjugate structure, however, computational problems render posterior simulation difficult and hinder the routine application of such nonparametric model augmentations. In this paper we present a scheme which overcomes this hurdle and allows the implementation of robust nonparametric model extensions with equal ease in nonconjugate models.

In this chapter we shall use models based on Dirichlet process prior distributions (Ferguson, 1973; Antoniak, 1974). Many alternative approaches are possible for the encompassing nonparametric model. Among the many models proposed for nonparametric Bayesian modelling in the recent literature are Polya trees (Lavine, 1992, 1994), Gaussian processes (O'Hagan, 1992; Angers and Delampady, 1992), beta processes (Hjort, 1990), beta-Stacy processes (Walker and Muliere, 1997), extended gamma processes (Dykstra and Laud, 1981), random Bernstein polynomials (Petrone, 1999a, 1999b). See Walker et al. (1999) for a recent review of these alternative forms of nonparametric Bayesian modelling.

Consider a generic Bayes model for a collection of n nominally identical problems with likelihood

$$y_i \overset{iid}{\sim} p_{\theta,\nu}(y_i), \quad i = 1, \dots, n, \tag{1}$$

and prior $\theta \sim G_0(\theta|\nu)$ and $\nu \sim H(\nu)$. In anticipation of the later generalization the parameter vector is partitioned into (θ, ν), where θ is the subvector of parameters with respect to which the model extension will be defined below. Model (1) could, for example, be a normal distribution with unknown location θ and variance ν. Inference from such a model is extremely restrictive in that a single parameter θ indexes the conditional distribution for each and every y_i. Estimation of an observation specific parameter - say θ_i, representing the mean of the conditional distribution for y_i in our simple example - is identical for every i since there is only a single θ. At the far extreme from model (1), we may write

$$y_i \overset{iid}{\sim} p_{\theta_i,\nu_i}(y_i), \quad i = 1, \dots, n, \tag{2}$$

and prior $\theta_i \sim G_0(\theta_i|\nu_i)$ and $\nu_i \sim H(\nu_i)$, creating n separate problems. Since the joint distribution on the n collections of parameters, θ_i, ν_i, y_i, forms a set of n independent distributions, inference is made independently in the n cases. This model does not permit any pooling of information across the n problems, leading to potentially poor inference.

We consider generalizations of (1) to

$$y_i \overset{iid}{\sim} \int p_{\theta,\nu}(y_i)\, dG(\theta), \quad G \sim DP\left(M\, G_0(\cdot|\nu)\right). \tag{3}$$

The original sampling model $p_{\theta,\nu}$ is replaced by a mixture over such models, with a mixing measure G. For example, we might replace a simple normal sampling model by a location mixture of normals. As a probability model for the random mixing measure we assume a Dirichlet process (DP) with base measure MG_0, where G_0 is a probability measure. See, for example, Antoniak (1974) or Ferguson (1973) for a definition and discussion of DPs. The model contains the original model (1) as a special case when G is a point mass. The DP prior puts nonzero prior probability on G being arbitrarily close to such a single point mass and implies that the point mass be a sample from G_0. The base measure of the DP need not be the same as the prior in the original parametric model, but this is a natural choice since it implies the same marginal distribution $p(y_i)$ as under (1). The model provides a nice alternative to (2), allowing us to pool information obtained from the entire collection of problems to make better inference for each individual problem.

The perspective of providing a flexible, nonparametric version of the parametric Bayes model motivated much early work in the area (see, for example, Susarla and van Ryzin, 1976; Kuo, 1983; MacEachern, 1988; Escobar, 1988). The flexibility of the nonparametric analysis both allows for an alternative to formal sensitivity analysis by comparing the fit of the parametric model and its elaboration and also provides a fresh look at the data with what can alternatively be considered a larger model.

For the sake of presentation it is convenient to consider the case of a parametric hierarchical model which is to be elaborated separately from the case of nonhierarchical models. To wit,

$$\begin{aligned} y_i &\overset{iid}{\sim} p_{\theta_i,\nu}(y_i), \\ \theta_i &\overset{iid}{\sim} G_0(\theta_i|\nu), \end{aligned} \tag{4}$$

with prior $\nu \sim H(\nu)$. The model is generalized by replacing the prior G_0 with a random distribution G:

$$\begin{aligned} y_i &\overset{iid}{\sim} p_{\theta_i,\nu}(y_i) \\ \theta_i \overset{iid}{\sim} G(\theta_i), &\quad G \sim DP\left(M\, G_0(\cdot|\nu)\right). \end{aligned} \tag{5}$$

As can easily be seen by marginalizing over θ_i in (5), model (5) is identical to (3). Following traditional terminology we refer to (5) as the mixture of Dirichlet process model (MDP). Given a MDP model it is often a matter of perspective whether it is seen as a generalization of a basic model (1) or a hierarchical model (4), although we believe the latter is the more common view in the literature. See Escobar and West (1998) for a recent summary of this perspective. Below, in examples (i) through (xii), we give examples of both.

In the rest of this chapter we will argue that Markov chain Monte Carlo (MCMC) posterior simulation in model (5), and thus in (3), can be easily implemented by adding just one additional (multinomial) sampling step to an MCMC scheme for the original model (4) or (1). Posterior inference under the augmented model (3) or (5) provides a basis for investigating model sensitivity and robustness.

Basu (2000) discusses the important related question of sensitivity to the prior specification in the DP mixture model. Basu considers sensitivity with respect to the total mass parameter M and the base measure G_0 and computes corresponding Gâteaux derivatives.

16.2 Nonparametric modelling and robustness

MDP models contribute to the study of robustness of Bayesian methods by highlighting the assumptions of a parametric Bayesian analysis and then by providing an easily implementable model extension that can provide additional robustness in a nearly automatic fashion. The robustness here is a robustness to the form of the "true" mixing distribution G, say G_T, as described by Ghoshal et al. (1999b). It is a robustness of both decision and loss for the motivating setting of many nominally identical problems with an overall loss equal to the sum of the n independent component losses, often referred to as the empirical Bayes problem. By robustness of the decision rule, we mean that the posterior action for the ith component problem tends to the posterior action that would be taken if G_T were known. By robustness of the loss, we mean that the posterior expected loss for the ith component problem, and also the preposterior expected loss for this problem, tends to the posterior expected loss if G_T were known. In the following paragraphs, we construct an argument which indicates that a nonparametric approach, as described in models (3) and (5), can enable us to perform inference as if we knew G_T.

Parametric approaches fail to provide robust inference in this setting because their support is restricted to a small set of distributions. That is, for a parametric model in which $y_i \sim p_{\theta_i}(y_i), \theta_i \sim G(\theta_i|\nu)$, and $\nu \sim p_\nu$, the mixing distribution G is restricted to have some parametric form, say $G_0(\nu)$. Both the prior and posterior distributions for the mixing distribution, G, have this restricted support. In fact, this support for G is typically so small that only very limited behavior is allowed for decision rules for estimating θ_i as a function of $y_1, \ldots, y_n$. The limitations on these decision rules ensure that, under typical loss functions, for G_T outside the support of the prior distribution, the posterior action and expected loss for the ith component problem under the parametric model do not converge to the posterior action and expected loss under the simple model where $\theta_i \sim G_T$ and $y_i|\theta_i \sim p_{\theta_i,\nu}$. Since this simple model represents the ideal for inference (as it is true model), the lack of convergence of decision rules and posterior expected loss to their ideal targets suggests that one does not do as well under the parametric model as one could by using a prior with larger support. The validity of this argument has been effectively demonstrated by Escobar (1994) for a particular prior, likelihood, loss combination.

Consistency results for the MDP extensions in (3) and (5) ensure that, under a variety of conditions, as $n \to \infty$, the posterior distribution of G converges in distribution to a degenerate distribution at the true mixing distribution, G_T. Ghoshal et al. (1999a, 1999b) and Wasserman (1998) provide sets of these conditions and illustrate their use for MDP models. With typical choices of likelihoods $p_{\theta,\nu}$, the marginal posterior distribution for each θ_i then converges to the posterior distribution for the model where $\theta_i \sim G_T$ and $y_i \sim p_{\theta_i,\nu}$—the ideal model for inference. This posterior distribution represents the best one can hope for. Convergence to this posterior distribution implies, under a broad range of loss functions, convergence of expected loss for the ith component problem to the posterior expected loss under perfect knowledge of G_T.

Thus, the MDP model extensions provide a means of gaining robustness, in both finite sample and asymptotic senses, to a relaxation of assumptions about G. It should be noted that a traditional robust Bayesian analysis does not solve this problem: as the prior distribution on G is varied over a class, the marginal posterior distribution on θ_i varies over a class. If the prior contains parametric models, the class of posterior distributions for $\theta_i|(y_1, \ldots, y_n)$ will typically not converge to a single distribution as $n \to \infty$, and one will be left with a range of inferences forever—an unpleasant thought when the simple method of MDP extension solves the problem.

Robustness concerns contribute to the study of nonparametric Bayesian methods by highlighting a weakness of many nonparametric model extensions. The central issue is what a single aberrant observation (or a small fraction of unusual observations) can do to the entire analysis. In the setting of many nominally identical problems, often referred to as the empirical Bayes problem, a single "bad," outlying observation, say y_i, may not only destroy inference for the corresponding θ_i, but can also destroy inference for the $\theta_j, j \neq i$. The paragraphs below describe how this behavior can work for MDP model extensions.

In the case of the MDP models (3) and (5), we have described the difference in spirit between the two models. This difference in spirit often translates into (3) representing a generalization of a nonhierarchical model and (5) representing a generalization of a hierarchical model. To illustrate, consider a simple model for normal means, with known variance. The conjugate, nonhierarchical model leading to (3) has $G_0(\cdot)$ a normal distribution with mean, say μ and variance τ. The simple MDP extension of this model replaces G_0 with $G \sim DP(MG_0)$. We imagine that the first observation, y_1, is aberrant, and consider behavior of the posterior distribution as $y_1 \to \infty$. In this simple model, the limit of posterior distributions tends to a state where θ_1 is placed in a component of the mixture all by itself and $\theta_2, \ldots, \theta_n$ have a distribution that matches the posterior under model (3) for the $n-1$ component problem involving only $\theta_2, \ldots, \theta_n$ and $y_2, \ldots, y_n$.

Thus, the aberrant observation induces no difficulty for inference regarding $\theta_j, j \neq 1$. For predictive inference for θ_{n+1}, there will be a bit of mass (of size $1/(M+n)$) spread over extremely large values, corresponding to θ_1.

In contrast to this naturally robust behavior, the corresponding model (5) exhibits a definite lack of robustness. The conjugate, hierarchical model leading to (5) includes a normal hyperprior on the mean, μ. The MDP extension retains this normal prior on μ. With this additional stage in the hierarchy, as $y_i \to \infty$, the center of the distribution of μ, given the data, gets pulled toward ∞ while the variance is bounded above. In turn, the mass of the base measure near the bulk of the data becomes negligible, leading the posterior to assign probability tending to one to the event that $\theta_2 = \ldots = \theta_n$. Worse yet, the posterior distribution for this common value of θ moves toward ∞ as y_1 grows, never converging to any limiting distribution. Thus, the aberrant observation destroys inference for $\theta_2, \ldots, \theta_n$ as well as for θ_1.

One promising solution to the occasional lack of robustness of MDP extensions along the lines of (5) is to first estimate hyperparameters in the parametric version of the model with a method robust to aberrant observations, and to then use these estimated hyperparameters to determine the base measure, formally working with a model in the style of (3). A second approach is to use a robust form for the base measure, in extreme cases even proceeding to improper base measures to remedy the situation. Care must be taken with both proper and improper base measures to ensure satisfactory performance.

The arguments presented above have focused on the empirical Bayes problem, although there is nothing special about this setting. Similar arguments can be constructed in much more varied settings. In fact, the analysis of the Old Faithful geyser data appearing later in this chapter is in the tradition of regression modelling.

16.3 Survey of MDP models

A number of models in the recent literature fit into the framework of (5). Recent versions of these models, and new developments, include those that follow. When likelihoods do not depend on certain parameters, the corresponding subscripts have been omitted. Most of these applications include priors on ν which have been omitted. Using the notation of (1) and (4), for each application we point out the corresponding G_0 and parameter θ or θ_i, respectively. Depending on what we think is the more natural perspective, we write $p_{\theta,\nu}$ as in (1), or $p_{\theta_i,\nu}$ as in (4). We use $N(x; m, S)$ to indicate that the random variable x follows a normal distribution with mean and variance (m, S). Also, we use $\text{Bern}(x;\ p)$, $\text{Bin}(x;\ n, \theta)$, $W(x;\ \nu, A)$, $\text{Ga}(x;\ a, b)$, $\text{Exp}(x;\ \lambda)$, $U(x; a, b)$, $\text{Dir}(x; \lambda)$ and $\text{Be}(x;\ a, b)$ to denote a Bernoulli, binomial, Wishart, gamma, exponential, uniform, Dirichlet and beta distribution, respectively. Our notation ignores distinctions between random variables and their realizations.

(i) Nonparametric regression: Müller et al. (1996) use $\theta_i = (\mu_i, \Sigma_i)$, $p_{\theta_i}(y_i) = N(\mu_i, \Sigma_i)$ and $G_0(\mu, \Sigma) = N(\mu; a, B)\, W(\Sigma^{-1}; s, S)$.

(ii) Density estimation: West et al. (1994) have $\theta_i = (\mu_i, \Sigma_i)$, $p_{\theta_i}(y_i) = N(y_i; \mu_i, \Sigma_i)$ and $G_0(\mu, \Sigma) = N(\mu; a, B)\, W(\Sigma^{-1}; s, S)$.
Gasparini's (1996) model can be reformulated as an MDP model with $p_{\theta_i,\nu}(y_i) = \mathrm{U}(y; \theta_i - \nu, \theta_i + \nu)$ and $G_0(\theta)$ a discrete measure on $\{a, a + 2\nu, a + 4\nu, \dots\}$.

(iii) Estimation of a monotone density. Brunner (1995) has $p_{\theta_i}(y_i) = \mathrm{U}(y; 0, \theta_i)$, where $G_0(\theta)$ is an arbitrary distribution on the positive half-line. Brunner and Lo (1989) use a similar model for estimation of a symmetric, unimodal density.

(iv) Hierarchical modelling: Escobar and West (1995) have $\theta_i = (\mu_i, \sigma_i)$, $p_{\theta_i}(y_i) = N(y_i; \mu_i, \sigma_i)$, and $G_0(\mu, \sigma) = Ga(\sigma^{-2}; s/2, S/2)\, N(\mu; m, \tau\sigma^2)$.
MacEachern (1994) uses $p_{\theta_i}(y_i) = N(y_i; \theta_i, \sigma^2)$.
Liu (1996) proceeds from (1), the nonhierarchical model, and uses $p_\theta(y_i) = \mathrm{Bin}(n_i, \theta)$, where $G_0(\theta) = \mathrm{Be}(a, b)$.

(v) Fixed and random effects modelling: Bush and MacEachern (1996) have $p_{\theta_i,\lambda}(y_i) = N(y_i; \lambda' x_i + \theta_i, \sigma_i^2)$ where $G_0(\theta) = N(m, \sigma^2)$ and x_i is a vector of covariates for observation i. Malec and Müller (1999) use a similar random effects model in the context of small area estimation.

(vi) Contingency tables: Quintana (1998) has $\theta_i = (p_{i1}, \dots, p_{il})$ and $p_{\theta_i}(n_i) = \mathrm{Multin}(n_i, p_i)$, with $G_0(p) = \mathrm{Dir}(p; \lambda)$;

(vii) Longitudinal data models: Müller and Rosner (1997) and Kleinman and Ibrahim (1998) use for patient-specific random effects z_i $p_{\mu,\Sigma}(z_i) = N(z_i;\, \mu, \Sigma)$, with $G_0(\mu) = N(m, S)$.

(viii) Estimating possibly nonstandard link functions: Erkanli et al. (1993) use $p_\theta(z_i) = N(z_i; \mu, 1)$ and $G_0(\mu) = N(\mu; m, \tau^2)$ for a latent variable z_i defined by $y_i = 1$ if $z_i < 0$ and $y_i = 0$ if $z_i \geq 0$.

(ix) Binary regression: Basu and Mukhopadhyay (1998) use $p_{\theta_i}(y_i) = \mathrm{Bern}(\sqrt{\theta_i}\, x_i'\beta)$ and $G_0(\theta) = G(\nu/2, \nu/2)$.

(x) Censored data: Doss' (1994) model for survival data and one of the proposed models in Gelfand and Kuo (1991) for dose-response data can be rewritten as:

$$p_{\theta_i}(y_i) = \begin{cases} 0 \text{ with prob. } 1 \text{ if } \theta_i > x_i, \\ 1 \text{ with prob. } 1 \text{ if } \theta_i \leq x_i, \end{cases}$$

for those data values that are right-censored. Left- and interval-censored data values have similarly defined likelihoods. Uncensored observations are absorbed into the base measure. Doss uses $G_0(\theta) = \text{Exp}(\mu)$, while Gelfand and Kuo take $G_0(\theta) = N(\theta; \mu, \sigma^2)$.

(xi) Survival analysis with covariates: Kuo and Mallick (1997) use the accelerated failure-time model where $p_{\mu_i,\sigma_i,\beta}(\log T_i) = N(\log T_i;\ \mu_i - \beta' x_i,\ \sigma_i)$ for failure times T_i, and - in one example - $G_0(\mu, \sigma) = \text{Exp}(1)\ \delta_{0.1}(\sigma)$, where $\delta_x(\cdot)$ is a point mass at x. Alternatively they consider a similar DP mixture on $v_i = T_i\ \exp(x_i \beta)$ instead of $w_i = \log T_i + \beta' x_i$.

(xii) Generalized linear models: Mukhopadhyay and Gelfand (1997) use $p_{\theta,\beta}(y_i) = f(y_i | \eta = \theta + x_i' \beta)$, where $f(y|\eta)$ is a generalized linear model with linear predictor η.

(xiii) Errors in variables models: Müller and Roeder (1997) use for the joint distribution of the missing covariate x_i and observed proxy w_i $p_{\mu,\Sigma}(w_i, x_i) = N(w_i, x_i;\ \mu, \Sigma)$, and $G_0(\mu) = N(m, S)$.

Other related models are used in Lavine and Mockus (1995), Kuo and Smith (1992) and Newton et al. (1996). Numerous other authors are currently working with models that fit into this MDP framework.

16.4 Gibbs sampling in conjugate MDP models

We briefly review Markov chain Monte Carlo schemes currently applied to estimate MDP models. Estimation of the MDP model (5) can be efficiently implemented by a Gibbs sampling scheme if $p_{\theta,\nu}$ and G_0 are conjugate (cf. Escobar and West, 1995, MacEachern, 1994, West et al., 1994, Bush and MacEachern, 1996). In MacEachern and Müller (1998), we define a model augmentation and outline a Markov chain Monte Carlo implementation which allows the use of nonconjugate pairs $p_{\theta,\nu}$ and G_0. The focus is on discussing the conceptual framework.

The next two sections summarize the discussion in MacEachern and Müller (1998) which is relevant for a practical implementation of a Gibbs sampling algorithm. It has the added benefit of providing an explicit description of the "complete model" algorithm which was trimmed from the published version of that manuscript. Building on this general discussion, we give specific Gibbs sampler algorithms suitable for a practical implementation.

A key feature of the DP is the almost sure discreteness of the random measure G which gives positive probability to some of the θ_is being equal. When the base measure G_0 is continuous with probability 1, the θ_is are equal only due to the discreteness inherent in the Dirichlet process, and not to the discreteness of G_0. In this case, write $\{\theta_1^*, \dots, \theta_k^*\}$ for the set of $k \leq n$ distinct elements in $\{\theta_1, \dots, \theta_n\}$. Thus θ is partitioned into k sets. Call this partitioning a configuration, and let $s_i = j$ iff $\theta_i = \theta_j^*$ denote configuration indicators. Also let n_j be the number of s_i equal to j, that is, the size of the jth element of the partition (also called the jth cluster).

A Gibbs sampling scheme to estimate MDP models is described by the following conditional distributions.

(i) Resampling s_i given all other parameters: We marginalize over θ_i and sample s_i from

$$\Pr(s_i = j|\theta_{-i}, s_{-i}, \nu, y) \propto \begin{cases} n_j^- \; p_{\theta_j^*,\nu}(y_i) & j = 1, \ldots, k^- \\ M \; \int p_{\theta,\nu}(y_i) \; dG_0(\theta) & j = k^- + 1, \end{cases} \tag{6}$$

Here, θ_{-i} denotes the vector $(\theta_1, \ldots, \theta_{i-1}, \theta_{i+1}, \ldots, \theta_n)$, n_j^- denotes the size of the jth cluster with θ_i removed from consideration (that is, $n_{s_i}^- = n_{s_i} - 1$ while for other j, $n_j^- = n_j$), and k^- denotes the number of clusters with θ_i removed from consideration. If $n_{s_i}^- = 0$, we relabel the remaining clusters $j = 1, \ldots, k^- = k - 1$.

After sampling s_i, redefine k accordingly, that is, set $k = k^-$ if $s_i \leq k^-$, and $k = k^- + 1$ if $s_i = k^- + 1$.

(ii) Resampling θ_j^* is straightforward. The posterior $p(\theta_j^*|s, \nu, y)$ is the same as in the simple Bayes model (4) with i going over all indices with $s_i = j$:

$$y_i \sim p_{\theta_j^*,\nu}(y_i), \; \theta_j^* \sim G_0(\theta_j^*|\nu)$$

for all i such that $s_i = j$.

(iii) Resampling M: If one wishes to express uncertainty about the total mass parameter, it can be included in the parameter vector and resampled in the MCMC simulation. West (1992) shows that if M is given a $\text{Ga}(a, b)$ hyperprior, it can be resampled by introducing an additional latent variable x with $p(x|k, M) = \text{Be}(M + 1, n)$ and

$$p(M|x, k) = \pi \; \text{Ga}\left(a + k, b - \log x\right) + (1 - \pi) \; \text{Ga}\left(a + k - 1, b - \log x\right),$$

where $\pi/(1 - \pi) = (a + k - 1)/n(b - \log x)$. Alternatively, uncertainty about the mass parameter can be expressed and then eliminated from the sampling scheme through a preintegration, as described in MacEachern (1998).

(iv) Resampling ν given all other parameters: The portion of the model involving ν is a conventional parametric model. Hence, conditioning on all other parameters leaves a standard Bayes model. Often, this will be of conjugate form and a standard generation will suffice.

Only step (i) and, if included, step (iii) go beyond the MCMC for the original parametric model. Step (i) is a multinomial draw and step (iii) is a gamma and a beta random variate generation. Steps (ii) and (iv) might require complicated posterior simulation, depending on the application. These steps remain almost unchanged.

16.5 Novel algorithms for nonconjugate MDP models

The Gibbs sampler described in Section 16.4 is practical only if $p_{\theta,\nu}$ and $G_0(\theta|\nu)$ are conjugate in θ, allowing analytic evaluation of $q_0 = \Pr(s_i = k^- + 1| \ldots)$ in equation (6). In many applications, however, a nonconjugate setup is required. West et al. (1994) present an algorithm for nonconjugate MDP models using an approximate evaluation of q_0. In MacEachern and Müller (1998), we propose a general framework which allows nonconjugate pairs G_0 and $p_{\theta,\nu}$. The scheme is based on a model augmentation introducing latent variables, $\{\theta^*_{k+1}, \ldots, \theta^*_n\}$, for up to n possible cluster locations. At any time, $n - k$ of the clusters are empty, that is, $n_j = 0$ for these j. For a detailed definition and discussion we refer to MacEachern and Müller (1998). Here we build on the conceptual framework described there to formulate a practical implementation of a Gibbs sampling scheme for continuous base measures G_0.

Alternative approaches for MCMC in nonconjugate models are described in Neal (1998) and Green and Richardson (1998) and in Walker and Damien (1998) and MacEachern (1998) for one-dimensional distributions. Neal (1998) proposes alternative algorithms using Metropolis Hastings-type moves to propose new configuration indicators s_i. Similar to Neal (1998), Green and Richardson (1998) exploit the relationship of the DP mixture model with a Dirichlet/multinomial allocation model and propose an algorithm based on split/merge moves. Walker and Damien (1998) use the auxiliary variable technique introduced in Damien et al. (1999), essentially avoiding evaluation of the integral in (6) by introducing a uniform latent variable u with $p(u_i|\theta_i) \sim U[0, p_{\theta_i,\nu}(y_i)]$. MacEachern (1998) suggests the use of adaptive rejection techniques for the special case of log-concavity in the complete conditional posterior distribution for θ_i.

We define two alternative algorithms, based on the "no-gaps" model and the "complete model" defined in MacEachern and Müller (1998). Choice of the algorithm depends on the particular application. As a guideline, if k is typically much smaller than n, and n is large, then we recommend the "no-gaps" algorithm. In other nonconjugate situations, we recommend the complete model.

16.5.1 No-gaps algorithm

Application of the "no-gaps" model results in the following changes of the Gibbs sampler steps (i) through (v) described in Section 16.4:

(i') If $n_{s_i} > 1$, then resample s_i from

$$\Pr(s_i = j|\theta^*, s_{-i}, \nu, y) \propto \begin{cases} n_j^- \; p_{\theta_j^*,\nu}(y_i) & j = 1, \ldots, k^- \\ \frac{M}{k^-+1} \; p_{\theta^*_{k^-+1},\nu}(y_i) & j = k^- + 1. \end{cases} \tag{7}$$

Here, k^- and n_j^- denote the number of distinct values in $\{\theta_1, \ldots, \theta_{i-1}, \theta_{i+1}, \ldots, \theta_n\}$ and the number of observations in cluster j after removing observation i, respectively.

If $n_{s_i} = 1$, then with probability $(k-1)/k$, leave s_i unchanged. With probability $1/k$, resample s_i from (7).

If a cluster j with $n_j = 1$ is removed by resampling s_i, then switch the labels of clusters j and k and decrement k by 1. But keep the old value of θ_j^* recorded as θ_{k+1}^*.

After repeating step (i') for $i = 1, \ldots, n$ to resample all indicators s_i, marginalize over $\theta_{k+1}^*, \ldots, \theta_n^*$ by simply dropping them from the simulation. Steps (ii), (iii) and (iv) are executed conditioning on $\theta_1^*, \ldots, \theta_k^*$ only. Before returning to step (i') in the next iteration, augment the θ^* vector again by sampling $\theta_{k+1}^*, \ldots, \theta_n^*$ from $G_0(\theta_j^*|\nu)$. Of course, this could be done when and as θ_j^* is needed in step (i') only, that is, θ_j^*, $j = k+1, \ldots, n$ need not be actually generated and kept in memory until they are needed in step (i').

16.5.2 The complete model algorithm

The complete conditional posterior for resampling s_i is given by

$$\Pr(s_i = j|\theta^*, s_{-i}, \nu, y) \propto \begin{cases} n_j^- \; p_{\theta_j^*,\nu}(y_i) & j = 1, \ldots, k^- \\ \frac{M}{n-k^-} \; p_{\theta_j^*,\nu}(y_i) & j = k^- + 1, \ldots, n, \end{cases} \tag{8}$$

where k^- and n_j^- are defined as before.

Again, after step (i") update k and relabel the clusters such that all "empty" clusters (with $n_j = 0$) have higher indices than the "non-empty" clusters (with $n_j > 0$).

After completing step (i") for $i = 1, \ldots, n$, marginalize over $\theta_{k+1}^*, \ldots, \theta_n^*$ by dropping them from the simulation and execute steps (ii) through (iv) conditioning on $\theta_1^*, \ldots, \theta_k^*$ only. Before returning to step (i"), augment the θ^* vector again by generating $\theta_j^* \sim G_0(\theta_j^*|\nu)$, $j = k+1, \ldots, n$.

16.6 Non-identifiability of the algorithms

The primitive notion of identifiability is easily described. An identifiable model has the property that, for every point in the parameter space, a different distribution is implied for the data that are to be collected in an experiment. Unfortunately, this notion becomes a bit fuzzy in the hierarchical Bayesian model, as just what is considered a parameter is open to several interpretations. Since several different parameterizations of a model, some of which may even be nested in others, can be considered identifiable, it is difficult to write a concise discussion of identifiability and its impact on fitting nonparametric Bayesian models. Nevertheless, the issue is important enough for fitting these models that we provide a brief discussion of the issue here.

We stress that identifiability is often important for interpretation of an analysis and that the question can appear in subtle forms in nonparametric Bayesian analysis. Newton et al. (1996) create a model so that a set of parameters that are useful for interpretation is identifiable and also explicitly appears in their model. Green and Richardson (1997) provide a focused treatment of identifiability in the context of finite mixture models of varying dimension. In general, the interpretation of a model is often tied to a particular parameterization at an intuitive level if not in mathematical terms. As an example, the two distinct routes to creation of the MDP model lead to different identifiable parameterizations of the model. While the eventual use of the model may follow from either generalization, for computational purposes, we need only ensure that the strategy used to fit the models allows us to make inference under either parameterization.

First, an example, to illustrate the variety of ways in which one can term models as identifiable or not in the context of the MDP model. The parameter for model (3) or (5) can be considered either G, or $(\theta_1, \ldots, \theta_p)$, or $(G, \theta_1, \ldots, \theta_p)$. Under relatively weak conditions on the likelihood, the first parameterization leads to a model that is identifiable in the sense that the joint distribution on y differs for each differing G. This parameterization is most naturally thought of as the generalization of model (3). The second model is identifiable in that the distribution for y differs for each differing vector θ. This parameterization is most naturally thought of as the generalization of model (5). The third model only becomes identifiable when one steps outside of the current experiment, as when performing a predictive analysis. The joint distribution of the data collected from the current experiment and a future observation, say y_{p+1} which depends on θ_{p+1} with $\theta_{p+1} \sim G$, depends both on G and on $\theta_1, \ldots, \theta_p$.

In terms of computation, the algorithms we have just described facilitate inference under any of the three parameterizations given above (see the next section for the treatment of predictive inference relevant to the third parameterization). At any stage in the Gibbs sampler, the stored parameters - the vector s and the vector θ^* - enable us to reconstruct the entire vector θ. Consequently, inference about the individual θ_i can be performed on the basis of the standard MCMC formulas. This handles inference for the second parameterization given above.

Inference for the first parameterization follows from the distribution of G given θ. Recalling the early result from Ferguson (1973), we know that the distribution of $G|\theta$ follows a Dirichlet process with parameter $\alpha + \sum_{i=1}^{p} \delta_{\theta_i}$. Subsequent inference about functionals against G can be made with any of the many varied tricks that have been described in the literature. The two main approaches for functionals that do not have tidy, closed-form expressions are either to pin down the random G at a collection of sites (the joint distribution of $G(x_1), \ldots, G(x_n)$ for $x_1 < \ldots < x_n$ follows from a Dirichlet distribution on the increments between successive values of the distribution function) or to approximate the countably infinite discrete distribution by a finite discrete distribution, perhaps of arbitrary size. The distribution on the finite approximation follows from Sethuraman's (1994) representation of the Dirichlet process and the rule for determining the (possibly random) number of components in the finite mixture. See, in particular, Tardella and Muliere (1998) for the ϵ-Dirichlet process and Gelfand and Kottas (1999) for the first sort of approximation. Guglielmi (1998) provides a means for calculating what are effectively exact values for linear functionals of Dirichlet processes.

The computational algorithms described in the preceding section rely on an additional non-identifiability in terms of the model that is written out for simulation. Instead of describing the value of the parameters $\theta_1, \ldots, \theta_p$ at any stage of the algorithm directly in terms of the θs, a latent structure is introduced. The vector (s, θ^*) contains all the information needed to reconstruct the vector θ through the relation $\theta_i = \theta^*_{s_i}$. There will be many vectors (s, θ^*) that result in the same value for the vector θ. We term these models non-identifiable because for any inference made from the posterior distribution for G and θ, the inference will not depend on the particular (s, θ^*) which produced this θ, G pair. The point that we wish to emphasize is that the model devised for computational purposes provides a finer scale of latent structure than does a model (3) or (5) elaboration.

The no-gaps algorithm and the complete model algorithm result from particular models for the latent structure. Each grouping of the θ_i into clusters corresponds to several vectors s. We have deliberately created non-identifiable models in order to improve the mixing/convergence of the Markov chain used to fit the models. In particular, the models are created by first writing an identifiable version of the model that leads to a one-to-one relationship between s and the grouping of the θ_i. This identifiable version of the model is then symmetrized by creating many labellings that correspond to that particular grouping of the θ_i and by apportioning the probability assigned to that grouping to each of the possible labellings.

For the no-gaps model, symmetrization proceeds by first labelling the, say k, groups $1, \ldots, k$. Next, all permutations of the labels $1, \ldots, k$ are considered for the group names. Each such permutation receives $1/k!$ of the probability assigned to that particular grouping of the θ_i.

For the complete model, symmetrization proceeds by first labelling the k groups $1, \ldots, k$ in an identifiable fashion. Next, all subsets of k distinct labels chosen from the integers $1, \ldots, n$ are considered as labellings of the groups. Each of the $n!/(n-k)!$ labellings receives an equal share of the probability assigned to that particular grouping of the θ_i.

The motivation behind the introduction of non-identifiable models for simulation as well as the idea behind symmetrization can be found in MacEachern (1996, 1998). West (1997) and Huerta and West (1999) use the technique to improve simulation in related models. For theory on the improvement that non-identifiable models can bring to MCMC simulation, see Meng and van Dyk (1999).

16.7 The predictive distributions

The posterior feature of greatest interest is often a predictive distribution. In the case of density estimation, the predictive distribution for a future observation is of direct interest. In the basic MDP model, the posterior predictive distribution is most easily found by returning from the *no-gaps* or *complete* model to the parameterization in terms of θ. Then the predictive distribution is given by $p(y_{n+1}|y) = \int\int p(y_{n+1}|\theta_{n+1})dp(\theta_{n+1}|\theta, y)dp(\theta|y)$. The inner integral reduces to an integral of $p(y_{n+1}|\theta_{n+1})$ against $(\sum n_j \delta_{\theta_j^*} + MG_0)/(M+n)$. The term involving G_0 may be evaluated as $M\tilde{\theta}$, where $\tilde{\theta}$ represents a new draw from G_0.

To obtain an estimate of the predictive distribution as the algorithm proceeds, we use an average over iterates of the resulting Markov chain. After each complete cycle of the algorithm, just after stage (ii), one has the estimate $1/T\sum_{t=1}^T p(y_{n+1}|\tilde{\theta}^t, \theta^t)$ when evaluation of the conditional distributions are feasible. Here θ^t refers to the imputed parameter vector θ after t iterations. When this evaluation is not feasible, after each iteration a value y_{n+1} can be generated, with the resulting estimator $1/T\sum_{t=1}^T y_{n+1}$.

16.8 Example

Azzalini and Bowman (1990) analyzed a data set concerning eruptions of the Old Faithful geyser in Yellowstone National Park in Wyoming. The data set records eruption durations and intervals between subsequent eruptions, collected continuously from August 1 until August 15, 1985. The data is available as data set `faithful` under R (Gentleman and Ihaka, 1997). Figure 1 (white dots) plots the data. Consider the inference problem of predicting duration (y) based on the waiting time since the previous eruption (x). A linear regression model (with improper priors) gives the posterior mean line shown Figure 1 (dotted straight line).

A look at the data already shows that the linear regression model is likely to be inappropriate. There are two clearly distinguishable clusters in the data. While a linear regression provides a good fit for each cluster separately, these two local regression lines are different and cannot be combined into one global linear regression model.

To investigate a more robust regression model we fit a DP mixture model

$$g(x,y) = \underbrace{\int N(\mu, V)\, dG(\mu, V)}_{\sum_h w_h\, N(\mu_h, V_h)}, \quad G \sim DP(M\, G_0), \tag{9}$$

with base measure $G_0(\mu, V^{-1}) = G_0(\mu)\, G_0(V^{-1})$, where $G_0(\mu)$ is a uniform distribution over the rectangle $[1, 6] \times [40, 100]$ and $G_0(V^{-1}) = W(s, S)$ is a Wishart distribution with a conjugate hyperprior on S. We include a gamma hyperprior $M \sim G(1, 1)$ for the total mass parameter. Except for the nonconjugate base measure, the model is as in Müller et al. (1996) (see item (i) above).

Implied in g is a regression curve $f(x) = E(y|x)$, which takes the form of a locally weighted mixture of linear regressions with each term $N(\mu_h, V_h)$ in the mixture defining a straight line which dominates for x close to μ_h. A (global) linear regression model is included as a special case when the mixture of normals reduces to a single bivariate normal, that is, when G is dominated by just one point mass. Figure 1 summarizes results from

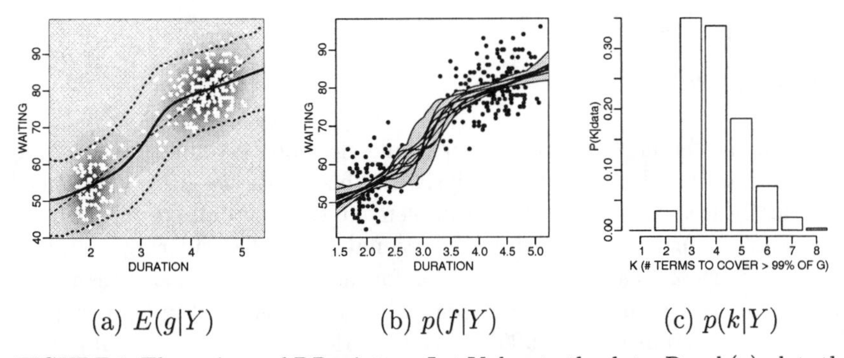

(a) $E(g|Y)$ (b) $p(f|Y)$ (c) $p(k|Y)$

FIGURE 1. The estimated DP mixture. Let Y denote the data. Panel (a) plots the estimated mixture of normal model $E(g|Y)$ (grey shades), the estimated regression curve $E(f|Y)$ (bold curve), 95% pointwise HPD (highest posterior density) intervals (dashed curves), and the posterior mean curve in a linear regression (dotted straight line). The white dots show the data points. Panel (b) plots some elements of $p(f|Y)$. The curves show 10 draws from the posterior distribution $p(f|Y)$ on the unknown regression curve f. The grey-shaded polygon shows the pointwise range over 100 draws from $p(f|Y)$. Panel (c) shows (approximately) $p(k|Y)$, where k is the smallest number of point masses in the random mixing measure G which cover at least 99% of the total mass. The special case $k = 1$ corresponds (up to the remaining 1% mass) to a single linear regression.

the analysis. Panel (a) compares the posterior estimated regression curve with a linear regression (posterior mean using improper priors). Not surprisingly, the two curves coincide around the centers of the two clusters and show the same overall trend, but differ in the interpolation between, and extrapolation beyond, the two data clusters. Panel (b) shows the range of a posteriori likely regression curves, which barely includes the linear regression line. In many problems the reported uncertainties are an important part of the inference. Note how the mixture model (9) reports increased uncertainty for the regression curve between the two data clusters. This is in contrast with the linear regression model, which would report least uncertainty around the overall mean, that is, exactly the opposite in this case. Panel (c) shows more inference related to how plausible a simple linear regression would be a posteriori. Let k denote the smallest number of terms in the normal mixture (9) which cover at least 99% of the total mass. Since the DP generates discrete measures with infinitely many point masses, there are always infinitely many terms in the mixture. But typically only few have appreciable weight a posteriori. Including only the k largest point masses accounts for 99% of the entire distribution. The figure shows the (approximate) posterior distribution of k. The near-zero posterior probability for $k = 1$ is to be expected given the obvious two clusters in the data.

16.9 Discussion

Inference in the encompassing nonparametric model provides an alternative to traditional sensitivity analysis. In this paper we have introduced practically feasible approaches to implement such model extensions with - at least in principle - minimal additional effort. Considering nonparametric extensions can also be used for formal robustness measures in the form of model diagnostics. Such approaches are considered in Carota et al. (1996) and Florens et al. (1996) using DP priors, in Berger and Guglielmi (1999) using Polya tree priors, and in Verdinelli and Wasserman (1998) using Gaussian processes.

The beauty of the nonparametric Bayesian sensitivity analysis/model elaboration is that it plays this dual role. For a sensitivity analysis technique, one can monitor the range of posterior inferences as the prior distribution is varied over a class of nonparametric priors; for a more general class of models, one can monitor summaries of the fit of the model as the prior varies over a class. In any realistic setting, the sensitivity analysis produces a range of inferences; often, the general class of models exhibits a substantially better fit than does the parametric model. Qin (1998) provides a prime example of this improvement in fit where, in a fairly complex model, the posterior standard deviations for several parameters are *smaller* under the nonparametric elaboration than under the parametric model. Such examples illustrate the benefit of a technique that plays both exploratory and confirmatory roles.

References

ANGERS, J.F. and DELAMPADY, M. (1992). Hierarchical Bayesian curve fitting and smoothing. *Canadian Journal of Statistics*, **20**, 35–49.

ANTONIAK, C.E. (1974). Mixtures of Dirichlet processes with applications to non-parametric problems. *The Annals of Statistics*, **2**, 1152–1174.

AZZALINI, A. and BOWMAN, A.W. (1990). A look at some data on the Old Faithful geyser. *Applied Statistics*, **39**, 357–365.

BASU, S. (2000). Bayesian robustness and Bayesian nonparametrics. In *Robust Bayesian Analysis* (D. Ríos Insua and F. Ruggeri, eds.). New York: Springer-Verlag.

BASU, S. and MUKHOPADHYAY, S. (1998). Binary response regression with normal scale mixture links. To appear in *Generalized Linear Models: A Bayesian Perspective* (D.K. Dey et al., eds.).

BERGER, J.O. and GUGLIELMI, A. (1999). Bayesian testing of a parametric model versus nonparametric alternatives. *Technical Report*, **99-04**, Institute of Statistics and Decision Sciences, Duke University.

BRUNNER, L.J. (1995). Using the Gibbs sampler to simulate from the Bayes estimate of a decreasing density. *Communications in Statistics A: Theory and Methods*, **24**, 215–226.

BRUNNER, L.J. and LO, A. (1989). Bayes methods for a symmetric unimodal density and its mode. *The Annals of Statistics*, **17**, 1550–1566.

BUSH, C.A. and MACEACHERN, S.N. (1996). A semiparametric Bayesian model for randomised block designs. *Biometrika*, **83**, 275–285.

CAROTA, C., PARMIGIANI, G. and POLSON, N.G. (1996). Diagnostic measures for model criticism. *Journal of the American Statistical Association*, **91**, 753–762.

DAMIEN, P., WAKEFIELD, J.C. and WALKER, S.G. (1999). Gibbs sampling for Bayesian non-conjugate and hierarchical models using auxiliary variables. *Journal of the Royal Statistical Society, Series B*, **61**, 331–344.

DOSS, H. (1994). Bayesian nonparametric estimation for incomplete data via successive substitution sampling. *The Annals of Statistics*, **22**, 1763–1786.

DYKSTRA, R.L. and LAUD, P. (1981). A Bayesian nonparametric approach to reliability. *The Annals of Statistics*, **9**, 356–367.

ERKANLI, A., STANGL, D.K. and MÜLLER, P. (1993). A Bayesian analysis of ordinal data using mixtures, *ASA Proceedings of the Section on Bayesian Statistical Science*, 51–56.

ESCOBAR, M. (1988). Estimating the means of several normal populations by estimating the distribution of the means. *Ph.D. thesis*, Yale University.

ESCOBAR, M.D. (1994). "Estimating normal means with a Dirichlet process prior," *Journal of the American Statistical Association*, **89**, 268–277.

ESCOBAR, M.D. and WEST, M. (1995). Bayesian density estimation and inference using mixtures. *Journal of the American Statistical Association*, **90**, 577–588.

ESCOBAR, M.D. and WEST, M. (1998). Computing nonparametric hierarchical models. In *Practical Nonparametric and Semiparametric Bayesian Statistics* (D. Dey, P. Müller and D. Sinha, eds.), 1–22. New York: Springer-Verlag.

FERGUSON, T.S. (1973). A Bayesian analysis of some nonparametric problems. *The Annals of Statistics*, **1**, 209–230.

FLORENS, J.P., RICHARD, J.F. and ROLIN, J.M. (1996). Bayesian encompassing specifications tests of a parametric model against a nonparametric alternative. *Technical Report*, Université Catholique de Louvain.

GASPARINI, M. (1996). Bayesian density estimation via Dirichlet density processes. *Journal of Nonparametric Statistics*, **6**, 355–366.

GELFAND, A. and KUO, L. (1991). Nonparametric Bayesian bioassay including ordered polytomous response. *Biometrika*, **78**, 657–666.

GELFAND, A.E. and Kottas, A. (1999). Full Bayesian inference for the nonparametric analysis of single and multiple sample problems. *Technical Report*, University of Connecticut.

GENTLEMAN, R. and IHAKA, R. (1997). *The R Manual*, University of Auckland.

GHOSHAL, S., GHOSH, J.K. and RAMAMOORTHI, R.V. (1999a). Consistency issues in Bayesian nonparametrics. In *Asymptotics, Nonparametrics and Time Series* (S. Ghosh, ed.), 639–667. Marcel Dekker.

GHOSHAL, S., GHOSH, J.K. and RAMAMOORTHI, R.V. (1999b). Consistent semiparametric Bayesian inference about a location parameter. To appear in *Journal of Statistical Planningand Inference.*

GREEN, P. and RICHARDSON, S. (1997). On Bayesian analysis of mixtures with an unknown number of components (Discussion: 758–792). *Journal of the Royal Statistical Society, Series B*, **59**, 731–758.

GREEN, P. and RICHARDSON, S. (1998). Modelling heterogeneity with and without the Dirichlet process. *Technical Report*, University of Bristol.

GUGLIELMI, A. (1998). Results with distribution functions of means of a Dirichlet process. *Unione Matematica Italiana*, **8**, **1A** (suppl.), 125–128.

HJORT, N.L. (1990). Nonparametric Bayes estimators based on beta processes in models for life history data. *The Annals of Statistics*, **18**, 1259–1294.

HUERTA, G. and WEST, M. (1999). Priors and component structures in autoregressive time series models. *Journal of the Royal Statistical Society, Series B*, **61**, 881–899.

KLEINMAN, K.P. and IBRAHIM, J.G. (1998). A semi-parametric Bayesian approach to the random effects model. *Biometrics*, **54**, 921–938.

KUO, L. (1983). Bayesian bioassay design. *The Annals of Statistics*, **11**, 886–895.

KUO, L. and MALLICK, B. (1997). Bayesian semiparametric inference for the accelerated failure time model. *Canadian Journal of Statistics*, **25**, 457–472.

KUO, L. and SMITH, A.F.M. (1992). Bayesian computations in survival models via the Gibbs sampler (discussion: 22–24). In *Survival Analysis: State of the Art* (J.P. Klein and P.K. Goel, eds.), 11–22 Dordrecht: Kluwer Academic Publishers.

LAVINE, M. (1992). Some aspects of Polya tree distributions for statistical modelling. *The Annals of Statistics*, **20**, 1203–1221.

LAVINE, M. (1994). More aspects of Polya tree distributions for statistical modelling. *The Annals of Statistics*, **22**, 1161–1176.

LAVINE, M. and MOCKUS, A. (1995). A nonparametric Bayes method for isotonic regression. *Journal of Statistical Planning and Inference*, **46**, 235–248.

LIU, J. (1996). Nonparametric hierarchical Bayes via sequential imputations. *The Annals of Statistics*, **24**, 911–930.

MACEACHERN, S.N. (1988). Sequential Bayesian bioassay design. Unpublished Ph.D. Dissertation. University of Minnesota.

MACEACHERN, S.N. (1994). Estimating normal means with a conjugate style Dirichlet process prior. *Communications in Statistics B: Simulation and Computation*, **23**, 727–741.

MACEACHERN, S.N. (1996). Identifiability and Markov chain Monte Carlo methods. Unpublished manuscript.

MACEACHERN, S.N. (1998). Computations for MDP models. In *Practical Nonparametric and Semiparametric Bayesian Statistics* (D. Dey, P. Müller and D. Sinha, eds.), 23–43. New York: Springer-Verlag.

MACEACHERN, S.N. and MÜLLER, P. (1998). Estimating mixture of Dirichlet process models. *Journal of Computational and Graphical Statistics*, **7**, 223–238.

MALEC, D. and MÜLLER, P. (1999). A Bayesian semi-parametric model for small area estimation. *Technical Report*, Institute of Statistics and Decision Sciences, Duke University, Durham, NC.

MENG, X.L. and VAN DYK, D.A. (1999). Seeking efficient data augmentation schemes via conditional and marginal augmentation. *Biometrika*, **86**, 301–320.

MUKHOPADHYAY, S. and GELFAND, A.E. (1997). Dirichlet process mixed generalized linear models. *Journal of the American Statistical Association*, **92**, 633–639.

MÜLLER, P., ERKANLI, A. and WEST, M. (1996). Bayesian curve fitting using multivariate normal mixtures. *Biometrika*, **83**, 67–79.

MÜLLER, P. and ROEDER, K. (1997). A Bayesian semiparametric model for case-control studies with errors in variables. *Biometrika*, **84**, 523–537.

MÜLLER, P. and ROSNER, G. (1997). A Bayesian population model with hierarchical mixture priors applied to blood count data. *Journal of the American Statistical Association*, **92**, 1279–1292.

NEAL, R.M. (1998). Markov chain sampling methods for Dirichlet process mixture models. *Journal of Computational and Graphical Statistics*, to appear.

NEWTON, M.A., CZADO, C. and CHAPPELL, R. (1996). Semiparametric Bayesian inference for binary regression. *Journal of the American Statistical Association*, **91**, 142–153.

O'HAGAN, A. (1992). Some Bayesian numerical analysis. In *Bayesian Statistics 4* (J.M. Bernardo, J.O. Berger, A.P. Dawid, and A.F.M. Smith, eds.), 355–363. Oxford: Clarendon Press.

PETRONE, S. (1999a). Bayesian density estimation using Bernstein polynomials. *Canadian Journal of Statistics*, **27**, 105–126.

PETRONE, S. (1999b). Random Bernstein polynomials. *Scandinavian Journal of Statistics*, **26**, 373–393.

QIN, L. (1998). Nonparametric Bayesian models for item response data. *Ph.D. dissertation*, The Ohio State University.

QUINTANA, F. (1998). Nonparametric Bayesian analysis for assessing homogeneity in $k \times l$ contingency tables with fixed right margin totals. *Journal of the American Statistical Association*, **93**, 1140–1149.

SETHURAMAN, J. (1994). A constructive definition of Dirichlet priors. *Statistica Sinica*, **4**, 639–650.

SUSARLA, J. and VAN RYZIN, J. (1976). Nonparametric Bayesian estimation of survival curves from incomplete observations. *Journal of the American Statistical Association*, **71**, 897–902.

TARDELLA, L. and MULIERE, P. (1998). Approximating distributions of random functionals of Ferguson-Dirichlet priors. *Canadian Journal of Statistics*, **26**, 283–297.

VERDINELLI, I. and WASSERMAN, L. (1998). Bayesian goodness of fit testing using infinite dimensional exponential families. *The Annals of Statistics*, **20**, 1203–1221.

WALKER, S. and DAMIEN, P. (1998). Sampling methods for Bayesian nonparametric inference involving stochastic processes. In *Practical Nonparametric and Semiparametric Bayesian Statistics* (D. Dey, P. Müller and D. Sinha, eds.), 243–254. New York: Springer-Verlag.

WALKER, S. and MULIERE, P. (1997). Beta-Stacy processes and a generalization of the Polya urn scheme. *The Annals of Statistics*, **25**, 1762–1780.

WALKER, S.G., DAMIEN, P., LAUD, P.W. and SMITH, A.F.M. (1999). Bayesian nonparametric inference for distributions and related functions (discussion: 510–527). *Journal of the Royal Statistical Society, Series B*, **61**, 485–509.

WASSERMAN, L. (1998). Asymptotic properties of nonparametric Bayesian procedures. In *Practical Nonparametric and Semiparametric Bayesian Statistics* (D. Dey, P. Müller and D. Sinha, eds.), 293–304. New York: Springer-Verlag.

WEST, M. (1992). Hyperparameter estimation in Dirichlet process mixture models. *Technical Report*, **92-A03**, Institute of Statistics and Decision Sciences, Duke University.

WEST, M. (1997). Hierarchical mixture models in neurological transmission analysis. *Journal of the American Statistical Association*, **92**, 587–606.

WEST, M., MÜLLER, P. and ESCOBAR, M. (1994). Hierarchical priors and mixture models, with application in regression and density estimation. In *Aspects of Uncertainty: A Tribute to D.V. Lindley* (A.F.M. Smith and P. Freeman, eds.), 363–386. New York: Wiley.

17

Sensitivity Analysis in IctNeo

Concha Bielza, Sixto Ríos-Insua, Manuel Gómez and Juan A. Fernández del Pozo

ABSTRACT IctNeo is a complex decision support system to manage neonatal jaundice, the situation in which bilirubin accumulates when the liver does not excrete it at a normal rate. This system finds a maximum expected utility treatment strategy based on an influence diagram (ID). Due to the computational intractability of such a large ID, IctNeo incorporates some procedures to the standard evaluation algorithm.

In this chapter we show how IctNeo conducts sensitivity analysis (SA). We start from the situation in which the doctor has provided the most appropriate value for each parameter (both probabilities and utilities), but has imprecision leading to upper and lower values of the parameter. We first deduce a relevant set of parameters on which to perform SA based on one-way SA tornado diagrams and some interviews with the doctors. This set basically includes probabilities of the related pathologies, certain injuries and costs, and some weights of the multiplicative utility function. Then, we study more carefully joint sensitivity of the parameters. We use a sensitivity indicator based upon the expected value of perfect information that considers simultaneously changes in the probability of a decision and in the magnitude of the model optimal value (Felli and Hazen, 1998). This measure is approximated via Monte Carlo simulation.

Key words: Decision analysis, influence diagram, imprecision, value of perfect information, neonatal jaundice.

17.1 IctNeo system

17.1.1 Neonatal jaundice

Neonatal jaundice is a common medical problem which arises in a healthy newborn because of the breakdown of excess red blood cells in his or her system. It produces a substance called bilirubin, providing the infant with a yellowish cast in the skin and eyes. Most babies have higher bilirubin levels than adults because they have extra oxygen-carrying red blood cells, and their young livers cannot metabolise the excess bilirubin.

More than 50% of newborns are visibly jaundiced during the first week of their life, therefore, it is important to differentiate between pathological and standard variations. Pathological jaundice may cause potentially serious central nervous system damages (including irreversible brain damage), related to the development of kernicterus (bilirubin-induced encephalopathy).

A low or moderate bilirubin level is not harmful, although the baby needs to be closely watched. When bilirubin levels increase too much, all the blood needs to be replaced by means of a risky procedure called *exchange transfusion*. The introduction of *phototherapy* in jaundice management, exposing the infants to special lights that break down excess bilirubin, reduced the number of blood exchanges treating lower bilirubin cases. Current recommendations try to balance out the risks of undertreatment and overtreatment (Newman and Maisels, 1992). However, the current protocol does not clearly delimit when it is best to start each treatment and which treatment to administer.

As a consequence of this difficulty among others, the Neonatology Service of Gregorio Marañón Hospital in Madrid suggested the development of a decision support system *IctNeo* to provide the doctors with an automated problem-solving tool as an aid for improving jaundice management. The problem is represented and solved by means of an influence diagram (Shachter, 1986).

17.1.2 Construction of IctNeo

IctNeo is implemented in C++ under Windows 95 and is driven by menus. A user-friendly interface allows for data entry of a patient already treated by the doctor. Then, we can compare the system recommendations and the doctor decisions in order to draw conclusions.

The development of the system has been very complex and time consuming. The structure of the diagram changed as the doctors and we understood better the model and its scope. These changes are a way of SA as well as an art, especially in complex problems; see Mahoney and Laskey (1996). The resultant model conditions in a decisive manner the remaining process and the difficulties we will face. For example, one of them was the definition of the decisions because of the existence of a number of constraints placed by doctors on the chain of treatment decisions (e.g., not to perform more than two exchange transfusions per full treatment, etc.); see Ríos-Insua et al. (1998) for the initial conception.

Then, probabilities and utilities are elicited. For probabilities, the main obstacles were the unavailability of both data bases and statistical studies for many uncertainties, distributions to be assigned as functions of too many parameters, and not much availability of the experts due to overwork. Subjective encoding, generalised noisy OR-gates as prototypes of uncertainty aggregation based on a causal model, and patience were basically the tools employed. For utilities, a multi-attribute utility function was assigned in a standard manner (Keeney and Raiffa, 1976) except to allow the DM for imprecise answers as a way of SA; see Gómez et al. (1999) for a detailed explanation.

The evaluation of the ID to obtain the optimal policy was even harder. The standard algorithm (Shachter, 1986), entailed an intractable ID due to the size of the set of nodes and arcs. IctNeo required 10^{16} storage positions to record all probabilities and expected utilities, thereby exceeding the capacity of every personal computer. We provided various procedures to alleviate this computational burden, proposing some heuristics to find good deletion sequences based on a genetic algorithm (Gómez and Bielza, 1999) and a graphical criterion of *d*-separation to identify and remove redundant arcs (Fagiuoli and Zaffalon, 1998).

Other ideas included postponing the computation of expected utilities at chance nodes until it is necessary and incorporating in the grammar script the knowledge about the decisions constraints, thus avoiding calculi of many expected utilities. Finally, should the problem be unmanageable even in this case, we could take advantage of the evidence propagation operation (Ezawa, 1998).

Bielza et al. (2000) explain all the issues that have been set out in this section, showing the final ID with 50 chance nodes, 5 decision nodes, 144 connective arcs, and 112 no-forgetting arcs.

17.1.3 The need for SA in IctNeo

Once the prototype IctNeo has been constructed, it is convenient to check the robustness to many elements embedded in it. Important questions are: (1) the influence and importance of each variable and its domain; (2) the suitability of probability and utility assignments; (3) the suitability of the time representation selected; (4) whether the hypothesis of OR-gates is held; (5) whether the hypothesis of the functional form of the utility function is held (see, e.g. Butler et al., 1997); (6) whether the system is biased in favour of certain objectives, and so on. SA facilitates answers to many of those questions, showing the extent to which the model represents the knowledge and where we must intervene to correct inaccurate features.

As is usually done in SA, instead of relying on point estimates for input parameters, we now recognise their possible imprecision and influence on the output of the model and study the relationship between changes in their values and changes induced in model output. We will label the model as sensitive to an input parameter whenever it shows large fluctuations in output in response to relatively small perturbations in the value of that parameter; see Kadane et al. (2000) for a conceptual framework.

Since we are interested in identifying a treatment strategy, we must distinguish between value sensitivity and decision sensitivity. Section 2 presents some procedures focused on value sensitivity, referred to as a change in the magnitude of the maximum expected utility of the problem in response to changes in the input parameters. It also reviews other methods that are not value-range methods. Section 3 implements a better sensitivity measure that simultaneously takes into account the changes both in the optimal value and in the preferred alternative proposed by the model. We end with some conclusions.

17.2 Basic sensitivity analyses

17.2.1 SA methods in influence diagrams

Although graphical methods for decision analysis have experimented with huge developments during the last decade (see, e.g., Bielza and Shenoy, 1999), SA tools for these more sophisticated models have not grown up accordingly. Our intention is to conduct SA not only with respect to the probabilities as many methods propose, but also with respect to the utilities.

First, there is a kind of *qualitative* SA for every decision-making problem. It refers to the fact of whether we are solving the right problem. In medical decision-making for instance, as Clemen (1996) points out, it can be the case of treating a symptom instead of a cause. The problem identification is essential, and for that we perform SA with respect to the ID structure. It basically consists of careful thought, introspection and many dialogues with the experts. Some authors try to formalise the idea by computing the so-called expected value of decision-model refinement. This refinement can be conceptual (variable domains), structural and even quantitative (probabilities and utilities); see Poh and Horvitz (1993). Other authors, like Horsch and Poole (1996), progressively change the ID structure by taking into account the computational costs to find the optimal policy and provide an iterative refinement of (suboptimal) policies for IDs.

As far as *quantitative* SA is concerned, we often find the standard methods applied in simple IDs, many from the robust Bayesian literature. We briefly review why these methods become unwieldy for more realistic problems.

Threshold proximity measures use decision maker beliefs of a threshold crossing as a proxy for decision sensitivity. For two-way SA, the parameter space formed from two parameters p, q is partitioned into regions, each one representing pairs of values (p, q) for which a particular alternative is optimal. Boundaries –thresholds– shared by two regions designate indifference between adjacent alternatives. The proximity of the baseline value (p_0, q_0) to neighbouring thresholds informs how sensitive the optimal alternative for reasonable perturbations in p_0 and q_0 is. However, it is difficult to construct and interpret graphical displays of these ideas for SAs involving three or more parameters. In real problems like ours, studying the sensitivity to more than two parameters will be a common practice.

One possibility is to base the likelihood of a threshold crossing on the calculation of the *distance* between the baseline value and the nearest threshold. But there are more problems related to finding the appropriate distance metric, which is the best way of jointly defining parameters, and so on (see, e.g., Ríos Insua and French, 1991).

Another possibility is to conduct *probabilistic* SA. Once the DM has assigned a probability distribution to each uncertain parameter, the sensitivity will be based on the probability of a threshold crossing rather than the distance to the threshold. The multiparametric case can be handled by using joint distributions over the parameters of interest. The drawback is that mathematical calculations can be quite cumbersome, although Monte Carlo methods can help to alleviate this complexity somewhat (see, e.g., Doubilet et al., 1985). The results allow us to observe the percentage of the time of each alternative is optimal and the percentage of the time each alternative has an expected value significantly greater (say, with differences in payoffs of 0.004) than those of the other strategies.

From a computational point of view, many SA measures reviewed in this book have not been conceived for complicated problems. Other approaches provide a *first* step towards a framework for SA in IDs to accommodate more realistic problems. Identifying nondominated alternatives can be viewed as a kind of SA to be applied early in an analysis to discard bad strategies. For example, Bielza et al. (1996) consider IDs in which probabilities and utilities have been only partially assessed, and an algorithm computes nondominated alternatives in a Pareto sense based on a parametric model with foundations in Ríos Insua (1992). However, the nondominated set gets too big while moving backwards in the diagram. Although some additional criteria are introduced to reduce that set, we prefer here not to apply this algorithm and use a simulation-based one like Felli and Hazen's (1998). This SA method is based on the expected value of perfect information (EVPI) and will be specified in the next section.

17.2.2 Difficulties of IctNeo influence diagram

We have already described why a detailed SA is difficult in big IDs with a lot of parameters to be examined. In fact, most papers in the literature deal with rather simple problems, with few nodes (especially few decision nodes, leading to almost nonsequential models) and not very complicated utility function structures. However, the jaundice problem presents the following difficulties to conducting usual SAs:

- Its big size implies too many parameter candidates to be examined.
- Probabilistic dependencies in the ID imply further parameters to be examined. For example, if chance node A is conditioned to nodes B, C and D, all of them being dichotomous, we have 2^3 parameters related to A, one for each combination of B, C and D.

- Domains of the (decision and random) variables do influence the number of parameters. In the previous example, if A has 5 possible outcomes, the number of parameters would be 4×2^3. IctNeo ID has variables with up to 8 outcomes.

- The utility function, although already structured with a multiplicative functional form (see Gómez et al., 1999), has 6 attributes. It means that there are 27 parameters: 3 parameters for each of the 6 component exponential utility functions; 2 scaling constants for each one of both additive utility functions; and 5 scaling constants for the multiplicative expression.

- The complicated structure of the diagram with so many decision and chance nodes interrelated implies a hard definition of a strategy itself and the existence of many strategies.

For those reasons and for explanation purposes, in this paper we reduce our initial ID to have only two decision nodes. Figure 1 shows the diagram to work with. It is enough to get an overview of the policy of our problem because it covers the first 48 hours of a baby's life, the most critical ones. The total number of nodes is 31, once we delete nodes related to the remaining decision nodes. Other information to suggest the size of the problem is 72 arcs, 1369 probability entries, the biggest table having 225 entries, and an initial table for the value node of size 5400. The evaluation of this ID takes approximately half an hour in a 200-MHz Pentium PC. The maximum size achieved during the process is 10^6 storage positions. The optimal strategy (the base-optimal one) involves two decision tables. It leads to 22 possible full two stage treatments for a fixed patient. The first decision depends on 9 variables, whereas at the second stage, it depends on these 9, on the first decision and on 3 other variables. Hence, it amounts to having an optimal strategy indicating what to do at both decisions for each combination of those variables in the tables. Taking into account the cardinal of each variable, the number of combinations is 207,360. Summarising the optimal policy content is then tough work. Yet, we think it seems reasonable, with most cases treated with the appropriate sessions of phototherapies, and only more serious cases corresponding to treatments with an exchange transfusion.

As an illustration, let us take an ill patient with the data given in Table 1. The optimal strategy at the first stage is to administer to the patient a 24-hour long session of phototherapy. Let us suppose that afterwards, the bilirubin concentration has decreased (bilirubin concentration 2 equals normal), the hemoglobin concentration is still the same and the baby is obviously older (age 2 is more than 24 hours). Then, IctNeo recommends exposing the neonate to another session of 24-hour long phototherapy, observe him and finally discharge him.

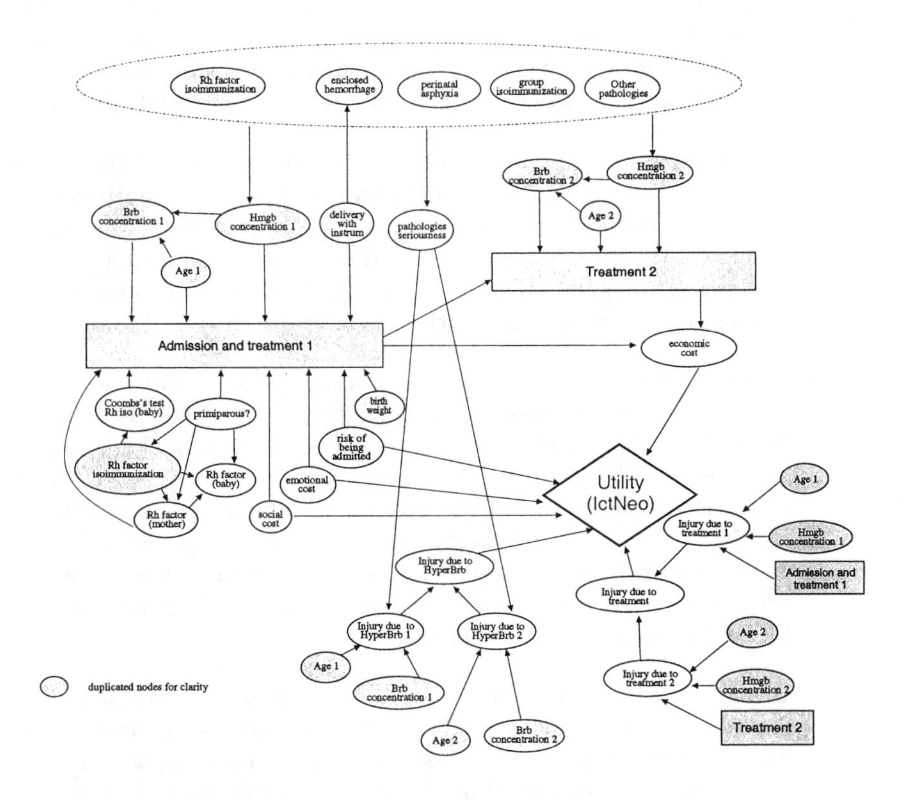

FIGURE 1. ID of the jaundice problem.

TABLE 1. A particular patient

Variable	Outcome
bilirubin concentration 1	high
hemoglobin concentration 1	medium
age 1	less than 12 hours
Coombs' test (for Rh isoimm.)	positive
delivery	without instruments
primiparous?	no
Rh factor (mother)	positive
birthweight	1000–1499 g
social cost	medium
emotional cost	medium
risk of being admitted	medium

17.2.3 Selection of parameters

We interviewed the doctors to find out the most important factors of the jaundice problem and the assignments they felt more uncomfortable with. Basically, the responses are probabilities of some pathologies, probabilities of the utility function attributes, and some weights of the utility function:

1. The probability of some pathologies: isoimmunization of Rh factor conditioned to mother being primiparous (ξ_1); isoimmunization of Rh factor conditioned to mother not being primiparous (ξ_2); isoimmunization of blood group (ξ_3); perinatal asphyxia (ξ_4); concealed hemorrhage (ξ_5) and other pathologies (ξ_6). Some of them have been converted to marginal distributions in order to get fewer parameters.

2. Random variable *emotional cost* has three possible outcomes (low, medium, high) with baseline probabilities .03, .85 and .12, respectively. We will take the mode as the value whose probability is parameter ξ_7 for SA purposes. Probability $1 - \xi_7$ will be proportionally distributed between the other two outcomes, that is, P(low emotional cost) $= .2(1 - \xi_7)$ and P(high emotional cost) $= .8(1 - \xi_7)$. Analogously, we proceed with *social cost* being ξ_8, the probability of a low social cost, currently at a baseline value of .90, and with ξ_9, probability of a low *risk of being admitted.*

3. *Economic cost* is conditioned to both decisions and can take four outcomes. Doctors consider that the most important costs to account for are obtained when the most essential therapies are applied. Therefore, we add two additional parameters: ξ_{10}, probability of a low cost given the patient receives a 24-hour long phototherapy session at the first stage and an exchange transfusion preceded and followed by a 6-hour long phototherapy session at the second stage; and ξ_{11}, probability of a high cost given the same first treatment and then an exchange transfusion preceded by a 6-hour long phototherapy session and followed by another session of 12 hours.

4. Another important attribute of the utility function is *injuries due to hyperbilirubinemia* at stage $i, i = 1, 2$ ($IHB1, IHB2$) with five outcomes and conditioned to three chance nodes: bilirubin concentration at stage $i, i = 1, 2$; pathology seriousness; and baby age at stage $i, i = 1, 2$. The dialogue with the doctors led us to choose four critical configurations of these three chance nodes: $A1$: high bilirubin concentration at stage 1, serious pathology, and age at stage 1 less than 12 hours old; $B1$: high bilirubin concentration at stage 1, *very* serious pathology, and age at stage 1 less than 12 hours old; and the same for stage 2 ($A2, B2$). Finally, we have four new parameters: ξ_{12} for the probability of the mode of $IHB1$ given $A1$, and ξ_{13} given $B1$,

ξ_{14} for the probability of the mode of $IHB2$ given $A2$, and ξ_{15} given $B2$.

5. As far as the utility function is concerned, it is deduced in Gómez et al. (1999) as multiplicative on four attributes - economic cost, socio-emotional cost, risk of being admitted and injuries (due to the treatment and to the hyperbilirubinemia) - obtaining four weights k_1, k_2, k_3 and k_4. They are our last parameters.

Each of the 19 parameters for which the doctors desired to perform SA requires a range along which the parameter will be varied. For ξ's, we will take the interval centred at the baseline value with radius 0.15. We think it is enough to cover many cases. For k's, we will take the intervals obtained in Gómez et al. (1999) where imprecise assignments when providing tradeoffs were allowed as a way of SA. These are narrower, the widest having a length of 0.18.

17.2.4 One-way sensitivity analysis

We start with a one-way SA. For that, we solve the problem, that is, we obtain the optimal strategy and its expected utility -the maximum expected utility of the problem– when one specified parameter is varied from the minimum value of its range to the maximum, with all other parameters held at their baseline values. We take 10 equally spaced values per interval. The same process is applied for each parameter.

The length of a bar representing the resulting maximum expected utility range will show us the extent to which the problem is sensitive to the specified parameter. All bars are arranged in a graph from the longest bar at the top to the shortest bar at the bottom. This kind of graph is usually called a *tornado diagram* (Howard, 1988). It is commonly used to decide in early analyses which variables might be left at their baseline values (as deterministic nodes) and which should be considered as random variables. As an example of this early analysis, we took the six attributes of the utility function. While five of them were kept at their base outcome therefore becoming deterministic nodes, the other variable was moved along all its possible outcomes. Results are in Figure 2, where utility values range from 0 to 1000. It reveals the need for considering all those nodes as chance nodes because of the wide variation of maximum expected utilities, although what is *wide* is critically tied to doctors' beliefs, as are many other SA measures.

It is more interesting to construct a tornado diagram at later stages of our modelling process, checking the sensitivity to more elaborate assignments, not only to variable outcomes. We now do it for the 19 parameters already explained and deemed more important by doctors. Figure 3 shows such a tornado diagram, displaying only those parameters whose sensitivity is nonnegligible. Maximum expected utility values range from 0 to 1000. The

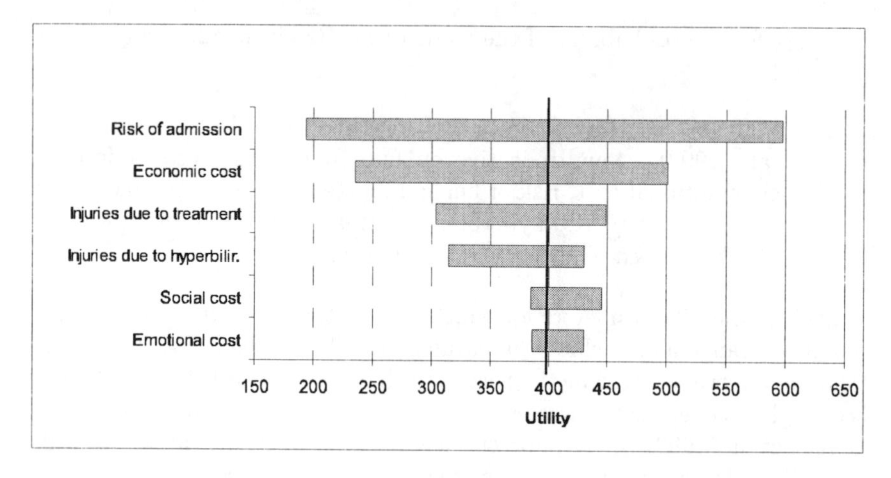

FIGURE 2. Deterministic tornado diagram for six variables.

vertical line is located at the base maximum expected utility (724.78), that is, that obtained with all parameters at their baseline values.

For instance, setting ξ_9 at 0.70 instead of 0.85, and keeping all other parameters at their baseline values, yields a maximum expected utility of 691.5 instead of 724.78, inducing a decrease of 33.28 in the maximum expected utility.

From this figure, we observe the importance of the scaling constants k_i. k_3 corresponds to the risk of being admitted and parameter ξ_9 is the probability of a low risk of being admitted. Therefore, both data inform about the sensitivity of the problem with respect to the risk of infections, and so forth, derived from the stay at the hospital, from the point of view of its uncertainty (ξ_9) and from its weight in the preferences (k_3).

The preferences of parents are measured by the inconvenience of visiting the baby every day (social cost) if she is at hospital and the interruption of parent–infant bonding (emotional cost). The associated scaling constant is k_2, that implies not very much sensitivity, compared to the other k's. The first cost is taken into account in ξ_8, which leads to more sensitivity than the emotional cost (ξ_7).

As far as pathologies are concerned, the doctors' uncertainty about the probability of perinatal asphyxia (ξ_4) and isoimmunization of blood group (ξ_3) seems important. The first one is related to oxygen problems (and thereby to bilirubin problems since red blood cells carry oxygen), while the second reveals incompatibility between mother and baby bloods. Also, other pathologies (ξ_6) stands out.

All these parameters would need to be considered more closely. We might expend effort to refine their assignment. On the contrary, the maximum expected utility seems very insensitive to probabilities related to isoimmunization of Rh factor, injuries due to hyperbilirubinemia, and economic

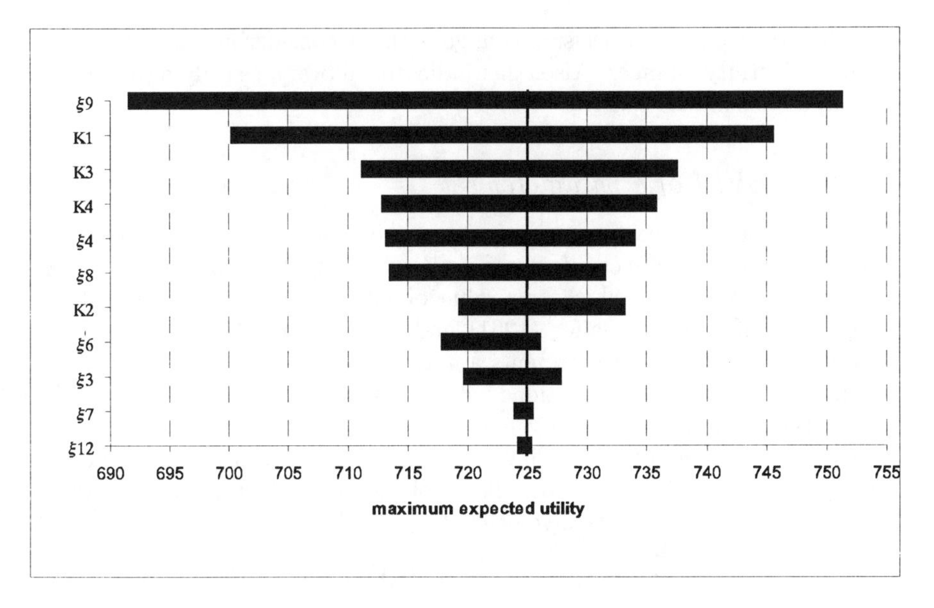

FIGURE 3. Tornado diagram for parameters.

cost. They could remain at their baseline values.

It would be interesting to draw in each bar whether the optimal strategy changes, by using a different colour, for example, to detect decision sensitivity. In our case, it is not very useful since there are too many changes because of having many strategies and they depend on many variables.

Tornado diagrams limit us to observe what happens only when one parameter changes at a time. Typically, we will want to explore the impact of several parameters at a time, finding out possible relationships among them. It is not rare, for example, to find a problem sensitive to its entire parameter set but not to any individual parameter (see, e.g., Ríos Insua, 1990). Two-way SAs are easy to perform and it would be interesting to consider the joint impact of changes in the two most critical parameters. However, we prefer to conduct SA using another measure, as done in the next section.

17.3 Sensitivity based on expected value of perfect information

Felli and Hazen (1998) introduce a new SA indicator based on the expected value of perfect information, and it is shown to be the proper way to proceed: methods usually encountered in medical decision-making literature, based on threshold proximity, probability of a threshold crossing, and entropy, can overstate problem sensitivity because of focusing only on the probability of a decision change without considering the associated expected utility changes. Also, this indicator allows for studying multiple parameters simultaneously.

17.3.1 EVPI of a parameter set

The concept of clairvoyance in decision problems led to the concept of expected value of perfect information (Howard, 1966). Some authors have discussed the way in which the value of information can be related to SA; see Raiffa and Schlaifer (1961) and Howard and Matheson (1983). Methodologically, EVPI represents a natural extension of probabilistic SA and, unlike the measures mentioned above, its calculation is consistent with the maximisation of expected value.

Consider parameter set ξ. The DM first solves the decision problem using ξ_0, the value he feels most likely, obtaining the alternative a_0 that maximises expected utility given that $\xi = \xi_0$. That is,

$$E(V_{a_0}|\xi = \xi_0) = \max_a E(V_a|\xi = \xi_0).$$

But ξ is uncertain and there is some risk in using only that value to determine the optimal alternative. With others value of ξ, some alternative other than a_0 could yield a higher expected utility. The difference is $\max_a E(V_a|\xi) - E(V_{a_0}|\xi)$. Averaging over all possible values of ξ, we obtain the EVPI on ξ:

$$EVPI(\xi) = E_\xi[\max_a E(V_a|\xi) - E(V_{a_0}|\xi)].$$

EVPI can be expressed mathematically as the product of both the probability of a change in the baseline decision due to variation in ξ and the payoff differential resulting from such a decision change, joining two sources of information DMs typically seek from SA (Felli and Hazen, 1998). EVPI represents the average improvement we could expect to gain over the payoff resulting from the selection of a_0 given perfect information on ξ prior to the time of making the decision. Note that the parentheses of EVPI expression is an opportunity loss or regret of the strategy a_0, for each ξ, so used in criteria for making decisions under uncertainty (e.g., minimax regret). However, EVPI has a hierarchical component since there is a probability distribution on (hyperparameters) ξ's, and all these regrets are averaged with respect to that distribution. Thus, EVPI is the expected opportunity loss suffered for not utilising perfect information about ξ.

The assignment of that distribution is carried out as in conventional SAs. For example, doctors may give a base value and upper and lower bounds for ξ, and then an approximation to the distribution, like Pearson-Tukey's, may be applied (Keefer and Bodily, 1983). Anyway, this task is a natural extension of the thought processes already employed in the problem modelling.

The higher the EVPI(ξ) is, the more sensitivity there is to ξ, where the comparisons of values are made in the same units as the problem payoffs (as opposed to the other measures mentioned above). Typically, the DM will specify a value V^* of EVPI such that parameter sets with EVPI greater that V^* will be labelled as sensitive. Felli and Hazen suggest V^* to be considered a *minimum significant improvement*, an accepted practice among Bayesian approaches to clinical trials (Spiegelhalter et al., 1994), that perfect information on the parameter set could yield over the payoff the DM expects to gain without that information.

17.3.2 Computing EVPI via Monte Carlo simulation

Since EVPI requires the computation of an expectation, it may be cumbersome with many parameters and complex parameter distributions. Therefore, we usually compute EVPI via Monte Carlo simulation.

Let $\xi = \xi_I \cup \xi_I^c$ be the set of parameters, with ξ_I the subset of interest. Let $E(V|\xi, A)$ be the expected utility as a function of ξ and an alternative A. Let $A^*(\xi_I)$ be the optimal decision as a function of ξ_I, and let A^* be the base-optimal alternative. Then,

$$EVPI(\xi_I) = E_\xi[E(V|\xi_I, \xi_I^c, A^*(\xi_I)) - E(V|\xi_I, \xi_I^c, A^*).]$$

The Monte Carlo simulation procedure generates N of ξ and averages the resulting differences in expected utility. The output is an approximation to EVPI(ξ_I). The general algorithm is simplified in case the utility function is linear in each individual parameter and all parameters are probabilistically independent. Since this is our case, as confirmed by the doctors, if $\bar{\xi}_I^c$ designates the mean value of ξ_I^c, our simulation procedure follows the steps:

`Set` $i = 1$
`While` $i < N$
  `1. Generate` ξ_I^i `according to its distribution`
  `2. Compute` $A^*(\xi_I^i)$ `and the improvement:`

$$\text{Diff}^i = E(V|\xi_I^i, \bar{\xi}_I^c, A^*(\xi_I^i)) - E(V|\xi_I^i, \bar{\xi}_I^c, A^*)$$

  `3. Set` $i = i + 1$
`Return` $\frac{1}{N}\sum_{i=1}^N \text{Diff}^i$

In our jaundice problem, we select the most outstanding parameters of the tornado diagram: $\xi_9, k_1, k_2, k_3, k_4, \xi_4, \xi_8$. The task of assigning a probability distribution for parameters under examination was not so involved since the doctors decided they should be uniform on the range already defined. Therefore, the mean values required in the simulation scheme coincide with the baseline values.

Since generating from a uniform is not difficult, the key step is step 2. To compute $A^*(\xi_I^i)$ we must solve the whole ID with the new values ξ_I^i, obtaining finally $E(V|\xi_I^i, \bar{\xi}_I^c, A^*(\xi_I^i))$ stored at the value node. To compute $E(V|\xi_I^i, \bar{\xi}_I^c, A^*)$, we must solve the whole ID with the new values ξ_I^i and slightly modify the standard algorithm to use only the chunks of the value node tables associated with the base-optimal strategy A^*.

We performed the analyses using a 200-MHz Pentium PC and the C++ random number generator. They were run within our system IctNeo. Table 2 summarises the results of our experiments with $N = 100$ and expected utilities varying from 0 to 1000. The computation of each row meant 12 hours processing.

TABLE 2. EVPI analysis of the jaundice problem

Parameter ξ_I	EVPI(ξ_I)
k_1	0.059
ξ_8	0.085
k_2	7.725
k_3	8.267
ξ_4	32.940
k_4	67.570
ξ_9	139.684
ξ_9, k_3	127.529
ξ_8, k_2	157.413
ξ_9, ξ_8, ξ_4	192.875
k_1, k_2, k_3, k_4	347.284
all parameters jointly	392.187

As an illustration of the EVPI meaning, let us take EVPI(ξ_9) $= 139.684$. The average benefit from switching alternatives in response to perfect foreknowledge on ξ_9 is 139.684, which represents a 19.27% improvement over the maximum expected utility 724.78 obtained with the base-optimal strategy (that is, $0.1927 \cdot 724.78 =$ EVPI(ξ_9)). Tornado diagram confirmed only a 3.7% improvement achieved when the maximum expected utility reached 751.41.

Concerning individual parameters, none of the first four EVPI values seems large enough to raise sensitivity concerns, all of them representing less than 1.2% of the base-optimal maximum expected utility. Therefore, we can label the problem insensitive to parameters k_1, k_2, k_3, ξ_8 singly. By looking at EVPI(ξ_4) and EVPI(k_4), we could say that the problem is possibly somewhat sensitive to parameters ξ_4 and k_4. Yet, as occurred in tornado diagram, ξ_9 is the individual parameter with the largest value of EVPI.

It is more interesting to study the joint sensitivity of various parameters. Reilly (1996) gives a clever way to identify groups of parameters which are jointly influential and identifies the particular linear combination of variables as being influential. However, his method does not work for sequential decision-making problems. We show here five of those studies; see Table 2. First, ξ_9 (a probability) and k_3 (a scaling constant) are varied concurrently because both refer to the risks of the stay at hospital, revealing an EVPI of 127.529. Second, ξ_8 and k_2 jointly give an EVPI of 157.413. Two studies follow, first with all the probabilities varied simultaneously, and then with all the scaling constants. Both indicate sensitivity, especially the latter.

When all parameters are allowed to vary concurrently, the EVPI for the entire parameter set is 392.187, concluding that the model is sensitive to variations in the entire parameter set.

Therefore, the jaundice problem is sensitive to its entire parameter set and to some joint variations of certain parameters, but it is not sensitive to most individual parameters. We might conclude that parameter interactions are important.

17.4 Conclusions

It is well known that no "optimal" SA procedure exists for decision analysis. Through the decision analysis cycle, SA is part of the modelling process, which is an art. Remember that we are accounting for expected utility changes accompanying decision changes. In this paper we have tried to show how to undertake SA in a big problem by means of several tools.

Most decision analysis software packages have built-in SA routines but they are too basic. For example, programs like TreeAge DATA, DPL, Decision Tool Suite, and Analytica, offer options of tornado diagrams with up to three parameters varying simultaneously, two-way threshold proximity measures, three-way threshold proximity measures (with animated region graphs), or SA on some specific parameters as the risk tolerance coefficient.

For that reason we have implemented the SA method described based on the EVPI. With EVPI we can draw conclusions concerning the relative importance of parameter sets based on direct consideration of the average marginal improvement attributable to a (hypothetical) gain of perfect information before the decision. EVPI is easily computed via simulation, being tractable even for large parameter sets. It is also easy to use and interpret. It has allowed us to declare the jaundice problem sensitive to some sets of parameters, meaning that it may indeed be worthwhile to put considerable effort in modelling the uncertainty related to those parameters.

In the next future, IctNeo will generate explanations of the results, providing doctors with considerable insight, thus serving as another way of SA. This is an ongoing issue for further research.

Acknowledgements

Research supported by Projects UPM A9902, CICYT HID98-0379-C02-2 and CAM 07T/0009/1997.

References

BIELZA, C., GÓMEZ, M., RÍOS-INSUA, S. and FERNANDEZ DEL POZO, J.A. (2000). Structural, elicitation and computational issues faced when solving complex decision problems with influence diagrams. *Computers and Operations Research* (to appear).

BIELZA, C., RÍOS INSUA, D. and RÍOS-INSUA, S. (1996). Influence diagrams under partial information. In *Bayesian Statistics 5* (J.M. Bernardo, J.O. Berger, A.P. Dawid and A.F.M. Smith, eds.), 491–497. Oxford: Oxford University Press.

BIELZA, C. and SHENOY, P.P. (1999). A comparison of graphical techniques for asymmetric decision problems. *Management Science*, **45**, 11, 1552–1569.

BUTLER, J., JIA, J. and DYER, J. (1997). Simulation techniques for the sensitivity analysis of multi-criteria decision models. *European Journal of Operational Research*, **103**, 531–546.

CLEMEN, R.T. (1996). *Making Hard Decisions*, 2nd ed. Belmont, CA: Duxbury.

DOUBILET, P., BEGG, C.B., WIENSTEIN, M.C., BRAUN, P. and MCNEIL, B.J. (1985). Probabilistic sensitivity analysis using Monte Carlo simulation. *Medical Decision Making*, **5**, 157–177.

EZAWA, K. (1998). Evidence propagation and value of evidence on influence diagrams. *Operations Research*, **46**, 1, 73–83.

FAGIUOLI, E. and ZAFFALON, M. (1998). A note about redundancy in influence diagrams. *International Journal of Approximate Reasoning*, **19**, 3-4, 231–246.

FELLI, J.C. and HAZEN, G.B. (1998). Sensitivity analysis and the expected value of perfect information. *Medical Decision Making*, **18**, 95–109.

GÓMEZ, M. and BIELZA, C. (1999). Node deletion sequences in influence diagrams by genetic algorithms. *Technical Report*, Department of Artificial Intelligence, Technical University of Madrid.

GÓMEZ, M., RÍOS-INSUA, S., BIELZA, C. and FERNANDEZ DEL POZO, J.A. (1999). Multi-attribute utility analysis in the IctNeo system. In *XIV-th International Conference on Multiple Criteria Decision Making* (Y.Y. Haimes and R. Steuer, eds.), Lecture Notes in Economics and Mathematical Systems. Berlin: Springer (to appear).

HORSCH, M.C. and POOLE, D. (1996). Flexible policy construction by information refinement. In *Uncertainty in Artificial Intelligence, Proceedings of the 12th Conference* (E. Horvitz and F. Jensen, eds.), 174-182. San Francisco, CA: Morgan Kaufmann.

HOWARD, R.A. (1966). Information value theory. *IEEE Transactions on Systems, Science and Cybernetics SCC*, **2**, 1, 22–26.

HOWARD, R.A. (1988). Decision analysis: practice and promise. *Management Science*, **34**, 679–695.

HOWARD, R.A. and MATHESON, J.E. (EDS.) (1983). *The Principles and Applications of Decision Analysis.* Palo Alto, CA: Strategic Decisions Group.

KADANE, J., SALINETTI, G. and SRINIVASAN, C. (2000). Stability of Bayes decisions and applications. In *Robust Bayesian Analysis* (D. Rios Insua and F. Ruggeri, eds.). New York: Springer-Verlag.

KEEFER, D.L. and BODILY, S.E. (1983). Three-point approximations for continuous random variables. *Management Science*, **29**, 5, 595–609.

KEENEY, R.L. and RAIFFA, H. (1976). *Decisions with Multiple Objectives: Preferences and Value Tradeoffs.* New York: Wiley.

MAHONEY, S.M. and LASKEY, K.B. (1996). Network engineering for large belief networks. In *Uncertainty in Artificial Intelligence, Proceedings of the 12th Conference*, (E. Horvitz and F. Jensen, eds.), 389–396. San Francisco, CA: Morgan Kaufmann.

NEWMAN, T.B. and MAISELS, M.J. (1992). Evaluation and treatment of jaundice in the term infant: a kinder, gentler approach. *Pediatrics*, **89**, 5, 809–818.

POH, K.L. and HORVITZ, E.J. (1993). Reasoning about the value of decision-model refinement: methods and application. In *Uncertainty in Artificial Intelligence, Proceedings of the 9th Conference* (D. Heckerman and A. Mamdani eds.), 174–182. San Mateo, CA: Morgan Kaufmann.

RAIFFA, H. and SCHLAIFFER, R. (1961). *Applied Statistical Decision Theory.* Cambridge, MA: Harvard University Press.

REILLY, T. (1996). Sensitivity analysis for dependent variables. *Technical Report*, University of Oregon, Eugene.

RÍOS INSUA, D. (1990). *Sensitivity Analysis in Multiobjective Decision Making*, Lecture Notes in Economics and Mathematical Systems 347. Berlin: Springer.

RÍOS INSUA, D. (1992). On the foundations of decision making under partial information. *Theory and Decision*, **33**, 83–100.

RÍOS INSUA, D. and FRENCH, S. (1991). A framework for sensitivity analysis in discrete multi-objective decision making. *European Journal of Operational Research*, **54**, 176–190.

RÍOS-INSUA, S., BIELZA, C., GÓMEZ, M., FERNANDEZ DEL POZO, J.A., SÁNCHEZ, M. and CABALLERO, S. (1998). An intelligent decision system for jaundice management in newborn babies. In *Applied Decision Analysis* (F.J. Girón ed.), 133–144. Norwell, MA: Kluwer.

SHACHTER, R.D. (1986). Evaluating influence diagrams. *Operations Research*, **34**, 6, 871–882.

SPIEGELHALTER, D.J., FREEDMAN, L.S. and PARMER, M.K.B. (1994). Bayesian approaches to randomized trials. *Journal of the Royal Statistical Society A*, **157**, 357–416.

18

Sensitivity of Replacement Priorities for Gas Pipeline Maintenance

Enrico Cagno, Franco Caron, Mauro Mancini and Fabrizio Ruggeri

ABSTRACT Knowledge management is a key issue in the industrial field because of the difficulties in both properly quantifying experts' judgments and integrating them with available historical data. Although many studies have been proposed on expert knowledge elicitation, new methods are still sought and practical applications are used to validate the effectiveness of the proposed ones.

We discuss some sensitivity issues in dealing with experts' judgments in a real case study about strategies for the preventive maintenance of low-pressure cast-iron pipelines in an urban gas distribution network. We are interested in replacement priorities, as determined by the failure rates of pipelines deployed under different conditions.

Many gas company experts have been interviewed using an ad hoc questionnaire, based on pairwise comparison of propensity to failure in different pipelines. Judgments have been combined and quantified using the analytic hierarchy process, which ensures a control on the consistency of the experts but has some drawbacks, as discussed in the paper. We cope with some of them by relaxing the assumptions, made in previous papers, about the prior distributions on the failure rates, and we study changes in replacement priorities under different choices of classes of priors.

Key words: Analytic hierarchy process, gas failures, reliability.

18.1 Gas pipeline replacement

The ability to assess effective replacement priorities in gas pipeline sections of an urban distribution network is very critical since replacements over a given planning horizon are limited by the gas company's budget. On the other side, a detonation could cause casualties and will have, in any case, a negative impact on the gas company's image. Therefore, accurate tools are needed to define priorities and improve service performance and safety.

A methodology based on interval judgment elicitation and Bayesian inference was proposed in Cagno et al. (1998, 2000) to assess the failure probability for each pipeline section, which we view as the priority index for the replacement policy.

Considering the very small number of failures every company actually experiences, effective evaluation of failure probabilities must be based on the integration of historical data and the knowledge of company experts. In order to guarantee reliable results, the robustness of the methodology must be particularly focused. This is the main goal of the current paper.

The gas distribution network considered in our case study was developed in a large urban area during the 1800s and thereby it is characterised by very different technical and environmental features (material, diameter of pipes, laying location, etc.). It consists of several thousand kilometers of pipelines in low, medium and high pressure. The main concerns about the replacement plan were related to the low-pressure network (20 mbar over the atmospheric pressure). For the low-pressure pipelines, several materials have been used to develop the network: traditional cast iron (CI), treated cast iron (TCI), spheroidal graphite cast iron (SGCI), steel (ST) and polyethylene (PE).

Since CI pipelines have a higher failure rate than other materials, even 10 times greater (see Cagno et al., 2000) and cover more than one fourth of the whole network, we paid more attention to this kind of material and studied it in more detail. The low-pressure CI network consists of about 6000 different sections of pipes whose lengths range from 3 to 250 meters for a total of 320 kilometers.

It should be noted that medium- and high-pressure steel pipelines are less critical since automatic devices stop gas flow if pressure decreases. Moreover, medium- and high-pressure pipelines are laid deeper in the ground, with more care about laying procedure, so they are less subject to accidental stress.

The main objective of the gas company is to progressively replace cast-iron pipelines with a different material, such as polyethylene. The foreseen time horizon to complete the replacement process is rather long since it is not possible to act simultaneously over the whole network. As a consequence, the company needs a criterion to decide which pipeline section must be changed first to guarantee safety and performance of the distribution network. The number of pipelines to be replaced in a given planning period depends on the available budget.

Using failure probability as an index of replacement priority and taking advantage of studies developed in other similar companies and available in literature (e.g., Bacon, 1988), we identified factors directly involved in the probability of failure for a low-pressure CI pipeline. Since an urban network has a considerable range of technical features, factors can be divided as shown in Table 1. A complete report on these parameters is rarely available, especially because in many countries distribution companies are responsible for a single urban network, and comparisons in different contexts, data sharing and reusing are difficult.

Intrinsic features of the pipeline	Factors concerning the laying of the pipeline	Environmental parameters
Thickness	Depth	Traffic characteristics
Diameter	Location	Intensity of underground services
Age	Ground characteristics	External temperature and moisture
	Type and state of the pavement	
	Laying techniques	

TABLE 1. Factors related to failure probability

Moreover, remarkable difficulties arise in collecting information about a single gas distribution network, since each network is generally laid over a long period of time and data registration methods have changed. Some information (such as age, depth, laying technique) is difficult to find because either it was poorly registered or it was recorded for other purposes (such as external temperature and moisture, intensity of underground services) which are not sufficiently related to the requirements deriving from a correct safety and reliability analysis.

For the case studied, in order to show the significance of the different factors, a multiple analysis of variance of the historical failure rate has been carried out in Cipolletti and Mancini (1997). Using archives of different bodies responsible for underground services, the length of the CI distribution network and the corresponding number of failures – over a significantly long period (10 years) – have been classified with reference to the above factors, giving the associated historical failure rate.

The subdivision into levels has been determined for each factor to maximise the factor's effects, avoiding nonsignificant ones. The available information led to division into two levels, even if a more detailed sub-division (3 or 4 levels) has been preferable, we avoided it because of the entailed loss of information due to data scattering.

The study identified diameter, laying depth and location as the discriminant factors. The other factors were not particularly important as they turned out to be either homogeneous for the analysed distribution network (e.g., the company had always used an embankment with the same characteristics during works; the external temperature and moisture have actually no effect on pipes as the ground has a strong insulating capacity, even with little laying depth; etc.) or fixed for a given level of the above-identified factors (e.g., due to production constraints, the thickness of a pipe is fixed for a given diameter). This selection of relevant factors was fully shared and validated by company experts. The identified levels of the selected factors are given in Table 2, where the notations "high" and "low" are qualitative rather than quantitative, thus more easily distinguishing two levels of the same factor and simplifying the reading of conclusions.

Factors	Low level	High level
Diameter	≤ 125 mm	> 125 mm
Laying depth	< 0.9 m	≥ 0.9 m
Laying location	Sidewalk	Street
Traffic	Light	Heavy

TABLE 2. Significant levels of factors related to failure

As already mentioned, the methodology currently applied to assess the replacement priority for each pipeline section requires significant improvement, since, at present, most gas companies use "demerit points cards" (e.g., the technical report of the British Gas Corporation; Bacon, 1988) to assess the tendency to failure, determining priority of maintenance and replacement intervention, for different sections of a gas pipeline by attempting to quantify the influence of some qualitative and quantitative parameter (diameter, laying depth, etc.). This tool has the advantage of highlighting the main factors strongly correlated to pipeline failures, but it also revealed several application problems.

The principal shortcoming of this tool is that in most cases setting the weights for each considered factor (and the choice of factors) is based only on previous empirical experience concerning other city centers, without any adjustment on the actual urban context. It is hardly worth saying that those weights (and also factors) can be misleading in correctly setting maintenance and replacement policies. Moreover, often an aggregate demerit index, obtained by adding the individual weights of the different factors, hides possible interactions between factors (e.g., diameter and laying depth), hence losing other important information and worsening an already critical situation.

Another shortcoming of this assessment method is that the tendency to failure of a given section of gas pipeline is considered independently from the length of the section and the given planning horizon. In a generic gas distribution network, the length of a homogeneous section may change remarkably, even more than 10 times. Since this kind of analysis is useful to decide the replacement of one section of a pipeline rather than another (not just to replace a kind of pipeline), it is very important to distinguish sections with the same characteristics but different lengths, as it is more likely that longer sections will fail rather than shorter ones.

18.2 Methodology

Following the data analysis, the most important conclusions in terms of guidelines for the study were (see Cagno et al., 1998, 2000) as follows:

1. The gas distribution sector is characterised by a remarkable shortage of data since not only does pipeline failure represent a rare event but also the available information is not always adequate to fulfill safety analysis requirements. This scarcity of data, along with some underlying "noise," suggested improving the information obtained from the company archives with experts' judgments by means of a Bayesian approach.

2. Since the structural failure of CI pipes is a rare phenomenon, it was felt appropriate to model failure events with a Poisson process. As a matter of fact, CI pipeline failure rate seems to be scarcely sensitive both to wear and to closeness to leak points (an assumption confirmed by available data, taking into account a useful life phase varying from 50 to 100 years), but mostly influenced by accidental stress. Due to these considerations, a time and space homogeneous Poisson process was used, with unit intensity (or failure rate) λ (i.e., per km and year).

In this way, it was possible also to take into account the length of each section of pipe and the given planning period, which are usually not considered within the demerit point cards approach. The evident importance of interactions between factors (e.g., diameter is significant if laying depth is low, but not if it is high) led to the abandonment of the additive technique, typical of the demerit point cards approach, which simply sums the effects of the factors, in favour of the determination of pipeline classes based on the combination of different levels of the most significant factors. This data organisation also facilitated the expression of experts' judgment.

18.2.1 *Subdivision in classes*

Table 3 shows the annual failure rate for the different combinations of factors, and highlights (by virtue of the partial values) both the influence of laying location (higher failure rates in street location branch) and the interactions between factors (e.g., the varying influence of laying depth as a function of diameter). Note that in the company's pristine way of assessing tendency to failure, the largest demerit point was given to small diameters and the laying location was one of the less important factors, so that it was clearly against the historical data analysis and was giving misleading results.

18.3 Bayesian approach

The first conclusion from the data analysis – that is, the shortage and the underlying noise of the historical data – was addressed in Cagno et al. (1998,

Factors	1	2	3	4	5	6	7	8
Diameter	L	L	L	L	H	H	H	H
Laying location	L	L	H	H	L	L	H	H
Laying depth	L	H	L	H	L	H	L	H
Annual failure rate	.072	.094	.177	.115	.067	.060	.131	.178

TABLE 3. Classes of pipelines, according to low/high (L/H) levels of factors

2000). They felt that historical data on failures should be integrated with the knowledge of company experts: the choice of the Bayesian approach was then natural.

Experts' opinions were collected using an ad hoc questionnaire and priors were obtained, as described in the next section. In Cagno et al. (1998) the experts' opinions were synthesised into one value for mean and variance of the prior distribution on each failure rate $\lambda_i, i = 1, 8$, computing their Bayes estimates (i.e., the posterior mean) using both a Gamma prior and a lognormal one. They found that both solutions (the former leading to closed expression for the estimators, whereas the latter is more accepted among engineers) led to the same ranking of the pipelines, based upon their propensity to failure.

Cagno et al. (2000) considered and compared two classes of Gamma priors with mean and variance in intervals around the synthesised values from the experts' opinions. The classes led to different, but similar, ranges on the posterior mean of each $\lambda_i, i = 1, 8$. Since the authors had the goal of ranking the classes, they needed a unique value for the failure rate of each class. They chose the one compatible with the "posterior regret Γ–minimax" criterion, presented in Ríos Insua et al. (1995) and Vidakovic (2000). As proved in Ríos Insua et al. (1995), the optimal estimator, according to the criterion, is given by the central value in the posterior range (under quadratic loss function).

18.4 Eliciting opinions: the experts' group and the pairwise comparison method

We present in detail the technique (analytic hierarchy process) which will be used later to combine the experts' judgments.

18.4.1 AHP as an elicitation method

The analytic hierarchy process (AHP) is a robust and flexible multi-attribute decision making (MADM) technique which formulates the decision problem in a hierarchical structure, prioritising both the evaluation criteria and the alternatives by pairwise comparisons. For a complete description of this methodology, the reader could refer to a number of specialised papers (e.g., Saaty, 1980, 1994).

The application of AHP to the pipelines replacement problem has been considered in Cagno et al. (1998, 2000), who deemed this approach as promising since it may give better decision support than scoring methods, (i.e. demerit point cards as in Bacon, 1988) mainly because the methodology is robust, due to the pairwise comparison process and the consistency check tool. The importance of each element is assessed through a sequential process of pairwise comparisons. It is common experience that estimates based on pairwise comparison are more reliable than single direct estimates. This is even more true, for both qualitative and quantitative factors, when insufficient data is available to make direct estimates. In addition, the elicitation based on redundant judgments, which characterises the standard application of AHP, is useful to check the consistency of the decision maker. The level of inconsistency can be calculated and, if it exceeds a given threshold, the expert may review judgments more carefully.

At each level of the hierarchy, a relative weight is given to a decision element (alternative/criterion) by comparing it with another element (alternative/criterion) at the same level with respect to their common adjacent element (criterion/super-criterion) at the upper level. The comparison takes the following form: "How important is element 1 when compared to element 2 with respect to the adjacent one at the upper level?" The decision maker can express his/her preference between each couple of elements verbally as equally preferred (or important or likely), moderately preferred, strongly preferred, very strongly preferred or extremely preferred. These descriptive preferences would then be translated into absolute numbers 1, 3, 5, 7 and 9, respectively, with 2, 4, 6, and 8 as intermediate values expressing a compromise between two successive qualitative judgments. The verbal scale used in AHP enables the decision maker to incorporate subjectivity, experience and knowledge in an intuitive and natural way. Pairwise comparisons are structured in matrices and the eigenvector method is used to derive the relative weights of the elements at each level with respect to the element in the adjacent upper level (Saaty, 1980). The overall weights of the decision criteria and alternatives are then determined by aggregating the weights through the hierarchy. This is done by following a top-down path through the hierarchy and multiplying the weights along each segment of the path. The outcome of this aggregation is a normalised vector of the overall weights of the alternatives.

The use of the AHP to model and analyse real-world problems can be made much easier using a software implementation of the method, such as Expert Choice Professional 9.5 (1999). It makes structuring and modifying the hierarchy simple and quick and eliminates tedious calculations.

As previously mentioned, in the deterministic version of AHP, judgments are expressed in a linguistic scale and every judgment is translated into a number. The result of the evaluation process is a list of priority indexes.

Nevertheless, the application of the deterministic AHP to a real-world decision making context has a main shortcoming: it makes no reference to the uncertainty which characterises the expert judgments.

Indeed, judgments are translated into numbers irrespective of the level of uncertainty which inevitably characterises the expert when he/she compares the elements. Analogous concerns may be extended to a group decision making process. A way to consider uncertainty within the deterministic version of AHP could be to apply sensitivity analysis and assess how sensitive the ranking of priority indexes is to variations in the single point judgments. However, sensitivity analysis *(in this context)* allows us to vary only one factor at a time (more factors would be changed in a scenario analysis) and does not provide a measure of the robustness of the result.

We shall present a probabilistic version of AHP which is able to include uncertainty of the expert judgments and thus provides a measure of the dispersion of each priority index around its mean value.

AHP can deal with decisional problems in both a deterministic and a probabilistic way, even if the vast majority of previous applications has used a deterministic approach. This characteristic is not exclusive of AHP, since most MADM techniques can be extended to include uncertainty.

In the probabilistic version, uncertainty in judgments can be accounted for by substituting the point estimates with interval judgments, that is, judgments which take values in a given range according to a specified probability distribution function; see Saaty and Vargas (1987). Interval judgments can represent the uncertainty of each individual involved in the decision process. While in the deterministic approach judgments are point estimates and the outcome is a comparison between the elements in terms of their final priority indexes, in the probabilistic approach judgments are random variables, as are the overall priority indexes, too. In addition to single-point priority indexes (assuming the mean value of the distribution of the final priority as the single-point estimate), relevant additional information is therefore available: a measure of the dispersion, such as the standard deviation, of final priorities around the single-point estimate.

When giving interval judgments, in order to take into account some uncertainties (Saaty and Vargas, 1998), any expert could answer, for example, to a pairwise comparison of the questionnaire "between moderately more probable and very strongly probable" (3–7) instead of "strongly probable" (5) (see Table 4). This methodology generated for each expert, when assessing the priority indexes, eight failure density functions rather than eight point estimates. It was found that, given the robustness of the eigenvalue method, the variance of the density functions was negligible in comparison with the mean value, giving evidence to the robustness of the methodology.

When some or all of the judgments in the decision hierarchy are treated as random variables, the application of any analytic method to assess the distribution of the final priority indexes generally becomes a theoretically and computationally prohibitive problem (Saaty and Vargas, 1987). As a

Intensity	Definition
1	Equally probable
3	Moderately more probable
5	Strongly more probable
7	Very strongly more probable
9	Absolutely more probable
2,4,6,8	Intermediate values

TABLE 4. Numerical opinion scale

consequence, it is easier and more practical to handle the uncertainty regarding judgments in AHP using a Monte Carlo simulation approach. At each replication, the point judgments are extracted from the corresponding probability distribution functions (interval judgments). A test evaluates the consistency of the set of judgments and, in case of unacceptable inconsistency (i.e., consistency index above a given threshold; see Saaty, 1980), the set of point judgments is re-extracted. Final priorities describing the importance of each element are derived from each set of point judgments, according to the standard procedure, the number of replications being sufficient to obtain a given level of confidence in the results.

The Monte Carlo approach can disregard the particular distribution used to describe the interval judgments; nevertheless, some recommendations are given. Although there are different ways of modeling individual judgments (Rosenbloom, 1996), the choice should reflect a precautionary approach in a safety-based problem. We therefore recommend that

- each expert is only asked a lower bound and an upper bound for his/her judgment;

- a uniform distribution is used in the interval between the lower and upper bounds (i.e., giving equal likelihood to all possible values in the interval judgment) for individual judgments.

This approach giving equal probability to all possible values included in an interval judgment guarantees the maximum safeguard of the residual uncertainty for individual judgments. Furthermore, it resolves the typical problem that experts are inclined to be excessively confident in their single-point estimates (i.e., the most likely) thus giving unreasonably strict ranges near the most likely value (Capen, 1976). In fact, following this AHP based approach, the probability distribution of final priorities is generally more thickened than the probability distributions of the interval judgments (Paulson and Zahir, 1995). This is especially true for a high number of elements to compare.

18.4.2 Case study

The first step in collecting experts' judgments was the identification of the sample to be interviewed. Three departments within the company were selected:

- pipeline design: responsible for designing the network structure (4 experts);
- emergency squad: responsible for locating failures (8 experts);
- operations: responsible for repairing broken pipelines (14 experts).

Although the experts (26) were extremely proficient and experienced, considerable difficulty emerged in determining the density function of failure rate for each pipeline class according to the classical Bayesian approach. It was felt inappropriate to ask experts to give a failure density function or a mean and a standard deviation. In fact, the interviewees were actually not able to say how many failures they expected to see on a kilometer of a given kind of pipe in a year (the situation became even more untenable, if possible, when they were asked to express the corresponding standard deviation or a maximum and minimum value). Therefore, each expert was asked to give just a propensity-to-failure index for each pipeline class. Thus, for each class, from the obtained index values (26 values from 26 experts) the mean and standard deviation were derived and subsequently the failure density function was obtained.

To obtain such a propensity-to-failure index, each expert was asked to pairwise compare the pipeline classes. In the pairwise comparison the judgment is the expression of the relation between two elements that is given, for greater simplicity, in a linguistic way (Saaty, 1980). The linguistic judgment scale is related to a numerical scale (Saaty's proposal has been used, see Table 4). Numerical judgments can be reported in a single matrix of pairwise comparisons (Table 5).

Pipes	1	2	3	4	5	6	7	8
1	1	3	3	3	1/6	1	1/6	3
2	1/3	1	1/4	2	1/6	1/2	1/5	1
3	1/3	4	1	1	1/4	1	1/6	2
4	1/3	1/2	1	1	1/5	1	1/5	1
5	6	6	4	5	1	4	4	5
6	1	2	1	1	1/4	1	1/6	1
7	6	5	6	5	1/4	6	1	4
8	1/3	1	1/2	1	1/5	1	1/4	1

TABLE 5. Example of an expert opinion matrix

Each judgment denotes the dominance of an element in the left column over an element in the first row. The results show the answer to the question: "If you have to lay two new CI pipes, in which of these pipeline classes, on the basis of your experience and knowledge, is a failure more probable, and by how much?" As an example, the number 5 at row 5 and column 8 means that the expert thinks that failures in class 8 are "strongly more probable" than in class 5 (see Table 4). To better use the expert's judgments, interviewed experts were invited to consider, with respect to the failure event, also all other factors (temperature, other underground services,etc.) not expressed in the classes but probably implicitly correlated in the experts' mind. This further underlines the importance of company knowledge collecting, recording and reusing.

Calculating the eigenvector associated to the maximum eigenvalue of an expert's opinion matrix, we obtain the vector of the weights of the i-class (w_i, which sum to 1) and, consequently, an estimate of the propensity to failure index for a given pipeline class as expressed by the experts.

The number of comparisons required to determine the weights for each class is redundant with respect to the strictly necessary figure (a single row of the matrix would be sufficient). However, this redundancy contributes significantly to increase the robustness of the method. In any case, the level of inconsistency can be calculated and, if it exceeds a given threshold, the expert may review judgments more carefully (Saaty, 1980).

The priority weight of the ith class (w_i) can be obtained by taking into account all the judgments (w_{ij}) given by the experts. Since the weights w_i represent for each class the experts' estimate of the propensity-to-failure index, a scale factor was sought to allow translation into the corresponding failure rate in order to integrate it with the value given by data records. In other words, the scale factor allows us to translate the propensity-to-failure index, stemming from the experts' subjective opinions, into the same "measure unit" of the failure rate, derived from the historical data.

Cagno et al. (1998, 2000) defined a scale factor (SF) as

$$\mathrm{SF} = \sum_{i=1}^{8} \lambda_i, \tag{1}$$

where λ_i is the historical failure rate of the ith class.

Since $\sum_{i=1}^{8} w_{ij} = 1$, where w_{ij} is the weight of the jth expert for the ith class, then λ_{ij} (the subjective estimate of the failure rate associated by the jth expert to the ith class) can be determined as $\lambda_{ij} = w_{ij}\mathrm{SF}$.

Once all λ_{ij}s were obtained, the mean and the variance of the empirical distribution of the failure rate within the group of experts for a given class were estimated, highlighting slight diversities in the experts' opinions on the different classes (diversities fully represented by the variance of the distribution). Those means and variances were used by Cagno et al. (1998, 2000) in their analyses.

18.5 Sensitivity analysis

Findings about failure propensity are well described in Cagno *et al.* (1998, 2000) and we refer to them; we just mention the importance of the laying location and the influence of the diameters for those pipelines laid under streets.

In this paper, we are dealing with sensitivity of ranking under changes in the prior specification. We relax some of the assumptions made in Cagno et al. (1998, 2000), mainly the ones subject to criticism, compute the ranges of the posterior mean of each $\lambda_i, i = 1, 8$, and consider all the possible rankings to check if the conclusions drawn in the above mentioned papers are robust with respect to the introduced changes.

18.5.1 Scale factor and variance from the AHP

One of the major drawbacks in the use of the AHP procedure is the need to multiply weights by the scale factor (1). The choice of the scale factor, as done in Cagno et al. (1998, 2000), to be equal to the sum of the historical failure rates of the classes may be questionable, at least from the philosophical viewpoint of considering data to assess prior probabilities. We assume that prior means of failure rates, as expressed by the experts, are still proportional to the weights obtained with the AHP, but their sum ranges in an interval containing SF. We keep the same (Gamma) distribution form for priors and the same variance obtained by the AHP. Therefore, we consider, for each class i, the family of priors

$$\Gamma = \{\Pi_c : \Pi_c \sim \mathcal{G}(c^2 m_i^2/\sigma_i^2, c m_i/\sigma_i^2), c_1 \leq c \leq c_2, \text{ for given } c_1 \leq c_2\},$$

where c is a positive constant, bound to the same interval for all the classes, and m_i and σ_i^2, $\forall i$, are the mean and the variance, respectively, obtained when multiplying the weights from AHP by (1). Note that the prior mean under Π_c is $\mathcal{E}^{\Pi_c}(\lambda_i) = c m_i$.

Suppose $c_1 = 0.5$ and $c_2 = 2$; thus we allow the prior mean to vary between half and double its value in Cagno et al. (1998, 2000). We consider the induced changes in ranking among the classes about their failure propensity (see the first two lines in Table 6). We observe that the findings confirm the previous ones: the pipelines laid under the street are more keen to failures, specially if they are right under the surface.

Another open problem involves the choice of the prior variance from the AHP. We assume that the prior means m_i of failure rates are the same as in Cagno et al. (2000), but the variances are the same for each class and allowed to vary in an interval. We keep the same (Gamma) distribution form for priors. Therefore, we consider, for each class i, the family of priors

$$\Gamma = \{\Pi_c : \Pi_c \sim \mathcal{G}(m_i^2/c, m_i/c), 0 < c_1 \leq c \leq c_2\},$$

Pipeline		1	2	3	4	5	6	7	8
Fixed variance	$c = 0.5$	6	4	1	5	7	8	2	3
Fixed variance	$c = 2$	5	6	1	3	8	7	2	4
Fixed mean	$c = 0.002$	6	5	1	3	7	8	2	4
Fixed mean	$c = 0.2$	6	5	1	4	7	8	3	2
Cagno et al.		6	5	1	3	7	8	2	4

TABLE 6. Sensitivity of ranking

Noting that each Π_c has mean m_i and variance c, with a choice of $c_1 = 0.002$ and $c_2 = 0.2$, we allow for different order of magnitude in the variance and include in the classes all the priors considered by Cagno et al. We consider the induced changes in ranking among the classes about their failure propensity (see the third and fourth lines in Table 6).

We can see that, despite allowing for sensible changes in the prior, the worst cases, that is, the ones with the largest changes in ranking (corresponding to c_1 and c_2 in both cases), are actually not very different from the one considered by Cagno et al. (2000). Therefore, we can see that relaxing, as done above, some questionable aspect does not change the ranking significantly.

18.5.2 Functional form of the prior

Another crucial aspect is the choice of the prior distribution. In Cagno et al. (1998) the Gamma distribution was substituted by a lognormal one, sharing the same mean and variance. It was found that no relevant changes were detected, neither in ranking nor in the estimates of the failure rates (see Table 7). Here we consider, for each class i, the family of all probability measures sharing the same mean m_i and variance σ_i^2 as in Cagno et al. (2000). We obtain the family of priors

$$\Gamma = \{\Pi : \mathcal{E}^{\Pi}(\lambda_i) = m_i, \mathcal{E}^{\Pi}(\lambda_i^2) = \sigma_i^2 + m_i^2\},$$

a special case of the generalised moment class, studied by Betrò et al. (1994) and presented in Betrò and Guglielmi (2000). Upper and lower bounds on the posterior means of the λ_is are given in Table 7. As in Cagno et al. (2000), we present the optimal estimates according to the "posterior regret Γ–minimax" criterion, well described in Vidakovic (2000). We compare our results with the ones in Cagno et al. (1998), for both Gamma and lognormal priors. We can see that the estimates of the failure rates are barely changed whereas the different ranking of classes 1, 2 and 8 can be explained with the very small difference of their estimates in both our analysis and Cagno et al.'s one. We add, for completeness, the estimate obtained with the hierarchical model studied by Masini (1998) and the maximum likelihood estimate. We observe that the former is, usually, between our estimates and the MLE's.

Pipeline	1	2	3	4	5	6	7	8
Lower	0.050	0.041	0.172	0.093	0.035	0.036	0.095	0.058
Upper	0.117	0.116	0.321	0.125	0.111	0.059	0.239	0.102
Minimax	0.083	0.078	0.247	0.109	0.073	0.047	0.167	0.080
Gamma	0.075	0.081	0.231	0.105	0.066	0.051	0.143	0.094
Lognormal	0.074	0.082	0.217	0.102	0.069	0.049	0.158	0.092
Hierarchical	0.074	0.085	0.170	0.160	0.066	0.064	0.136	0.142
MLE	0.072	0.094	0.177	0.115	0.067	0.060	0.131	0.178

TABLE 7. Ranges for fixed mean and variance

18.6 Discussion

Cagno et al. (1998, 2000) conclude their papers by reporting that the company management was satisfied with their methodology, based upon integration of (relatively scarce) historical data and experts' opinions. Their conclusions are slightly different from those obtained by looking at the observed failure rate of each class, sometimes affected by shortage of data. As an example, consider the pipelines in class 8: there were only 3 failures over 2.813 Km (out of a network longer than 300 Km) and, according to the MLE (see Table 7), this class (with large diameter, laid deeply under the street) was the one most subject to failures. The conclusion was very different when considering experts' judgments.

Cagno et al. found that laying location is the factor mainly affecting failures; in fact, classes 3, 4, 7 and 8 are the ones with the largest failure rates. Among these classes, 3 and 7 (corresponding to pipelines not deeply laid) were the worst. Therefore, the replacement strategy should aim to substitute pipelines under the street and laid near its surface.

Their conclusions should be reinforced, since we have shown that the replacement priorities, as set by the estimates of the failure rates of each class, are quite insensitive to the changes in the prior we have considered.

We have presented the AHP method to assess prior probabilities. We have suggested a robust Bayesian approach to enhance some strengths (highlighted in the paper) of the method to cope with some problems involved in experts' elicitation. Nevertheless, we deem that more work is needed to exploit the AHP properly.

Finally, we want to mention two works which might "robustify" inferences in our context, even if they were developed mainly for nonhomogeneous Poisson processes. Saccuman (1998) considered failure rates dependent on covariates, instead of considering separately the failure rates in the classes determined by the values of the covariates. Masini (1998) considered a hierarchical model and applied it to the same gas data we have been considering in this paper (see her results in Table 7).

References

BACON, J.F. (1988). *King report January 1988*. London: British Gas Corporation.

BETRÒ, B. and GUGLIELMI, A. (2000). Methods for global prior robustness under generalized moment conditions. In *Robust Bayesian Analysis* (D. Ríos Insua and F. Ruggeri, eds.). New York: Springer-Verlag.

BETRÒ, B., MĘCZARSKI, M. and RUGGERI, F. (1994). Robust Bayesian analysis under generalized moments conditions. *Journal of Statistical Planning and Inference*, **41**, 257–266.

CAGNO, E., CARON, F., MANCINI, M. and RUGGERI F. (1998). On the use of a robust methodology for the assessment of the probability of failure in an urban gas pipe network. In *Safety and Reliability, vol. 2* (S. Lydersen, G.K. Hansen and H.A. Sandtorv, eds.), 913–919, Rotterdam: Balkema.

CAGNO, E., CARON, F., MANCINI, M. and RUGGERI F. (2000). Using AHP in determining prior distributions on gas pipeline failures in a robust Bayesian approach. *Reliability Engineering and System Safety*, **67**, 275–284.

CAPEN, E.C. (1976). The difficulty of assessing uncertainty. *Journal of Petroleum Technology*, 843–850.

CIPOLLETTI, D. and MANCINI, M. (1997). Valutazione della probabilità di rottura di tubazioni interrate per la distribuzione di gas in ambito metropolitano. Un approccio bayesiano robusto. *B.Sc. Dissertation*, Dipartimento di Ingegneria Meccanica, Politecnico di Milano.

EXPERT CHOICE PROFESSIONAL 9.5 (1999). *Expert Choice*. Pittsburgh: Expert Choice Inc.

MASINI, L. (1998). Un approccio bayesiano gerarchico nell'analisi dell'affidabilità dei sistemi riparabili. *B.Sc. Dissertation*, Dipartimento di Matematica, Università degli Studi di Milano.

PAULSON, D. and ZAHIR, S. (1995). Consequences of uncertainty in the analytic hierarchy process: a simulation approach. *European Journal of Operational Research*, **87**, 45–56.

RÍOS INSUA, D., RUGGERI, F. and VIDAKOVIC B. (1995). Some results on posterior regret Γ–minimax estimation. *Statistics and Decision*, **13**, 315–331.

ROSENBLOOM, E. S. (1996). A Probabilistic Interpretation of the Final Rankings in AHP. *European Journal of Operational Research*, **96**, 371–378.

SAATY, T. L. (1980). *The Analytic Hierarchy Process.* New York: Mc Graw-Hill.

SAATY, T. L. (1994). *Fundamentals of Decision Making and Priority Theory with the Analytic Hierarchy Process.* Pittsburgh: RWS Publications.

SAATY, T. L. and VARGAS L.G. (1987). Uncertainty and Rank Order in the analytic hierarchy process. *European Journal of Operational Research*, **32**, 107–117.

SAATY, T. L. and VARGAS L.G. (1998). Diagnosis with dependent symptoms: Bayes theorem and the analytic hierarchy process. *Operations Research*, **46**, 491–502.

SACCUMAN, E. (1998). Modelli che incorporano covariate nell'analisi bayesiana dell'affidabilità di sistemi riparabili. *B.Sc. Dissertation*, Dipartimento di Matematica, Università degli Studi di Milano.

VIDAKOVIC, B. (2000). Γ–minimax: a paradigm for conservative robust Bayesians. In *Robust Bayesian Analysis* (D. Ríos Insua and F. Ruggeri, eds.). New York: Springer-Verlag.

19

Robust Bayesian Analysis in Medical and Epidemiological Settings

Bradley P. Carlin and María-Eglée Pérez

ABSTRACT Many medical and epidemiological professionals cite their distaste for informative priors as a prime reason for their ongoing aversion to Bayesian methods. In this paper we attempt to ease these concerns by investigating Bayesian robustness in such settings. Past attempts in this regard have demonstrated either the range of results possible using a certain class of priors ("forward robustness") or the range of priors leading to a particular result ("backward robustness"). Application areas of particular interest include longitudinal data studies, clinical trial monitoring, survival analysis, and spatial epidemiology. After a brief review of the relevant methodology we consider two specific application areas. First, in the context of AIDS clinical trials we analyze a dataset that compared the effectiveness of the drug pyrimethamine versus a placebo in preventing toxoplasmic encephalitis. Our method uses nonparametric classes of prior distributions which attempt to model prior neutrality regarding the effect of the treatment. The resulting classes of prior distributions are reasonably wide, so that a clear conclusion emerging therefrom should be regarded as convincing by a broad group of potential consumers. Turning to spatial disease mapping, we investigate the impact of changes in the "heterogeneity plus clustering" priors commonly used to model excess spatial variation. In particular, we use the notion of Bayesian learning about the proportion of excess variability due to clustering to see whether a prior can be determined that offers a "fair" prior balance between these two components while exerting little influence on the posterior.

Key words: AIDS, clinical trials, disease mapping, Markov chain Monte Carlo methods, prior neutrality, prior partitioning, spatial statistics.

19.1 Introduction

Over the last decade or so, the expansion of Bayesian methods into biostatistical practice in general (and medicine and epidemiology, in particular) has been substantial. This expansion has been due in large part to the advent of Markov chain Monte Carlo (MCMC) methods and associated generalist software (e.g., the BUGS language; Spiegelhalter et al., 1995a), but also to increasing recognition by practitioners that traditional, frequentist techniques were inadequate. As such, in this chapter we make no attempt at an exhaustive review, but rather merely seek to elucidate some of the ways in which robust Bayesian methods and practices have found application in the fields of medicine and epidemiology. Similarly, since many previous chapters of this volume have carefully elucidated the necessary key methodological tools, we do not review them here, but instead focus only on those tools most relevant in our context.

Naturally the most common approach to Bayesian robustness in biomedical settings is through *sensitivity analysis*, wherein we simply make various modifications to a particular modeling assumption (say, some aspect of the prior) and recompute the posterior quantities of interest. If our resulting interpretations or decisions are essentially unaffected by this change, we say the Bayesian model is robust with respect to the assumption in question. This "forwards" approach to robustness is conceptually and implementationally simple, but it does not free us from careful development of the original prior, which must still be regarded as a reasonable baseline. Unlike many applied settings (e.g., business decision making), in biomedical work this approach can be impractical, since the prior beliefs and vested interests of the potential consumers of our analysis may be very broad. For example, the results of a Bayesian clinical trial analysis might ultimately be read by doctors in clinical practice, epidemiologists, government regulatory workers, legislators, members of the media, and of course, patients and patient advocate groups. These groups are likely to have widely divergent opinions as to what constitutes "reasonable" prior opinion, ranging from quite optimistic (e.g., a clinician who has seen a few patients respond well to the drug being tested) to rather pessimistic (e.g., a regulatory worker who has seen many similar drugs emerge as ineffective). What is needed is a method for communicating the robustness of our conclusions to *any* prior input a consumer deems appropriate.

Carlin and Louis (1996a) suggest an alternate, "backwards" approach to this problem. Suppose that, rather than fix the prior and compute the posterior distribution, we fix the posterior (or set of posterior distributions) that produce a given conclusion, and determine which prior inputs are consistent with this desired result, given the observed data. The reader would then be free to determine whether the outcome was reasonable according to whether the prior class that produced it was consistent with his or her own prior beliefs. Carlin and Louis (1996a) refer to this approach simply as *prior partitioning* since it subdivides the prior class based on possible outcomes, though it is important to remember that such partitions also depend on the data and the decision to be reached.

To illustrate the basic idea, consider the point null testing scenario $H_0 : \theta = \theta_0$ versus $H_1 : \theta \neq \theta_0$. Without loss of generality, set $\theta_0 = 0$. Suppose our data x has density $f(x|\theta)$, where θ is an unknown scalar parameter.

Let π represent the prior probability of H_0, and $G(\theta)$ the prior cumulative distribution function (cdf) of θ conditional on $\{\theta \neq 0\}$. The complete prior cdf for θ is then $F(\theta) = \pi I_{[0,\infty)}(\theta) + (1-\pi)G(\theta)$, where I_S is the indicator function of the set S. Hence the posterior probability of the null hypothesis is

$$P_G(\theta = 0|x) = \frac{\pi f(x|0)}{\pi f(x|0) + (1-\pi)\int f(x|\theta)dG(\theta)} \,. \tag{1}$$

Prior partitioning seeks to characterize the G for which this probability is less than or equal to some small probability $p \in (0, 1/2)$, in which case we reject the null hypothesis. (Similarly, we could also seek the G leading to $P_G(\theta \neq 0|x) \leq p$, in which we would reject H_1.) Elementary calculations show that characterizing this class of priors $\{G\}$ is equivalent to characterizing the set $\mathcal{H}_c$, defined as

$$\mathcal{H}_c = \left\{ G : \int f(x|\theta)dG(\theta) \geq c = \frac{1-p}{p}\frac{\pi}{1-\pi} f(x|0) \right\} \,. \tag{2}$$

Carlin and Louis (1996a) establish results regarding the features of $\mathcal{H}_c$ and then use these results to obtain sufficient conditions for $\mathcal{H}_c$ to be nonempty for classes of priors that satisfy various moment and percentile restrictions. The latter are somewhat more useful, since percentiles and tail areas of the conditional prior G are transform-equivariant, and Chaloner et al. (1993) have found that elicitees are most comfortable describing their opinions through a "best guess" (mean, median or mode) and a few relatively extreme percentiles (say, the 5th and the 95th).

Sargent and Carlin (1996) extend this general approach to the case of an interval null hypothesis, that is, $H_0 : \theta \in [\theta_L, \theta_U]$ versus $H_1 : \theta \notin [\theta_L, \theta_U]$. This formulation is useful in the context of clinical trial monitoring, where $[\theta_L, \theta_U]$ is thought of as an *indifference zone*, within which we are indifferent as to the use of treatment or placebo. For example, if positive values of θ indicated superiority of a treatment associated with increased costs or toxicities, we might take $\theta_U > 0$ (instead of merely $\theta_U = 0$), thus insisting on a "clinically significant" benefit to offset the treatment's costs. Let π again denote the prior probability of H_0, and let $g(\theta)$ now correspond to the prior density of θ given $\theta \notin [\theta_L, \theta_U]$. If we make the simplifying assumption of a uniform prior over the indifference zone, the complete prior density for θ may be written as

$$p(\theta) = \frac{\pi}{\theta_U - \theta_L} I_{[\theta_L, \theta_U]}(\theta) + (1-\pi)g(\theta). \tag{3}$$

Sargent and Carlin (1996) derive expressions similar to (1) and (2) under the nonparametric percentile restrictions mentioned above. However, these rather weak restrictions lead to prior classes that, while plausible, are often too broad for practical use. As such, we might consider a sequence of increasingly tight restrictions on the shape and smoothness of permissible priors, which in turn enable increasingly informative results. For example, we might retain the mixture form (3), but now restrict $g(\theta)$ to some particular parametric family. Carlin and Sargent (1996) refer to such a prior as "semiparametric" since the parametric form for g does not cover the indifference zone $[\theta_L, \theta_U]$, although since we have adopted another parametric form over this range (the uniform) one might argue that "biparametric" or simply "mixture" would be better names.

In the remainder of our paper we consider two biomedical settings where it is important that any Bayesian method be not only robust, but also "prior neutral," in the sense that the informative content of any prior used should be symmetric with respect to the null and the alternative. First, Section 19.2 considers a new approach to the Bayesian analysis of clinical trials data, illustrating with an example from a recent AIDS clinical trial. Section 19.3 then turns to robust Bayesian methods for spatial disease mapping, illustrating with the well-known Scottish lip cancer data of Clayton and Kaldor (1987). Finally, Section 19.4 discusses our findings and presents some possible avenues for future research.

19.2 Clinical trial monitoring and analysis

Phase III clinical trials are large-scale trials designed to identify patients who are best treated with a drug whose safety and effectiveness have already been reasonably well established. In order to ensure scientific integrity and ethical conduct, decisions concerning whether or not to continue a clinical trial based on the accumulated data are made by an independent group of statisticians, clinicians and ethicists who form the trial's *data and safety monitoring board*, or DSMB. These boards have been a standard part of clinical trials practice (and NIH policy) since the early 1970s. As a result, the trial's statisticians require efficient algorithms for computing posterior summaries of quantities of interest to the DSMB.

From a Bayesian point of view, this also implies the need to select a prior distribution (or a set of prior distributions) on which all further analyses will be based. Kass and Greenhouse (1989) point out that, "randomization is ethically justifiable when a *cautious reasonable skeptic* would be unwilling to state a preference in favor of either the treatment or the control." If we assume that it is ethical to conduct a clinical trial, then an appropriate baseline prior for its monitoring and analysis might be one that reflects the opinions of a cautious, reasonable skeptic. Here one might naturally think of a reference prior, but this might not add sufficient "skepticism," as we now describe.

Suppose that the effect of the drug is summarized by means of a parameter θ, such that $\theta < 0$ if patients treated with the drug perform better than those receiving placebo, and $\theta > 0$ otherwise. $\theta = 0$ means that both treatments (drug and placebo) have the same behavior with respect to the outcome of interest (take, for example, the log-odds of the probabilities of survival for both groups). A posterior quantity of interest for the DSMB might be the posterior probability that $\theta < 0$ given the accumulated data, $P(\theta < 0 \mid \mathbf{x})$. Following the approach of Section 19.1, if this probability is either smaller than p or greater than $1-p$, the trial should be stopped. (Note that the precise selection of p *implicitly* determines a utility function for this problem; see Berry and Ho, 1988, Stangl, 1995, and Carlin et al., 1998, for Bayesian methods in clinical trial analysis that determine the utility *explicitly*.) Our objective in this section is proposing classes of priors for θ which we believe reflect prior neutrality with respect to the treatment and the control and which are wide enough so that a clear conclusion emerging therefrom should be regarded as convincing by most observers.

In a similar situation arising in an age-discrimination trial, Kadane (1990) addresses the problem of representing prior neutrality, stating symmetry around zero and unimodality as reasonable features for a neutral prior for θ to have, and choosing the normal family with mean zero and standard deviations 1, 2, 4 and ∞ for calculating posterior quantities. Pérez and Pericchi (1994) suggested other neutral and almost neutral priors for the same problem, but these reflect only a discrete class of priors, and so do not satisfactorily address the issue of robustness in the inference. Though Pérez and Pericchi explored classes of prior distributions, these classes don't articulate well with the idea of prior neutrality.

A very natural class for representing the opinion of a cautious, reasonable skeptic is the class of all unimodal priors symmetric around 0. However, Kadane et al. (1999) show that this class turns out to be too broad, in the sense that it leads to trivial bounds (one being the prior value 1/2) for $P(\theta < 0 \mid \mathbf{x})$, provided the likelihood $f(x|\theta)$ goes to 0 exponentially fast as $|\theta| \to \infty$. After finding similar behavior in several subsets of the class of unimodal priors symmetric around zero (the ones obtained by fixing its height at 0, fixing one of its quantiles, or bounding its variance from below), Kadane et al. conclude that a subclass of all unimodal priors symmetric around zero can lead to a nontrivial bound only if it avoids putting too much probability close to zero and avoids allowing too much probability to be put on extremely high and low values of θ.

For achieving such a class, Kadane et al. focus on predictive distributions. Berger (1994) points out that the predictive distribution is in fact the likelihood of the prior (for a fixed likelihood), and a limitation of some robust Bayesian analyses is that robustness might be missing due to the inclusion of priors which have a very low (predictive) likelihood. In other words, lack of robustness might be caused by priors which are ruled out by the data. As the set of possible data outcomes involved in the age-discrimination trial is discrete and finite, Kadane et al. state that neutrality might be considered in terms of not being too surprised at any way the data might come out. More formally, suppose that the prior is $\pi(\theta)$, and $f_i(\theta)$

is the likelihood corresponding to a hypothetical observed value i, where i runs over the set of all possible data outcomes I. Let

$$g_\pi(i) = \int_{-\infty}^{\infty} f_i(\theta)\pi(\theta)d\theta. \tag{4}$$

Then the restricted *neutral prior class* A can be defined as

$$\begin{aligned} A \quad = \{\pi(\theta): \quad & \pi(\theta) \text{ is unimodal and symmetric around } 0, \\ & \text{and } g_\pi(i) \geq \varepsilon \text{ for all } i \in I\} \end{aligned} \tag{5}$$

The parameter ε of this class is then the minimum prior predictive probability of the possible data. The idea of this class is that it constrains the neutral arbitrator to have probability at least $\varepsilon > 0$ on each possible data point. In other words, only priors which have a nonnegligible likelihood, for all possible data, are allowed in this neutrality class. Kadane et al. determine restrictions on ε for getting a nonempty class which leads to nontrivial posterior bounds for $P(\theta < 0|x)$, and show how to find those bounds in the case where I is a discrete finite set.

We now extend these ideas to the case where the set of possible data outcomes I is uncountably infinite. To handle this problem, we replace the true sample space I by a representative grid of plausible hypothetical values. To illustrate this, suppose $f_i(\theta)$ can be reasonably well approximated by a normal density $\phi(\theta|m, s)$ having mean $m \in M$ and standard deviation $s \in S$, where M and S are discrete finite sets of grid points corresponding to plausible likelihoods. Define

$$g_\pi(m, s) = \int_{-\infty}^{\infty} \phi(\theta|m, s)\pi(\theta)d\theta. \tag{6}$$

A neutral prior class similar to the class A in (5) can then be defined as

$$\begin{aligned} \Gamma_\varepsilon \quad = \{\pi(\theta): \quad & \pi(\theta) \text{ is unimodal and symmetric around } 0, \\ & \text{and } g_\pi(m, s) \geq \varepsilon \text{ for all } m \in M, s \in S\}. \end{aligned}$$

This is a class of neutral priors such that none of them is going to be ruled out by data leading to parameters (m, s), $m \in M$, $s \in S$ for the normal approximation to the likelihood of θ.

Bounds for $P(\theta < 0|\mathbf{x})$ over the class Γ_ε can be obtained using the following procedure, similar to the one indicated in Kadane et al. (1999). Using Khinchine's representation, the set of priors in Γ_ε can be expressed as the set of distribution functions F satisfying $\pi(\theta) = \int_\theta^\infty (1/a)dF(a)$, where F is a distribution function in the set

$$\mathcal{F} = \left\{ F(\cdot): \int_0^\infty dF(a) = \frac{1}{2}, \int_0^\infty f^*_{m,s}(a)dF(a) \geq \varepsilon > 0, \ \forall\, m \in M,\ s \in S \right\}$$

and

$$\begin{aligned} f^*_{m,s}(a) &= \frac{1}{a}\int_0^a (\phi(-\theta|m,s) + \phi(\theta|m,s))d\theta \\ &= \frac{1}{a}\left[\Phi\left(\frac{a-m}{s}\right) - \Phi\left(\frac{-a-m}{s}\right)\right] \end{aligned}$$

for $a > 0$. Here, $\Phi(.)$ is the distribution function of a standard normal variable, and $f^*_{m,s}(0)$ is defined by continuity. The quantity of interest is the probability that $\theta < 0$ when the observed data is $\mathbf{x}$ and the prior is $\pi(\theta) \in \Gamma_\varepsilon$. This can be rewritten as

$$P^\pi(\theta < 0 \mid \mathbf{x}) = \left[1 + \frac{\int_{\theta \geq 0} \phi(\theta|\mu_\mathbf{x}, \sigma_\mathbf{x})\pi(\theta)d\theta}{\int_{\theta \geq 0} \phi(-\theta|\mu_\mathbf{x}, \sigma_\mathbf{x})\pi(\theta)d\theta}\right]^{-1}, \qquad (7)$$

where $\phi(\theta|\mu_\mathbf{x}, \sigma_\mathbf{x})$ is the normal approximation to the likelihood with parameters $\mu_\mathbf{x}$ and $\sigma_\mathbf{x}$ corresponding to the observed data $\mathbf{x}$. Equation (7) can equivalently be written as

$$P^F(\theta < 0 \mid \mathbf{x}) = \left[1 + \frac{\int_0^\infty f^1_\mathbf{x}(a)dF(a)}{\int_0^\infty f^2_\mathbf{x}(a)dF(a)}\right]^{-1}, \qquad (8)$$

where

$$\begin{aligned} f^1_\mathbf{x}(a) &= \tfrac{1}{a}\int_0^a \phi(\theta|\mu_\mathbf{x}, \sigma_\mathbf{x})d\theta = \tfrac{1}{a}\left[\Phi\left(\tfrac{a-\mu_\mathbf{x}}{\sigma_\mathbf{x}}\right) - \Phi\left(-\tfrac{\mu_\mathbf{x}}{\sigma_\mathbf{x}}\right)\right], \\ f^2_\mathbf{x}(a) &= \tfrac{1}{a}\int_0^a \phi(-\theta|\mu_\mathbf{x}, \sigma_\mathbf{x})d\theta = \tfrac{1}{a}\left[\Phi\left(-\tfrac{\mu_\mathbf{x}}{\sigma_\mathbf{x}}\right) - \Phi\left(\tfrac{-a-\mu_\mathbf{x}}{\sigma_\mathbf{x}}\right)\right], \end{aligned}$$

and $F(\cdot) \in \mathcal{F}$. Then the supremum of the posterior probability that $\theta < 0$ can be written as

$$\sup_{\pi \epsilon A} P^\pi(\theta < 0 \mid \mathbf{x}) = \sup_{F \epsilon \mathcal{F}} P^F(\theta < 0 \mid \mathbf{x}) = \left[1 + \inf_{F \in \mathcal{F}} \frac{\int_0^\infty f^1_\mathbf{x}(a)dF(a)}{\int_0^\infty f^2_\mathbf{x}(a)dF(a)}\right]^{-1}. \qquad (9)$$

By the linearization algorithm (Lavine et al., 1993), the infimum in (9) is the unique solution λ_0 of the equation in λ:

$$\inf_{F \in \mathcal{F}} \int_0^\infty \left[f^1_\mathbf{x}(a) - \lambda f^2_\mathbf{x}(a)\right] dF(a) = 0. \qquad (10)$$

Once λ_0 has been found, $\sup_{\pi \in \Gamma_\varepsilon} P^\pi(\theta < 0 \mid \mathbf{x}) = (1 + \lambda_0)^{-1}$. Using Kemperman (1987) (see also Salinetti (1994) and Liseo *et al.* (1996)), we can rewrite equation (10) as

$$0 = \sup_{\substack{d_{m,s}\geq 0\\ m\in M\\ s\in S}} \left\{ \varepsilon \sum_{\substack{m\in M\\ s\in S}} d_{m,s} + \inf_{F\in\mathcal{F}_0} \int_0^\infty \left[f^1_{\mathbf{x}}(a) - \lambda f^2_{\mathbf{x}}(a) - \sum_{\substack{m\in M\\ s\in S}} d_{m,s} f^*_{m,s}(a) \right] dF(a) \right\}, \tag{11}$$

where $\mathcal{F}_0$ is the class $\{F(\cdot) : \int_0^\infty dF(a) = 1/2\}$.

This last equation has important consequences. First, the internal infimum occurs at an F that puts all its probability at a single point a. This means that the extremum will occur at a single uniform distribution for θ. Thus

$$\inf_{F\epsilon\mathcal{F}_0} \int_0^\infty \left[f^1_{\mathbf{x}}(a) - \lambda f^2_{\mathbf{x}}(a) - \sum_{\substack{m\in M\\ s\in S}} d_{m,s} f^*_{m,s}(a) \right] dF(a)$$
$$= \inf_{a\geq 0} \frac{1}{2} \left[f^1_{\mathbf{x}}(a) - \lambda f^2_{\mathbf{x}}(a) - \sum_{\substack{m\in M\\ s\in S}} d_{m,s} f^*_{m,s}(a) \right],$$

which permits reduction of (11) to

$$\sup_{\substack{d_{m,s}\geq 0\\ m\in M\\ s\in S}} \inf_{a\geq 0} \left\{ f^1_{\mathbf{x}}(a) - \lambda f^2_{\mathbf{x}}(a) - \sum_{\substack{m\in M\\ s\in S}} (f^*_{m,s}(a) - 2\varepsilon) d_{m,s} \right\} = 0. \tag{12}$$

¿From (12), if for any pair (m, s), $f^*_{m,s}(a) < 2\varepsilon$, allowing the corresponding $d_{m,s}$ to go to infinity results in a sup of infinity, so (12) cannot be satisfied. Hence the supremum is attained when $d_{m,s} = 0$ for all (m, s). Thus (12) further simplifies to finding a value λ_0 of λ such that

$$\inf_{a\geq 0} \left\{ (f^1_{\mathbf{x}}(a) - \lambda f^2_{\mathbf{x}}(a)) I_{\{a\geq 0: f^*_{m,s}(a)\geq 2\varepsilon, \forall m\in M,\ s\in S\}}(a) \right\} = 0. \tag{13}$$

Now there are two cases to be considered separately. Since the supremum of $P^\pi(\theta < 0 \mid \mathbf{x})$ corresponds to a small value of λ, find the value of a for which $f^1_{\mathbf{x}}(a)/f^2_{\mathbf{x}}(a)$ is a minimum. If that value of a satisfies the constraint $\min_{\substack{m\in M\\ s\in S}} f^*_{m,s}(a) > 2\varepsilon$, then the supremum has been found. If not, then the constraint is binding. In this case, because $f^1_{\mathbf{x}}(a) - \lambda f^2_{\mathbf{x}}(a)$ is continuous in a, the infimum in (13) occurs when $f^*_{m,s}(a) = 2\varepsilon$ for some $m \in M$, $s \in S$.

Thus the search for a solution of (13) in the second case can be made at the points a at which

$$\min_{\substack{m \in M \\ s \in S}} f^*_{m,s}(a) = 2\varepsilon, \tag{14}$$

and then $\lambda(a) = f^1_{\mathbf{x}}(a)/f^2_{\mathbf{x}}(a)$. If there are several points a satisfying (14), the smallest $\lambda(a)$ in the set corresponds to the infimum in (9). This can be accomplished by a one-dimensional search over possible values a. Finally, to find $\inf_{\pi \in \Gamma_\varepsilon} P^\pi(\theta < 0 \mid \mathbf{x})$, we simply reverse the roles of inf and sup in (9). This can be done by reversing the roles of $f^1_{\mathbf{x}}(a)$ and $f^2_{\mathbf{x}}(a)$ in each of the subsequent formulas.

We now illustrate the approach in the specific context of the Community Programs for Clinical Research on AIDS toxoplasmic encephalitis (TE) prophylaxis trial. When the degree of immune damage becomes sufficiently severe, an HIV-infected person may develop a specific subset of more than 20 infections, several cancers, a variety of neurological abnormalities including severe declines in mental function, and wasting. Among the most ominous infections is encephalitis due to *Toxoplasma gondii*. This infection is the cause of death in approximately 50% of persons who develop it and the median survival is approximately six months. Additional clinical and immunological background concerning TE is provided in the review paper by Carlin et al. (1995).

Our study is a double-blind randomized TE prophylaxis trial comparing the drug pyrimethamine to placebo; a previous robust Bayesian analysis of this trial can be found in Carlin and Sargent (1996). All patients entered into the study had either an AIDS-defining illness or a CD4 count (a blood-borne measurement for which higher levels indicate a healthier immune system) of less than 200. In addition, all had a positive titre for *Toxoplasma gondii* and were therefore at risk for TE. As described in the report by Jacobson et al. (1994), the trial's data and safety monitoring board met on three occasions after the start of the trial in September of 1990 to assess its progress and determine whether it should continue or not. These three meetings analyzed the data available as of the file closing dates 1/15/91, 7/31/91, and 12/31/91, respectively. At its final meeting, the board recommended stopping the trial based on an informal stochastic curtailment rule: using classical significance tests, the pyrimethamine group had not shown significantly fewer TE events up to that time, and due to the low TE rate a significant difference was judged unlikely to emerge in the future. An increase in the number of deaths in the pyrimethamine group was also noted, but this was not a stated reason for the discontinuation of the trial (although subsequent follow-up confirmed this mortality increase). The recommendation to terminate the study was conditional on the agreement of the protocol chairperson after unblinding and review of the data. As a result, the trial did not actually stop until 3/30/92, when patients were instructed to discontinue their study medication.

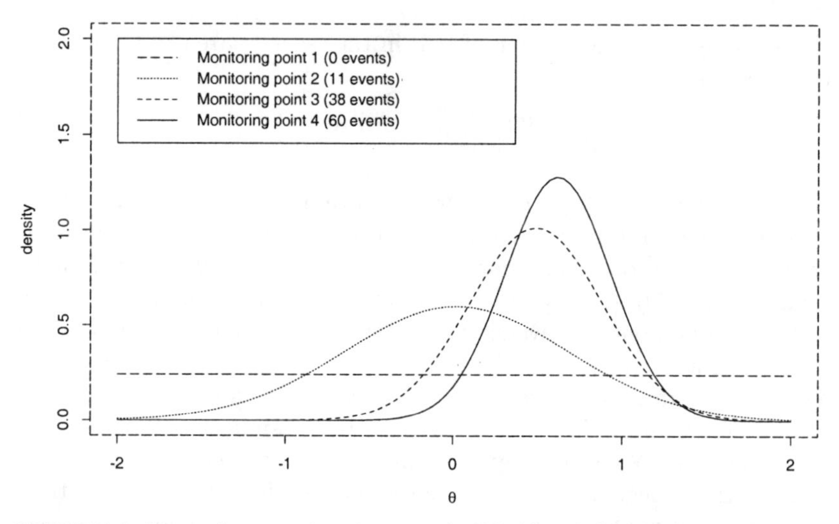

FIGURE 1. Normal approximations to the likelihood for the treatment effect, TE trial data. Endpoint is TE or death; covariate is baseline CD4 count.

In a Bayesian reanalysis of this data, Carlin et al. (1993) employed a proportional hazards likelihood using the time from randomization until development of TE or death as the response variable. Specifically, their model used two covariates for each patient: baseline CD4 cell count, and a treatment effect indicator (1 for active drug, 0 for placebo). Denoting the parameters which correspond to these two covariates as β and θ, respectively, we obtain a marginal partial likelihood for θ by numerically integrating β out of the Cox partial likelihood. Negative values of θ correspond to an efficacious treatment; the relatively low support for these values at the final two monitoring points suggests that perhaps the trial should be stopped and the treatment rejected. Normal approximations to the standardized Cox partial likelihood at each of the four data monitoring points are shown in Figure 1. We will base our inference for θ on these normal approximations.

Each outcome of the trial produces different values for the parameters of the normal approximation. Stating that a cautious, reasonable skeptic shouldn't be too surprised at any data outcome is equivalent to saying that he or she won't be too surprised at observing a wide set of parameters for this approximation. For this analysis we defined our likelihood grids as follows: for means, $M = \{-2.0, -1.9, -1.8, \ldots, 1.8, 1.9, 2.0\}$; for standard deviations, $S = \{0.1, 0.2, \ldots, 2.0\}$. Note that $\theta = -2.0$ corresponds to a reduction of 86% in hazard for the treatment relative to control, significantly greater than the target reduction of 50% specified in the trial protocol. In a similar way, $\theta = 2$ indicates that the hazard rate in the control group is 86% lower than the hazard for treatment group. So, this class states that a cautious, reasonable skeptic shouldn't be surprised by results that support

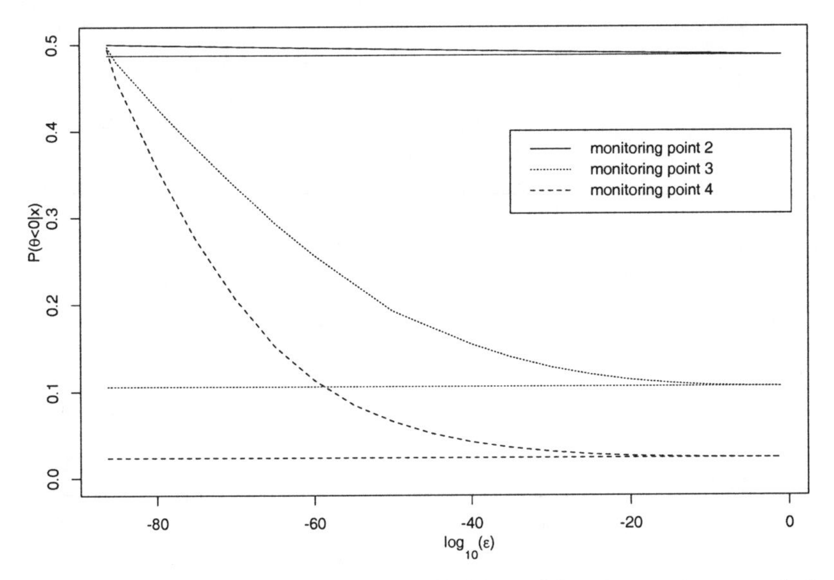

FIGURE 2. Bounds for the posterior probability of $\theta < 0$ over the neutral class Γ_ε for different values of ε, TE trial data.

either the treatment or the placebo. Different grids can be defined on the basis of further design considerations.

Figure 2 shows the bounds for $P^\pi(\theta < 0|\mathbf{x}_i)$, where $\mathbf{x}_i$ is the data collected at monitoring point i, $i = 2, 3, 4$ and $\pi \in \Gamma_\varepsilon$, for different values of $\log_{10}(\varepsilon)$. At monitoring points 3 and 4, these bounds are very near 0 even for very small values of ε, suggesting the need to stop the trial.

Other posterior probabilities are also of interest, especially if the treatment is preferred only if it reduces the hazard rate by some meaningful amount. Following Carlin et al. (1993), we can concentrate on $P(\theta < \theta_L|\mathbf{x})$, where $\theta_L = \log(0.75) = -0.288$ (corresponding to a reduction of at least 25% in relative hazard). Bounds for this posterior probability over the class Γ_ε can be obtained in a similar way and are shown in Figure 3. The message of this graph is even clearer: regardless of the value of ε, the trial should be stopped at the penultimate monitoring point.

19.3 Spatial epidemiology

Bayes and empirical Bayes models for spatial data aggregated by geographic region have begun to see widespread use in spatial epidemiology, especially in the creation of disease maps. Developed by Clayton and Kaldor (1987) and refined by Besag et al. (1991), these models typically assume the observed disease count in region i, Y_i, has a Poisson distribution with mean $E_i e^{\mu_i}$, where E_i is an expected disease count (perhaps obtained via refer-

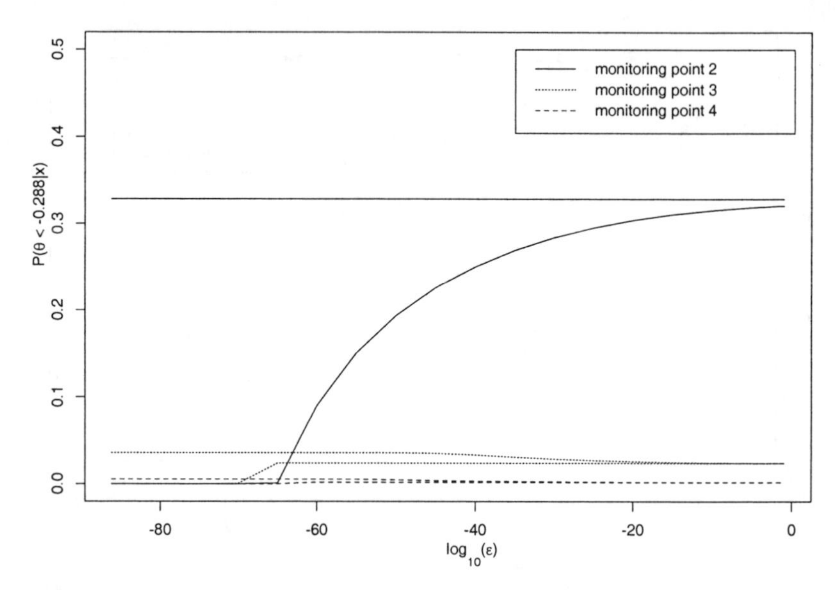

FIGURE 3. Bounds for the posterior probability of $\theta < -0.2880$ over the neutral class Γ_ε for different values of ε, TE trial data.

ence to an external standard table) and μ_i is a log-relative risk of disease, modeled linearly as

$$\mu_i = \mathbf{x}_i'\boldsymbol{\beta} + \theta_i + \phi_i,\ i = 1, \ldots, I. \tag{15}$$

Here the $\mathbf{x}_i$ are explanatory spatial covariates, while $\boldsymbol{\beta}$ is a vector of fixed effects. The θ_i capture *heterogeneity* among the regions via the mixture specification $\theta_i \overset{iid}{\sim} N(0\,,\,1/\tau_h)$, while the ϕ_i capture regional *clustering* by assuming that

$$\phi_i \mid \phi_{j \neq i} \sim N(\bar{\phi}_i\,,\,1/(n_i\tau_c)), \tag{16}$$

where n_i is the number of "neighbors" of region i, and $\bar{\phi}_i = n_i^{-1}\Sigma_{j\in\partial_i}\,\phi_j$ with ∂_i denoting the neighbor set of region i. The usual assumption is that regions are neighbors if and only if they are adjacent on the map, though other (e.g., distance-based) modifications are often considered. This distribution for $\boldsymbol{\phi} \equiv \{\phi_i\}$ is called an *intrinsically* or *conditionally autoregressive* specification, which for brevity we typically write in compact notation as $\boldsymbol{\phi} \sim CAR(\tau_c)$.

Model (15) with the CAR prior formulation has several quirks that cloud its complete specification. First, the CAR prior is translation invariant, since an arbitrary constant could be added to all of the ϕ_i without changing the joint probability specification. This necessitates the addition of a identifiability-preserving constraint (say, $\sum_{i=1}^{I}\phi_i = 0$), which is awkward theoretically but easy to implement "on the fly" during an MCMC algorithm by recentering the $\phi_i^{(g)}$ samples around their own mean at the end of each iteration g. Even with this correction to the prior, only the sum of the two random effects, $\eta_i \equiv \theta_i + \phi_i$, is identified by the datapoint Y_i, so the effective dimension of the full model is often much smaller than the actual parameter count.

This identifiability problem is in some sense a non-issue for Bayesians, since as observed by Besag et al. (1995) and others, even under improper priors for the θ_i and ϕ_i, MCMC algorithms may still operate on the resulting overparametrized space, with convergence still obtaining for the *proper embedded posterior* (that is, the lower-dimensional parameter vector having a unique integrable posterior distribution). Of course, warnings of apparent but false convergence of parameters having improper posteriors due to prior impropriety (e.g., Casella, 1996) are relevant here; the analyst must understand which parameters are identified and which are not. Still, Gelfand and Sahu (1999) show that noninformative priors are often optimal in such settings, producing immediate convergence for the well-identified embedded subset. Unfortunately, such an approach is less attractive here due to our genuine interest in the random effects $\boldsymbol{\theta}$ and $\boldsymbol{\phi}$, and in particular in the proportion of excess variability due to clustering, since this may help us identify missing covariates which vary spatially. As a specific measure of this proportion, Best et al. (1999) define the quantity

$$\psi = \frac{sd(\boldsymbol{\phi})}{sd(\boldsymbol{\theta}) + sd(\boldsymbol{\phi})}\,, \tag{17}$$

where $sd(\cdot)$ is the empirical marginal standard deviation of the random effect vector in question. A posterior for ψ concentrated near 1 suggests most of the excess variation (i.e., that not explained by the covariates $\mathbf{x}_i$) is due to spatial clustering, while a posterior concentrated near 0 suggests most of this variation is mere unstructured heterogeneity. This genuine interest in the tradeoff between $\boldsymbol{\theta}$ and $\boldsymbol{\phi}$ forces these authors into a search for proper yet vague priors for these two components – a task complicated by the fact that the prior for the former is specified marginally, while that for the latter is specified conditionally. In the remainder of this section, we investigate whether a prior can be determined that offers a fair prior balance between heterogeneity and clustering, while remaining minimally informative. We can then check whether this in fact enables robust analyses within this fair class.

Eberly and Carlin (2000) show that Bayesian learning about ψ is indeed possible (i.e., that its prior and posterior can be determined and shown to be distinct). As such, the class of priors for which $\psi \approx 1/2$ seems an appropriate fair class to which we may restrict our attention. To proceed with this line of inquiry, we reconsider the Scottish lip cancer data originally presented by Clayton and Kaldor (1987) and reanalyzed by many others since. This dataset provides observed and expected cases of lip cancer in the 56 districts of Scotland for 1975–1980; the expected cases are based on MLEs of the age effects in a simple multiplicative risk model and are

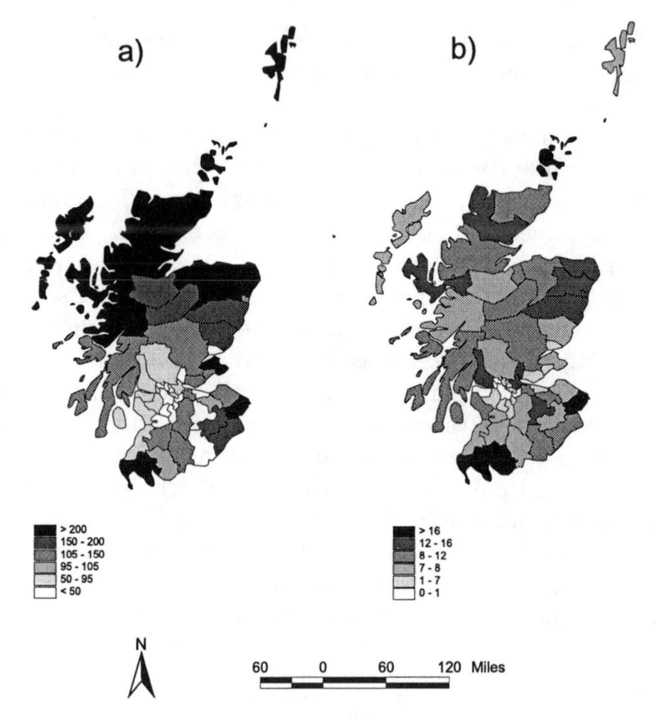

FIGURE 4. Scotland lip cancer data: a) crude standardized mortality ratios (observed/expected $\times$ 100); b) AFF covariate values.

thought of as fixed and known. For each district i we also have one covariate x_i (the percentage of the population engaged in agriculture, fishing or forestry, or AFF) and a list of which other districts j are adjacent to i. The raw data and the AFF covariate are mapped in Figure 4.

Since Gibbs sampler code for analyzing these data and model is readily available as an example in the `BUGS` software package (Spiegelhalter et al., 1995b), we use this language to carry out our investigation. The newest version of `BUGS` for Windows, `WinBUGS 1.2`, automatically imposes the sum-to-zero constraint $\sum_{i=1}^{I} \phi_i = 0$ numerically by recentering the ϕ_i samples around their own mean at the end of each iteration (Best et al., 1999). All older versions of the program do not, which in turn prohibits the inclusion of an intercept term in the log-relative risk model (15). Note that neither approach solves the Bayesian identifiability problem with the ϕ_i due to the continued presence of the covariate coefficient β and the θ_i.

In order to specify a fair prior balance between heterogeneity and clustering, we must first make a connection between the prior precision parameters τ_c and τ_h, and then see what prior for ψ they induce. Regarding the first problem, the primary difficulty is obviously that τ_c is a *conditional* prior precision, while τ_h is a precision in a standard, marginal prior specification. An investigation by Bernardinelli et al. (1995) suggests that the marginal

	posterior for ψ			posterior for β		
τ_c, τ_h	mean	sd	l1acf	mean	sd	l1acf
1, 1.81	.54	.041	.64	.42	.21	.97
0.1, 0.181	.50	.032	.64	–.022	.59	.99
0.001, 0.00181	.49	.017	.22	.043	3.00	.99
	posterior for η_1			posterior for η_{56}		
τ_c, τ_h	mean	sd	l1acf	mean	sd	l1acf
1, 1.81	1.05	.46	.46	–1.31	.72	.06
0.1, 0.181	1.82	.99	.86	–2.85	1.85	.02
0.001, 0.00181	1.73	4.81	.95	–2.20	1.62	–.04

TABLE 1. Posterior summaries for spatial model with fixed values for τ_c and τ_h, Scotland lip cancer data; "sd" denotes standard deviation while "l1acf" denotes lag 1 sample autocorrelation. In each case, the prior for ψ has mean .50 and standard deviation .059.

standard deviation of ϕ_i is roughly proportional to the corresponding conditional expression; that is, from (16) we have $sd(\phi_i) \approx 1/(K\sqrt{n_i \tau_c})$, where the authors propose $K = .7$ as a plausible variance inflation factor. A sensible "fair" prior might then be one which equates this expression to that for the marginal standard deviation of the θ_i, $1/\sqrt{\tau_h}$. Replacing n_i by $\bar{n} = 264/56 = 4.71$, the average number of neighbors across the map of Scotland, we obtain the rule of thumb

$$\tau_h = K^2 \bar{n} \tau_c. \tag{18}$$

To check the accuracy of this formula, we employ a more direct approach made possible by `WinBUGS 1.2`. Freely available over the web at `http://www.mrc-bsu.cam.ac.uk/bugs/`, this program allows direct sampling from the centered version of the CAR prior (that is, the version incorporating the sum-to-zero constraint) via its `car.normal` function. Running the `WinBUGS` code for a simplified version of our model that does not include the data produces draws $\psi^{(g)}$ via equation (17), hence an estimate of the induced prior for ψ. (All of our simulations use a single sampling chain, run for a 1000-iteration burn-in period followed by a 10,000-iteration "production" period.) After a bit of experimentation, we discovered that setting $K = .62$ in equation (18) produced ψ samples having empirical mean (and median) .50 and standard deviation .059, a suitably fair specification.

Note that the induced prior for ψ does not depend on the τ_c value selected, since ψ measures only the amount of excess variability due to clustering *relative to* the total amount present; changing the scale of the θ_i and ϕ_i does not affect ψ provided (18) still holds. Table 1 investigates the robustness of our conclusions under three such priors, namely, those obtained by setting $\tau_c = 1$, 0.1, and 0.001 in (18), respectively. We see that the posterior for ψ is rather robust to these changes, remaining close to 0.50 albeit with a slightly decreasing posterior standard deviation ("sd" in

priors for τ_c, τ_h	posterior for ψ mean	sd	l1acf	posterior for β mean	sd	l1acf
G(1.0, 1.0), G(3.2761, 1.81)	.57	.058	.80	.43	.17	.94
G(.1, .1), G(.32761, .181)	.65	.073	.89	.41	.14	.92
G(.1, .1), G(.001, .001)	.82	.10	.98	.38	.13	.91
	posterior for η_1			posterior for η_{56}		
priors for τ_c, τ_h	mean	sd	l1acf	mean	sd	l1acf
G(1.0, 1.0), G(3.2761, 1.81)	.92	.40	.33	−.96	.52	.12
G(.1, .1), G(.32761, .181)	.89	.36	.28	−.79	.41	.17
G(.1, .1), G(.001, .001)	.90	.34	.31	−.70	.35	.21

TABLE 2. Posterior summaries for spatial model with Gamma hyperpriors for τ_c and τ_h, Scotland lip cancer data; "sd" denotes standard deviation while "l1acf" denotes lag 1 sample autocorrelation.

the table). However, the picture is less reassuring for the other parameters summarized (the AFF covariate, β, and the sums of the random effects for the counties with the highest and lowest observed rates, η_1 and η_{56}). Vaguer priors appear to lead to larger random effects and a corresponding collapse in the significance of the AFF covariate. Worse, the very high lag 1 sample autocorrelations ("l1acf") for β suggest that the posterior summaries for this parameter are likely to be quite unreliable; far more Monte Carlo samples (or perhaps thinning or batching techniques; see Carlin and Louis, 1996b, p. 194) would be required to obtain effective sample sizes large enough for reliable inference.

The lack of robustness and convergence problems evident in Table 1 are not terribly surprising, given that fixing τ_c and τ_h at small values essentially precludes borrowing of strength across counties in an overparametrized model setting where such borrowing is badly needed. Indeed, hyperpriors for these two parameters are commonly used in practice, in order to reduce the prior specification burden and allow the data to play a bigger role in determining the posterior. However, adding these hyperpriors also complicates specification of a fair prior balance between heterogeneity and clustering. To see this, consider the usual conjugate hyperprior specification

$$\tau_c \sim \text{Gamma}(a_c, b_c) \quad \text{and} \quad \tau_h \sim \text{Gamma}(a_h, b_h).$$

If we require $E(\tau_c) = 1.0$ and $\text{Var}(\tau_c) = \sigma_c^2$, then we must take $a_c = b_c = 1/\sigma_c^2$. If we then follow (18) and similarly insist $E(\tau_h) = 1.81$ and $\text{Var}(\tau_h) = \sigma_h^2$, it follows that $a_h = (1.81)^2/\sigma_h^2$ and $b_h = 1.81/\sigma_h^2$. Clearly, taking σ_c^2 and σ_h^2 very small would essentially reproduce the first line in Table 1. The first two lines of Table 2 summarize results obtained from two nontrivial hyperpriors of this form, the first setting $\sigma_c^2 = \sigma_h^2 = 1$, and the second setting $\sigma_c^2 = \sigma_h^2 = 10$. These two hyperpriors actually do *not* produce priors for ψ having means of .50, but rather .62 and .99, respectively, suggesting that our moment-matching approach is not sufficient to ensure our previous definition of fairness. The third line of Table 2 reports results obtained under the specification recommended by Best et al. (1999), namely $a_c = b_c = 0.1$ and $a_h = b_h = 0.001$ (that is, hyperpriors having mean 1 and variance 10 and 1000, respectively). This specification actually leads to a ψ prior mean of .09, so it is not particularly fair either. Obviously, we could continue searching for induced ψ priors centered near .50 via ad-hoc experimentation, but we instead move on to the summarization of results under these three since they are reasonably encouraging even in the absence of a rigorously imposed prior-fairness constraint.

The results in Table 2 are to some extent the "mirror image" of those in Table 1, since the results for ψ are disappointing but those for the remaining parameters are reassuring. Under all three priors, it appears that the excess variability in the data is mostly due to clustering ($E(\psi|\mathbf{y}) > .50$), but the posterior distribution for ψ does not seem robust to changes in its prior. In fact, $E(\psi|\mathbf{y})$ is actually largest under the prior for which $E(\psi)$ was the smallest! Apparently the extreme one-tailed shapes of our hyperpriors for τ_c and τ_h can make both the prior and posterior distributions for ψ difficult to anticipate and interpret.

Despite these difficulties, we do see a reasonable degree of robustness for the remaining parameters. While autocorrelation in the β chain is still fairly high, it is reduced to a level enabling fairly precise estimation (and in particular, a 95% credible interval excluding 0 under all three priors). The random effect sums are much more modest than those in Table 1 and are well estimated from essentially uncorrelated MCMC chains.

19.4 Discussion and future directions

In this paper we have investigated the use of robust Bayesian methods in medical research, focusing on clinical trials and spatial disease mapping, two application areas where the methods are increasingly popular. Our methods emphasized use of priors which are not only minimally informative, but also neutral or fair in the sense that they do not favor any particular hypothesis (drug versus placebo, unstructured heterogeneity versus spatial clustering).

Our findings in both cases suggest a bright future for practitioners seeking to apply Bayesian methods in the absence of overly informative or unfair priors. In the clinical trial setting of Section 19.2, the bounds we obtain seem helpful, especially in Figure 3, where the range produced is quite narrow and nearly constant across ε for monitoring points 3 and 4. Our results in the spatial epidemiology setting of Section 19.3 are similarly encouraging, and consistent with those of Gelfand and Sahu (1999) and Eberly and Carlin (2000). That is, they indicate that well-identified subsets of the parameter space tend to converge quickly and can be robustly estimated under vague prior specifications, but less well-identified parameters (such as the θ_i, ϕ_i and ψ) may converge poorly and in any case produce posterior estimates that are difficult to interpret. To the extent that spatial epidemiologists and statisticians wish to focus on such quantities, there would appear to be much work remaining to do in the specification of fair, informative priors for ψ and the determination of more efficient algorithms.

In many model settings, Bayesian inferences will depend on the precise form of the prior selected for variance components. This is somewhat surprising, since such components are typically thought of as nuisance parameters, and all priors contemplated for them are typically minimally informative in some sense. Still, the results for ψ in Table 2 reemphasize this point. While the Gamma(ϵ, ϵ) prior (that is, having mean 1 but variance $1/\epsilon$) currently seems to enjoy widespread use, recent work by Hodges and Sargent (1998) and Natarajan and Kass (2000) shows that such priors can actually have significant impact on the resulting posterior distributions. And while this prior is proper, it is nearly improper for suitably small ϵ, potentially leading to MCMC convergence failure – or worse, the appearance of MCMC convergence when in fact the joint posterior is also improper. More work is needed to determine priors for variance components that have minimal impact on the resulting posterior while still allowing MCMC algorithms with acceptable convergence properties. Alternatively, Carlin and Louis (2000) suggest that reverting to an empirical Bayes approach (that is, replacing an unknown variance component by a point estimate, rather than attempting to pick a hyperprior for it) may well produce an estimated posterior that produces improved estimates while at the same time is safer to use and easier to obtain.

Acknowledgements

The research of the first author was supported in part by National Institute of Allergy and Infectious Diseases (NIAID) Grant R01-AI41966 and by National Institute of Environmental Health Sciences (NIEHS) Grant 1-R01-ES07750, while that of the second author by the National Council for Science and Technology of Venezuela (CONICIT) Grant G97-000592. This research was completed during the second author's sabbatical visit to the Division of Biostatistics at the University of Minnesota. The authors are grateful to Drs. Lynn Eberly and Erin Conlon for their assistance in the development of Section 19.3.

References

BERGER, J.O. (1994). An overview of robust Bayesian analysis (with discussion). *Test*, **3**, 5–124.

BERNARDINELLI, L., CLAYTON, D.G. and MONTOMOLI, C. (1995). Bayesian estimates of disease maps: How important are priors? *Statistics in Medicine*, **14**, 2411–2431.

BERRY, D.A. and HO, C.-H. (1988). One-sided sequential stopping boundaries for clinical trials: a decision-theoretic approach. *Biometrics*, **44**, 219–227.

BESAG, J., GREEN, P., HIGDON, D. and MENGERSEN, K. (1995). Bayesian computation and stochastic systems (with discussion). *Statistical Science*, **10**, 3–66.

BESAG, J., YORK, J.C. and MOLLIÉ, A. (1991). Bayesian image restoration, with two applications in spatial statistics (with discussion). *Annals of the Institute of Statistical Mathematics*, **43**, 1–59.

BEST, N.G., WALLER, L.A., THOMAS, A., CONLON, E.M. and ARNOLD, R.A. (1999). Bayesian models for spatially correlated disease and exposure data (with discussion). In *Bayesian Statistics 6* (J.M. Bernardo, J.O. Berger, A.P. Dawid and A.F.M. Smith, eds.), 131–156. Oxford: Oxford University Press.

CARLIN, B.P., CHALONER, K., CHURCH, T., LOUIS, T.A. and MATTS, J.P. (1993). Bayesian approaches for monitoring clinical trials with an application to toxoplasmic encephalitis prophylaxis. *The Statistician*, **42**, 355–367.

CARLIN, B.P., CHALONER, K., LOUIS, T.A. and RHAME, F.S. (1995). Elicitation, monitoring, and analysis for an AIDS clinical trial (with discussion). In *Case Studies in Bayesian Statistics, Volume II* (C. Gatsonis, J.S. Hodges, R.E. Kass and N.D. Singpurwalla, eds.), 48–89. New York: Springer-Verlag.

CARLIN, B.P., KADANE, J.B. and GELFAND, A.E. (1998). Approaches for optimal sequential decision analysis in clinical trials. *Biometrics*, **54**, 964–975.

CARLIN, B.P. and LOUIS, T.A. (1996a). Identifying prior distributions that produce specific decisions, with application to monitoring clinical trials. In *Bayesian Analysis in Statistics and Econometrics: Essays in Honor of Arnold Zellner* (D. Berry, K. Chaloner and J. Geweke, eds.), 493–503. New York: Wiley.

CARLIN, B.P. and LOUIS, T.A. (1996b). *Bayes and Empirical Bayes Methods for Data Analysis.* Boca Raton, FL: Chapman and Hall/CRC Press.

CARLIN, B.P. and LOUIS, T.A. (2000). Empirical Bayes: past, present, and future. To appear in *Journal of the American Statistical Association.*

CARLIN, B.P. and SARGENT, D. (1996). Robust Bayesian approaches for clinical trial monitoring. *Statistics in Medicine,* **15**, 1093–1106.

CASELLA, G. (1996). Statistical inference and Monte Carlo algorithms (with discussion). *Test,* **5**, 249–344.

CHALONER, K., CHURCH, T., LOUIS, T.A. and MATTS, J.P. (1993). Graphical elicitation of a prior distribution for a clinical trial. *The Statistician,* **42**, 341–353.

CLAYTON, D.G. and KALDOR, J. (1987). Empirical Bayes estimates of age-standardized relative risks for use in disease mapping. *Biometrics,* **43**, 671–681.

EBERLY, L.E. and CARLIN, B.P. (2000). Identifiability and convergence issues for Markov chain Monte Carlo fitting of spatial models. To appear in *Statistics in Medicine.*

GELFAND, A.E. and SAHU, S.K. (1999). Identifiability, improper priors, and Gibbs sampling for generalized linear models. *Journal of the American Statistical Association,* **94**, 247–253.

HODGES, J.S. and SARGENT, D.J. (1998). Counting degrees of freedom in hierarchical and other richly-parameterised models. *Technical report* **98-004**, Division of Biostatistics, University of Minnesota.

JACOBSON, M.A., BESCH, C.L., CHILD, C., HAFNER, R., MATTS, J.P., MUTH, K., WENTWORTH, D.N., NEATON, J.D., ABRAMS, D., RIMLAND, D., PEREZ, G., GRANT, I.H., SARAVOLATZ, L.D., BROWN, L.S., DEYTON, L., and THE TERRY BEIRN COMMUNITY PROGRAMS FOR CLINICAL RESEARCH ON AIDS (1994). Primary prophylaxis with pyrimethamine for toxoplasmic encephalitis in patients with advanced human immunodeficiency virus disease: results of a randomized trial. *Journal Infectious Diseases,* **169**, 384–394.

KADANE, J.B. (1990). A statistical analysis of adverse impact of employer decisions. *Journal of the American Statistical Association,* **85**, 925–933.

KADANE, J.B., MORENO, E., PÉREZ, M.-E. and PERICCHI, L.R. (1999). Applying nonparametric robust Bayesian analysis to non-opinionated judicial neutrality. *Technical Report*, **695**, Department of Statistics, Carnegie Mellon University.

KASS, R. and GREENHOUSE, J. (1989). Comment on "Investigating therapies of potentially great benefit: ECMO" by J. Ware. *Statistical Science*, **4**, 31–317.

KEMPERMAN, J. (1987). Geometry of the moment problem. *Proceedings of Symposia in Applied Mathematics*, **37**, 16–53.

LAVINE, M., WASSERMAN, L. and WOLPERT, R. (1993). Linearization of Bayesian robustness problems. *Journal of Statistical Planning and Inference*, **37**, 307–316.

LISEO, B., MORENO, E., and SALINETTI, G. (1996). Bayesian robustness of the class with given marginals: an approach based on moment theory (with discussion). In *Bayesian Robustness*, IMS Lecture Notes – Monograph Series, **29** (J.O. Berger et al., eds.), 101–118. Hayward, CA: Institute of Mathematical Statistics.

NATARAJAN, R. and KASS, R.E. (2000). Reference Bayesian methods for generalized linear mixed models. To appear in *Journal of the American Statistical Association.*

PÉREZ, M.-E. and PERICCHI, R. (1994). A case study on the Bayesian Analysis of 2 × 2 tables with all margins fixed. *Revista Brasileira de Probabilidade e Estatistica*, **1**, 27–37.

SALINETTI, G. (1994). Discussion of "An overview of robust Bayesian analysis," by J.O. Berger, *Test*, **3**, 1–125.

SARGENT, D.J. and CARLIN, B.P. (1996). Robust Bayesian design and analysis of clinical trials via prior partitioning (with discussion). In *Bayesian Robustness*, IMS Lecture Notes – Monograph Series, **29**, (J.O. Berger et al., eds.), pp. 175-193, Hayward, CA: Institute of Mathematical Statistics.

SPIEGELHALTER, D.J., THOMAS, A., BEST, N. and GILKS, W.R.(1995a). BUGS: Bayesian inference using Gibbs sampling, Version 0.50. *Technical report*, Medical Research Council Biostatistics Unit, Institute of Public Health, Cambridge University.

SPIEGELHALTER, D.J., THOMAS, A., BEST, N. and GILKS, W.R.(1995b). BUGS examples, Version 0.50. *Technical report*, Medical Research Council Biostatistics Unit, Institute of Public Health, Cambridge University.

STANGL, D.K. (1995). Prediction and decision making using Bayesian hierarchical models. *Statistics in Medicine*, **14**, 2173–2190.

20

A Robust Version of the Dynamic Linear Model with an Economic Application

Juan Miguel Marín

ABSTRACT In dynamic linear models it is often necessary to consider a more robust model than the normal one because of the appearance of outliers. Here, we consider that errors and parameters that follow a multivariate exponential power distribution. In the univariate version, this distribution has been successfully applied to robustify statistical procedures. In this chapter, a robust version of the standard normal dynamic linear model, the exponential power dynamic linear model, is introduced and applied to study the temporal relationship between the activity rate and the unemployment rate in the community of Valencia, Spain.

Key words: Exponential power dynamic linear model, outliers.

20.1 Introduction

The mathematical modelling of time-series processes may be based on the general framework of dynamic linear models. The term *dynamic* refers to the time-dependent nature of such processes. Dynamic linear models (DLM) were introduced from the Bayesian point of view by Harrison and Stevens (1976). They are commonly employed by control engineers and have been successfully used in many areas like signal processing in aerospace tracking or underwater sonar, and statistical quality control.

Standard dynamic linear models, as described by Harrison and Stevens (1976), assume normal distributions for errors and parameters. However, there are many situations where such assumption of normality is not realistic, even quite arbitrary, though the hypothesis of symmetry may be appropriate. Furthermore, the standard model may be sensitive to outliers in data, and its predictions may not be adequate. The robustness literature, see for example Berger (1994), consequently suggests that using flat-tailed priors and models tends to be much more robust than using the normal priors and models. In the case of dynamic linear models, several distributions have been employed to deal with robustness issues in predictions, including mixtures of normal distributions (Girón et al., 1989), multivariate t

distributions (Meinhold and Singpurwalla, 1989) and elliptically contoured distributions (Girón and Rojano, 1994).

In contrast with most chapters in this book, we shall be concerned with robustness issues having to do with the model rather than the prior (see West, 1996, and Corradi and Mealli, 1996, for other ideas).

In this chapter, we introduce another robustified, more general version of the standard dynamic linear model, assuming a multivariate exponential power distribution for the parameters and errors of the model. This distribution is a multidimensional generalization of the standard unidimensional power exponential distribution. The unidimensional exponential power distribution has been used in many studies relating to robust procedures (see Box and Tiao, 1973). Its multivariate version may be used to robustify many multivariate statistical procedures. Specifically, the multivariate exponential power distribution exhibits a broad range of symmetric forms and includes many important distributions, including normal, multivariate double exponential and multivariate uniform distributions, with heavier or lighter tails than the normal distribution (Gómez et al., 1998).

The exponential power dynamic linear model (EPDLM) is more robust than the standard (DLM) because it does not assume normality of errors and parameters and may adapt to the appearance of outliers in data. Furthermore, the standard independence error assumption may be replaced with the weaker assumption of uncorrelated errors.

We shall use some properties of elliptically contoured distributions and multidimensional exponential power distributions. A complete exposition on elliptically contoured and multivariate exponential power distributions may be found in Fang et al. (1990), Gómez et al. (1998) and Marín (1998).

We illustrate the ideas with an application of an EPDLM to study the temporal relationship between two economic quantities, the *activity rate* (Y_t) and the *unemployment rate* (x_t), in the community of Valencia, Spain. We do not impose normality on errors and parameters, but only assume symmetry of the distribution and uncorrelation among parameters and errors.

20.2 Definition and properties of the exponential power dynamic linear model

DLMs are described through the next two relations, called *the observation equation* and *state equation*, respectively, for $t = 1, \ldots, n$:

$$\begin{aligned} Y_t &= F_t'\theta_t + v_t, \\ \theta_t &= G_t\theta_{t-1} + w_t, \end{aligned}$$

where, for each t, θ_t is an r-dimensional state variable vector and Y_t is an s-dimensional vector of observations; v_t and w_t are error vectors with dimensions s and r, respectively; F_t is a fixed $r \times s$ matrix of regressors and G_t is a fixed $r \times r$ matrix relating two consecutive states. A detailed exposition of DLMs can be found in West and Harrison (1997).

For each $t = 1, \ldots, n$, we shall denote by $\mathbf{Y}_t = (Y_1', \ldots, Y_t')'$ the vector of all observations until time t, its dimension being $s \cdot t$; the scalar $d_t = r + (n-t)(r+s)$ and $H_t = \left(\theta_t', v_{t+1}', w_{t+1}', \ldots, v_n', w_n'\right)'$, whose dimension is d_t. We shall suppose that prior distribution of H_0 is a multivariate exponential power distribution.

This distribution is a multidimensional generalization of the standard unidimensional power exponential distribution (see Box and Tiao, 1973) and includes, among others, the multivariate normal, the multivariate double exponential and the multivariate uniform distributions (Gómez et al., 1998). The advantage of using this distribution in robustness studies is that it includes, besides the normal distribution, a broad range of multivariate symmetrical distributions with flatter or heavier tails than the normal distribution.

We shall say that $X \sim PE_n(\mu, \Sigma, \beta)$ if X has a multivariate exponential power distribution, with parameters μ, Σ, β. Its density function is

$$f(x; \mu, \Sigma, \beta) = \frac{n\Gamma(\frac{n}{2})}{\pi^{\frac{n}{2}}\Gamma(1+\frac{n}{2\beta})2^{1+\frac{n}{2\beta}}} |\Sigma|^{-\frac{1}{2}} \exp\left(-\frac{1}{2}\left[(x-\mu)'\Sigma^{-1}(x-\mu)\right]^{\beta}\right)$$

where $\mu \in \Re^n$, Σ is an $(n \times n)$ definite positive symmetric matrix, and $\beta \in (0, \infty)$. Note, for example, that for $\beta = 1$ we have a normal distribution; for $\beta = 0.5$ we have the double exponential distribution; when $\beta \to 0$ we approximate a flat improper prior; when $\beta \to \infty$ we approximate a uniform distribution on an appropriate elliptical form region.

In words, this family is a particular case of elliptically contoured distributions whose density functions are constant over ellipsoids. We shall say that $X \sim E_n(\mu, \Sigma, g)$ if X has an elliptically contoured distribution, with parameters μ, Σ, g. Its density function is

$$f(x; \mu, \Sigma, g) = \frac{\Gamma\left(\frac{n}{2}\right)}{\pi^{\frac{n}{2}} \int_0^\infty t^{\frac{n}{2}-1} g(t) dt} |\Sigma|^{-\frac{1}{2}} g\left((x-\mu)'\Sigma^{-1}(x-\mu)\right)$$

where $\mu \in \Re^n$, Σ is a positive definite $(n \times n)$ matrix and g is a nonnegative Lebesgue measurable function on $[0, \infty)$ such that $\int_0^\infty t^{\frac{n}{2}-1} g(t) dt < \infty$. Note that the exponential power distribution is an elliptically contoured distribution with $g(t) = \exp\left\{-1/2t^{\beta}\right\}$ (Gómez et al., 1998).

We shall provide first the basic definition of an EPDLM.

Definition 20.2.1. *An exponential power dynamic linear model is defined as a dynamic linear model where the initial vector* $H_0 = (\theta_0', v_1', w_1', \ldots, v_n', w_n')'$, *whose dimension is* $d_0 = r + n(r+s)$, *has an exponential power distribution:*

$$H_0 \sim PE_{d_0}\left(\mu_0^H, \Sigma_0^H, \beta\right)$$

where $\mu_0^H = \left(m_0', 0'_{(d_0-r)\times 1}\right)'$, *with* $m_0 \in \Re^r$; Σ_0^H *is a block-diagonal matrix whose diagonal components are symmetric positive definite matrices* C_0, V_1, $W_1, \ldots, V_n, W_n$, *of orders* $r, s, r, \ldots, s, r$, *respectively, and* $\beta \in (0, \infty)$.

The EPDLM is a generalization of the standard normal model DLM, when $\beta = 1$.

The next result shows that the updated distributions of vectors θ_t, θ_{t+k} and Y_{t+k}, conditional on $\mathbf{Y}_t$, are elliptically contoured, where the corresponding functional parameters g are related to the functional parameter of the power exponential distribution. The proof may be found in Marín (1998). Note the resemblance of the EPDLM recursions with those of the standard DLM. Specifically, posterior and predictive means coincide. However, and this is very important from a robust point of view, posterior and predictive variances will differ, just like the multivariate kurtosis measure.

Theorem 20.2.1. *For each possible value* $\mathbf{y}_t = (y_1', \ldots, y_t')'$ *of* $\mathbf{Y}_t$, *the following statements hold:*

(i) For each $t = 1, \ldots, n-1$ *and each* $k = 0, 1, \ldots, n-t$, *the distribution of* θ_{t+k} *conditional on* $\mathbf{Y}_t = \mathbf{y}_t$ *is*

$$\theta_{t+k} \mid \mathbf{Y}_t = \mathbf{y}_t \sim E_r\left(\mu_{t,k}^\theta, \Sigma_{t,k}^\theta, g_t^\theta\right)$$

where $\mu_{t,k}^\theta$ *and* $\Sigma_{t,k}^\theta$ *are recursively given by*

$$\begin{aligned} \mu_{t,k}^\theta &= G_{t+k}\mu_{t,k-1}^\theta, \\ \Sigma_{t,k}^\theta &= G_{t+k}\Sigma_{t,k-1}^\theta G_{t+k}' + W_{t+k}, \end{aligned}$$

with $\mu_{t,0}^\theta = m_t$, $\Sigma_{t,0}^\theta = C_t$,

$$g_t^\theta(z) = \int_0^\infty w^{\frac{d_t + rt - r}{2} - 1} \exp\left\{-\frac{1}{2}(z + q_t + w)^\beta\right\} dw,$$

and m_t, C_t *and* q_t *are recursively defined by*

$$\begin{aligned} m_t &= G_t m_{t-1} + R_t F_t Q_t^{-1} e_t', \\ C_t &= R_t - R_t F_t Q_t^{-1} F_t' R_t, \\ q_t &= q_{t-1} + e_t' Q_t^{-1} e_t, \end{aligned}$$

and

$$\begin{aligned} R_t &= G_t C_{t-1} G_t' + W_t, \\ Q_t &= F_t' R_t F_t + V_t, \\ e_t &= y_t - F_t' G_t m_{t-1} \end{aligned}$$

and $q_0 = 0$.

(ii) For each $t = 1, \ldots, n-1$ and for each $k =, 1, \ldots, n-t$, the distribution of Y_{t+k} conditional on $\mathbf{Y}_t = \mathbf{y}_t$ is

$$Y_{t+k} \mid \mathbf{Y}_t = \mathbf{y}_t \sim E_s\left(\mu^Y_{t,k}, \Sigma^Y_{t,k}, g^Y_t\right)$$

where

$$\begin{aligned}
\mu^Y_{t,k} &= F'_{t+k}\mu^\theta_{t,k}, \\
\Sigma^Y_{t,k} &= F'_{t+k}\Sigma^\theta_{t,k}F_{t+k} + V_{t+k}, \\
g^Y_t(z) &= \int_0^\infty w^{\frac{d_t+rt-s}{2}-1} \exp\left\{-\frac{1}{2}(z + q_t + w)^\beta\right\} dw.
\end{aligned}$$

In the next corollary (see Marín, 1998), we provide several features of the previous vectors, where $\gamma_2[X]$ is the multivariate kurtosis measure defined by Mardia et al. (1979) as

$$\gamma_2[X] = E\left[\left((X - E[X])' Var[X]^{-1}(X - E[X])\right)^2\right].$$

Corollary 20.2.1. *Under the hypotheses and notation in Theorem ??, let us denote*

$$I\left(\frac{d_t + rt}{2}\right) = \int_0^\infty w^{\frac{d_t+rt}{2}} \exp\left\{-\frac{1}{2}(q_t + w)^\beta\right\} dw. \tag{1}$$

Then, the following statements hold:

(i) The mean vector, covariance matrix and kurtosis measure of θ_{t+k} conditional on $\mathbf{Y}_t = \mathbf{y}_t$ are, respectively,

$$\begin{aligned}
E[\theta_{t+k} \mid \mathbf{Y}_t = \mathbf{y}_t] &= \mu^\theta_{t,k}, \\
\mathrm{Var}[\theta_{t+k} \mid \mathbf{Y}_t = \mathbf{y}_t] &= \frac{1}{d_t+rt}\frac{I\left(\frac{d_t+rt}{2}\right)}{I\left(\frac{d_t+rt}{2} - 1\right)}\Sigma^\theta_{t,k}, \\
\gamma_2[\theta_{t+k} \mid \mathbf{Y}_t = \mathbf{y}_t] &= \frac{r(r+2)(d_t+rt)}{d_t+rt+2}\frac{\left(I\left(\frac{d_t+rt}{2} + 1\right)\right)\left(I\left(\frac{d_t+rt}{2} - 1\right)\right)}{\left(I\left(\frac{d_t+rt}{2}\right)\right)^2}.
\end{aligned}$$

(ii) The mean vector, covariance matrix and kurtosis measure of Y_{t+k} conditional on $\mathbf{Y}_t = \mathbf{y}_t$ are

$$\begin{aligned}
E[Y_{t+k} \mid \mathbf{Y}_t = \mathbf{y}_t] &= \mu^Y_{t,k}, \\
\mathrm{Var}[Y_{t+k} \mid \mathbf{Y}_t = \mathbf{y}_t] &= \frac{1}{d_t+rt}\frac{I\left(\frac{d_t+rt}{2}\right)}{I\left(\frac{d_t+rt}{2} - 1\right)}\Sigma^Y_{t,k}, \\
\gamma_2[Y_{t+k} \mid \mathbf{Y}_t = \mathbf{y}_t] &= \frac{s(s+2)(d_t+rt)}{d_t+rt+2}\frac{\left(I\left(\frac{d_t+rt}{2} + 1\right)\right)\left(I\left(\frac{d_t+rt}{2} - 1\right)\right)}{\left(I\left(\frac{d_t+rt}{2}\right)\right)^2}.
\end{aligned}$$

The updated probabilistic characteristics of all vectors depend on one-dimensional integrals with form (1) that may be computed by numerical integration.

Observe that all distributions of θ_t, θ_{t+k} and Y_{t+k} are elliptically contoured, where the functional parameter g has form

$$g(z) = \int_0^\infty w^{\frac{m}{2}-1} \exp\left\{-\frac{1}{2}(z+q+w)^\beta\right\} dw,$$

where $m \in \mathcal{N}$, $q \geq 0$ and $\beta \in (0,\infty)$. This function is similar to that of the exponential power distribution $g(z) = \exp\left\{-1/2z^\beta\right\}$. This might suggest approximating the actual functional parameter by that of the exponential power. However, the computational gain is not important, and we shall therefore use the actual functional parameter for forecast. However, we may be interested in providing an approximation of $g(z)$. For that we shall use the exponential power distribution $PE_d\left(\mu^*, \Sigma^*, \beta^*\right)$, which is closest to the elliptically contoured distribution $E_d(\mu,\Sigma,g)$, in the sense of, say, minimizing the Kullback Leibler divergence between the obtained elliptically contoured distribution $E_d(\mu,\Sigma,g)$ and the exponential power distribution $PE_d\left(\mu^*, \Sigma^*, \beta^*\right)$. This implies a considerable computational effort; alternatively, we shall use the moments method. Hence, as the means, covariance matrices and kurtosis of all vectors are easy to compute, we consider the exponential power distribution with the same probabilistic characteristics as the obtained elliptically contoured distribution. Then, we choose the values of μ^*, Σ^* and β^* which solve the system

$$\left\{\begin{array}{c} \mu^* = E \\ \\ \dfrac{2^{\frac{1}{\beta^*}}\Gamma\left(\frac{d+2}{2\beta^*}\right)}{d\Gamma\left(\frac{d}{2\beta^*}\right)}\Sigma^* = \mathrm{Var} \\ \\ d^2\dfrac{\Gamma\left(\frac{d+4}{2\beta^*}\right)\Gamma\left(\frac{d}{2\beta^*}\right)}{\left(\Gamma\left(\frac{d+2}{2\beta^*}\right)\right)^2} = \gamma_2 \end{array}\right\}, \tag{2}$$

where E, Var and γ_2 are the mean vector, covariance matrix and kurtosis measure of the distribution $E_d(\mu,\Sigma,g)$.

20.3 Temporal relations between two economic quantities

We illustrate the use of the EPDLM to study the relations between the *activity* (Y_t) and *unemployment rates* (x_t) in Valencia, Spain. The data

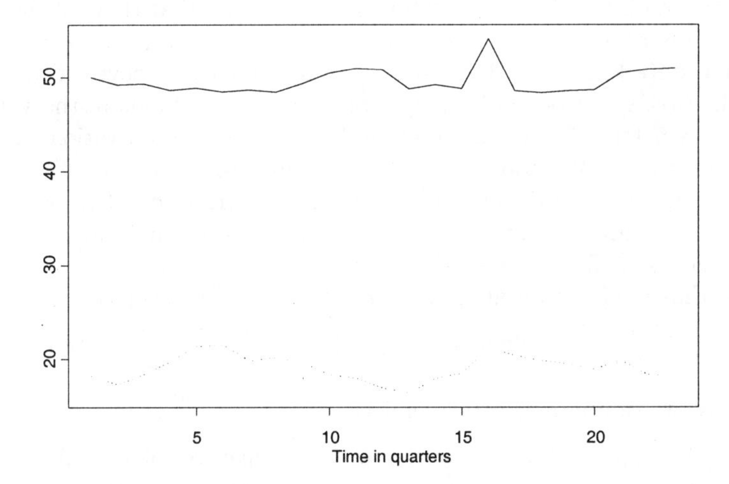

FIGURE 1. Quarterly observations ranging from 1983 to 1988 of the activity rate and unemployment rate in Valencia, Spain.

consist of 23 pairs (y_t, x_t) of quarterly observations from the years 1983 to 1988 and are shown in Table 1. The observations are taken from Girón et al. (1989).

Table 1. Quarterly observations from the years 1983 to 1988

Activ.	Unemp.	Activ.	Unemp.
50.00	18.19	48.67	19.81
48.75	16.37	48.57	19.48
49.21	17.28	48.41	20.12
49.20	17.91	48.67	18.86
49.30	18.36	49.34	19.61
48.79	18.48	50.46	19.83
48.62	19.74	50.44	18.32
54.14	21.14	50.83	18.31
48.85	21.36	50.93	17.97
48.55	20.37	50.97	18.10
48.45	21.55	50.80	16.92
48.33	19.79		

We assume a linear relation between both rates with intercept and slope varying along time. Hence, we choose $F_t' = (1, x_t)$. We shall use an EPDLM; this model is more robust than the standard DLM because it may cope with outliers in the observations. Furthermore, we assume the less restrictive and general condition that errors are uncorrelated and that they follow a symmetric distribution. Then, we assume that $H_0 = (\theta_0, v_1, \omega_1, \ldots, v_n, \omega_n)'$, of dimension$(r + n(r+s))$, has a multivariate exponential power distribution.

The first seven pairs of observations are used to simulate the values of the parameters of prior distribution H_0. The other observations are used for forecasting. We adopt $G_t = I$, as there are no reasons to assume a systematic trend in the evolution of θ_t. The distribution of H_0 is adjusted using the regression line between Y_t and x_t and simulating a sequence of values of θ_t from consecutive pairs (x_t, y_t), (x_{t+1}, y_{t+1}). Hence, we assume that the multivariate power distribution of H_0 has prior mean vector $m_0 = \begin{pmatrix} 50.78 \\ -0.10 \end{pmatrix}$, scale matrix $C_0 = \begin{pmatrix} 780.73 & -41.59 \\ -41.59 & 2.22 \end{pmatrix}$ and $\beta = 2.00$.

Finally, we compute $V_t = 0.26$ and $W_t = \begin{pmatrix} 0.35 & 0 \\ 0 & 1.19 \end{pmatrix}$.

We then update the distribution of θ_t and features of Y_t. Table 2 shows the mean vector, the covariance matrix of θ_t and the value β of the closest exponential power distribution, in the sense of (2), to the elliptically contoured distribution obtained for θ_t.

Table 2. Updated Distribution of θ_t

t	Mean	Covariance M.		β	t	Mean	Covariance M.		β
8	14.85	61.00	−2.89	1.03	16	19.23	81.05	−4.30	1.05
	1.86	−2.89	1.37			1.56	−4.19	0.23	
9	21.53	75.02	−3.51	1.04	17	19.67	82.09	−4.19	1.05
	1.28	−3.51	0.16			1.51	−4.19	0.21	
10	20.52	77.64	−3.81	1.05	18	19.82	83.14	−4.19	1.05
	1.38	−3.81	1.87			1.55	−4.19	0.21	
11	21.77	78.10	−3.62	1.04	19	18.86	82.99	−4.53	1.05
	1.24	−3.62	0.17			1.72	−4.53	0.25	
12	20.32	77.96	−3.94	1.04	20	18.31	84.44	−4.61	1.05
	1.42	−3.94	0.20			1.78	−4.61	0.25	
13	19.81	79.13	−3.99	1.05	21	18.01	85.55	−4.76	1.06
	1.46	−3.99	0.20			1.83	−4.76	0.27	
14	19.50	80.02	−4.11	1.05	22	18.10	86.79	−4.80	1.06
	1.49	−4.11	0.21			1.82	−4.80	0.27	
15	20.21	80.81	−4.02	1.05	23	17.36	86.95	−5.14	1.06
	1.40	−4.02	0.20			1.98	−5.14	0.30	

We therefore obtain how the regression lines vary for each quarter. Hence, we may study the relationship between the activity (Y_t) and the unemployment rates (x_t). Observe that the slope is positive for all quarters and the covariance between the independent terms and the slopes is negative.

Therefore, it seems that when the activity rate increases, the unemployment rate increases as well. It appears, also, that an inverse relationship exists between the slope of such dependence and the intercept.

The parameter β, corresponding to the closest exponential power distribution, is slightly greater than 1; then, the distribution of θ_t for each t seems to deviate slightly from the normal: there is some evidence of somewhat more platykurtic distributions, namely, symmetrical distributions with lower tails than multivariate normal distribution.

Table 3 shows the updated mean and standard deviation of Y_t, as well as the prediction error.

Table 3. Updated Characteristics of $\mathbf{Y}_t$

t	Mean	S.D.	Error	t	Mean	S.D.	Error
8	48.75	1.46	5.39	16	46.65	0.61	2.03
9	54.53	0.22	−5.68	17	49.84	0.43	−0.50
10	47.59	0.46	0.96	18	49.67	0.26	0.79
11	50.17	0.56	−1.72	19	48.13	0.73	2.31
12	46.27	0.76	2.06	20	50.42	0.24	0.41
13	48.36	0.23	0.31	21	50.23	0.30	0.70
14	48.19	0.28	0.38	22	51.17	0.26	−0.20
15	49.53	0.38	−1.11	23	48.83	0.65	1.97

Table 4 shows the 95% confidence intervals for Y_t. They are slightly shorter from those obtained with the standard DLM.

Table 4. 95% Confidence Intervals for $\mathbf{Y}_t$

t	8	9	10	11
C.I. Y_t	(46.06, 51.44)	(54.11, 54.96)	(46.72, 48.46)	(49.10, 51.24)
t	12	13	14	15
C.I. Y_t	(44.84, 47.71)	(47.91, 48.80)	(47.67, 48.71)	(48.81, 50.24)
t	16	17	18	19
C.I. Y_t	(45.49, 47.80)	(49.02, 50.66)	(49.18, 50.17)	(46.74, 49.51)
t	20	21	22	23
C.I. Y_t	(49.96, 50.88)	(49.67, 50.78)	(50.68, 51.65)	(47.60, 50.06)

Errors have no trend and standard deviations are in some cases large; it may suggest that the prior value 0.26 chosen for V_t is too small. Nevertheless, the results show that the exponential power model provides suitable predictions and the distributions obtained each time are more platykurtic than the normal distribution. If the standard normal model is used, we obtain less adjusted predictions. Furthermore, as this is a robust version it

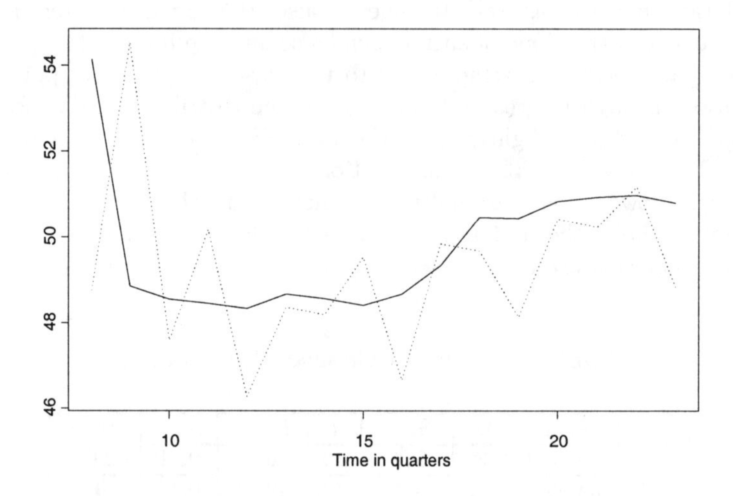

FIGURE 2. Predictions (dotted line) and actual observations (solid line) of Y_t for successive periods.

includes also, as a possible model, the standard normal one. But, in this case, the β parameters of the power exponential distributions obtained are different than 1, therefore, the adjusted model is slightly different than the standard normal model.

20.4 Discussion

We have introduced a robust version of the standard DLM, based on the multivariate exponential power distribution. Its main advantage stems from making fewer assumptions (just symmetry instead of normality, uncorrelation instead of independence) and it may potentially cope with outliers.

There are several possible extensions. For example, we could introduce a prior distribution over the parameter β and by considering its posterior, test the hypothesis $\beta = 1$.

Acknowledgements

I am grateful to the discussions and comments by two referees and David Ríos Insua.

References

BERGER, J.O. (1994). An overview of robust Bayesian analysis (with discussion). *Test*, **3**, 5 - 124.

BOX, G. and TIAO, G. (1973). *Bayesian Inference in Statistical Analysis.* Reading, MA: Addison-Wesley.

CORRADI, F. and MEALLI, F. (1996). Nonparametric specification of error terms in dynamic models. In *Bayesian Robustness* (J.O. Berger et al., eds.) IMS Lecture Notes, Vol. 29, 293 - 304. Hayward: IMS.

FANG, K., KOTZ, S. and NG, K. (1990). *Symmetric Multivariate and Related Distributions.* New York: Chapman and Hall.

GIRÓN, F.J., MARTÍNEZ, M.L. and ROJANO, J.C. (1989). Modelos lineales dinámicos y mixturas de distribuciones. *Estadística Española*, **31**, 165 - 206.

GIRÓN, F.J. and ROJANO, J.C. (1994). Bayesian Kalman filtering with elliptically contoured errors. *Biometrika*, **80**, 390 - 395.

GÓMEZ, E., GÓMEZ-VILLEGAS, M.A. and MARÍN, J.M. (1998). A multivariate generalization of the power exponential family of distributions. *Communications in Statistics, Theory and Methods*, **27**, 589 - 600.

HARRISON, P.J. and STEVENS, C.F. (1976). Bayesian forecasting (with discussion). *Journal of the Royal Statistical Society, B*, **38**, 205 - 247.

MARDIA, K.V., KENT, J.T. and BIBBY, J.M. (1979). *Multivariate Analysis.* London: Academic Press.

MARÍN, J.M. (1998). Continuous elliptical and power exponential distributions with applications to dynamic linear models. *Ph.D. Dissertation*, Universidad Complutense de Madrid.

MEINHOLD, R.J. and SINGPURWALLA, N.Z. (1989). Robustification of Kalman filter models. *Journal of American Statistical Association*, **84**, 479 - 486.

WEST, M. (1996). Modeling and robustness issues in Bayesian time series analysis. In *Bayesian Robustness* (J.O. Berger et al., eds.) IMS Lecture Notes, Vol. 29, 231 - 245. Hayward: IMS.

WEST, M. and HARRISON, J. (1997). *Bayesian forecasting and dynamic models.* New York: Springer-Verlag.

21

Prior Robustness in Some Common Types of Software Reliability Model

Simon P. Wilson and Michael P. Wiper

ABSTRACT We investigate prior sensitivity to predictions of software reliability made with two well-known software reliability models; one based on a nonhomogeneous Poisson process and the other a time series. A mixture of formal (global) and informal sensitivity approaches is used. We demonstrate that while inference based on the first of these models does not seem too sensitive to the prior input, inference from the time series model does exhibit considerable prior sensitivity, even when the sample of observed data is quite large.

Key words: Bayesian statistics, software reliability, Jelinski-Moranda model, autoregressive models, robustness.

21.1 Introduction

Software reliability can be defined as the probability of the failure-free operation of a computer program in a specified environment for a specified period of time, see Musa and Okumoto (1987). A failure is the result of a fault in the program. Many stochastic models for software reliability have been proposed, using several different concepts from the classical reliability literature, time series and the theory of stochastic processes; see Singpurwalla and Wilson (1999) for a recent review.

In this chapter, we investigate the sensitivity of software reliability predictions to prior specification. This is done for two well-known software reliability models. We illustrate the use of some prior robustness techniques applied to these models. We conclude that prior sensitivity does exist for these two models, particularly when one is estimating the reliability function.

Software reliability models are defined in one of two ways. Either the sequence of successive failure times $T_1, T_2, \ldots$ or a model for the number of faults $N(t)$ to have been discovered by time t is specified. Most models in the former case specify the failure rate for each successive time, and most models in the latter use some form of nonhomogeneous Poisson process.

Another approach to modelling successive failure times is to use a time series. Most, but by no means all, software reliability models will fall into one of these categories. Both the failure rate and the Poisson process models are special cases of the self-exciting Poisson process (Chen and Singpurwalla, 1997). Thus, from a probability modelling point of view, it is perhaps better to make a distinction between two classes of model: self-exciting Poisson processes and time series.

A variety of problems in the software production cycle can be addressed with these models. The most common application is in the software testing phase of this cycle, just prior to release of the code to the user, when the software is subject to various inputs, the objective being to detect and remove faults. Given this, it is hoped that as more faults are discovered, the software will become more reliable. The idea of increasing reliability is to be found in almost every software reliability model.

Software reliability models can be fitted to failure time data from the testing, and predictions on future failure times made. More formally, in conjunction with decision theory, decisions on whether to continue testing and, if so, for how long can be made; see for example Dalal and Mallows (1986), Singpurwalla (1991) and Wiper et al. (1998). In most applications the model is used to make reliability predictions; estimation of model parameters is not usually an objective in itself.

Bayesian approaches to inference and prediction were used even early on in the development of software reliability models; see Littlewood and Verall (1973). As with many statistical applications, flexibility in prior specification had to wait until suitable computational tools were developed. Now that the use of such methods is routine, the question of prior robustness can be more fully investigated. Thorough reviews are given by Berger (1994) and now this book.

In this chapter, we investigate prior sensitivity of predictions, specifically the time to next failure, for two models. The first is a self-exciting point process, the well known model of Jelinski and Moranda (1972), which was one of the earliest software reliability models to be developed. The second is a time series, the autoregressive model of Singpurwalla and Soyer (1985).

The chapter is organised as follows. Section 21.2 investigates prior sensitivity for the Jelinski-Moranda model. Section 21.3 does likewise for the autoregressive model of Singpurwalla and Soyer. Section 21.4 completes the chapter with some concluding remarks.

21.2 The Jelinski-Moranda model

Assume that $T_1, T_2, \ldots$ are the successive times between failures of a piece of software under testing. Then the Jelinski-Moranda (JM) model assumes that

$$T_i|N, \phi \sim \mathcal{E}\left((N-i+1)\phi\right),$$

namely, that intra-failure times are exponentially distributed. The model is basically assuming that there are N faults in the software initially and that after each one is discovered, it is perfectly corrected. The parameter ϕ relates to the size of the faults.

Assume that we observe m failure times $t_1, \ldots, t_m$. Then the likelihood function is given by

$$\frac{N!}{(N-m)!}\phi^m \exp\left(-\left[(N+1)m\bar{t} - \sum_{i=1}^{m} it_i\right]\phi\right).$$

Maximum likelihood inference has been carried out for this model, but problems have been pointed out by Forman and Singpurwalla (1977) and Joe and Reid (1985), who suggested that maximum likelihood estimates of N can sometimes be unreasonable. It is interesting to see if we have problems with Bayesian inference under non-informative priors, which are often assumed to be a robust choice.

Here, the Jeffreys prior for this problem (assuming independence of N and ϕ) would be given by

$$f(N, \phi) \propto \frac{1}{\phi}.$$

However, given the data, the marginal posterior for N would now be of the form

$$\begin{aligned} f(N|t_1, \ldots, t_m) &\propto \frac{N!}{(N-m)!}\left((N+1)m\bar{t} - \sum_{i=1}^{m} it_i\right)^{-m}, \qquad N \geq m, \\ &= o(1) \qquad \text{as } N \to \infty, \end{aligned}$$

which is improper. A globally uniform prior $f(N, \phi) \propto 1$ can also be shown to give an improper posterior. Thus, prior information is needed to make Bayesian inference for this model.

The standard approach to Bayesian inference for this model has been to consider a Poisson prior distribution for N, say

$$P(N = n) = \frac{\lambda^n e^{-\lambda}}{n!}, \tag{1}$$

and a gamma prior for ϕ

$$f(\phi) = \frac{\beta^\alpha}{\Gamma(\alpha)}\phi^{\alpha-1}\exp(-\beta\phi); \tag{2}$$

see, for example, Dalal and Mallows (1986) and Keiller et al. (1983). Given these priors, we can derive the joint and marginal posterior distributions. See, for example Wiper et al. (1998).

Often, interest in reliability modelling is concerned with estimation of the time to the next failure. It is to be noted that under the JM model, given any prior distribution for N defined on the positive integers, the expected time to next failure does not exist, as there is a nonnull posterior probability that there are no faults remaining in the software, and hence that the time to next failure will be infinite. Thus, we can only estimate the time to next failure conditional on there being some faults remaining in the software. However, we can estimate the expected reliability $R(T)$ of the software at any time T after the last observed failure. The reliability function after m failures have been observed is given by

$$R(T) = e^{(N-m)\phi T},$$

and given the priors (1,2) on N and ϕ, the posterior expected reliability may be calculated as

$$\begin{aligned} E[R(T)|\mathbf{t}] &= \frac{1}{D}\sum_{N=m}^{\infty}\frac{\lambda^N}{(N-m)!}\Bigg(\beta+(N+1)m\bar{t} \\ &\qquad -\sum_{i=1}^{m} it_i + (N-m)T\Bigg)^{-(\alpha+m)} \end{aligned} \tag{3}$$

where

$$D = \sum_{N=m}^{\infty}\frac{\lambda^N}{(N-m)!}\left(\beta+(N+1)m\bar{t}-\sum_{i=1}^{m} it_i\right)^{-(\alpha+m)}.$$

The summations may be approximated by truncation at a sufficiently large value of N.

Given the structure of the model, posterior estimates of the number of faults remaining in the software, $N-m$, will also be of interest. A similar form to that of Equation (3) may be derived for the posterior expectation of $N-m$ given the earlier priors.

In the following subsection, we will undertake a Bayesian analysis for a real set of software failure data using the priors given in (1) and (2). In subsections 21.2.2 and 21.2.3 we shall extend the analysis to situations of various contamination classes around the original prior.

21.2.1 Data analysis given usual prior distributions

We consider a data set given by Littlewood (1989, Table 2). These data consist of execution times of a piece of software code in 100ths of a second

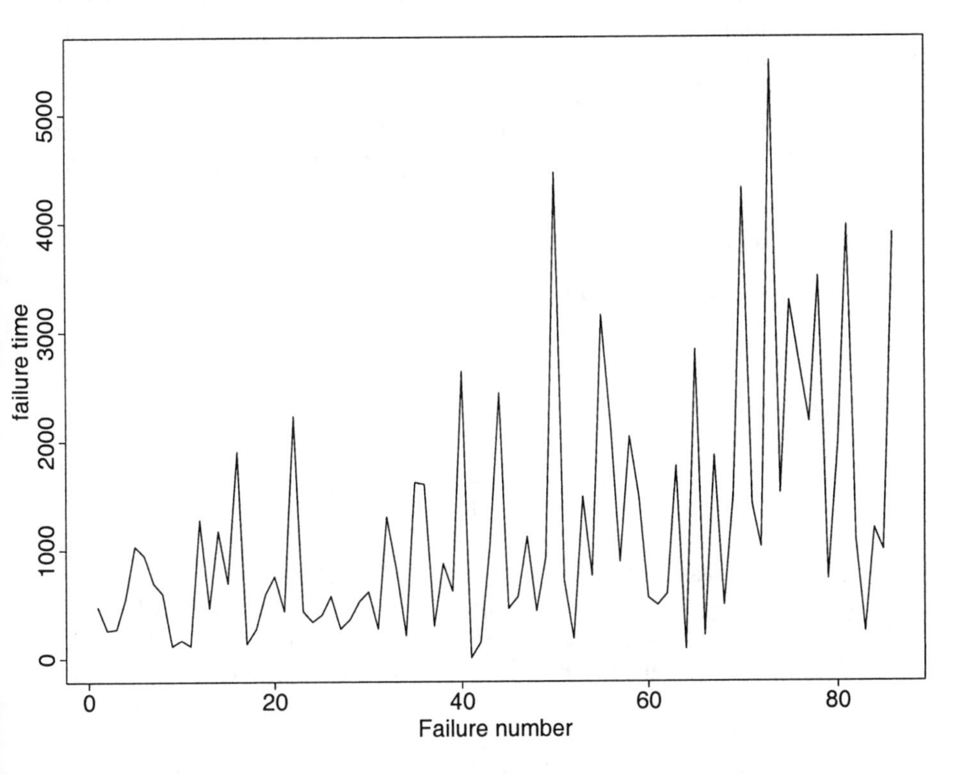

FIGURE 1. Successive times to failure for the Littlewood data.

between successive failures. In total, there are 86 observed failures. The data are illustrated in Figure 1. As can be seen, there is some evidence of reliability growth here, as times between the later failures are, on average, greater than times between the earlier failures. Thus it may be reasonable to apply the Jelinski-Moranda model to this data.

We will suppose that a Poisson(100) prior is elicited for the number of faults N and that a (fairly uninformative) exponential(.0001) prior is used for ϕ. This could reflect the opinions of an expert who had fairly strong knowledge that N was between 70 and 130 but had little idea of the average operation time that would be taken before a fault was discovered.

Given these data, the posterior mean of N is given by $E[N|\text{data}] = 103.89$ (maximum likelihood estimate: mle = 106) so that it is expected that there are approximately 18 bugs remaining in the software. The posterior median time to next failure is estimated to be 2440 100ths of a second (mle = 2177) and the expected time to failure, assuming that there are still some faults remaining, is about 3765 100ths of a second (mle = 3141). Thus, in this case, the Bayesian conclusions are slightly more optimistic than the maximum likelihood estimates.

In the following subsections, we will consider two classes of priors based on modifications of our original priors. First we will undertake informal sensitivity analysis by contaminating the Poisson prior for N with a long tailed prior distribution, and second, we will undertake a global sensitivity analysis by embedding our initial priors within a density bounded class. See Gustafson (2000) for a discussion of informal, global and local sensitivity.

21.2.2 ϵ-contamination with a long tailed prior

Consider the class of mixture priors

$$\Gamma = \{(1-\epsilon)f(N) + \epsilon g(N)\},$$

where $f(N)$ is the Poisson($\lambda = 100$) prior given in Equation (1) and

$$g(N) \propto \frac{1}{\left(\left(N - 100 + \frac{1}{2}\right)^2 + \delta\right)}.$$

We study this class for $0 \leq \epsilon < 1$ and $100 \leq \delta \leq 1,000,000$.

The prior $g(N)$ may be thought of as a discrete analogue of a Cauchy distribution. It is bimodal at 99 and 100 (as is the Poisson) and is long tailed, having no moments. For $\delta \approx 100$, the prior is fairly close to the Poisson prior for values of N close to 100. Larger values of δ give a more disperse prior distribution.

Figure 2 shows the expected numbers of errors remaining in the software after observing the data, for $0 < \epsilon < 1$ and for various different elections of the scale parameter δ.

It can be seen from Figure 2 that for $\epsilon < .2$, the ϵ–contamination makes very little difference to the estimated number of faults, although for large ϵ and δ, the difference becomes considerable.

In Figure 3, we show the posterior estimated reliability function for $\epsilon = .2$ (left-hand side) and for $\epsilon = 1$ (right-hand side) and various values of δ. In each diagram, the expected reliability curve for the uncontaminated prior distribution is given as the solid line and the contaminated priors are short dashes for $\delta = 10^2$, medium dashes for $\delta = 10^3$ and long dashes for $\delta = 10^6$, as marked on the right-hand diagram.

It can be seen that contamination with $\epsilon = .2$ has virtually no effect on the posterior expected reliability function, although when $\epsilon = 1$, the posterior reliability can be somewhat different for large δ. In particular, for $\epsilon = .2$, the estimated median time to failure varies between about 2410 and 2440 given the different contaminating distributions, whereas, in the case $\epsilon = 1$, the estimated median is as low as 1960 in the most extreme case of $\delta = 10^6$.

Note that in all cases, the expected reliability function goes to zero at approximately $t = 22,000$ (or $\log t \approx 10$). We shall see in Section 21.3 that

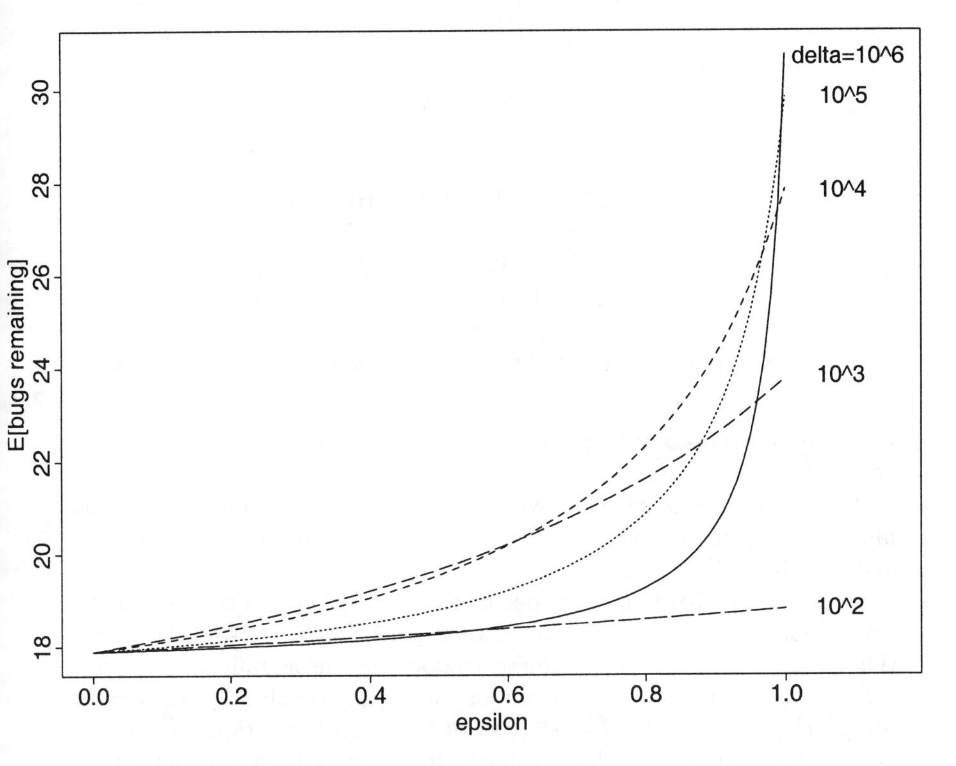

FIGURE 2. Expected number of faults remaining in the software given an ϵ–contaminated prior for N.

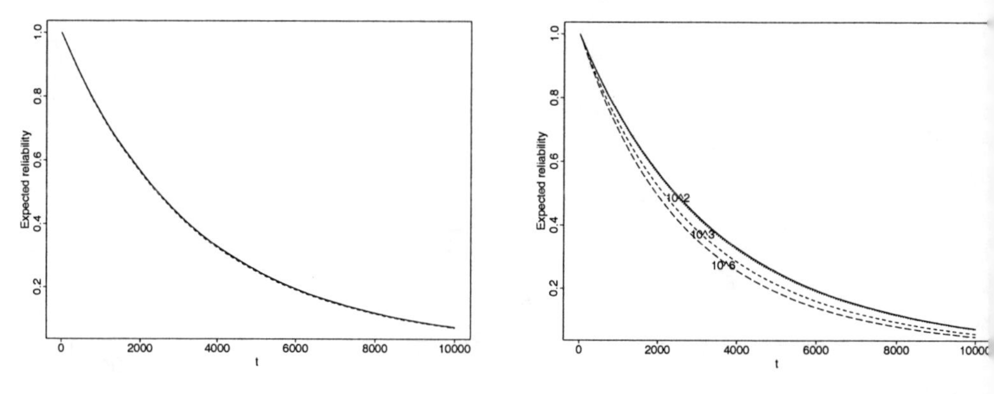

FIGURE 3. Expected reliability function given $\epsilon = .2$ (lhs) and $\epsilon = 1$ (rhs).

		δ				
		10^2	10^3	10^4	10^5	10^6
	.2	3757	3726	3737	3752	3760
ϵ	.6	3737	3606	3625	3693	3735
	1	3714	3360	3168	3116	3106

TABLE 1. Expected time to next failure for various values of ϵ and δ

when an alternative basic model is used, the predicted reliability can be very different.

We can also calculate the expected times to failure assuming that some faults remain in the software. These are tabulated for various values of ϵ and δ in Table 1.

Table 1 illustrates that the expected time to next failure does not change very much from the uncontaminated value (3765) for $\epsilon = .2$ and even $\epsilon = .6$, though for large ϵ and δ, the difference does become significant.

Given the results of this subsection, we might conclude that inference seems relatively robust to small and medium contaminations of the prior for N. It would appear that posterior inference is largely dependent on the prior mode, with the variability of the prior distribution being less important. In the following subsection we will assess the effects on inference about the expected number of bugs remaining in the software when we also consider sensitivity of the prior density for ϕ.

21.2.3 Density bounded priors preserving independence between N and ϕ

The density band class of prior distributions is one of the most popular classes of priors considered in the robustness literature. See, for example,

		ϵ_2			
		0	.1	.5	.9
ϵ_1	0	17.89	18.29	20.17	24.18
		17.89	17.48	15.72	12.57
	.1	18.42	18.84	20.75	24.76
		17.36	16.96	15.23	12.15
	.5	20.90	21.35	23.38	27.52
		15.11	14.74	13.18	10.39
	.9	26.22	26.70	28.85	33.11
		11.12	10.81	9.53	7.36

TABLE 2. Upper (upper number) and lower (lower number) bounds for numbers of bugs remaining for various ϵ_1 and ϵ_2.

the references in Berger (1994). This class is of form

$$\Gamma = \{\pi : L(N, \phi) < \pi(N, \phi) < U(N, \phi)\},$$

where $L(\cdot)$ and $U(\cdot)$ are lower and upper generalised density bounds.

One problem with this class is that for multivariate models, without the use of extra constraints, it can be very wide. Suppose that in our situation, we are willing to consider N and ϕ as independent; a reasonable assumption since N represents the number of faults in the software and ϕ the average size, or propensity for discovery of a fault. This is also the model we have assumed in subsection 21.2.1.

If we now consider the related class

$$\begin{aligned} \Gamma_{IB} \quad = \quad & \{\pi : \pi(N, \phi) = \pi_1(N)\pi_2(\phi), \;\; L_1(N) < \pi_1(N) < U_1(N), \\ & L_2(\phi) < \pi_2(\phi) < U_2(\phi)\}, \end{aligned}$$

where $L_1(\cdot)$ ($U_1(\cdot)$) and $L_2(\cdot)$ ($U_2(\cdot)$) are lower (upper) generalised bounds for N and ϕ, respectively, then Berger and Moreno (1994) illustrate methods that may be used to assess the posterior sensitivity of various quantities.

Given our base priors $\phi \sim \mathcal{G}(\alpha, \beta)$ and $N \sim \mathcal{P}(\lambda)$, we consider the density bands

$$\begin{aligned} L_1(N) = (1-\epsilon_1)f(N) \quad & \leq \pi_1(N) \leq \quad U_1(N) = (1+\epsilon_1)f(N), \\ L_2(\phi) = (1-\epsilon_2)f(\phi) \quad & \leq \pi_2(\phi) \leq \quad U_2(\phi) = (1+\epsilon_2)f(\phi), \end{aligned}$$

where $f(N)$ and $f(\phi)$ are the Poisson(100) and exponential(.0001) priors given earlier.

Here, we shall consider the posterior sensitivity of the expected number of bugs remaining, $E[N|\text{data}] - 86$ to this class of priors. In Table 2 we illustrate the upper and lower bounds for this expectation for various values of ϵ_1 and ϵ_2.

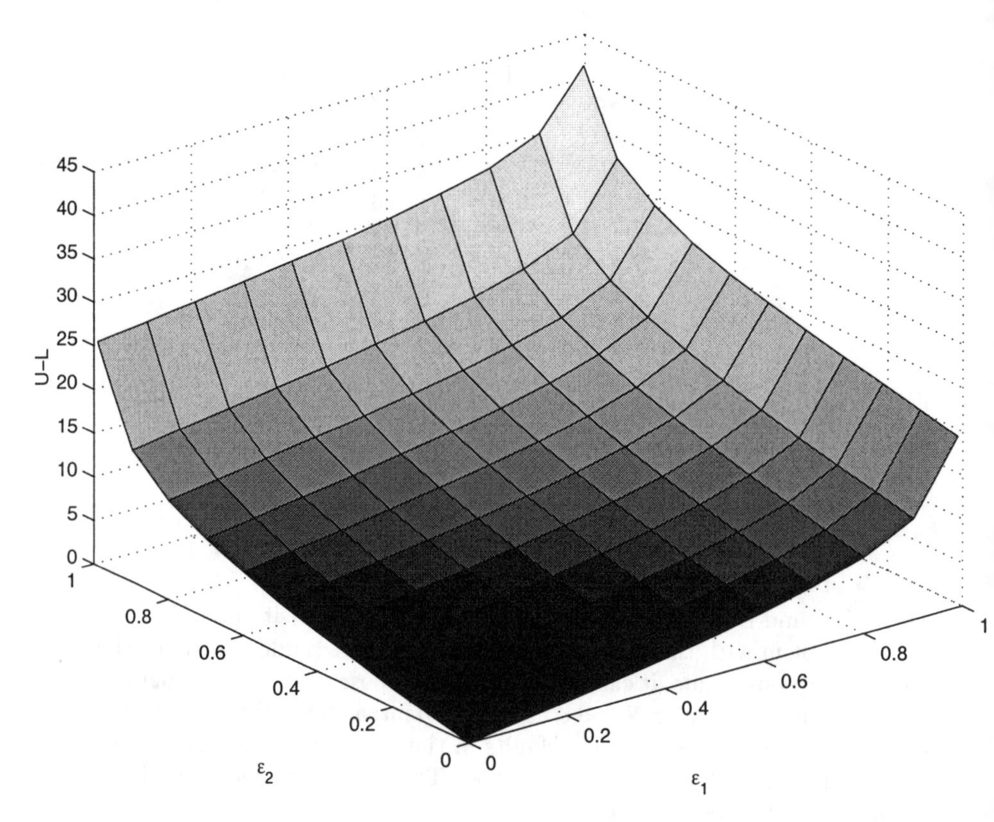

FIGURE 4. Difference between upper and lower bounds for $0 \leq \epsilon_1, \epsilon_2 \leq 0.99$.

In Figure 4 we illustrate how the difference between these bounds varies over various values of ϵ_1 and ϵ_2. Note that for $\epsilon_1, \epsilon_2 > 0.99$, the difference increases very rapidly.

Table 2 indicates that for small ϵ_1 and ϵ_2, there is relatively small variation in the expected number of bugs remaining, between about 17 and 19 for $\epsilon_1 = \epsilon_2 = .1$. For large contaminations, the range in this expectation increases greatly. There appears to be slightly greater marginal sensitivity to the prior for N than to that for ϕ, as can be seen by comparing the results for $\epsilon_1 = 0$ with those for $\epsilon_2 = 0$. From Figure 4, we can also see that for $\epsilon_1, \epsilon_2 < 0.5$, the difference between the upper and lower bounds is less than 10. For $\epsilon_1, \epsilon_2 < 0.2$, the maximum difference is below 4.

These results seem to back up those of the previous section. It seems that the Poisson/gamma prior elected here is a reasonably robust choice. In the following section we consider one of the most popular time-series models for software reliability.

21.3 The autoregressive model of Singpurwalla and Soyer

Singpurwalla and Soyer (1985) have proposed a Gaussian autoregressive process for log failure times. Let $Y_i = \log(T_i)$, then assume

$$Y_{i+1} = \theta_i Y_i + \epsilon_i, \tag{4}$$

where $\epsilon_i \sim N(0, \sigma_1^2)$ and we assume a reasonably robust inverse χ_1^2 for σ_1^2. We consider three prior structures for the θ_i, the first and third of which appear in Singpurwalla and Soyer (1985):

1. Exchangeable prior, where $\theta_i \sim N(\lambda, \sigma_2^2)$. We assume a robust inverse χ_1^2 for σ_2^2 and specify a prior for λ.

2. Multivariate Gaussian, with $\theta = (\theta_1, \dots, \theta_N) \sim MVN(\lambda 1_N, \sigma_2^2 \Sigma)$, where Σ is a first-order band matrix with $\Sigma_{ii} = 1$, $\Sigma_{i,i+1} = \rho$ and $\Sigma_{ij} = 0$ otherwise (i.e., $\mathrm{Corr}(\theta_i, \theta_{i+1}) = \rho$). An inverse χ_1^2 is assumed for σ_2^2, ρ is assumed known and a prior is specified for λ. This is a generalisation of the first prior, which is obtained with $\rho = 0$.

3. Autoregressive prior, where $\theta_i = \theta_{i-1} + w_i$ and $w_i \sim N(0, w^2)$, and a prior is specified for w^2. This implies a Gaussian Kalman filter model for the Y_i.

All posterior distributions, under any of these prior structures, are available by Gibbs sampling. Using the output from the Gibbs sampler, we can obtain posterior predictive distributions for the time to next failure.

We analyse the same data set as was used in the last section. We first see that it is important that a prior be specified on the model parameters, rather than assuming some of them known. We also see that the prior structure has considerable effect on predictions of the next failure time, even when one tries to pick the priors to be quite robust. For example, in Table 3 we show selected percentiles and the mean of the posterior predictive distribution for the next log time to failure under five priors on the θ_i:

1. Exchangeable prior with a non-central t_1 of mean 1 on λ (equivalent to the multivariate Gaussian prior with $\rho = 0$);

2. Multivariate Gaussian prior with a non-central t_1 of mean 1 on λ and a modest positive correlation of $\rho = 0.5$;

3. Multivariate Gaussian prior with a non-central t_1 of mean 1 on λ and a small negative correlation of $\rho = -0.2$;

4. Autoregressive prior with inverse χ_1^2 prior on the w^2 (i.e., a robust t_1 on the w_i);

5. Autoregressive prior with $w^2 = 1$, that is, the same as the mean of w^2 in prior 4.

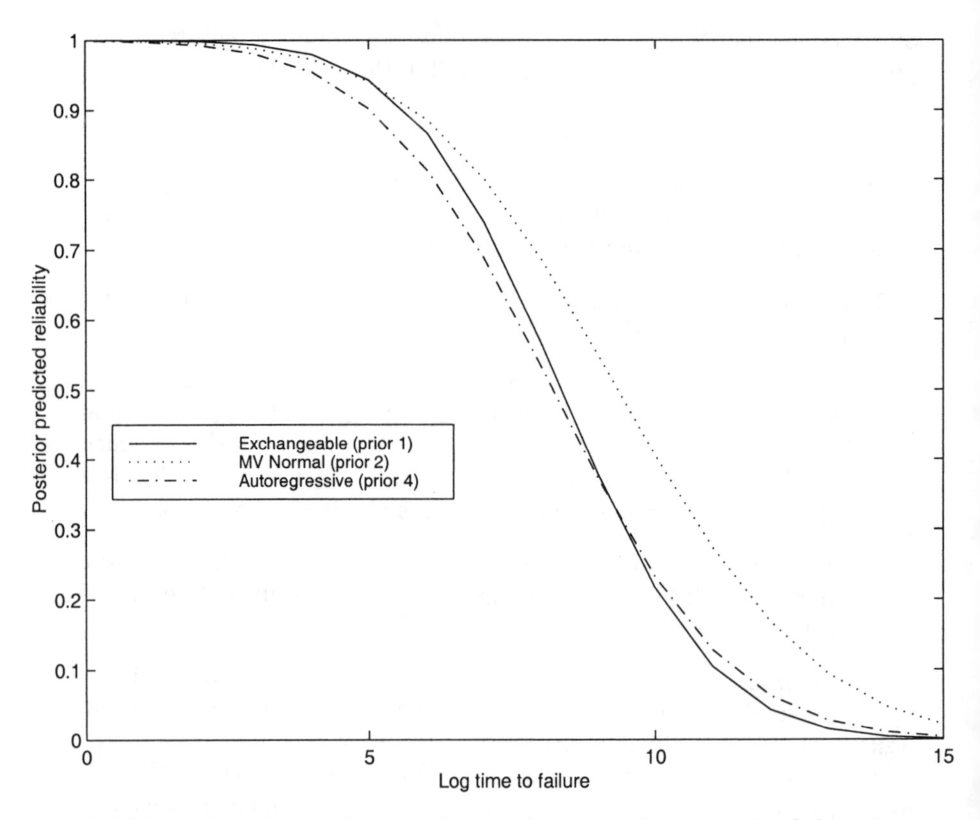

FIGURE 5. Posterior predictive reliability functions of the next log failure time under priors 1, 2 and 4 (as defined in Table 3) for the Littlewood data.

Figure 5 shows the posterior reliability function of the next log time to failure for priors 1, 2 and 4. Comparison of priors 1, 2 and 4 shows that picking what would often be considered robust priors for the three different prior structures gives substantially different results, even on the log time scale and even for a reasonably large data set as is used here. Looking at priors 1, 2 and 3, we see that the posterior mean under the multivariate Gaussian model is robust to the specification of ρ, whereas the reliability function is not. Comparison of priors 4 and 5, both autoregressive priors, shows the importance of assuming w^2 random; fixing $w^2 = 1$ gives a much wider predictive distribution in contrast to allowing it to be distributed with a mean of 1 a priori. However, fixing w^2 to be the posterior mean of w^2 from prior 4 (which is 0.05) yields a predictive distribution almost identical to that obtained by prior 4. Similarly, comparison of priors 2 and 3 shows that the correlation parameter value has considerable effect on the

	Prior Structure on θ_i	Percentiles					Mean
		2.5	25	50	75	97.5	
1	Exchangeable, $\lambda \sim t_1(1)$	3.0	5.3	8.1	9.5	11.7	8.34
2	MV Gaussian, $\lambda \sim t_1(1)$, $\rho = 0.5$	2.0	5.4	8.0	9.2	12.4	8.20
3	MV Gaussian, $\lambda \sim t_1(1)$, $\rho = -0.2$	3.0	6.7	8.2	9.5	11.7	8.37
4	Autoregressive, $w^2 \sim \chi_1^2$	2.0	6.4	9.0	11.3	13.6	9.34
5	Autoregressive, $w^2 = 1$	−8.1	3.1	8.4	14.6	25.2	9.86

TABLE 3. Percentiles and mean for the posterior predictive distribution of next log failure time under different priors for the autoregressive model of Singpurwalla and Soyer and the Littlewood data.

uncertainty in the prediction, and so ideally should also be considered an unknown parameter.

As we have seen that robustness is an important issue here, it might be useful to consider a more robust (prior and) model structure by replacing the assumption of normality with a more general distributional structure. One possibility within the time series framework is examined in Marin (2000).

21.4 Conclusions

In this paper, we have examined sensitivity in Bayesian inference for two well known software reliability models. For the Jelinski-Moranda model, we have seen that small changes to the usual priors have relatively little effect on posterior inference concerning the number of faults remaining, or the posterior reliability. However, medium and large contaminations can cause differences in some posterior estimates. In particular, Table 2 shows that there is some sensitivity to the prior in the estimated number of faults remaining.

For the autoregressive model we have shown that inference concerning both failure time and the reliability function can be very sensitive to the choice of prior distribution. In particular, introducing correlation into the prior distribution can have a strong effect. Also, inference given the autoregressive prior model and the simple exchangeable Gaussian model can be very different.

One thing we have not considered in this article is model sensitivity. Clearly, with the proliferation of different software reliability models available, this is a subject of great importance. It has been observed before that many seemingly similar classical and Bayesian models can lead to very different reliability and failure time estimates. See, for example, Keiller et al. (1983). We should also note that these differences can be very important when models are being used to decide upon testing times for software, see Wiper et al. (1998).

It is interesting to note that the predicted reliability functions generated by the two models are very different and that the choice of model would appear to have a much stronger influence on the posterior inference than the prior. The analyses given by Littlewood (1989) also indicate that models can lead to very different (Bayesian and classical) inference based on the same data. Furthermore, we should mention that prior robustness should not be taken as any indicator of model quality. Although here we have seen that inference from the Jelinski-Moranda model is relatively robust, there are a number of articles, both theoretical (Singpurwalla and Wilson 1994) and empirical (Adams 1984), which point out important defects in this approach. On the contrary, the Singpurwalla and Soyer model has been shown to give quite good predictions in practice: see Singpurwalla and Soyer (1985), Singpurwalla (1992). Model choice should be based on predictive power; see, for example, Littlewood (1989).

Finally, we also note that there have been few examples of real prior elicitation in the software reliability literature. Exceptions illustrating elicitation with two Poisson process models are Campodónico and Singpurwalla (1994) and McDaid (1998). We predict that robust methods will become more important in this field when subjective, expert priors are introduced into analyses on a more regular basis.

References

ADAMS, E.N. (1984). Optimizing preventive service of software products. *IBM Journal of Research and Development*, **28**.

BERGER, J.O. (1994). An overview of robust Bayesian analysis. *Test*, **3**, 5–124.

BERGER, J.O. and MORENO, E. (1994). Bayesian robustness in bidimensional models: prior independence. *Journal of Statistical Planning and Inference*, **40**, 161–178.

CAMPODÓNICO, S. and SINGPURWALLA, N.D. (1994). A Bayesian analysis of the logarithmic-Poisson execution time model based on expert opinion and failure data. *IEEE Transactions in Software Engineering*, **SE-20**, 677–683.

CHEN, Y. and SINGPURWALLA, N.D. (1997). Unification of software reliability models via self-exciting Poisson processes. *Advances in Applied Probability*, **29**, 337–352.

DALAL, S.R. and MALLOWS, C.L. (1986). When should one stop testing software? *Journal of the American Statistical Association*, **83**, 872–879.

FORMAN, E.H. and SINGPURWALLA, N.D. (1977). An empirical stopping rule for debugging and testing computer software. *Journal of the American Statistical Association*, **72**, 750–757.

GUSTAFSON, P. (2000). Local robustness of posterior quantities. In *Robust Bayesian Analysis* (D. Ríos Insua and F. Ruggeri, eds.). New York: Springer-Verlag.

JELINSKI, Z. and MORANDA, P. (1972). Software reliability research. In *Statistical Computer Performance Evaluation* (W. Freiberger, ed.), 465–484. New York: Academic Press.

JOE, H. and REID, N. (1985). On the software reliability models of Jelinski Moranda and Littlewood. *IEEE Transactions on Reliability*, **R-34**, 216–218.

KEILLER, P.A., LITTLEWOOD, B., MILLER, D.R. and SOFER, A. (1983). A comparison of software reliability predictions. *Digest FTCS*, **13**, 128–134.

LITTLEWOOD, B. (1989). Forecasting software reliability. *Lecture Notes in Computer Science*, **341**. Berlin: Springer-Verlag.

LITTLEWOOD, B. and VERALL, J.L. (1973). A Bayesian reliability growth model for computer software. *Applied Statistics*, **22**, 332–346.

MARIN, J.M. (2000). Case study of a robust dynamic linear model. In *Robust Bayesian Analysis* (D. Ríos Insua and F. Ruggeri, eds.). New York: Springer-Verlag.

MCDAID, K. (1998). Deciding how long to test software. *Ph.D. Dissertation*, The University of Dublin, Trinity College.

MUSA, J.D. and OKUMOTO, K. (1987). A logarithmic Poisson execution time model for software reliability measurement. *Proceedings of the 7th International Conference on Software Engineering, Orlando*, 230–237.

SINGPURWALLA, N.D. (1991). Determining an optimal time for testing and debugging software. *IEEE Transactions in Software Engineering*, **SE-17**, 313–319.

SINGPURWALLA, N.D. (1992). Non-homogeneous autoregressive processes for tracking (software) reliability growth, and their Bayesian analysis. *Journal of the Royal Statistical Society*, **54**, 145–156.

SINGPURWALLA, N.D. and SOYER, R. (1985). Assessing (software) reliability growth using a random coefficient autoregressive process and its ramifications. *IEEE Transactions in Software Engineering*, **SE-11**, 12: 1456–1464.

SINGPURWALLA, N.D. and WILSON, S.P. (1994). Software reliability modeling. *International Statistical Review*, **62**, 289–317.

SINGPURWALLA, N.D. and WILSON, S.P. (1999). *Statistical Methods in Software Engineering: Reliability and Risk.* New York: Springer-Verlag.

WIPER, M.P., RÍOS INSUA, D. and HIERONS, R.M. (1998). Bayesian inference and optimal release times for two software failure models. *Revista de la Real Academia de Ciencias Exactas, Fisicas y Naturales (España)*, **92**, 323–328.

Bibliography

ABRAHAM, C. (1999). Asymptotic limit of the Bayes actions set derived from a class of loss functions. *Tech. Report*, ENSAM INRIA.

ABRAHAM, C. and DAURES, J.P. (1998). Global robustness with respect to the loss function and the prior. *Tech. Report*, ENSAM INRIA.

ABRAHAM, C. and DAURES, J.P. (1999). Analytic approximation of the interval of Bayes actions derived from a class of loss functions. *Test*, **8**, 129–145.

ALBERT, J., DELAMPADY, M. and POLASEK, W. (1991). A class of distributions for robustness studies. *J. Statist. Plan. Inference*, **28**, 291–304.

ANGERS, J.F. (1996). Protection against outliers using a symmetric stable law prior. In *Bayesian Robustness* (J.O. Berger et al., eds). IMS Lecture Notes - Monograph Series, **29**. Hayward: IMS.

ANGERS, J.F. and BERGER, J.O. (1986). The Stein effect and Bayesian analysis: a reexamination. *Comm. Statist. Theory Methods*, **15**, 2005–2023.

ANGERS, J.F. and BERGER, J.O. (1991). Robust hierarchical Bayes estimation of exchangeable means. *Canad. J. Statist.*, **19**, 39–56.

ANGERS, J.F., MACGIBBON, B. and WANG, S. (1998). A robust Bayesian approach to the estimation of intra-block exchangeable normal means with applications. *Sankhyā, Ser. A*, **60**, 198–213.

BASU, S. (1994). Variations of posterior expectations for symmetric unimodal priors in a distribution band. *Sankhyā, Ser. A*, **56**, 320–334.

BASU, S. (1995). Ranges of posterior probability over a distribution band. *J. Statist. Plan. Inference*, **44**, 149–166.

BASU, S. (1996a). Local sensitivity, functional derivatives and nonlinear posterior quantities. *Statist. Decisions*, **14**, 405–418.

BASU, S. (1996b). Bayesian hypothesis testing using posterior density ratios, *Statist. Probab. Lett.*, **30**, 79–86.

BASU, S. (1996c). A new look at Bayesian point null hypothesis testing. *Sankhyā, Ser. A*, **58**, 292–310.

BASU, S. (1999). Posterior sensitivity to the sampling distribution and the prior: more than one observation. *Ann. Inst. Statist. Math.*, **51**, 499–513.

BASU, S. (2000a). Uniform stability of posteriors, *Statist. Probab. Lett.*, **46**, 53–58.

BASU, S. (2000b). Bayesian robustness and Bayesian nonparametrics. In *Robust Bayesian Analysis* (D. Ríos Insua and F. Ruggeri, eds.). New York: Springer-Verlag.

BASU, S. and DASGUPTA, A. (1995). Robust Bayesian analysis with distribution bands. *Statist. Decisions*, **13**, 333–349.

BASU, S., JAMMALAMADAKA, S.R. and LIU, W. (1996). Local posterior robustness with parametric priors: maximum and average sensitivity. In *Maximum Entropy and Bayesian Methods* (G. Heidlbreder, ed.). Dordrecht: Kluwer.

BASU, S., JAMMALAMADAKA, S.R. and LIU, W. (1998). Stability, uniform stability and qualitative robustness of posterior distributions and posterior quantities. *J. Statist. Plan. Inference*, **71**, 151–162.

BAYARRI, M.J. and BERGER, J.O. (1993). Robust Bayesian bounds for outlier detection. In *Recent Advances in Statistics and Probability* (J.P. Vilaplana and M.L. Puri, eds). Ah Zeist: VSP.

BAYARRI, M.J. and BERGER, J.O. (1994). Applications and limitations of robust Bayesian bounds and type II MLE. In *Statistical Decision Theory and Related Topics V* (S.S. Gupta and J.O. Berger, eds.). New York: Springer-Verlag.

BAYARRI, M.J. and BERGER, J.O. (1998). Robust Bayesian analysis of selection models. *Ann. Statist.*, **26**, 645–659.

BERGER, J.O. (1982a). Estimation in continuous exponential families: Bayesian estimation subject to risk restrictions and inadmissibility results. In *Statistical Decision Theory and Related Topics III,* (S.S. Gupta and J.O. Berger, eds.), **1**. New York: Academic Press.

BERGER, J.O. (1982b). Bayesian robustness and the Stein effect. *J. Amer. Statist. Assoc.*, **77**, 358–368.

BERGER, J.O. (1984). The robust Bayesian viewpoint (with discussion). In *Robustness of Bayesian Analysis* (J. Kadane, ed.). Amsterdam: North-Holland.

BERGER, J.O. (1985). *Statistical Decision Theory and Bayesian Analysis.* New York: Springer Verlag.

BERGER, J.O. (1990). Robust Bayesian analysis: sensitivity to the prior. *J. Statist. Plan. Inference*, **25**, 303–328.

BERGER, J.O. (1992). A comparison of minimal Bayesian test of precise hypotheses. *Rassegna di Metodi Statistici ed Applicazioni*, **7**, Bologna: Pitagora Editrice.

BERGER, J.O. (1994). An overview of robust Bayesian analysis (with discussion). *Test*, **3**, 5–59.

BERGER, J.O. and BERLINER, L.M. (1986). Robust Bayes and empirical Bayes analysis with ε-contaminated priors. *Ann. Statist.*, **14**, 461–486.

BERGER, J.O., BETRÒ, B., MORENO, E., PERICCHI, L.R., RUGGERI, F., SALINETTI, G. and WASSERMAN, L. (1996). *Bayesian Robustness*, IMS Lecture Notes - Monograph Series, **29**. Hayward: IMS.

BERGER, J.O. and CHEN, M.H. (1993). Determining retirement patterns: prediction for a multinomial distribution with constrained parameter space. *Statistician*, **42**, 427–443.

BERGER, J.O. and DELAMPADY, M. (1987). Testing precise hypotheses (with discussion). *Statist. Sci.*, **2**, 317–352.

BERGER, J.O. and JEFFERYS, W. (1992). The application of robust Bayesian analysis to hypothesis testing and Occam's razor. *J. Ital. Statist. Society*, **1**, 17–32.

BERGER, J.O. and MORENO, E. (1994). Bayesian robustness in bidimensional models: prior independence. *J. Statist. Plan. Inference*, **40**, 161–176.

BERGER, J.O. and MORTERA, J. (1991). Interpreting the stars in precise hypothesis testing. *Int. Statist. Review*, **59**, 337–353.

BERGER, J.O. and MORTERA, J. (1994). Robust Bayesian hypothesis testing in the presence of nuisance parameters. *J. Statist. Plan. Inference*, **40**, 357–373.

BERGER, J.O. and O'HAGAN, A. (1988). Ranges of posterior probabilities for unimodal priors with specified quantiles. In *Bayesian Statistics 3* (J.M. Bernardo et al., eds.). Oxford: Oxford University Press.

BERGER, J.O., RÍOS INSUA, D. and RUGGERI, F. (2000). Bayesian robustness. In *Robust Bayesian Analysis* (D. Ríos Insua and F. Ruggeri, eds.). New York: Springer-Verlag.

BERGER, J.O. and SELLKE, T. (1987). Testing a point null hypothesis: the irreconcilability of significance levels and evidence (with discussion). *J. Amer. Statist. Assoc.*, **82**, 112–122.

BERGER, R. (1979). Gamma-minimax robustness of Bayes rules. *Comm. Statist. Theory Methods*, **8**, 543–560.

BERLINER, M. (1985). A decision theoretic structure for robust Bayesian analysis with the applications to the estimation of a multivariate normal mean. In *Bayesian Statistics 2* (J.M. Bernardo et al., eds.). Oxford: Oxford University Press.

BERLINER, M. and GOEL, P. (1990). Incorporating partial prior information: ranges of posterior probabilities. In *Bayesian and Likelihood Methods in Statistics and Econometrics: Essays in Honor of George A. Barnard* (S. Geisser et al, eds.). Amsterdam: North-Holland.

BERTOLINO, F. and RACUGNO, W. (1994). Robust Bayesian analysis in analysis of variance and χ^2-tests by using marginal likelihoods. *Statistician*, **43**, 191–201.

BETRÒ, B. (1999). The accelerated central cutting plane algorithm in the numerical treatment of Bayesian global prior robustness problems. *Tech. Report*, **99.17**, CNR-IAMI.

BETRÒ, B. and GUGLIELMI, A. (1994). An algorithm for robust Bayesian analysis under generalized moment conditions. *Tech. Report*, **94.6**, CNR-IAMI.

BETRÒ, B. and GUGLIELMI, A. (1996). Numerical robust Bayesian analysis under generalized moment conditions. In *Bayesian Robustness* (J.O. Berger et al., eds). IMS Lecture Notes - Monograph Series, **29**. Hayward: IMS.

BETRÒ, B. and GUGLIELMI, A. (2000). Methods for global prior robustness under generalized moment conditions. In *Robust Bayesian Analysis* (D. Ríos Insua and F. Ruggeri, eds.). New York: Springer-Verlag.

BETRÒ B., GUGLIELMI, A. and ROSSI, F. (1996). Robust Bayesian analysis for the power law process. *ASA 1996 Proc. Sec. Bay. Statist. Sc.*, 288–291. Alexandria: ASA.

BETRÒ, B., MĘCZARSKI, M. and RUGGERI, F. (1994). Robust Bayesian analysis under generalized moments conditions. *J. Statist. Plan. Inference*, **41**, 257–266.

BETRÒ, B. and RUGGERI, F. (1992). Conditional Γ-minimax actions under convex losses. *Comm. Statist. Theory Methods*, **21**, 1051–1066.

BIELZA, C., MARTIN, J. and RÍOS INSUA, D. (2000). Approximating nondominated sets in continuous problems. *Tech. Report*, Technical University of Madrid.

BIELZA, C., RÍOS INSUA, D. and RÍOS-INSUA, S. (1996). Influence diagrams under partial information. In *Bayesian Statistics 5* (J.M. Bernardo et al., eds.). Oxford: Oxford University Press.

BIELZA, C., RÍOS-INSUA, S., GÓMEZ, M. and FERNÁNDEZ DEL POZO, J.A. (2000). Sensitivity analysis in IctNeo. In *Robust Bayesian Analysis* (D. Ríos Insua and F. Ruggeri, eds.). New York: Springer-Verlag.

BIRMIWAL, L.R. and DEY, D.K. (1993). Measuring local influence of posterior features under contaminated classes of priors. *Statist. Decisions*, **11**, 377–390.

BORATYŃSKA, A. (1994). Robustness in Bayesian statistical models (Polish). *Appl. Math.*, **37**, 67–106.

BORATYŃSKA, A. (1996). On Bayesian robustness with the ε-contamination class of priors. *Statist. Probab. Lett.*, **26**, 323–328.

BORATYŃSKA, A. (1997). Stability of Bayesian inference in exponential families. *Statist. Probab. Lett.*, **36**, 173–178.

BORATYŃSKA, A. and DROZDOWICZ, M. (1998). Robust Bayesian estimation in normal model with asymmetric loss function. *Tech. Report*, Institute of Mathematics, University of Warsaw.

BORATYŃSKA, A. and MĘCZARSKI, M. (1994). Robust Bayesian estimation in the one-dimensional normal model. *Statist. Decisions*, **12**, 221–230.

BORATYŃSKA, A. and ZIELIŃSKI, R. (1991). Infinitesimal Bayes robustness in the Kolmogorov and the Lévy metrics. *Tech. Report*, Institute of Mathematics, University of Warsaw.

BORATYŃSKA, A. and ZIELIŃSKI, R. (1993). Bayes robustness via the Kolmogorov metric. *Appl. Math.*, **22**, 139–143.

BOSE, S. (1990). Bayesian robustness with shape-constrained priors and mixtures of priors. *Ph.D. Dissertation*, Purdue University.

BOSE, S. (1994a). Bayesian robustness with mixture classes of priors. *Ann. Statist.*, **22**, 652–667.

BOSE, S. (1994b). Bayesian robustness with more than one class of contaminations. *J. Statist. Plan. Inference*, **40**, 177–187.

BOX, G.E.P. (1980). Sampling and Bayes' inference in scientific modelling and robustness (with discussion). *J. Roy. Statist. Soc. Ser. A*, **143**, 383–430.

BOX, G.E.P. and TIAO, G. (1992). *Bayesian Inference in Statistical Analysis.* New York: Wiley.

BROWN, L.D. (1975). Estimation with incompletely specified loss functions. *J. Amer. Statist. Assoc.*, **70**, 417–427.

BUTLER, J., JIA, J. and DYER, J. (1997). Simulation techniques for the sensitivity analysis of multi-criteria decision models. *Eur. J. Oper. Res.*, **103**, 531–546.

CAGNO, E., CARON, F., MANCINI, M. and RUGGERI, F. (1998). On the use of a robust methodology for the assessment of the probability of failure in an urban gas pipe network. In *Safety and Reliability, vol. 2* (S. Lydersen et al, eds.). Rotterdam: Balkema.

CAGNO, E., CARON, F., MANCINI, M. and RUGGERI, F. (2000a). Using AHP in determining prior distributions on gas pipeline failures in a robust Bayesian approach. *Reliab. Engin. System Safety*, **67**, 275–284.

CAGNO, E., CARON, F., MANCINI, M. and RUGGERI, F. (2000b). Sensitivity of replacement priorities for gas pipeline maintenance. In *Robust Bayesian Analysis* (D. Ríos Insua and F. Ruggeri, eds.). New York: Springer-Verlag.

CANO, J.A. (1993). Robustness of the posterior mean in normal hierarchical models. *Comm. Statist. Theory Methods*, **22**, 1999–2014.

CANO, J.A., HERNÁNDEZ, A. and MORENO, E. (1985). Posterior measures under partial prior information. *Statistica*, **2**, 219–230.

CARLIN, B.P., CHALONER, K.M., LOUIS, T.A. and RHAME, F.S. (1993). Elicitation, monitoring, and analysis for an AIDS clinical trial. In *Case Studies in Bayesian Statistics V* (C. Gatsonis et al., eds.). New York: Springer-Verlag.

CARLIN, B.P. and LOUIS, T.A. (1996). Identifying prior distributions that produce specific decisions, with application to monitoring clinical trials. In *Bayesian Analysis in Statistics and Econometrics: Essays in Honor of Arnold Zellner* (D. Berry et al., eds.). New York: Wiley.

CARLIN, B.P. and PÉREZ, M.E. (2000). Robust Bayesian analysis in medical and epidemiological settings. In *Robust Bayesian Analysis* (D. Ríos Insua and F. Ruggeri, eds.). New York: Springer-Verlag.

CARLIN, B.P. and SARGENT, D. (1996). Robust Bayesian approaches for clinical trial monitoring. *Statist. Med.*, **15**, 1093–1106.

CAROTA, C. (1996). Local robustness of Bayes factors for nonparametric alternatives. In *Bayesian Robustness* (J.O. Berger et al., eds). IMS Lecture Notes - Monograph Series, **29**. Hayward: IMS.

CAROTA, C. and RUGGERI, F. (1994). Robust Bayesian analysis given priors on partition sets. *Test*, **3**, 73–86.

CASELLA, G. and BERGER, R. (1987). Reconciling Bayesian and frequentist evidence in the one-sided testing problem (with discussion). *J. Amer. Statist. Assoc.*, **82**, 106–111.

CHATURVEDI, A. (1996). Robust Bayesian analysis of the linear regression model. *J. Statist. Plan. Inference*, **50**, 175–186.

CHUANG, D.T. (1984). Further theory of stable decisions. In *Robustness of Bayesian Analysis* (J. Kadane, ed.). New York: North Holland.

CLARKE, B. and GUSTAFSON, P. (1998). On the overall sensitivity of the posterior distribution to its inputs. *J. Statist. Plan. Inference*, **71**, 137–150.

CONIGLIANI C., DALL'AGLIO, M., PERONE PACIFICO, M. and SALINETTI, G. (1994). Robust statistics: from classical to Bayesian analysis. *Metron*, **LII**, 89–109.

CONIGLIANI, C. AND O'HAGAN, A. (2000). Sensitivity of the fractional Bayes factor to prior distributions. To appear in *Canad. J. Statist.*.

CORRADI, F. and MEALLI, F. (1996). Non parametric specification of error terms in dynamic models. In *Bayesian Robustness* (J.O. Berger et al., eds). IMS Lecture Notes - Monograph Series, **29**. Hayward: IMS.

COZMAN, F. (1997). Robustness analysis of Bayesian networks with local convex sets of distributions. In *Proc. Conf. Uncertainty Artif. Intell.*

COZMAN, F. (1999). Calculation of posterior bounds given convex sets of prior probability measures and likelihood functions. *J. Comput. Graph. Statist.*, **8**, 824–838.

CUEVAS, A. (1984). A Bayesian definition of qualitative robustness (Spanish). *Trab. Estad. Invest. Oper.*, **35**, 170–186.

CUEVAS, A. (1988a). Qualitative robustness in abstract inference. *J. Statist. Plan. Inference*, **18**, 277–289.

CUEVAS, A. (1988b). On stability in a general decision problem. *Statistica*, **48**, 9–14.

CUEVAS, A. and SANZ, P. (1988). On differentiability properties of Bayes operators. In *Bayesian Statistics 3* (J.M. Bernardo et al., eds.). Oxford: Oxford University Press.

CUEVAS, A. and SANZ, P. (1990). Qualitative robustness in confidence region (Spanish). *Trab. Estad.*, **5**, 9–22.

DALL'AGLIO, M. (1995). Problema dei momenti e programmazione lineare semi-infinita nella robustezza bayesiana. *Ph.D. Dissertation*, Dipartimento di Statistica, Probabilità e Statistiche Applicate, Università "La Sapienza", Roma.

DALL'AGLIO, M. and SALINETTI, G. (1994). Bayesian robustness, moment problem and semi-infinite linear programming. *Tech. Report*, Dipartimento di Statistica, Probabilità e Statistiche Applicate, Università "La Sapienza", Roma.

DASGUPTA, A. (1991). Diameter and volume minimizing confidence sets in Bayes and classical problems. *Ann. Statist.*, **19**, 1225–1243.

DASGUPTA, A. and BOSE, A. (1988). Γ-minimax and restricted-risk Bayes estimation of multiple Poisson means under ϵ-contaminations of the subjective prior. *Statist. Decisions*, **6**, 311–341.

DASGUPTA, A. and DELAMPADY, M. (1990). Bayesian testing with symmetric and unimodal priors. *Tech. Report*, **90–47**, Department of Statistics, Purdue University.

DASGUPTA, A. and MUKHOPADHYAY, S. (1994). Uniform and subuniform posterior robustness: the sample size problem. *J. Statist. Plan. Inference*, **40**, 189–204.

DASGUPTA, A. and RUBIN, H. (1988). Bayesian estimation subject to minimaxity of the mean of a multivariate normal distribution in the case of a common unknown variance: a case for Bayesian robustness. In *Statistical Decision Theory and Related Topics IV* (S.S. Gupta and J.O. Berger, eds.), **1**. New York: Springer-Verlag.

DASGUPTA, A. and STUDDEN, W.J. (1988a). Robust Bayesian analysis in normal linear models with many parameters. *Tech. Report*, **88–14**, Department of Statistics, Purdue University.

DASGUPTA, A. and STUDDEN, W.J. (1988b). Frequentist behavior of robust Bayes procedures: new applications of the Lehmann-Wald minimax theory to a novel geometric game. *Tech. Report*, **88–36C**, Department of Statistics, Purdue University.

DASGUPTA, A. and STUDDEN, W.J. (1989). Frequentist behavior of robust Bayes estimates of normal means. *Statist. Decisions*, **7**, 333–361.

DASGUPTA, A. and STUDDEN, W.J. (1991). Robust Bayesian experimental designs in normal linear models. *Ann. Statist.*, **19**, 1244–1256.

DE LA HORRA, J. and FERNÁNDEZ, C. (1993). Bayesian analysis under ε-contaminated priors: a trade-off between robustness and precision. *J. Statist. Plan. Inference*, **38**, 13–30.

DE LA HORRA, J. and FERNÁNDEZ, C. (1994). Bayesian robustness of credible regions in the presence of nuisance parameters. *Comm. Statist. Theory Methods*, **23**, 689–699.

DE LA HORRA, J. and FERNÁNDEZ, C. (1995). Sensitivity to prior independence via Farlie-Gumbel-Morgenstern model. *Comm. Statist. Theory Methods*, **24**, 987–996.

DELAMPADY, M. (1989a). Lower bounds on Bayes factors for interval null hypotheses. *J. Amer. Statist. Assoc.*, **84**, 120–124.

DELAMPADY, M. (1989b). Lower bounds on Bayes factors for invariant testing situations. *J. Multivariate Anal.*, **28**, 227–246.

DELAMPADY, M. (1992). Bayesian robustness for elliptical distributions. *Rebrape*, **6**, 97–119.

DELAMPADY, M. and BERGER, J.O. (1990). Lower bounds on Bayes factors for multinomial and chi-squared tests of fit. *Ann. Statist.*, **18**, 1295–1316.

DELAMPADY, M. and DEY, D.K. (1994). Bayesian robustness for multiparameter problems. *J. Statist. Plan. Inference*, **40**, 375–382.

DEROBERTIS, L.(1978). The use of partial prior knowledge in Bayesian inference. *Ph.D. Dissertation*, Yale University.

DEROBERTIS, L. and HARTIGAN, J.A.(1981). Bayesian inference using intervals of measures. *Ann. Statist.*, **1**, 235–244.

De Santis, F. and Spezzaferri, F. (1996). Comparing hierarchical models using Bayes factor and fractional Bayes factor: a robust analysis. In *Bayesian Robustness* (J.O. Berger et al., eds). IMS Lecture Notes - Monograph Series, **29**. Hayward: IMS.

De Santis, F. and Spezzaferri, F. (1997). Alternative Bayes factors for model selection. *Canad. J. Statist.*, **25**, 503–515.

De Santis, F. and Spezzaferri, F. (1999). Methods for default and robust Bayesian model comparison: the fractional Bayes factor approach. *Int. Statist. Review*, **67**, 3, 1–20.

Dey, D.K. and Birmiwal, L.R. (1994). Robust Bayesian analysis using entropy and divergence measures. *Statist. Probab. Lett.*, **20**, 287–294.

Dey, D.K., Ghosh, S.K. and Lou, K. (1996). On local sensitivity measures in Bayesian analysis (with discussion). In *Bayesian Robustness* (J.O. Berger et al., eds). IMS Lecture Notes - Monograph Series, **29**. Hayward: IMS.

Dey, D., Lou, K. and Bose, S. (1998). A Bayesian approach to loss robustness. *Statist. Decisions*, **16**, 65–87.

Dey, D. and Micheas, A. (2000). Ranges of posterior expected losses and ϵ-robust actions. In *Robust Bayesian Analysis* (D. Ríos Insua and F. Ruggeri, eds.). New York: Springer-Verlag.

Drummey, K.W.(1991). Robust Bayesian estimation in the normal, gamma, and binomial probability models: a computational approach. *Ph.D. Dissertation*, University of Maryland Baltimore County.

Eichenauer-Herrmann, J. and Ickstadt, K. (1992). Minimax estimators for a bounded location parameter. *Metrika*, **39**, 227–237.

Eichenauer-Herrmann, J. and Ickstadt, K.(1993). A saddle point characterization for classes of priors with shape-restricted densities. *Statist. Decisions*, **11**, 175–179.

Fandom Noubiap, R. and Seidel, W. (1998). A minimax algorithm for calculating optimal tests under generalized moment conditions. *Tech. Report*, **85**, DP in Statistics and Quantitative Economics, Universitat der Bundeswehr, Hamburg.

Felli, J.C. and Hazen, G.B. (1998). Sensitivity analysis and the expected value of perfect information. *Medical Decision Making*, **18**, 95–109.

Fernández, C., Osiewalski, J. and Steel, M.F.J. (1997). Classical and Bayesian inference robustness in multivariate regression models. *J. Amer. Statist. Assoc.*, **92**, 1434–1444.

Fishburn P., Murphy A. and Isaacs H. (1967). Sensitivity of decision to probability estimation errors: a reexamination. *Oper. Res.*, **15**, 254–267.

Fortini, S. and Ruggeri, F. (1994a). Concentration function and Bayesian robustness. *J. Statist. Plan. Inference*, **40**, 205–220.

Fortini, S. and Ruggeri, F. (1994b). On defining neighbourhoods of measures through the concentration function. *Sankhyā, Ser. A*, **56**, 444–457.

Fortini, S. and Ruggeri, F. (1995). Concentration function and sensitivity to the prior. *J. Ital. Statist. Society*, **4**, 283–297.

FORTINI, S. and RUGGERI, F. (1997). Differential properties of the concentration function, *Sankhyā, Ser. A*, **59**, 345–364.

FORTINI, S. and RUGGERI, F. (2000). On the use of the concentration function in Bayesian robustness. In *Robust Bayesian Analysis* (D. Ríos Insua and F. Ruggeri, eds.). New York: Springer-Verlag.

FRENCH, S. and RÍOS INSUA, D. (2000). *Statistical Decision Theory.* London: Arnold.

GAIVORONSKI, A., MORASSUTTO, M., SILANI, S. and STELLA, F. (1996). Numerical techniques for solving estimation problems on robust Bayesian networks. In *Bayesian Robustness* (J.O. Berger et al., eds). IMS Lecture Notes - Monograph Series, **29**. Hayward: IMS.

GEISSER, S.(1992). Bayesian perturbation diagnostics and robustness. In *Bayesian Analysis in Statistics and Econometrics* (P.K. Goel and N.S. Iyengar, eds.). New York: Springer-Verlag.

GELFAND, A. and DEY, D.(1991). On Bayesian robustness of contaminated classes of priors. *Statist. Decisions*, **9**, 63–80.

GEWEKE, J. (1999). Simulation methods for model criticism and robustness analysis (with discussion). In *Bayesian Statistics 6* (J.M. Bernardo et al., eds.). Oxford: Clarendon Press.

GEWEKE, J. and PETRELLA, L. (1998). Prior density ratio class in Econometrics. *J. Bus. Econom. Statist.*, **16**, 469–478.

GHOSH, J.K., GHOSAL, S. and SAMANTA, T.(1994). Stability and convergence of the posterior in non-regular problems. In *Statistical Decision Theory and Related Topics V* (S.S. Gupta and J.O. Berger, eds.). New York: Springer-Verlag.

GHOSH, M. and KIM, D.H.(1996). Bayes estimation of the finite population mean under heavy-tailed priors. *Calcutta Statist. Assoc. Bull.*, **46**, 181–195.

GHOSH, S.K.(2000). Measures of model uncertainty to simultaneous perturbations in both the prior and the likelihood. To appear in *J. Statist. Plan. Inference.*

GHOSH, S.K. and DEY, D.(1994). Sensitivity diagnostics and robustness issues in Bayesian inference. *Tech. Report*, **94–30**, Department of Statistics, University of Connecticut.

GILBOA, I. and SCHMEIDLER, D. (1989). Maxmin expected utility with non-unique prior. *J. Math. Econom.*, **18**, 141–153.

GIRON, F.J. and RÍOS, S. (1980). Quasi Bayesian behaviour: a more realistic approach to decision making?. In *Bayesian Statistics* (J.M. Bernardo et al., eds.). Valencia: Valencia University Press.

GODSILL, S.J. and RAYNER, P.J.W.(1996). Robust treatment of impulsive noise in speech and audio signals. In *Bayesian Robustness* (J.O. Berger et al., eds). IMS Lecture Notes - Monograph Series, **29**. Hayward: IMS.

GOLDSTEIN, M. (1982). Contamination Distributions. *Ann. Statist.*, **10**, 174–183.

GOLDSTEIN, M. and WOOFF, D.A.(1994). Robustness measures for Bayes linear analyses. *J. Statist. Plan. Inference*, **40**, 261–277.

GÓMEZ DENIZ, E., HERNÁNDEZ BASTIDA, A. and VAZQUEZ POLO, F.J.(1998). A sensitivity analysis of the pricing process in general insurance (Spanish). *Estudios de Economìa Aplicada*, **9**, 19–34.

GÓMEZ DENIZ, E., HERNÁNDEZ BASTIDA, A. and VAZQUEZ POLO, F.J.(1999a). Robustness analysis of Bayesian models for auditing: the independence between error quantity and rate (Spanish). *Estudios de Economìa Aplicada*, **11**, 101–120.

GÓMEZ DENIZ, E., HERNÁNDEZ BASTIDA, A. and VAZQUEZ POLO, F.J.(1999b). The Esscher premium principle in risk theory: a Bayesian sensitivity study. *Insurance Math. Econom.*, **25**, 387–395.

GÓMEZ DENIZ, E., HERNÁNDEZ BASTIDA, A. and VAZQUEZ POLO, F.J.(1999c). The variance premium principle: a Bayesian robustness analysis. To appear in *Actuarial Research Clearing House*, **1**.

GÓMEZ DENIZ, E., HERNÁNDEZ BASTIDA, A. and VAZQUEZ POLO, F.J.(2000). Robust Bayesian premium principles in actuarial science. To appear in *Statistician*.

GOMEZ-VILLEGAS, M. and SANZ, L.(2000). ε-contaminated priors in testing point null hypothesis: a procedure to determine the prior probability. *Statist. Probab. Lett.*, **47**, 53–60.

GOOD, I.J. (1952). Rational decisions. *J. Roy. Statist. Soc. Ser. B*, **14,** 107–114.

GOOD, I.J. (1959). Could a machine make probability judgments? *Computers and Automation*, **8**, 14–16 and 24–26.

GOOD, I.J. (1961). Discussion of C.A.B. Smith: Consistency in statistical inference and decision. *J. Roy. Statist. Soc. Ser. B*, **23**, 28–29.

GOOD, I.J.(1983a). *Good Thinking: The Foundations of Probability and Its Applications*. Minneapolis: University of Minnesota Press.

GOOD, I.J.(1983b). The robustness of a hierarchical model for multinomials and contingency tables. In *Scientific Inference, Data Analysis, and Robustness* (G.E.P. Box et al., eds.). New York: Academic Press.

GOOD, I.J. and CROOK, J.F.(1987). The robustness and sensitivity of the mixed-Dirichlet Bayesian test for "independence" in contingency tables. *Ann. Statist.*, **15**, 694–711.

GOUTIS, C. (1994). Ranges of posterior measures for some classes of priors with specified moments. *Int. Statist. Review*, **62**, 245–357.

GREENHOUSE, J. and WASSERMAN, L.(1995). Robust Bayesian methods for monitoring clinical trials. *Statistics in Medicine*, **14**, 1379–1391.

GREENHOUSE, J. and WASSERMAN, L.(1996). A practical robust method for Bayesian model selection: a case study in the analysis of clinical trials (with discussion). In *Bayesian Robustness* (J.O. Berger et al., eds). IMS Lecture Notes - Monograph Series, **29**. Hayward: IMS.

GUSTAFSON P. (1994). Local sensitivity of posterior expectations. *Ph.D. Dissertation*, Department of Statistics, Carnegie-Mellon University.

GUSTAFSON, P. (1996a). Local sensitivity of inferences to prior marginals. *J. Amer. Statist. Assoc.*, **91**, 774–781.

GUSTAFSON, P. (1996b). Local sensitivity of posterior expectations. *Ann. Statist.*, **24**, 174–195.

GUSTAFSON, P. (1996c). Aspects of Bayesian robustness in hierarchical models (with discussion). In *Bayesian Robustness* (J.O. Berger et al., eds). IMS Lecture Notes - Monograph Series, **29**. Hayward: IMS.

GUSTAFSON, P. (1996d). Model influence functions based on mixtures. *Canad. J. Statist.*, **24**, 535–548.

GUSTAFSON, P. (1996e). Robustness considerations in Bayesian analysis. *Statistical Methods in Medical Research*, **5**, 357–373.

GUSTAFSON, P. (1996f). The effect of mixing-distribution misspecification in conjugate mixture models. *Canad. J. Statist.*, **24**, 307–318.

GUSTAFSON, P. (2000). Local robustness of posterior quantities. In *Robust Bayesian Analysis* (D. Ríos Insua and F. Ruggeri, eds.). New York: Springer-Verlag.

GUSTAFSON P., SRINIVASAN C. and WASSERMAN L. (1996). Local sensitivity. In *Bayesian Statistics 5* (J.M. Bernardo et al., eds.). Oxford: Oxford University Press.

GUSTAFSON, P. and WASSERMAN, L. (1995). Local sensitivity diagnostics for Bayesian inference. *Ann. Statist.*, **23**, 2153–2167.

HARTIGAN, J.A.(1983). *Bayes Theory.* New York: Springer-Verlag.

HILL, S. and SPALL, J.C.(1994). Sensitivity of a Bayesian analysis to the prior distribution. *IEEE Trans. Sys. Man Cybern.*, **24**, 216–221.

HWANG, J.T. (1985). Universal domination and stochastic domination: estimation simultaneously under a broad class of loss functions. *Ann. Statist.*, **13**, 295–314.

ICKSTADT, K.(1992). Gamma-minimax estimators with respect to unimodal priors. In *Operations Research '91* (P. Gritzmann et al., eds.). Heidelberg: Physica-Verlag.

JEFFERYS, W. and BERGER, J.(1992). Ockham's razor and Bayesian analysis. *American Scientist*, **80**, 64–72.

KADANE, J.(1984). *Robustness of Bayesian Analysis.* Amsterdam: North-Holland.

KADANE, J.(1994). An application of robust Bayesian analysis to a medical experiment. *J. Statist. Plan. Inference*, **40**, 221–232.

KADANE, J. and CHUANG, D.T.(1978). Stable decision problems. *Ann. Statist.*, **6**, 1095–1110.

KADANE, J., MORENO, E., PÉREZ, M.-E. and PERICCHI, L.R. (1999). Applying nonparametric robust Bayesian analysis to non-opinionated judicial neutrality. *Tech. Report*, **695**, Department of Statistics, Carnegie Mellon University.

KADANE, J., SALINETTI, G. and SRINIVASAN, C. (2000). Stability of Bayes decisions and applications. In *Robust Bayesian Analysis* (D. Ríos Insua and F. Ruggeri, eds.). New York: Springer-Verlag.

KADANE, J. and SRINIVASAN, C. (1996). Bayesian robustness and stability (with discussion). In *Bayesian Robustness* (J.O. Berger et al., eds). IMS Lecture Notes - Monograph Series, **29**. Hayward: IMS.

KADANE, J. and SRINIVASAN, C. (1998). Bayes decision problems and stability. *Sankhyā, Ser. A*, **60**, 383–404.

KADANE, J. and WASSERMAN, L. (1996). Symmetric, coherent, Choquet capacities. *Ann. Statist.*, **24**, 1250–1264.

KASS, R.E. and WASSERMAN, L.(1996). The selection of prior distributions by formal rules. *J. Amer. Statist. Assoc.*, **91**, 1343–1370.

KIM, D.H.(1998). Bayesian robustness in small area estimation. *Statist. Decisions*, **16**, 89–103.

KOUZNETSOV, V.P.(1991). *Interval Statistical Models.* Moscow: Moscow Radio and Communication.

KUO, L. and DEY, D. (1990). On the admissibility of the linear estimators of the Poisson mean using LINEX loss functions. *Statist. Decisions*, **8**, 201–210.

LAVINE, M. (1988). Prior influence in Bayesian statistics. *Tech. Report*, **88–06**, ISDS, Duke University.

LAVINE, M.(1989). The boon of dimensionality: how to be a sensitive multidimensional Bayesian. *Tech. Report*, **89–14**, ISDS, Duke University.

LAVINE, M.(1991a). Sensitivity in Bayesian statistics: the prior and the likelihood. *J. Amer. Statist. Assoc.*, **86**, 396–399.

LAVINE, M.(1991b). An approach to robust Bayesian analysis with multidimensional spaces. *J. Amer. Statist. Assoc.*, **86**, 400–403.

LAVINE, M.(1992a). A note on bounding Monte Carlo variances. *Comm. Statist. Theory Methods*, **21**, 2855–2860.

LAVINE, M.(1992b). Local predictive influence in Bayesian linear models with conjugate priors. *Comm. Statist. Simulation Comput.*, **21**, 269–283.

LAVINE, M.(1994). An approach to evaluating sensitivity in Bayesian regression analysis. *J. Statist. Plan. Inference*, **40**, 233–244.

LAVINE, M., PERONE PACIFICO, M., SALINETTI, G. and TARDELLA, G. (2000). Linearization techniques in Bayesian robustness. In *Robust Bayesian Analysis* (D. Ríos Insua and F. Ruggeri, eds.). New York: Springer-Verlag.

LAVINE, M., WASSERMAN, L. and WOLPERT, R.L.(1991). Bayesian inference with specified prior marginals. *J. Amer. Statist. Assoc.*, **86**, 964–971.

LEAMER, E.E.(1982). Sets of posterior means with bounded variance prior. *Econometrica*, **50**, 725–736.

LEVY, H. (1992). Stochastic dominance and expected utility analysis: survey and analysis. *Management Science*, **38**, 555–593.

LI, Y. and SAXENA, K.M.L.(1995). Optimal robust Bayesian estimation. *J. Statist. Plan. Inference*, **46**, 365–380.

LISEO, B.(1994). New perspectives in Bayesian robustness (Italian). *Proc. Ital. Statist. Soc.*, 127–138.

LISEO, B.(1995). A note on the posterior mean robustness for location parameters. *Metron*, **LIII**, 29–33.

LISEO, B.(1996). A note on the concept of robust likelihoods. *Metron*, **LIV**, 25–38.

LISEO, B. (2000). Robustness issues in Bayesian model selection. In *Robust Bayesian Analysis* (D. Ríos Insua and F. Ruggeri, eds.). New York: Springer-Verlag.

LISEO, B., MORENO, E., and SALINETTI, G. (1996). Bayesian robustness of the class with given marginals: an approach based on moment theory (with discussion). In *Bayesian Robustness* (J.O. Berger et al., eds). IMS Lecture Notes - Monograph Series, **29**. Hayward: IMS.

LISEO, B., PETRELLA, L. and SALINETTI, G. (1996). Bayesian robustness: an interactive approach. In *Bayesian Statistics 5* (J.M. Bernardo et al., eds.). Oxford: Oxford University Press.

LISEO, B., PETRELLA, L. and SALINETTI, G. (1993). Block unimodality for multivariate Bayesian robustness. *J. Ital. Statist. Society*, **2**, 55–71.

LIU, Y.H. and YANG, M.C.(1997). Posterior robustness in simultaneous estimation problem with exchangeable contaminated priors. *J. Statist. Plan. Inference*, **65**, 129–143.

MADANSKI, A.(1990). Bayesian analysis with incompletely specified prior distributions. In *Bayesian and Likelihood Methods in Statistics and Econometrics* (S. Geisser et al, eds.). Amsterdam: North-Holland.

MACEACHERN, S.N. and MÜLLER, P. (2000). Sensitivity analysis by MCMC in encompassing Dirichlet process mixture models. In *Robust Bayesian Analysis* (D. Ríos Insua and F. Ruggeri, eds.). New York: Springer-Verlag.

MAKOV, U.E. (1994). Some aspects of Bayesian loss robustness. *J. Statist. Plan. Inference*, **38**, 359–370.

MAKOV, U.E. (1995). Loss robustness via Fisher-weighted squared-error loss function. *Insurance Math. Econom.*, **16**, 1–6.

MARÍN, J.M. (2000). A robust version of the dynamic linear model with an economic application. In *Robust Bayesian Analysis* (D. Ríos Insua and F. Ruggeri, eds.). New York: Springer-Verlag.

MARTEL ESCOBAR, M., HERNÁNDEZ BASTIDA, A. and VAZQUEZ POLO, F.J. (2000). Testing independence hypothesis in biparametric auditing models using Bayesian robustness methodology. To appear in *J. Inter-American Statist. Inst.*.

MARTÍN, J. (1995). Robustez Bayesiana con imprecision en preferencias y en creencias. *Ph.D. Dissertation*, Technical University of Madrid.

MARTÍN, J. and ARIAS, J.P. (2000). Computing the efficient set in Bayesian decision problems. In *Robust Bayesian Analysis* (D. Ríos Insua and F. Ruggeri, eds.). New York: Springer-Verlag.

MARTÍN, J. and RÍOS INSUA, D. (1996). Local sensitivity analysis in Bayesian decision theory (with discussion). In *Bayesian Robustness* (J.O. Berger et al., eds). IMS Lecture Notes - Monograph Series, **29**. Hayward: IMS.

MARTÍN, J., RÍOS INSUA, D. and RUGGERI, F. (1996). Checking dominance for several classes of loss functions. *ASA 1996 Proc. Sec. Bay. Statist. Sc.*, 176–179. Alexandria: ASA.

MARTÍN, J., RÍOS INSUA, D. and RUGGERI, F. (1998). Issues in Bayesian loss robustness. *Sankhyā, Ser. A*, **60**, 405–417.

MARYAK, J.L. and SPALL, J.C. (1986). On the sensitivity of posterior to prior in Bayesian analysis. *ASA 1986 Proc. Sec. Bus. Econ. Statist. Sc.*, 131–134. Alexandria: ASA.

MARYAK, J.L. and SPALL, J.C. (1987). Conditions for the insensitivity of the Bayesian posterior distribution to the choice of prior distribution. *Statist. Probab. Lett.*, **5**, 401–407.

MAZZALI, C. and RUGGERI, F. (1998). Bayesian analysis of failure count data from repairable systems. *Tech. Report*, **98.19**, CNR-IAMI.

MCCULLOCH, R. (1989). Local model influence. *J. Amer. Statist. Assoc.*, **84**, 473–478.

MĘCZARSKI, M.(1993a). Stable Bayesian estimation in the Poisson model: a nonlinear problem. *Zesz. Nauk. Politech.*, **687**, 83–91.

MĘCZARSKI, M. (1993b). Stability and conditional gamma-minimaxity in Bayesian inference. *Appl. Math.*, **22**, 117–122.

MĘCZARSKI, M. (1998). *Robustness Problems in Bayesian Statistical Analysis* (Polish). Monograph Series **446**. Warszawa: Publishing House of Warsaw School of Economics.

MĘCZARSKI, M. and ZIELIŃSKI, R.(1991). Stability of the Bayesian estimator of the Poisson mean under the inexactly specified gamma prior. *Statist. Probab. Lett.*, **12**, 329–333.

MĘCZARSKI, M. and ZIELIŃSKI, R.(1997a). Stability of the posterior mean in linear models - an admissibility property of D-optimum and E-optimum designs. *Statist. Probab. Lett.*, **33**, 117–123.

MĘCZARSKI, M. and ZIELIŃSKI, R.(1997b). Bayes optimal stopping of a homogeneous Poisson process under LINEX loss function and variation in the prior. *Appl. Math.*, **24**, 457–463.

MENG, Q. and SIVAGANESAN, S. (1996). Local sensitivity of density bounded priors. *Statist. Probab. Lett.*, **27**, 163–169.

MICHEAS, A. and DEY, D. (1999). On measuring loss robustness using maximum a posteriori estimate. *Tech. Report*, **99–12**, Department of Statistics, University of Connecticut.

MORENO, E. (1997). Bayes factors for intrinsic and fractional priors in nested models: Bayesian robustness. In *L_1-Statistical Procedures and Related Topics* (Y. Dodge, ed.). IMS Lecture Notes - Monograph Series, **31**. Hayward: IMS.

MORENO, E. (2000). Global Bayesian robustness for some classes of prior distributions. In *Robust Bayesian Analysis* (D. Ríos Insua and F. Ruggeri, eds.). New York: Springer-Verlag.

MORENO, E. and CANO, J.A.(1989). Testing a point null hypothesis: asymptotic robust Bayesian analysis with respect to the priors given on a subsigma field. *Int. Statist. Review*, **57**, 221–232.

MORENO, E. and CANO, J.A.(1991). Robust Bayesian analysis for ε-contaminations partially known. *J. Roy. Statist. Soc. Ser. B*, **53**, 143–155.

MORENO, E. and CANO, J.A.(1992). Classes of bidimensional priors specified on a collection of sets: Bayesian robustness. *J. Statist. Plan. Inference*, **46**, 325–334.

MORENO, E. and GIRÓN, J. (1994). Prior elicitation and Bayesian robustness in a noisy Poisson process. *ASA 1994 Proc. Sec. Bay. Statist. Sc.*, 217–220. Alexandria: ASA.

MORENO, E. and GIRÓN, J. (1998). Estimating with incomplete count data: a Bayesian approach. *J. Statist. Plan. Inference*, **66**, 147–159.

MORENO, E. and GONZÁLEZ, A. (1990). Empirical Bayes analysis for ε-contaminated priors with shape and quantile constraints. *Rebrape*, **4**, 177–200.

MORENO, E., MARTÍNEZ, C. and CANO, J.A. (1996). Local robustness and influence for contamination classes of prior distributions (with discussion). In *Bayesian Robustness* (J.O. Berger et al., eds). IMS Lecture Notes - Monograph Series, **29**. Hayward: IMS.

MORENO, E. and PERICCHI, L.R.(1990). Sensitivity of the Bayesian analysis to the priors: structural contaminations with specified quantiles of parametric families. *Actas III Cong. Latinoamericano Probab. Estad. Mat.*, 143–158.

MORENO, E. and PERICCHI, L.R.(1991). Robust Bayesian analysis for ε-contaminations with shape and quantile constraints. *Proc. Fifth Inter. Symp. on Applied Stochastic Models and Data Analysis*. Singapore: World Scientific.

MORENO, E. and PERICCHI, L.R.(1992a). Subjetivismo sin dogmatismo: análisis Bayesiano robusto (with discussion). *Estadistica Española*, **34**, 5–60.

MORENO, E. and PERICCHI, L.R.(1992b). Bands of probability measures: a robust Bayesian analysis. In *Bayesian Statistics 4* (J.M. Bernardo et al., eds.). Oxford: Oxford University Press.

MORENO, E. and PERICCHI, L.R.(1993a). Bayesian robustness for hierarchical ε-contamination models. *J. Statist. Plan. Inference*, **37**, 159–168.

MORENO, E. and PERICCHI, L.R.(1993b). Precise measurement theory: robust Bayesian analysis. *Tech. Report*, Department of Statistics, Universidad de Granada.

MORENO, E. and PERICCHI, L.R.(1993c). Prior assessments for bands of probability measures: empirical Bayes analysis. *Test*, **2**, 101–110.

MORENO, E. and PERICCHI, L.R.(1993d). On ϵ-contaminated priors with quantile and piece-wise unimodality constraints. *Comm. Statist. Theory Methods*, **22**, 1963–1978.

MORENO, E., PERICCHI, L.R. and KADANE, J.(1998). A robust Bayesian look of the theory of precise measurement. In *Decision Science and Technology: Reflections on the Contributions of Ward Edwards* (J. Shanteau et al, eds.). Norwell, MA: Kluwer.

MOSKOWITZ, H.(1992). Multiple-criteria robust interactive decision analysis for optimizing public policies. *Eur. J. Oper. Res.*, **56**, 219–236.

MÜLLER, A.(1996). Optimal selection from distributions with unknown parameters: robustness of Bayesian models. *Math. Methods Oper. Res.*, **44**, 371–386.

MUSIO, M. (1997). Measuring local sensitivity in Bayesian analysis: a general setting. *Tech. Report*, Université de Haute Alsace.

NAU, R. (1992). Indeterminate probabilities on finite sets, *Ann. Statist.*, **20**, 4, 1737–1767.

NAU, R. (1995). The shape of incomplete preferences. *Tech. Report*, Duke University.

O'HAGAN, A.(1994). Robust modelling for asset management. *J. Statist. Plan. Inference*, **40**, 245–259.

O'HAGAN, A. and BERGER, J.O.(1988). Ranges of posterior probabilities for quasi-unimodal priors with specified quantiles. *J. Amer. Statist. Assoc.*, **83**, 503–508.

OSIEWALSKI, J. and STEEL, M.F.(1993a). Robust Bayesian inference in ℓ_q-spherical models. *Biometrika*, **80**, 456–460.

OSIEWALSKI, J. and STEEL, M.F.(1993b). Robust Bayesian inference in elliptical regression models. *J. Econometrics*, **57**, 345–363.

PEÑA, D. and ZAMAR, R. (1996). On Bayesian robustness: an asymptotic approach. In *Robust Statistics, Data Analysis, and Computer Intensive Methods*, (H. Rieder, ed.). New York: Springer-Verlag.

PEÑA, D. and ZAMAR, R. (1997). A simple diagnostic tool for local prior sensitivity. *Statist. Probab. Lett.*, **36**, 205–212.

PÉREZ, M.-E. and PERICCHI, R. (1994). A case study on the Bayesian analysis of 2×2 tables with all margins fixed. *Rebrape*, **8**, 27–37.

PERICCHI, L.R. and NAZARET, W.(1988). On being imprecise at the higher levels of a hierarchical linear model. In *Bayesian Statistics 3* (J.M. Bernardo et al., eds.). Oxford: Oxford University Press.

PERICCHI, L.R. and PEREZ, M.E. (1994). Posterior robustness with more than one sampling model. *J. Statist. Plan. Inference*, **40**, 279–291.

PERICCHI, L.R. and WALLEY, P.(1990). One-sided hypothesis testing with near-ignorance priors. *Rebrape*, **4**, 69–82.

PERICCHI, L.R. and WALLEY, P.(1991). Robust Bayesian credible intervals and prior ignorance. *Internat. Statist. Rev.*, **58**, 1–23.

PERONE PACIFICO, M., SALINETTI, G. and TARDELLA, L. (1994). Fractional optimization in Bayesian robustness. *Tech. Report*, **A 23**, Dipartimento di Statistica, Probabilità e Statistiche Applicate, Università "La Sapienza", Roma.

PERONE PACIFICO, M., SALINETTI, G. and TARDELLA, L. (1996). Bayesian robustness on constrained density band classes. *Test*, **5**, 395–409.

PERONE PACIFICO, M., SALINETTI, G. and TARDELLA, L. (1998). A note on the geometry of Bayesian global and local robustness. *J. Statist. Plan. Inference*, **69**, 51–64.

PIERCE, D. and FOLKS, J.L. (1969). Sensitivity of Bayes procedures to the prior distribution. *Oper. Res.*, **17**, 344–350.

POLASEK, W.(1985). Sensitivity analysis for general and hierarchical linear regression models. In *Bayesian Inference and Decision Techniques with Applications*, (P.K. Goel and A. Zellner, eds.). Amsterdam: North-Holland.

POLASEK, W.(1985). Joint sensitivity analysis for covariance matrices in Bayesian linear regression. In *Recent Advances in Statistics and Probability*, (J.P. Vilaplana and M.L. Puri, eds.). Ah Zeist: VSP.

POLASEK, W. and PÖTZELBERGER, K.(1988). Robust Bayesian analysis in hierarchical models. In *Bayesian Statistics 3* (J.M. Bernardo et al., eds.). Oxford: Oxford University Press.

POLASEK, W. and PÖTZELBERGER, K.(1994). Robust Bayesian methods in simple ANOVA models. *J. Statist. Plan. Inference*, **40**, 295–311.

PÖTZELBERGER, K. and POLASEK, W.(1991). Robust HPD-regions in Bayesian regression models. *Econometrica*, **59**, 1581–1590.

RAMSAY, J.O. and NOVICK, M.R. (1980). PLU robust Bayesian decision. Theory: point estimation. *J. Amer. Statist. Assoc.*, **75**, 901–907.

REGOLI, G.(1994). Qualitative probabilities in the elicitation process (Italian). *Proc. Ital. Statist. Soc.*, 153–165.

REGOLI, G. (1996). Comparative probability and robustness. In *Bayesian Robustness* (J.O. Berger et al., eds). IMS Lecture Notes - Monograph Series, **29**. Hayward: IMS.

REILLY, T. (1996). Sensitivity analysis for dependent variables. *Tech. Report*, University of Oregon, Eugene.

RÍOS, S. and GIRÓN, F.J.(1980). Quasi-Bayesian behavior: a more realistic approach to decision making? In *Bayesian Statistics* (J.M. Bernardo et al., eds.). Valencia: Valencia University Press.

RÍOS INSUA, D.(1990). *Sensitivity Analysis in Multiobjective Decision Making*. New York: Springer-Verlag.

RÍOS INSUA, D. (1991). On the foundations of decision making under partial information. *Theory Decis.*, **33**, 83–100.

RÍOS INSUA, D.(1992). Foundations for a robust theory of decision making: the simple case. *Test*, **1**, 69–78.

RÍOS INSUA, D. and CRIADO, R. (2000). Topics on the foundations of robust Bayesian analysis. In *Robust Bayesian Analysis* (D. Ríos Insua and F. Ruggeri, eds.). New York: Springer-Verlag.

RÍOS INSUA, D. and FRENCH, S. (1991). A framework for sensitivity analysis in discrete multi-objective decision making. *Eur. J. Oper. Res.*, **54**, 176–190.

RÍOS INSUA, D. and MARTÍN, J. (1994a). Robustness issues under precise beliefs and preferences. *J. Statist. Plan. Inference*, **40**, 383–389.

RÍOS INSUA, D. and MARTÍN, J. (1994b) On the foundations of robust decision making. In *Decision Theory and Decision Analysis: Trends and Challenges*, (S. Ríos, ed.). Cambridge: Kluwer.

RÍOS INSUA, D. and MARTÍN, J. (1995). On the foundations of robust Bayesian statistics. *Tech. Report*, Universidad Politecnica de Madrid.

RÍOS INSUA, D., MARTÍN, J., PROLL, L., FRENCH, S. and SALHI, A. (1997). Sensitivity analysis in statistical decision theory: a decision analytic view. *J. Statist. Comput. Simulation*, **57**, 197–218.

RÍOS INSUA, D. and RUGGERI, F. (2000). *Robust Bayesian Analysis*. New York: Springer-Verlag.

RÍOS INSUA, D., RUGGERI, F. and MARTÍN, J. (2000). Bayesian sensitivity analysis: a review. To appear in *Handbook on Sensitivity Analysis* (A. Saltelli et al., eds.). New York: Wiley.

RÍOS INSUA, D., RUGGERI, F. and VIDAKOVIC B. (1995). Some results on posterior regret Γ-minimax estimation. *Statist. Decisions*, **13**, 315–331.

RÍOS INSUA, S., MARTÍN, J., RÍOS INSUA, D. and RUGGERI, F. (1999). Bayesian forecasting for accident proneness evaluation. *Scand. Actuar. J.*, **99**, 134–156.

RUGGERI, F.(1990). Posterior ranges of functions of parameters under priors with specified quantiles. *Comm. Statist. Theory Methods*, **19**, 127–144.

RUGGERI, F.(1991). Robust Bayesian analysis given a lower bound on the probability of a set. *Comm. Statist. Theory Methods*, **20**, 1881–1891.

RUGGERI, F.(1992). Bounds on the prior probability of a set and robust Bayesian analysis. *Theory Probab. Appl.*, **37**, 358–359.

RUGGERI, F.(1993). Robust Bayesian analysis given bounds on the probability of a set. *Comm. Statist. Theory Methods*, **22**, 2983–2998.

RUGGERI, F. (1994a). Nonparametric Bayesian robustness. *Tech. Report*, **94.8**, CNR-IAMI.

RUGGERI, F. (1994b). Local and global sensitivity under some classes of priors. In *Recent Advances in Statistics and Probability* (J.P. Vilaplana and M.L. Puri, eds.). Ah Zeist: VSP.

RUGGERI, F. (1999). Robust Bayesian and Bayesian decision theoretic wavelet shrinkage. In *Robust Bayesian Analysis* (P. Müller and B. Vidakovic eds.). New York: Springer-Verlag.

RUGGERI, F. (2000). Sensitivity issues in the Bayesian analysis of the reliability of repairable systems. To appear in *Proc. MMR'2000 (Mathematical Methods in Reliability)*.

RUGGERI, F. and SIVAGANESAN, S. (2000). On a global sensitivity measure for Bayesian inference. To appear in *Sankhyā, Ser. A*.

RUGGERI, F. and WASSERMAN, L.(1993). Infinitesimal sensitivity of posterior distributions. *Canad. J. Statist.*, **21**, 195–203.

RUGGERI, F. and WASSERMAN, L.A. (1995). Density based classes of priors: infinitesimal properties and approximations. *J. Statist. Plan. Inference*, **46**, 311–324.

RUKHIN, A.L. (1993). Influence of the prior distribution on the risk of a Bayes rule. *J. Theoret. Probab.*, **6**, 71–87.

SALINETTI, G. (1994). Stability of Bayesian decisions. *J. Statist. Plan. Inference*, **40**, 313–320.

SALTELLI, A., CHAN, K. and SCOTT, M. (2000). *Handbook of Sensitivity Analysis*. New York: Wiley.

SANSÓ, B., MORENO, E. and PERICCHI, L.R. (1996). On the robustness of the intrinsic Bayes factor for nested models (with discussion). In *Bayesian Robustness* (J.O. Berger et al., eds). IMS Lecture Notes - Monograph Series, **29**. Hayward: IMS.

SANSÓ, B. and PERICCHI, L.R.(1992). Near ignorance classes of log-concave priors for the location model. *Test*, **1**, 39–46.

SARGENT, D.J. and CARLIN, B.P. (1996). Robust Bayesian design and analysis of clinical trials via prior partitioning (with discussion). In *Bayesian Robustness* (J.O. Berger et al., eds). IMS Lecture Notes - Monograph Series, **29**. Hayward: IMS.

SEIDENFELD, T., SCHERVISH, M. and KADANE, J. (1992). A representation of partially ordered preferences. *Ann. Statist.*, **23**, 2168–2217.

SEIDENFELD, T. and WASSERMAN, L. (1993). Dilation for sets of probabilities. *Ann. Statist.*, **21**, 1139–1154.

SHYAMALKUMAR, N.D. (1996a). Bayesian robustness with respect to the likelihood. *Tech. Report*, **96-23**, Department of Statistics, Purdue University.

SHYAMALKUMAR, N.D. (1996b). Bayesian robustness with asymptotically nice classes of priors. *Tech. Report*, **96-22**, Department of Statistics, Purdue University.

SHYAMALKUMAR, N.D. (2000). Likelihood robustness. In *Robust Bayesian Analysis* (D. Ríos Insua and F. Ruggeri, eds.). New York: Springer-Verlag.

SIVAGANESAN, S.(1988). Range of posterior measures for priors with arbitrary contaminations. *Comm. Statist. Theory Methods*, **17**, 1591–1612.

SIVAGANESAN, S.(1989). Sensitivity of posterior mean to unimodality preserving contaminations. *Statist. Decisions*, **7**, 77–93.

SIVAGANESAN, S.(1990). Sensitivity of some standard Bayesian estimates to prior uncertainty - a comparison. *J. Statist. Plan. Inference*, **27**, 85–103.

SIVAGANESAN, S.(1991). Sensitivity of some posterior summaries when the prior is unimodal with specified quantiles. *Canad. J. Statist.*, **19**, 57–65.

SIVAGANESAN, S.(1992). An evaluation of robustness in binomial empirical Bayes testing. In *Bayesian Statistics 4* (J.M. Bernardo et al., eds.). Oxford: Oxford University Press.

SIVAGANESAN, S.(1993a). Range of the posterior probability of an interval for priors with unimodality preserving contaminations. *Ann. Inst. Statist. Math.*, **45**, 187–199.

SIVAGANESAN, S. (1993b). Robust Bayesian diagnostics. *J. Statist. Plan. Inference*, **35**, 171–188.

SIVAGANESAN, S.(1993c). Optimal robust sets for a density bounded class. *Tech. Report*, Department of Mathematical Statistics, University of Cincinnati.

SIVAGANESAN, S.(1994). Bounds on posterior expectations for density bounded classes with constant bandwidth. *J. Statist. Plan. Inference*, **40**, 331–343.

SIVAGANESAN, S. (1995). Multi-dimensional priors: global and local robustness. *Tech. Report*, Department of Mathematical Statistics, University of Cincinnati.

SIVAGANESAN, S. (1996). Asymptotics of some local and global robustness measures (with discussion). In *Bayesian Robustness* (J.O. Berger et al., eds). IMS Lecture Notes - Monograph Series, **29**. Hayward: IMS.

SIVAGANESAN, S. (1997). Optimal robust credible sets for density-bounded priors. *J. Statist. Plan. Inference*, **63**, 9–17.

SIVAGANESAN, S. (1999). A likelihood based robust Bayesian summary. *Statist. Probab. Lett.*, **43**, 5–12.

SIVAGANESAN, S. (2000). Global and local robustness approaches: uses and limitations. In *Robust Bayesian Analysis* (D. Ríos Insua and F. Ruggeri, eds.). New York: Springer-Verlag.

SIVAGANESAN, S. and BERGER, J.O.(1989). Ranges of posterior measures for priors with unimodal contaminations. *Ann. Statist.*, **17**, 868–889.

SIVAGANESAN, S. and BERGER, J.O.(1993). Robust Bayesian analysis of the binomial empirical Bayes problem. *Canad. J. Statist.*, **21**, 107–119.

SIVAGANESAN, S., BERLINER, L.M. and BERGER, J.O.(1993). Optimal robust credible sets for contaminated priors. *Statist. Probab. Lett.*, **18**, 383–388.

SMITH, A.F.M.(1983). Bayesian approaches to outliers and robustness. In *Specifying Statistical Models* (J.P. Florens et al., eds.). New York: Springer-Verlag.

SMITH, J.E. (1995). Generalized Chebychev Inequalities: theory and applications in decision analysis. *Oper. Res.*, **43**, 807–825.

SRINIVASAN, C. and TRUSCZCYNSKA, H.(1990a). Approximation to the range of a ratio linear posterior quantity based on Fréchet derivative. *Tech. Report*, **289**, Department of Statistics, University of Kentucky.

SRINIVASAN, C. and TRUSCZCYNSKA, H.(1990b). On the ranges of posterior quantities. *Tech. Report*, **294**, Department of Statistics, University of Kentucky.

STONE, M. (1963). Robustness of nonideal decision procedures. *J. Amer. Statist. Assoc.*, **58**, 480–486.

VAZQUEZ POLO, F.J. and HERNÁNDEZ BASTIDA, A.(1995). Behavior of the posterior error rate with partial prior information in auditing. *J. Appl. Statist.*, **22**, 469–476.

VIDAKOVIC, B.(1992). A study of properties of computationally simple rules in estimation problems. *Ph.D. Dissertation*, Purdue University.

VIDAKOVIC, B. (2000). Γ-minimax: a paradigm for conservative robust Bayesians. In *Robust Bayesian Analysis* (D. Ríos Insua and F. Ruggeri, eds.). New York: Springer-Verlag.

VIDAKOVIC, B. and DASGUPTA, A. (1996). Efficiency of linear rules for estimating a bounded normal mean. *Sankhyā, Ser. A*, **58**, 81–100.

VIDAKOVIC, B. and RUGGERI, F. (1999). Expansion estimation by Bayes rules. *J. Statist. Plan. Inference*, **79**, 223–235.

WALLEY, P.(1991). *Statistical Reasoning with Imprecise Probabilities*. London: Chapman and Hall.

WASSERMAN, L.(1989). A robust Bayesian interpretation of likelihood regions. *Ann. Statist.*, **17**, 1387–1393.

WASSERMAN, L.(1990a). Prior envelopes based on belief functions. *Ann. Statist.*, **18**, 454–464.

WASSERMAN, L.(1990b). Belief functions and statistical inference. *Canad. J. Statist.*, **18**, 183–196.

WASSERMAN, L.(1992a). Recent methodological advances in robust Bayesian inference. In *Bayesian Statistics 4* (J.M. Bernardo et al., eds.). Oxford: Oxford University Press.

WASSERMAN, L.(1992b). Invariance properties of density ratio priors. *Ann. Statist.*, **20**, 2177–2182.

WASSERMAN, L. (1996). The conflict between improper priors and robustness. *J. Statist. Plan. Inference*, **52**, 1–15.

WASSERMAN, L. and KADANE, J.(1990). Bayes' theorem for Choquet capacities. *Ann. Statist.*, **18**, 1328–1339.

WASSERMAN, L. and KADANE, J.(1992a). Computing bounds on expectations. *J. Amer. Statist. Assoc.*, **87**, 516–522.

WASSERMAN, L. and KADANE, J.(1992b). Symmetric upper probabilities. *Ann. Statist.*, **20**, 1720–1736.

WASSERMAN, L. and KADANE, J.(1994). Permutation invariant upper and lower probabilities. In *Statistical Decision Theory and Related Topics V* (S.S. Gupta and J.O. Berger, eds.). New York: Springer-Verlag.

WASSERMAN, L., LAVINE, M. and WOLPERT, R.L.(1993). Linearization of Bayesian robustness problems. *J. Statist. Plan. Inference*, **37**, 307–316.

WASSERMAN, L. and SEIDENFELD, T.(1994). The dilation phenomenon in robust Bayesian inference. *J. Statist. Plan. Inference*, **40**, 345–356.

WATSON, S.R. (1974). On Bayesian inference with incompletely specified prior distributions. *Biometrika*, **61**, 193–196.

WEISS, R. (1996). An approach to Bayesian sensitivity analysis. *J. Roy. Statist. Soc. Ser. B*, **58**, 739–750.

WEST, M.(1996). Modeling and robustness issues in Bayesian time series analysis (with discussion). In *Bayesian Robustness* (J.O. Berger et al., eds). IMS Lecture Notes - Monograph Series, **29**. Hayward: IMS.

WHITE, D.J. (1982). *Optimality and Efficiency.* New York:Wiley.

WILSON, S.P. and WIPER, M.P.(2000). Prior robustness in some common types of software reliability model. In *Robust Bayesian Analysis* (D. Ríos Insua and F. Ruggeri, eds.). New York: Springer-Verlag.

WOLPERT, R.L. and LAVINE, M.(1996). Markov random field priors for univariate density estimation (with discussion). In *Bayesian Robustness* (J.O. Berger et al., eds). IMS Lecture Notes - Monograph Series, **29**. Hayward: IMS.

ZEN, M.M. and DASGUPTA, A.(1993). Estimating a binomial parameter: is robust Bayes real Bayes? *Statist. Decisions*, **11**, 37–60.

Lecture Notes in Statistics

For information about Volumes 1 to 78, please contact Springer-Verlag

Vol. 79: N. Schmitz, Optimal Sequentially Planned Decision Procedures. xii, 209 pages, 1992.

Vol. 80: M. Fligner, J. Verducci (Eds.), Probability Models and Statistical Analyses for Ranking Data. xxii, 306 pages, 1992.

Vol. 81: P. Spirtes, C. Glymour, R. Scheines, Causation, Prediction, and Search. xxiii, 526 pages, 1993.

Vol. 82: A. Korostelev and A. Tsybakov, Minimax Theory of Image Reconstruction. xii, 268 pages, 1993.

Vol. 83: C. Gatsonis, J. Hodges, R. Kass, N. Singpurwalla (Editors), Case Studies in Bayesian Statistics. xii, 437 pages, 1993.

Vol. 84: S. Yamada, Pivotal Measures in Statistical Experiments and Sufficiency. vii, 129 pages, 1994.

Vol. 85: P. Doukhan, Mixing: Properties and Examples. xi, 142 pages, 1994.

Vol. 86: W. Vach, Logistic Regression with Missing Values in the Covariates. xi, 139 pages, 1994.

Vol. 87: J. Müller, Lectures on Random Voronoi Tessellations.vii, 134 pages, 1994.

Vol. 88: J. E. Kolassa, Series Approximation Methods in Statistics. Second Edition, ix, 183 pages, 1997.

Vol. 89: P. Cheeseman, R.W. Oldford (Editors), Selecting Models From Data: AI and Statistics IV. xii, 487 pages, 1994.

Vol. 90: A. Csenki, Dependability for Systems with a Partitioned State Space: Markov and Semi-Markov Theory and Computational Implementation. x, 241 pages, 1994.

Vol. 91: J.D. Malley, Statistical Applications of Jordan Algebras. viii, 101 pages, 1994.

Vol. 92: M. Eerola, Probabilistic Causality in Longitudinal Studies. vii, 133 pages, 1994.

Vol. 93: Bernard Van Cutsem (Editor), Classification and Dissimilarity Analysis. xiv, 238 pages, 1994.

Vol. 94: Jane F. Gentleman and G.A. Whitmore (Editors), Case Studies in Data Analysis. viii, 262 pages, 1994.

Vol. 95: Shelemyahu Zacks, Stochastic Visibility in Random Fields. x, 175 pages, 1994.

Vol. 96: Ibrahim Rahimov, Random Sums and Branching Stochastic Processes. viii, 195 pages, 1995.

Vol. 97: R. Szekli, Stochastic Ordering and Dependence in Applied Probability. viii, 194 pages, 1995.

Vol. 98: Philippe Barbe and Patrice Bertail, The Weighted Bootstrap. viii, 230 pages, 1995.

Vol. 99: C.C. Heyde (Editor), Branching Processes: Proceedings of the First World Congress. viii, 185 pages, 1995.

Vol. 100: Wlodzimierz Bryc, The Normal Distribution: Characterizations with Applications. viii, 139 pages, 1995.

Vol. 101: H.H. Andersen, M.Højbjerre, D. Sørensen, P.S.Eriksen, Linear and Graphical Models: for the Multivariate Complex Normal Distribution. x, 184 pages, 1995.

Vol. 102: A.M. Mathai, Serge B. Provost, Takesi Hayakawa, Bilinear Forms and Zonal Polynomials. x, 378 pages, 1995.

Vol. 103: Anestis Antoniadis and Georges Oppenheim (Editors), Wavelets and Statistics. vi, 411 pages, 1995.

Vol. 104: Gilg U.H. Seeber, Brian J. Francis, Reinhold Hatzinger, Gabriele Steckel-Berger (Editors), Statistical Modelling: 10th International Workshop, Innsbruck, July 10-14th 1995. x, 327 pages, 1995.

Vol. 105: Constantine Gatsonis, James S. Hodges, Robert E. Kass, Nozer D. Singpurwalla(Editors), Case Studies in Bayesian Statistics, Volume II. x, 354 pages, 1995.

Vol. 106: Harald Niederreiter, Peter Jau-Shyong Shiue (Editors), Monte Carlo and Quasi-Monte Carlo Methods in Scientific Computing. xiv, 372 pages, 1995.

Vol. 107: Masafumi Akahira, Kei Takeuchi, Non-Regular Statistical Estimation. vii, 183 pages, 1995.

Vol. 108: Wesley L. Schaible (Editor), Indirect Estimators in U.S. Federal Programs. viii, 195 pages, 1995.

Vol. 109: Helmut Rieder (Editor), Robust Statistics, Data Analysis, and Computer Intensive Methods. xiv, 427 pages, 1996.

Vol. 110: D. Bosq, Nonparametric Statistics for Stochastic Processes. xii, 169 pages, 1996.

Vol. 111: Leon Willenborg, Ton de Waal, Statistical Disclosure Control in Practice. xiv, 152 pages, 1996.

Vol. 112: Doug Fischer, Hans-J. Lenz (Editors), Learning from Data. xii, 450 pages, 1996.

Vol. 113: Rainer Schwabe, Optimum Designs for Multi-Factor Models. viii, 124 pages, 1996.

Vol. 114: C.C. Heyde, Yu. V. Prohorov, R. Pyke, and S. T. Rachev (Editors), Athens Conference on Applied Probability and Time Series Analysis Volume I: Applied Probability In Honor of J.M. Gani. viii, 424 pages, 1996.

Vol. 115: P.M. Robinson, M. Rosenblatt (Editors), Athens Conference on Applied Probability and Time Series Analysis Volume II: Time Series Analysis In Memory of E.J. Hannan. viii, 448 pages, 1996.

Vol. 116: Genshiro Kitagawa and Will Gersch, Smoothness Priors Analysis of Time Series. x, 261 pages, 1996.

Vol. 117: Paul Glasserman, Karl Sigman, David D. Yao (Editors), Stochastic Networks. xii, 298, 1996.

Vol. 118: Radford M. Neal, Bayesian Learning for Neural Networks. xv, 183, 1996.

Vol. 119: Masanao Aoki, Arthur M. Havenner, Applications of Computer Aided Time Series Modeling. ix, 329 pages, 1997.

Vol. 120: Maia Berkane, Latent Variable Modeling and Applications to Causality. vi, 288 pages, 1997.

Vol. 121: Constantine Gatsonis, James S. Hodges, Robert E. Kass, Robert McCulloch, Peter Rossi, Nozer D. Singpurwalla (Editors), Case Studies in Bayesian Statistics, Volume III. xvi, 487 pages, 1997.

Vol. 122: Timothy G. Gregoire, David R. Brillinger, Peter J. Diggle, Estelle Russek-Cohen, William G. Warren, Russell D. Wolfinger (Editors), Modeling Longitudinal and Spatially Correlated Data. x, 402 pages, 1997.

Vol. 123: D. Y. Lin and T. R. Fleming (Editors), Proceedings of the First Seattle Symposium in Biostatistics: Survival Analysis. xiii, 308 pages, 1997.

Vol. 124: Christine H. Müller, Robust Planning and Analysis of Experiments. x, 234 pages, 1997.

Vol. 125: Valerii V. Fedorov and Peter Hackl, Model-oriented Design of Experiments. viii, 117 pages, 1997.

Vol. 126: Geert Verbeke and Geert Molenberghs, Linear Mixed Models in Practice: A SAS-Oriented Approach. xiii, 306 pages, 1997.

Vol. 127: Harald Niederreiter, Peter Hellekalek, Gerhard Larcher, and Peter Zinterhof (Editors), Monte Carlo and Quasi-Monte Carlo Methods 1996, xii, 448 pages, 1997.

Vol. 128: L. Accardi and C.C. Heyde (Editors), Probability Towards 2000, x, 356 pages, 1998.

Vol. 129: Wolfgang Härdle, Gerard Kerkyacharian, Dominique Picard, and Alexander Tsybakov, Wavelets, Approximation, and Statistical Applications, xvi, 265 pages, 1998.

Vol. 130: Bo-Cheng Wei, Exponential Family Nonlinear Models, ix, 240 pages, 1998.

Vol. 131: Joel L. Horowitz, Semiparametric Methods in Econometrics, ix, 204 pages, 1998.

Vol. 132: Douglas Nychka, Walter W. Piegorsch, and Lawrence H. Cox (Editors), Case Studies in Environmental Statistics, viii, 200 pages, 1998.

Vol. 133: Dipak Dey, Peter Müller, and Debajyoti Sinha (Editors), Practical Nonparametric and Semiparametric Bayesian Statistics, xv, 408 pages, 1998.

Vol. 134: Yu. A. Kutoyants, Statistical Inference For Spatial Poisson Processes, vii, 284 pages, 1998.

Vol. 135: Christian P. Robert, Discretization and MCMC Convergence Assessment, x, 192 pages, 1998.

Vol. 136: Gregory C. Reinsel, Raja P. Velu, Multivariate Reduced-Rank Regression, xiii, 272 pages, 1998.

Vol. 137: V. Seshadri, The Inverse Gaussian Distribution: Statistical Theory and Applications, xi, 360 pages, 1998.

Vol. 138: Peter Hellekalek, Gerhard Larcher (Editors), Random and Quasi-Random Point Sets, xi, 352 pages, 1998.

Vol. 139: Roger B. Nelsen, An Introduction to Copulas, xi, 232 pages, 1999.

Vol. 140: Constantine Gatsonis, Robert E. Kass, Bradley Carlin, Alicia Carriquiry, Andrew Gelman, Isabella Verdinelli, Mike West (Editors), Case Studies in Bayesian Statistics, Volume IV, xvi, 456 pages, 1999.

Vol. 141: Peter Müller, Brani Vidakovic (Editors), Bayesian Inference in Wavelet Based Models, xi, 394 pages, 1999.

Vol. 142: György Terdik, Bilinear Stochastic Models and Related Problems of Nonlinear Time Series Analysis: A Frequency Domain Approach, xi, 258 pages, 1999.

Vol. 143: Russell Barton, Graphical Methods for the Design of Experiments, x, 208 pages, 1999.

Vol. 144: L. Mark Berliner, Douglas Nychka, and Timothy Hoar (Editors), Case Studies in Statistics and the Atmospheric Sciences, x, 208 pages, 1999.

Vol. 145: James H. Matis and Thomas R. Kiffe, Stochastic Population Models, viii, 220 pages, 2000.

Vol. 146: Wim Schoutens, Stochastic Processes and Orthogonal Polynomials, xiv, 163 pages, 2000.

Vol. 147: Jürgen Franke, Wolfgang Härdle, and Gerhard Stahl, Measuring Risk in Complex Stochastic Systems, xvi, 272 pages, 2000.

Vol. 148: S.E. Ahmed and Nancy Reid, Empirical Bayes and Likelihood Inference, x, 200 pages, 2000.

Vol. 149: D. Bosq, Linear Processes in Function Spaces: Theory and Applications, xv, 296 pages, 2000.

Vol. 150: Tadeusz Caliński and Sanpei Kageyama, Block Designs: A Randomization Approach, Volume I: Analysis, ix, 313 pages, 2000.

Vol. 151: Håkan Andersson and Tom Britton, Stochastic Epidemic Models and Their Statistical Analysis: ix, 152 pages, 2000.

Vol. 152: David Ríos Insua and Fabrizio Ruggeri, Robust Bayesian Analysis: xiii, 435 pages, 2000.